# Utilisation and Management of Medicinal Plants

## — *Volume 2* —

*The Editor*

**Dr. Vijay Kumar Gupta,** Ph.D., FLS, London (born 1953-) Chief Scientist, CSIR- Indian Institute of Integrative Medicine, Jammu, India. He did his M.Sc. (1975) and Ph.D. (1979) in Zoology both from University of Jammu, Jammu-India. His research capabilities are substantiated by his excellent work on histopathology, ecology and reproductive biology of fishes, turtles, birds and mammals, which has already got recognition in India and abroad.

Dr. Gupta has to his credit more than 100 scientific publications and review articles which have appeared in internationally recognized Indian and foreign journals. Founder fellow, life member and office bearer of many national societies, academies and associations. He has successfully completed a number of research/consultancy projects funded by government, private and multinational agencies. His current areas of interest are histopathology, toxicology, pre-clinical safety pharmacology, reproductive efficacy studies of laboratory animals and biodiversity.

He is the Series Editor of the recently published multi-volume set of books, "**Comprehensive Bioactive Natural Products (Vols. 1-8)**", published by M/S Studium Press, LLC, USA. He is also Editor-in-Chief of the books, "**Utilisation and Management of Medicinal Plants**", "**Medicinal Plants: Phytochemistry, Pharmacology and Therapeutics (Vols. 1 and 2)**", "**Traditional and Folk Herbal Medicine**", "**Natural Products: Research Reviews**", "**Bioactive Phytochemicals: Perspectives for Modern Medicine**", "**Perspectives in Animal Ecology and Reproduction (Vols. 1-9)**" and "**Animal Diversity, Natural History and Conservation (Vols. 1-3)**". The Editor-in-Chief of the American Biographical Institute, USA, has appointed him as *Consulting Editor* of *The Contemporary Who's Who*. Dr. Gupta also appointed as Nominee for the *Committee for the Purpose of Control and Supervision of Experiments on Animals* (CPCSEA, Govt. of India). The *Linnaean Society of London, U.K.* has awarded fellowship to him in November 2009 in recognition of his contribution towards the cultivation of knowledge in Science of Natural History. Recently, Modern Scientific Press, USA has nominated Dr. Gupta as the Editor of the *International Journal of Traditional and Natural Medicine*.

**Dr. Anpurna Kaul** (born 1956-) working as Scientist in Pharmacology Division of, Indian Institute of Integrative Medicine (CSIR), Jammu, India. She has completed her Ph.D. in 1996 from University of Jammu, Jammu, India. Her field of research is immune-pharmacology.

Dr. Kaul has 35 research publications in national and international journals, 12 patents and also presented several research papers in various symposia/conferences. She has screened more than 4000 plant extracts/fractions and pure compounds on immune system by *in vivo* and *in vitro* methods. She has also actively participated in Indo-Malaysian research project as member and visited Malaysia in 2007 to attend *Women's Health and Asian Traditional Medicine* Conference.

# Utilisation and Management of Medicinal Plants

## — Volume 2 —

*Editor-in-Chief*
**V.K. Gupta**
*Chief Scientist*
*CSIR-Indian Institute of Integrative Medicine*
*Canal Road, Jammu – 180 001,*
*India*

*Editor*
**A. Kaul**
*Scientist*
*Department of Pharmacology*
*CSIR-Indian Institute of Integrative Medicine*
*Canal Road, Jammu – 180 001,*
*India*

2014
# Daya Publishing House®
*A Division of*
# Astral International Pvt. Ltd.
New Delhi – 110 002

| | | |
|---|---|---|
| *Published by* | : | **Daya Publishing House**® |
| | | A Division of |
| | | **Astral International Pvt. Ltd.** |
| | | – ISO 9001:2008 Certified Company – |
| | | 4760-61/23, Ansari Road, Darya Ganj |
| | | New Delhi-110 002 |
| | | Ph. 011-43549197, 23278134 |
| | | E-mail: info@astralint.com |
| | | Website: www.astralint.com |
| *Laser Typesetting* | : | **Classic Computer Services**, Delhi - 110 035 |
| *Printed at* | : | **Replika Press Pvt. Ltd.** |

PRINTED IN INDIA

# Editorial Board

**Prof. Yukihiro Shoyama**

*Faculty of Pharmaceutical Science,*
*Nagasaki International University,*
*His Ten Bosh, Sasebo, Nagasaki, JAPAN*
*Tel: +81-956-20-5653*
*E-mail: shoyama@niu.ac.jp*

**Prof. Ian Fraser Pryme**

*Department of Biomedicine,*
*University of Bergen,*
*Jonas Lies vei 91, N-5009 Bergen, NORWAY*
*Tel: +47-55-586438*
*Fax: +47-55-586360*
*E-mail: ian.pryme@biomed.uib.no*

**Dr. George Qian Li**

*Herbal Medicines Research and Education Centre,*
*Faculty of Pharmacy, The University of Sydney,*
*NSW 2006, AUSTRALIA*
*Tel: 612 9351*
*Fax: 9351 8638*
*E-mail: george.li@sydney.edu.au*

**Dr. Julia Serkedjieva**

*Institute of Microbiology,*
*Bulgarian Academy of Sciences,*
*26, Academician Georgy Bonchev St.,*
*1113 Sofia, BULGARIA*
*Tel: +359 2 979 31 85*
*Fax: +359 2 870 01 09*
*E-mail: jserkedjieva@microbio.bas.bg*

**Hawthorn Campus**

John Street Hawthorn
Victoria 3122 Australia

PO Box 218 Hawthorn
Victoria 3122 Australia

Telephone +61 3 9214 800
Facsimile +61 3 9819 5454
www.swinburne.edu.au

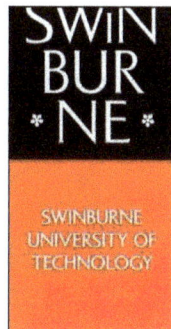

SWIN
BUR
* NE *

SWINBURNE
UNIVERSITY OF
TECHNOLOGY

# Foreword

Medicinal preparations derived from plants and other natural sources have been widely used throughout human history, and have been recorded in the ancient texts of India, China, the Middle East and Europe, Despite the development and growth of the pharmaceutical industry in the twentieth century, plants remain the major source of medicines for a large proportion of the human population, particularly in the developing world. Indeed, many of the medicines commonly used today have their origins as traditional medicines. The natural products derived from plants have proven to be an abundant source of biologically active compounds, many of which have been the basis for the development of new lead chemicals for pharmaceuticals.

Herbal medicine is globally accepted as a valid alternative therapeutic system and has been recognised as such by the World Health Organisation. Nevertheless, as our understanding of drug mechanisms, pharmacokinetics, phannacodynamics and drug-drug interactions increases, it is essential that modern scientific investigations be undertaken to rigorously validate the claims of traditional medicine and verify the safety and efficacy of such medicines. Given that herbal medicines are often used in conjunction with prescription drugs in modern society, a relevant health and safety concern is the potential interaction between plant extracts and drugs.

The book *Utilisation and Management of Medicinal Plants* is the second volume in the series and brings together a comprehensive collection of articles which present recent findings in the areas of traditional medicine, phytochemistry and phytomedicine. The topics covered are varied and have broad appeal across numerous disciplines including medicine, agriculture, industry and biotechnology.

I offer my congratulations to the editor and contributing authors for their immense efforts in producing this book. The volume will be of great value to students, researchers and practitioners engaged in the study and utilisation of medicinal plants.

**Enzo Palombo**

*Associate Professor and Director*
*Environment and Biotechnology Centre*
*Swinburne University of Technology*
*AUSTRALIA*

# Preface

The use of plants for treating diseases is as old as the human species. Medicinal plants continue to play a central role in the healthcare system of large proportions of the world's population. The history reveals that medicinal plants have traditionally served as man's most important weapon against pathogens. People use traditional, natural remedies for curing different types of disease and even in the next millennium the herbal medical practice will forge as important remedy to treat various disorders. This is particularly true in developing countries, where herbal medicine has a long and uninterrupted history of use. Recognition and development of the medicinal and economic benefits of these plants are on the increase in both developing and industrialized nations.

Medicinal plants belong to the earliest known health care products that have been used by mankind. Over three quarters of the world population rely on the use of traditional medicines for their primary health care needs. Medicinal plants are the backbone of all major systems of medicine. Progress in medicinal plant research has undergone phenomenal growth in last few decades. Day by day medicinal plant research is gaining popularity because of its lesser side effects, lesser price and broad spectrum activity. Being immense value to the mankind it is the need of the time to increase the safety, effectiveness and efficacy of novel as well as current medicinal plants. Vigorous research on medicinal plants is going on to divulge the secret treasure present in plants as pharmacologically and therapeutically active phytochemicals. Therefore, medicinal plants are treated as a subject of serious study and are undergoing intense research all over the world. The knowledge has spread in India and worldwide at a high pace, covering not only important and promising plants but also thousands of lesser investigated and uninvestigated medicinal plants. This has led to an unprecedented growth and multi disciplinary published information widely scattered across large number of periodicals, conferences, proceeding reports etc. However for effective use and fruitful management of this accumulated scientific knowledge which is being continuously generated at a greater pace, a need has been felt for its systematic compilation in a form that can be comprehended and used.

It was in this backdrop, the present volume of the book series (Volume 2) has been designed in such a way as to serve the purpose of the students, researchers and practitioners engaged in the study and utilization of medicinal plants. The book includes twenty six original research and review articles written by eminent researchers from within India and abroad. The lead articles included in this volume are: Anti-venom Activity of Medicinal Plants; Phytotherapeutic Potentialities and Efficacy of *Cymbopogon citratus*; The Antimalarial Potential of Medicinal Plants; Histochemical Studies on Selected Medicinally Important Primitive and Advanced Ferns; *Phyllanthus niruri* Linn.: A Versatile Herb; Process Management in Herbal Medicine; Plants of Holy Quran; An Overview of *Pimpinella anisum* L.; Wound Healing Activity of *Achyranthes aspera* L.; Chemoprofiling and Antioxidant Activities of Shilajit; Antioxidative Potential of *Marchantia* sp; Economic Botany of the Amazonian Copaiba Oil; Use of Timbó to Control Agriculture Pests; Essential Oils -Their Chemistry, Extraction Techniques, Quality Control and Utilization; Extraction Methods and Health Benefits of Nutraceuticals; Biotechnological Products of Amazonian Fruits; Medicinal and Aromatic Plants-Diversity, Conservation and Future Prospects; Pharmacognostic Evaluation and Antibacterial Activity of *Prosopis cineraria* Leaf; Antibacterial Effect of Rhizobial Exo-Polysaccharide Against some Multidrug Resistance Human Pathogens; Bioassay-guided Fractionation for the Isolation of the Antibacterial Compounds from *Loranthus* species; Processing of *Aloe vera* Leaf; A Study of Photo Damage of Hair and Its Protection Using *Aloe vera*; Anti-Proliferative Potential and Induction of Apoptosis from *Catharanthus roseus*; Pharmacognosy Review of *Emblica officinalis*.

The first volume of this book series entitled, "*Utilisation and Management of Medicinal Plants*" has already been submitted to general perusal in 2010 and it was a big success. This volume like the previous one shall definitely prove very useful to the researchers who are interested in seeking knowledge/information pertaining to the utilisation and management of medicinal plants. We are thankful to the publisher M/S Astral International Pvt. Ltd., New Delhi for publishing the book well in time without compromising the quality of the contents and the printing. Contributions from our esteemed contributors are also greatly acknowledged and errors if any are owned by the editors.

*Dr. V.K. Gupta*

# Contents

Utilisation and Management of Medicinal Plants Vol. 2 (2014)    *Pages* 1–61
*Editor-in-Chief:* V.K. Gupta
*Published by:* DAYA PUBLISHING HOUSE, NEW DELHI

# 1

# Anti-venom Activity of Medicinal Plants from South America

Eduardo Dellacassa[1]*, Ana M. Torres[2],
Gabriela A.L. Ricciardi[2], Francisco J. Camargo[2],
Sara G. Tressens[2] and Armando I.A. Ricciardi[2]

## ABSTRACT

*The use of plants to subdue or reverse the effects of snakebite has long been recognized. Plant extracts were widely used as therapy for snakebite by South American traditional healers, and especially in tropical regions where plant resources are diverse and plentiful. Several medicinal plants are also believed to have been used as a source of snakebite antidote. Reference to the use of plants as part of early indigenous ethnomedical practices appears in traditional drug recipes recovered from chronicles of Spanish explorers and includes evidence that methods were passed on orally through generations. The most frequently encountered poisonous snake genera found in regions of South America include: Bothrops (jarara/ lanceheads), Crotalus (casabel/rattlesnakes), Lachesis (surucucú/bushmasters) and Micrurus (coral). Coral snakes have very powerful venom but confrontation with these snakes is infrequent because of their quiet and secretive character. A bibliographic revision helped us identify more than one hundred seventy plants with anti-venom activity in South America. Most are identified using their common names, sometimes making it difficult to determine the formal taxonomic name. Furthermore, the absence of available references or other documentation to verify or even designate a proper scientific name continues to pose a problem. There have been*

1    Cátedra de Farmacognosia y Productos Naturales, Departamento de Química Orgánica, Facultad de Química, 11800-Montevideo, Uruguay

2    Laboratorio Dr. G.A. Fester, Facultad de Ciencias Exactas y Naturales y Agrimensura, Universidad Nacional del Nordeste, Corrientes, 3400, Argentina

*    *Corresponding author*: E-mail: edellac@fq.edu.uy

*numerous attempts to study and characterize the anti-venom activity in native plants. We show that modern methods that provide unequivocal identification of active compounds together with in vitro and in vivo assays have enabled both evaluation and validation of ancestral knowledge. Among the pharmacologically active secondary metabolites isolated from plants, flavonoids are most frequently cited inhibiting phospholipases, lipoxygenases and metalloproteases.*

*Keywords*: South American medicinal plants, Bothrops, Anti-venom activity.

## Introduction

Effective treatment for bites and stings of venomous animals like snakes, spiders and scorpions has always been of great concern, and certainly remains so for persons in isolated rural areas where modern health care is not available. In the writings of colonial Spanish and even in modern texts, there are frequent references to the application of plant materials to treat bites and stings for a range of venomous animals. One may read that amazing and wonderful cures were possible using information gained, compiled, refined and enriched over the many years.

The oldest relevant South American record documenting the healing properties of plants was in 1711, "*Materia Médica Misionera por el hermano Pedro de Montenegro*". This manuscript describes and references the virtues and healing or curative properties of ca. 108 plants using their Spanish and vernacular names (Guarani or Tupi), the areas in which they are found and instructions for use. Specific instructions for 16 of these plants are provided for treatment of snakebites (Ricciardi *et al.*, 1996).

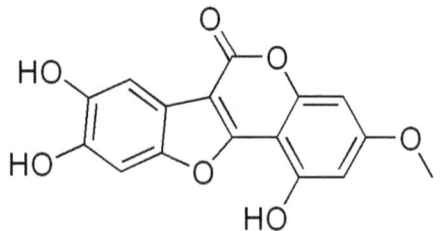

A number of other compounds with anti-venom activity were also reported. Among such compounds was the prenylated pterocarpan, (-)-edunol, isolated from the Fabaceae, *Brongniartia podalyrioides* Kunth. and *B. intermediate*: with structure and properties similar to the cabenegrines (Reyes-Chilpa *et al.*, 1994). Yet another was a compound whose activity was first recognized in studies with mice, *ar*-turmerone, a sesquiterpene from *Curcuma longa* (Zingiberaceae). This compound was shown to have anti-venom activity against venom of both *B. jararaca* and *Crotalus durissus terrificus* (Ferreira *et al.*, 1992). Wedelolactone is another compound isolated from *Eclipta prostrata* that antagonizes the poisonous effects of crotalid myotoxin, and in which an enzyme inhibitory mechaniam is likely (Melo, 1997; Melo *et al.*, 1993; Melo and Ownby, 1999).

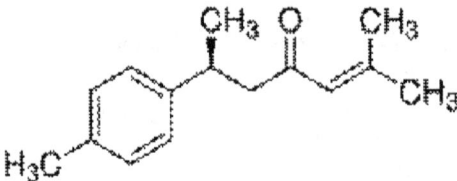

**ar-turmerone**                                   **wedelolactone**

The increasing number of publications examining *in vitro* and *in vivo* interaction between extracts or isolated components of plants and snake venoms encourage review and discussion of the issue of inhibition or neutralization of the toxicity of these poisons by extracts, oils or various natural products.

Even today in regions where modern medical facilities are not available or easily accessible or affordable, traditional medicine remains as the primary form of health-care. This type of health care also remains primarily based on ancestral knowledge. A clearer and deeper understanding of accumulated knowledge serving as a basis for these traditional methods should further validate potential usefulness. An understanding of mechanism of action for attenuation of venom toxicity will guide future research to further reduce ineffectiveness or danger (Farnsworth, 1988; Posey, 2002).

## Snakebites

Risk of being bitten or stung by a snake constitutes a serious and frequent threat in South America. The incidence of snakebite for different regions of the world is summarized in Table 1.1. Both the incidence of poisonous bites and resulting number of deaths is high in Latin America (López Sáez and Perez Soto, 2009):

Table 1.1

| Region | Population | N° Snakebites (x10⁸ inhabitants) | N° Poisonings | N ° Deads |
|---|---|---|---|---|
| Europe | 700 | 25,000 | 8,000 | 30 |
| Near East | 160 | 20,000 | 15,000 | 100 |
| North America | 270 | 45,000 | 6,500 | 15 |
| Latin America | 400 | 300,000 | 150,000 | 5,000 |
| Africa | 750 | 1,000,000 | 500,000 | 20,000 |
| Asia | 3,000 | 4,000,000 | 2,000,000 | 100,000 |
| Oceania | 20 | 10,000 | 3,000 | 200 |
| Total | 5,300 | 5,400,000 | 2,682,500 | 125,345 |

The poisonous action of venom is due to toxins and enzymes including: cardiotoxins (cytotoxins), disintegrins, hemotoxins, hemolysins, myotoxins, necrotoxins, nephrotoxins, neurotoxins, L-amino acid oxidase, phospholipases, adenesine triphosphatase, acetylcholinesterase, phosphodiesterases, proteases (hemorrhage-promoting) monoesterases (5'-nucleotidase), metalloproteinases, hyaluronidase, peptidase, deoxiribonucleasas, fibrin and fibrinogenolysinases, *L*-arginine ester hydrolase, coaglutinins, lectins.

The species of venomous snakes found in South America belong mainly to two families, Viperidae (solenoglyphous teeth, Figure 1.1A) and Elapidae (proteroglyphous teeth, Figure 1.1B). Genera frequently found within these two large families include *Bothrops*, *Crotalus*, *Lachesis* and *Micrurus*, and different species are found for each of these genera according to region.

## Historical Background

As noted above, chronicles of colonial times describing aboriginal medicine or ethnomedicine in South America make numerous references to the use of plants for treatment of snakebites. The oldest such document in 1711 was that of Hermano Pedro de Montenegro of the Compa ía de Jesús, and consisted of many handwritten sheets. The plant names and descriptions of that time are not always easy to equate with the most likely botanical species recognized today. Ricciardi *et al.* (1996) were able to identify some of the species referred to those having properties effective against snakebite in these documents: *Cyperus sesquiflorus, Cissampelos pareira (C. glaberrima), Sidastrum paniculatum, Asclepias mellodora (A. campestris), Euphorbia dichotoma, Dorstenia brasiliensis, Agrimonia eupatoria, Aristolochia rotunda, Gomphrena tuberous, Pilocarpus pennatifolius.*

**Figure 1.1A**: Dental structure of solenoglyphous teeth.

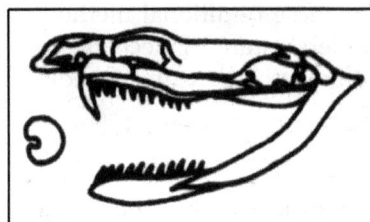

**Figure 1.1B**: Dental structure of proteroglyphous teeth.

Other medicinal plant historians include Antonio Ruiz de Montoya (1585-1652), Buenaventura Suárez (1679-1750) and Andreas Thevet who published his observations on medical practices of the Guarani in Brazil, *Les singularités de la France antarctique autrement nommée Amérique et des plusieurs terres et des decouvertes de notre temps*, reprinted in Paris in 1878. Jose Acosta (ca. 1539-1600) studied American flora and recorded his observations in *De natura novi orbis* which was later published in Salamanca in 1588. There were also numerous revisions of this work such as, *Historia Natural y Moral de las Indias* (Pedro Lozano, 1697-1752) a writer, historian and Spanish ecclesiastic who wrote on Chaco Valley. Like many American works of this period, the focus was mainly ethnographic detail. Noteworthy however, chapters on rivers and quality of land have extensive discussion of regional medicinal plants.

## Plants Historically Considered Effective against Snakebite

This review aims to present and update information from current literature on traditional knowledge for snakebite treatment from plants. Different authors have given detailed accounts on various snakebite treatments found in plants (Reyes-Chilpa and Jiménez Estrada, 1995; Mors *et al.*, 2000; Ricciardi, 2005; López Sáez *et al.*, 2009; Makhija and Khamar, 2010; Abhijit and Jitendra, 2012). These reports provided a comprehensive view of the ability of a given snakebite treatment to relieve one or more complex symptoms such as pain, bleeding, swelling, infection, in some cases for more than one poison. Knowledge and use of plants for snakebite treatment is very old. Medical treatment by the Ayurvedic in India (centuries before the Christian era) includes 211 plants. The use of plants in Mexico preceded colonization by Spanish, and in the 16th century Francisco Hernandez was able to document 119 newly described plants in his *Historia de las plantas de la nueva Espa a* (History of the New Spain plants). Also, in the 16th century Fray Bernardino de Sahagún cites the use of picietl or tobacco. Ethnic groups like the Colorados, Cayapa and Coaquier of northern Ecuador used 40 species of Gesneriaceae, and the Colorados and Cayapa also used 11 species of Polypodial and 7 species of Piperaceae.

Reyes-Chilpa and Jiménez Estrada (1995), in writing about the Tepehuanos of Chihuahua noted that: "the indians rarely open wounds caused by snakebites. To cure the bite they depend on the application of specific mixtures prepared with differents plant materials."

Otero *et al.* (2000a, 2000b, 2000c), describe the use of more than 77 plants in Colombia in the form of beverages prepared by infusion, as alcohol extracts (30-38 per cent v/v) or by mixing various plant materials with rum. Other methods included decoction or maceration in water for topical or localized use. Esteso (1985) in his work *Ofidismo en la República Argentina* (Snakebite in Argentina) citing folklore, almost always referred to the statements of a respectable elderly such as a grandfather, whose knowledge passed from family to family.

López Sáez *et al.* (2009) cite use of the genus: *Aristolochia* (22 species), *Ficus* (5 species in India besides the common fig), Araceae (26), *Amaranthus* (4), *Rawvolfia* (7), *Amorphophallus* (4), Araceae, *Heliotropium* (5, Heliotropiaceae), Asteraceae (*Eupatorium, Mikania, Vernonia*), *Terminalia* (5, Combretaceae), *Ipomoea* (6, Convolvulaceae), *Euphorbia* and *Phyllantus* (9 and 5, Euphorbiaceae), *Cassia* (6, Fabaceae), *Strychnos* (4, Loganiaceae), *Piper* (8, Piperaceae), *Zanthoxylum* (6, Rutaceae), *Solanum* (6, Solanaceae) and *Clerodendrum* (6, Verbenaceae). Also many ornamental or food plants are reported as having activity against snakebite: mango (*Mangifera indica*), litchi (*Litchi chinensis*), saffron (*Crocus sativus*), papaya (*Carica papaya*), chickpea (*Cicer arietinum*), nutmeg (*Myristica fragrans*), pepper (*Capsicum annuum*), longan (*Euphorbia longan*), castor (*Ricinus communis*), garlic and onion (*Allium cepa, A. sativum*), sweet potato (*Ipomoea batatas*), persimmon (*Diospyros kaki*), leaves of artichoke (*Cynara scolymus*), Brazilian oleander (*Nerium oleander*) in the Middle East, leaves of blood flower (*Asclepias curassavica*) in Central America, achiote (*Bixa orellana*) in India and the Philippines, roots of cassava (*Manihot esculenta*), the seeds of cacao (*Theobroma cacao*) in South America, coconut (*Cocos nucifera*), and many species in India such as sunflower (*Helianthus annuus*), pomegranate (*Punica granatum*), grape (*Vitis vinifera*) and cider (*Citrus medica*).

The following list of genera and species of South American plants have been reported as having antisnake activity:

## Genus *Aloysia*

*Aloysia citriodora* Palau (Verbenaceae), syn. *Aloysia triphylla* (L'Hér.) Britton, syn. *A. citriodora* Ortega ex Pers., *hom. illeg., Lippia citriodora* (Lam.) Kunth. *comb. illeg., L. triphylla* (L'Hér.) Kuntze, *Verbena triphylla* L´Hér., *Zapania citriodora* Lam. is a very common species that grows in South America (Zuloaga and Morrone, 1999) (Cabrera, 1993).

### Common Names

Lemon verbena, cedrón, hierba luisa (Jozamí and Mu oz, 1982); cerdón, cedrón de Castilla (Martínez Crovetto, 1961); in Brazil: salvia limao, erva cidreira (González Torres, 1997; Manfred, 1977).

## Background

This species is widely used in ethnomedicine for preparation of teas or infusions with the leaves, flowers and young stems for treatment of a range of disorders including: indigestion, diarrhoea, vomiting, anxiety, nervous affections, hysteria (Cáceres, 1996), asthma, cough, fever (Bassols and Gurni, 1996). Leaf infusions are used as stimulant (Fester *et al.*, 1961); also as antimalarial, antineuralgic and tonic (Bassols and Gurni, 1996), (Gupta, 1995), (Toursarkissian, 1980); anti-inflammatory, analgesic, antipyretic, tonic and stimulant (Oliva *et al.*, 2010). Infusions were also reported as being antimicrobial and antimycotic (Sartoratto *et al.*, 2004; Oksay *et al.*, 2005); and antioxidant (Stashenko *et al.*, 2003). The essential oil is also used in perfumery and in the flavor industry for liquors and foods (Cáceres, 1996). In Argentina, the plant is an official drug in the Farmacopea Argentina and is included in the Código Alimentario Argentino. In Europe its use in perfumes is restricted because it was reported to be photo sensitizing (Bandoni, 2000). The plant is mentioned as having anti-venom activity only in a few reports, and these reports lack scientific evidence or confirmation of activity (Manfred, 1977; Duke *et al.*, 2009).

## Chemical Constituents

The essential oil is composed mainly of: 1,8-cineole (15 per cent), limonene (1 per cent), a-pinene (0.5 per cent), citral a (37 per cent), citral b (22 per cent), geraniol (11 per cent), linalool (1 per cent) (Terblanché *et al.*, 1996); the essential oil of flowers (from Córdoba, Argentina): limonene (9 per cent), myrcene (2 per cent), a-pinene (1 per cent), a-thujone (17 per cent), mircenone (31 per cent), camphor (5 per cent), carvone (2 per cent), lippifoli-1(6)-en-5-one (africanone) (9 per cent) and spathulenol (3 per cent) (Zygadlo *et al.*, 1995). Neral (12 per cent), geranial (17 per cent), limonene (22-38 per cent) mainly *levo* isomer (98 per cent *l*-limonene and 2 per cent *d*-limonene) is found but is variable depending on stage of plant growth, as well as sabinene (6-15 per cent) mainly *dextro* isomer (99 per cent *d*-sabinene) and a sesquiterpene fraction (10-30 per cent) with caryophyllene, bicyclogermacrene and caryophyllene oxide and other compounds (Ricciardi *et al.*, 2011). Leaf infusions contain 400 mg/L verbascoside and 100 mg/L luteolin 7-diglucuronide; the same infusion yields 51 per cent of essential oil containing 77 per cent citral, far more than the citral content in leaves (41 per cent) (Carnat *et al.*, 1999).

Essential oil in plants from Brazil have: citral (59 per cent), geraniol (11 per cent), eucalyptol (15 per cent) and linalool (1 per cent); material from France: citral (38 per cent), geraniol (6 per cent), nerol (5 per cent), limonene (4 per cent); from Portugal: limonene (23 per cent) and citral (18 per cent) (Bandoni, 2000). The yield of essential oil of leaves and flowers from Guatemala representing 0.5–2 per cent (w/w) consisted of: cineol, citral (20–39 per cent), *l*-limonene (10–15 per cent), sesquiterpenes (β-caryopyhyllene, isocaryophyllene, 2,6-β-caryophyllene oxide) (40–45 per cent), linalool (3.5–11 per cent), *d*-verbenone, aldehydes and ketones (photocitral isomers, *l*-carvone, traces of furfural and kobusone) phenolic acids, acetic and valerianic acid, flavonoids, hydrolysable tannins, pyrrol, flavones and alkaloids (Cáceres, 1996).

## Properties Tested

SDS-PAGE was done as part of an *in vitro* screening method to determine the anti-venom activity of all essential oils and extracts of aerial parts of the plant from three different geographical regions (Paso de la Patria, Laguna Brava and San Luis del Palmar, Corrientes, Argentina). The volatile samples showed a substantial modification of the electrophoretic profile of the venom of *B. diporus* which was evident by the disappearance of the phospholipases protein bands (18 kDa) (Camargo *et al.*, 2011). On the other hand, aqueous, alcohol or hexane extracts of aerial plant parts showed no activity by SDS-PAGE at the doses tested (Cáceres Wenzel *et al.*, 2012).

*In vitro* SDS-PAGE screening was done to determine the anti-venom activity of all essential oils of aerial parts of the plant. All essential oils of aerial parts of the plant from three different geographical origins (Paso de la Patria, Laguna Brava and San Luis del Palmar, Corrientes, Argentina) shows a substantial modification of the electrophoretic profile of *B. diporus* venom with disappearance of the phospholipases protein bands (18 kDa) (Camargo *et al.*, 2011). On the other hand, aqueous, alcohol or hexane extracts of aerial plant parts showed no activity by SDS-PAGE at the doses tested (Cáceres Wenzel *et al.*, 2012).

# Genus *Aristolochia*

Among species acknowledged as being useful against snakebite, and usually considered weeds, are those assigned with relative certainty to be in *Aristolochia*. Treatments for both internal and external use have been developed from material from different species and are well known to travellers who frequent areas with snake risk as well as naturalists.

## Background

☆ Bonpland and Azara: Prescribed treatments developed using materials from species of this genus have been considered in some cases to be unfailing in effectiveness against snakebite.

☆ Gonzalez Torres (1997): Recipes for mixtures of powdered root material for addition to beverage or wine or that can be applied directly in a compress for snakebites.

☆ *A. fimbriata* in infusions and compresses; *A. gibertii*; *A. labiosa*; *A. macroura* as antidote against cobra venom; *A. serpentaria* whose roots can act as snake repellent as well as having healing properties for snakebite and *A. triangularis* also acts against cobra bites (Ricciardi, 2005).

## Chemical Constituents

Generally *Aristolochia* species are characterized by the presence of essential oil, but also presence of aristolochic acid (ACA), a phenanthrene-substituted carboxylic acid with ACA I, a nitroderivative. This compound is usually accompanied by aristolactams, or by germacranolides: aristolone, with ACA and *nor*aristolochic acid, in *A. clematitis*, whereas in *A. rotunda* and *A. longa*, aristolochic acid II (ACA II). Determination of ACAs is done by capillar electrophoresis (Priestap *et al.*, 2003; Ioset *et al.*, 2002).

## Properties Tested

ACA inhibits inflammation and edema, is associated with relieving pain and functions as a non-competitive inhibitor of PLA2. ACA was found to form a weak 1:1 complex with PLA2 resulting in an apparent increase in alpha helical secondary structure but with no measurable effect on tertiary structure (Vishwanath *et al.*, 1985; Vishwanath *et al.*, 1987a; Vishwanath *et al.*, 1987b; Vishwanath *et al.*, 1987c; Moreno, 1993; Houghton *et al.*, 1993; Pereira *et al.*, 1994 and 1992).

**Aristolochic Acid**

ACA was more recently found to have a high-affinity for PLA2 from *Vipera russelli* to provide a structural basis for inhibition of arachidonic acid biosynthesis leading to production of prostaglandins. The structure involved two crystallographically independent molecules of PLA2 as an asymmetric dimer with ACA bound specifically to only one of the two polypeptides, molecules A and B. ACA could bind to molecule A only while the binding site of molecule B is empty. Binding was associated with a change in conformational characteristics of W31 that resulted in a narrowing or occlusion of a hydrophobic channel serving as a passage to active site residues (Chandra *et al.*, 2002).

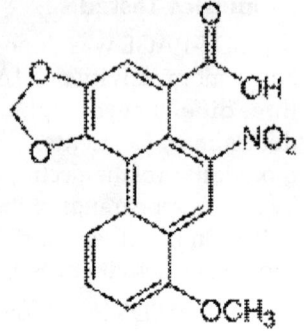

## Toxicity

ACA is an irritant of mucus membranes and at high doses can cause respiratory paralysis. It is also nephrotoxic and carcinogenic (mutagenic). Oral administration or intraperitomeal injection of ACA to rats or mice resulted in stomach carcinomas and kidney fibrosis, and in rabbits a low incidence of renal tumors. Subcutaneous injection of ACA to rats caused urothelial kidney tumors and malignant fibrohistiocytic sarcomas at the site of injection. ACA I and II were shown to be mutagenic in several systems. A mixture of the two acids administered to rats was found to be so highly carcinogenic in a study in Germany that Aristolochiaceae plant materials were eliminated from homeopathic products.

ACA passed in maternal milk or increasing intravenous dosage of the acids can result in renal toxicity, nephrosis, renal failure and even death (DeSmet, 1995). These observations underscore the importance of only administering ACA topically. Persons most adversely affected (as nephropathologies) by ACA s were those known to have regular dietary supplements containing material from *Aristolochia* containing ACAs (IARC, 2002).

*Aristolochia angustifolia*, Cham. (Aristolochiaceae); syn. *A. angustifolia* Cham. var. *guaranitica* Ahumada (Martínez Crovetto, 1961).

## Common Names

Cipo mil homens (González Tórres, 1997), luz del campo (Martínez Crovetto, 1961).

*Aristolochia argentina* Griseb. (Aristolochiaceae)

## Common Name

Charruga (Domínguez, 1928)

## Chemical Constituents

Essential oil is found throughout the whole plant with principal compounds: *cis* and *trans*-β-ocimene; linalool; (3*E*,5*Z*)-1,3,5-undecatriene; (*E,E*)-1,3,5-undecatriene; β-elemene; β-caryophyllene; (*E*)-β-farnesene; germacrene D; spathulenol, bicyclogermacrene (Priestap *et al.*, 2003b); lactams (Crohare *et al.*, 1974); seven aristolactams (Priestap, 1985a), two carboxyaristolactams and two hydroxymethyl aristolactams (Priestap, 1985b); specifically in roots: ARA, ARA I and II methyl esters (Priestap, 1982); other ARAs: (Priestap, 1987), argentilactone (Priestap *et al.*, 1977). Argentilactone is a mayor constituent of essential oil (57-89 per cent), and is allergenic as well as being a skin irritant. Plant extracts are also cytotoxic (Mongelli *et al.*, 2000).

*Aristolochia elegans* Mast. (Aristolochiaceae), syn. *A. hassleriana* Chodat; *A. littoralis* auct. (Parodi *et al.*, 1988).

## Background

Used in Mexico against snake venoms. The species has anti-inflammatory activity like other *Aristolochia*, and inhibits phospholipases consistent with attenuating action of mediators of inflammation and pain (Hutt and Houghton, 1998). This species was also shown to have anticholinergic activity (Rastrelli *et al.*, 1997).

## Chemical Constituents

In leaves, sesquiterpene hydrocarbons, β-caryophyllene, isocaryophllene and bicyclogermacrene, (in roots and stems) oxygenated sesquiterpenes: (*E*)-nerolidol (Vila *et al.*, 1997); *ent*-kaurane-16α,17-diol, (-)-cubebin, α-methylcubebin, methylcubebin, β-(-)-hinokinin, (-)-53-methoxyhinokinin, (-)-kobusin (Habib and El-Sebakhy, 1981).

Non-volatile compounds reported in leaves: (-)-(*R,R*)-7'-*O*-methylcuspidatin (El-Sebakhy and Waterman, 1994), aristololide, aristogin C and two porphyrins: aristophylls A and B (Wu *et al.*, 2000); diterpenoids as in the case of *ent*-16β,17-epoxykauran, *ent*-kaur-15-en-17-ol and *ent*-15β,16-epoxykauran-17-ol; pericampy linone-A, corydaldine, thalifoline, northalifoline, *N*-methylcorydaldine (El-Sebakhy *et al.*, 1989). In roots and stems: 4-methoxy-3, 42-oxydibenzoic acid, tetralones: aristelegone A, B, C and D; pericampylinone-A, E; (Wu *et al.*, 2000), lignans (Wu *et al.*, 2002); aristolactams and benzoyl alkaloids (Shi *et al.*, 2004); aristolochic acids, amides, aporphines, benzenoidsand steroids.

## Properties Tested

Hexane and methanol extracts had anti-toxic activity against venom of Mexican scorpion *Centruroides limpidus limpidus* in rats, and *in vitro* inhibition of concentration-dependent venom induced contractions of ileum (Jiménez-Ferrer *et al.*, 2005).

Treatment of snakebite venom with aqueous and alcoholic extracts of leaves, roots and stems of plants from Corrientes, Argentina resulted in slight variations in the intensity of bands by SDS-PAGE electrophoresis, and did not affect hemolytic activity. The procoagulant activity was inhibited by alcoholic extracts of leaves (31 per cent, 1: 20) and roots (42 per cent, 1: 40). Alcoholic extracts of leaves (1: 120), stems (1: 120) and roots (1: 80) also inhibited casein proteolytic activity of *B. diporus* venom (Torres *et al.*, 2012).

*Aristolochia eriantha* (Aristolochiaceae)

**Common Names**

Cipo-mil-homens (González Torres, 1997)

*Aristolochia fimbriata Cham.* (Aristolochiaceae) syn. *A. ciliata* Hook, *A. ciliosa* Benth

**Common Names**

Patito, ipe-mí, flor de patito, pajarito (Iturralde, 1925).

*Aristolochia gibertii Hook.* (Aristolochiaceae)

**Common Names**

Ysypó mil hombres, contrayerba, ypemí, patito (Gonzalez Torres, 1997; Toursarkissian, 1980; Martinez Crovetto, 1961 Parodi, 1988).

**Properties Tested**

Methanol extracts showed enzymatic and non-enzymatic protection against lipid peroxidation in experiments using a free radical generating system on mice membranes (Velazquez *et al.*, 2003).

Alcohol (17 per cent) and aqueous (35 per cent) extracts of leaves inhibited procoagulant activity in *B. diporus* venom. Aqueous extracts of leaves were analyzed for ability to bind/inactivate this venom *in vitro* by SDS-PAGE. The extract caused the disappearance of two bands: 57.5 kDa (hemorrhagin NHFb, hemorrhagic metalloprotease) and 52.5 kDa. The intensities of other bands were greatly reduced, one of 28.2 kDa botroalternin (type C lectin inhibitor of thrombin) and botrocetin (platelet conglutinin) and 17 kDa (Phospholipase A2, myotoxin) (Torres *et al.*, 2004). Finally, extracts showed no inhibition of venom hemolytic activity, and alcohol extracts of: leaves (1: 150), stems (1: 150) and root (1: 100) inhibited casein proteolytic activity in B. diporus venom (Torres *et al.*, 2012a).

*Aristolochia macroura* Ortega (Aristolochiaceae); syn. *A. appendiculata* Vell, *A. caudata* Booth ex Lindl., *A. caudata* Parodi, *A. macroura* Gomez *var. subtrifida* Duchtr, *A. trilobata* Lindl.

**Common Names**

Nil hombres (Domínguez, 1928; Martinez Crovetto, 1961).

**Background**

Useful against cobra venom (Ahumada, 1978).

*Aristolochia triangularis Cham.* (Aristolochiaceae) syn. *A. antihysterica* Mart. ex Duch., *A. sellowiana* (Klotzsch) Duch., *A. paraguariensis* D. Parodi, *A. salpinx* Mast, *A. triangularis* Cham. *var. salpinx* (Mast.) Hauman.

## Common Names

Mil hombres, ypemí, jarrinha, cipó cobra, papo-de-Peru (Gonzalez Torres, 1997; Toursarkissian, 1980; Dominguez, 1928; Parodi and Dimitri, 1988; Di Stasi *et al.,* 1989). ACA was found in roots.

## Background

Commonly used against cobra bites (Oliveira Simoes *et al.*, 1998), and for the cytotoxic effects of this plant see Mongelli *et al.* (2000).

*Aristolochia trilobata* **L.;** (Aristolochiaceae)

## Common Names

Jarrinha, mil hombres, papo de Peru, (Di Stasi *et al.,* 1989)

# Genus *Asclepias*

*Asclepias mellodora* A. St.-Hil. (Asclepiadaceae), syn. *Asclepias campestris* Decne.

## Common Names

Pasto de vibora, mboi ka'a (Dominguez, 1928; Gonzalez Torres, 1997; Toursarkissian, 1980; Ricciardi *et al.,* 1996; Amorin, 1988; Burkart, 1979; Marzocca, 1997).

## Chemical Constituents

Desglucouzarin, the main cardanolide glycoside (Petricic, 1966); vincetoxin and asclepiadin (cardenolide), soluble in alcohol but sparingly soluble in ether; the glycoside yields asclepin after hydrolysis as a first derivative.

## Background

The plant material is crushed or chewed to release active ingredients, and applied directly to snakebite wounds (Montenegro, 1711).

## Properties Tested

Contains cardiotoxic cardenolides which inhibit myocardial $Na^+/K^+$ ATPase affecting conductivity and myocardial contractility. Alcohol extract of the aerial parts of *A. mellodora* inhibits proteolytic activity (Ricciardi Verrastro *et al.,* 2012).

*Asclepias curassavica* **L.**

The alcohol extract of leaves inhibits hemolytic activity of *B. diporus* venom *in vitro* (Torres *et al.,* 2007) and hexane extract of aerial parts inhibits the procoagulant activity (Torres *et al.,* 2008). Also, both alcoholic and hexanic extracts inhibit proteolytic activity of venom (Ricciardi Verrastro *et al.,* 2012).

# Genus *Bidens*

*Bidens pilosa* L. (Asteraceae) annual grass of two recognized varieties: *B. pilosa* L. var. *minor* (Blume) Sherff and *B. pilosa* L. var *pilosa*, the latter having a greater geographic distribution. Syn.: *B. chilensis* DC, *B. hirsuta* Nutt., *B. pilosa* var. *minor* (Blume) Sherff., *B. pilosa* var. *radiata* Sch. Bip.

## Chemical Constituents

Essential oil has limonene, borneol, germacrenes, cadinols, D-muurolol, 3-propyl-3-(2,4,5-trimethoxy) benzyloxy-pentan-2,4-dione, diterpenoids: phytol, phytylheptanoate, triterpenoids: squalene (Zulueta *et al.*, 1995), β-amyrin, lupeol, lupeylacetate, purines: caffeine and derivatives (Ogawa and Sashida, 1992); coumarins: esculetin; flavonoids: quercetin, isoquercitrin, luteolin, friedelin, friedelan 3-β-ol, (Geissberger and Sequin, 1991; Wang *et al.*, 1997), 5-*O*-methylhoslundin (Sarker *et al.*, 2000), steroids, stigmasterol (in leaves, Zulueta *et al.*, 1995), β-sitosterol, acetylenes: 2-*O*-β-D-glucosyl trideca-11-(*E*)-ene-3,5,7,9,-tetrain-1,2-diol (PA-1) (Pereira *et al.*, 1999), 2-β-D-glucopyranosyloxy-1-hydroxy-5(*E*)-tridecene-7,9,11-triyne and 3-β-D-glucopyranosyloxy-1-hydroxy-6(*E*)-tetradecene-8,10,12-triyne (Ubillas *et al.*, 2000), β-D-glucopyranosyloxy-3-hydroxy-6-(*E*)-tetradecen-8,10,12-triyne (Alvarez *et al.*, 1996); phenylheptatriyne (Geissberger and Séquin, 1991); trideca-2,12-dien-4,6,8,10-tetrain-1-ol; trideca-3,11-dien-5,7,9-triine-1,2-diol; trideca-5-en-7,9,11-triine-3-ol; (Sarg *et al.*, 1991), in the root: 3,3'-dimethyl ether 7-*O*-α-L-rhamnopyranosyl-(1 → 6)-β-D-glucopyranoside and quercetin 3,3'-dimethyl ether 7-*O*-β-D-glucopyranoside (Brand o *et al.*, 1998). Caffeine (Sarker *et al.*, 2000), acids: succinic, vanillic, capric, lauric, palmitic, palmitoleic, elaidic behenic, linoleic, linolenic, myristic, phytic; in leaves: elaidic and behenic acids (in leaves: Zulueta *et al.*, 1995); carotenoid pigments: xanthophyll; tannins (Chippaux *et al.*, 1997).

## Properties Tested

Aerial parts are used for anti-venom treatments. The polyacetylenes (Alvarez *et al.*, 1996; Brand o *et al.*, 1997), linoleic and linolenic acids as well as flavonoids and friedelin and friedelan 3-ol are known to be anti-inflammatory (Geissberger and Séquin, 1991; Pereira *et al.*, 1999), plant extracts exhibit prostaglandin biosynthesis inhibitory activity (an inflammatory process) (Chippaux *et al.*, 1997). Intravenously and intraperitoneally treatment neutralizes the effect of Cameroon elapid snake, *Dendroaspis jameson*, venom, rich in neurotoxin agonists and antagonists of acetylcholine and whose toxicity is not dependent on the route of inoculation in rats. Such treatment also counteracts the action of other poisons and has anti-malarial activity (Oliveira *et al.*, 2004). The phenylheptatriyne is photoactive but unlike furanocoumarins, generates DNA crosslinks when UV irradiated (Wat *et al.*, 1979).

# Genus *Baccharis*

*Baccharis trimera* (Less) DC (Asteraceae)

## Common Names

Carqueja

## Background

Used to treat liver ailments, rheumatism, diabetes as well as kidney, liver and digestive disorders. The active component is a neo-clerodane diterpenoid: 7a-hydroxy-3,13-clerodadiene-16,15: 18,19-diolide, $C_{20}H_{28}O_5$.

## Properties Tested

Januário *et al.* (2004) observed that clerodane isolated from *B. trimera* inhibited hemorrhagic activity of *Bothrops* crude venom 70 per cent on fibrinogen in an approximate ratio 1: 10 w/w poison: clerodane. Also, metalloprotease activity, isolated from *B. jararacussú* venom, on casein at a ratio of 1: 10 protease/clerodane, was inhibited 80 per cent for Class I, and 89 per cent for Class II type metalloproteases. The strong hemolytic and proteolytic *Bothrops* venom inhibitory activity of clerodanes, as well as the beneficial effects on induced edema accompanying myotoxicity, indicates that continuing efforts to further develop snakebite treatments based on these natural products shows much promise as an alternative to treatment with heterologous serum.

# Genus *Boerhavia*

*Boerhavia diffusa* L. var. *diffusa* (Nyctaginaceae), syn. *B. paniculata* Rich., *B. coccinea* var. *paniculata* (Rich.) Moscoso.

## Common Names

Tangará (Gonzalez Torres, 1997), erva Tostao (in Brazil), in Venezuela, hierba de boca (Rodriguez, 1980).

## Background

In some regions treatment against snake bite involves taking tea of leaves and roots and applying locally as a compress after cooking.

*Boerhavia diffusa* L. var. *leiocarpa* (Heimerl) Adams (Nyctaginaceae), syn. *B. paniculata* var. *leiocarpa* (Heimerl) Heimerl; *B. diffusa* var. *paniculata* D. Parodi, nom. nud.; *B. coccinea* var. *leiocarpa* (Heimerl) Standl.

*Boerhavia repens* L. (Nyctaginaceae), syn. *B. adscendens* Willd., *B. coccinea* Mill, *B. caribaea* Jacq., *B. diffusa* L., *B. paniculata* Rich. syn. *Boerhavia diffusa* L. var. *diffusa*.

## Common Names

Tangará (Gonzalez Torres, 1997), in Venezuela, hierba de boca (Rodriguez, 1980), grows in tropical South America.

# Genus *Brunfelsia*

*Brunfelsia uniflora* (Pohl) D. Don; (Solanaceae), syn. *Franciscea uniflora* Pohl.

## Common Names

Jazmín del Paraguay, azucena, manacá, mercurio vegetal, Santa Maria. Half-height shrub, native to the Amazon, it is used as an ornamental because of its flowers.

## Background

Used to treat snakebites in South America. The root has been used in Europe to treat snakebites.

## Chemical Constituents

The essential oil consists of limonene, myrcene, ocimene, α-terpineol, terpinolene, geraniol, linalool, α-ionone, β-cyclocitral, β-bisabolene, elemol, nerolidol, farnesol, β-safranal, β-eudesmol, lavandulal, benzylbenzoate, isobutylsalicylate, benzylsalicylate, β-damascenone, 2-ethylfuran, *n*-heptane, *n*-octane, *n*-decane, *n*-pentadecane, *n*-heptadecane, *n*-hexadecane, *n*-nonadecane, *n*-pentacosane, heneicosane, tricosane, palmitic acid, myristic acid, linoleic acid, linolenic acid, pentadecanoic acid, neophytadiene (Maestri and Guzmán, 1995). Roots and bark have tropane alkaloids (lymphatic system stimulants) throughout the plant: mandragorine (central nervous system stimulant and anticholinergic) coumarins: esculetin, a furanocoumarin: scopoletin (analgesic and anti-inflammatory *in vitro*, induces psychopharmacological effects) (Ruppelt *et al.*, 1991), scopoline and other constituents: brunfelsene, lignans and sapogenins.

## Properties Tested

Oral administration of plant infusions have been shown to have analgesic and anti-inflammatory activities (Ruppelt *et al.*, 1990).

# Genus *Casearia*

*Casearia sylvestris* Sw (Flacourtiaceae) syn. *C. parviflora* Willd., *C. punctata* Spreng., *C. samyda* (Gaert) DC; *Samyda parviflora* L. *S. sylvestris* (Sw) Poir., *Anavinga samyda*.

## Common Names

Burro kaá, cafe-bravo, cafeiillo, corta-lengua, guaçatonga. Background: Decoction of leaf applied topically and taken internally is used by natives of the Amazon and Bolivia for snakebite (Borges *et al.*, 2000; Raslan *et al.*, 2002).

## Chemical Constituents

In leaves clerodane diterpenoids: casearvestrine A, B, and C, (Oberlies *et al.*, 2002); diterpenes: casearines G, S and T, (De Carvalho *et al.*, 1998; Morita *et al.*, 1991) and casearines A-F (Itokawa *et al.*, 1990); iridoids: 1-β-hydroxy-dihydrocornine; 1-α-hydroxy-dihydrocornine; α-gardiol; β-gardiol; plumericine; isoplumericine; 11-O-trans-caffeoylteucrein; ester derivatives: 2-methyl-4-hydroxy-butyl-caffeoate; 3-methyl-4-hydroxy-butyl-caffeoato; amides: N-[7-(3′,4′-methylenedioxyphenyl)-2Z,4Z-heptadienoyl] pyrrolidine and a triterpene: viburgenin. *Casearia* clerodanes I, II, III, IV, V and VI, polysaccharides and sterols sitosterol and stigmasterol, lapachol and saponins (Bolzani *et al.*, 1999).

## Properties Tested

Tests done *in vitro* have shown the plant to have analgesic and anti-inflammatory properties (Ruppelt *et al.*, 1990). Aqueous extracts of leaves were able to: neutralize hemorrhagic activity form *B. asper*, *B. jararacussu*, *B. neuwiedi* and *B. pirajai*, as well as two metalloproteinases from *B. asper* and at various levels proteolytic activity on casein in venom of *B. neuwiedi*; partially protect α-fibrinogen from degradation by venom of *B. jararacussu*; reduce the increase in plasma clotting times caused by venom of *B. jararacussu*, *B. moojeni* and *B. neuwiedi* (Borges *et al.*, 2001); inhibit Class I, II and III $PLA_2$ activity of various snake venoms, with having the greatest effect on Class II activity; inhibit myotoxic, anticoagulant and edema inducing activity in various venoms from *B. moojeni* and *B. jararacussu* (Borges *et al.*, 2000). Hydroalcoholic extracts of leaves or essential oil inhibits acute edema in mice caused by *B. alternatus*. Overall *C. sylvestris* is an excellent source of $PLA_2$ inhibitor, which is primarily responsible for the toxicity of lancehead venom. Plant extracts can counteract lethal doses of venom of vipers (mainly *Bothrops*) and bees.

Some *Casearia* have antitumor and cytotoxic activity (Itokawa *et al.*, 1988); anticancer activity, inhibit HIV replication or antibiotic activity on *Bacillus cereus* and *B. subtilis* (Itokawa *et al.*, 1990).

### *Casearia mariquitensis* (Flacourtiaceae)

Aqueous extracts of leaves of *C. mariquitensis*, a plant found in Brazilian open pastures, neutralizes hematologic dysfunctions induced by the crude venom or by neuwiedase, 22 kDa class PI metalloproteinase from venom of South American *B. neuwiedi pauloensis*. Incubation of erythrocytes with venom and extract at a ratio of 1: 10 (w/w, venom/extract), was generally effective but did not affect the decrease in platelet levels induced by the venom. Plasma fibrinogen concentration decreased ca. 36 per cent and 83 per cent when 0.6 $LD_{50}$ crude venom or neuwiedase, respectively, were i.p. injected in mice, and the aqueous extract could inhibit this effect. The beta fibrinogen chain was not degraded by crude venom or neuwiedase. The pulmonary hemorrhage induced by 0.6 $LD_{50}$ i.v. injected neuwiedase was completely inhibited when pre-incubated with extract at a ratio of 1: 10 (w/w, toxin/extract) (Izidoro *et al.*, 2003).

# Genus *Chaptalia*

*Chaptalia nutans* (L.) Pol. (Asteraceae), syn. *C. nutans* (L.) Hemsl. comb. superfl., *C. subcordata* Greene, *Leria nutans* (L.) D.C., *Tusilago nutans* L.

### Common Names

Peludilla, pelosilla, cerraja, lígua-de-vaca (Lemos de Arruga Camargo, 1999). Plant is found across north-central Argentina.

### Chemical Constituents

The aerial parts: prunasin (*d* form of mandelonitrile glucoside), parasorbic acid, 5-methyl-3α-hydroxivalerolactone and coumarins: 4-*O*-β-glucopyranosyl-5-

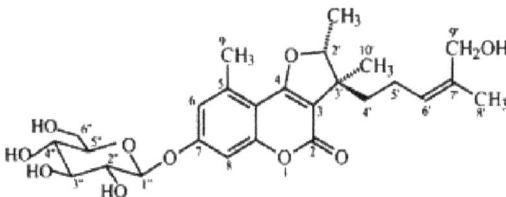

**7-*O*-β-D-glucopyranosyl-nutanocumarin**

methylcoumarin. The roots: nutanocoumarin, 9'-O-β-glucopyranosyl-nutanocoumarin, and 7-O-β-D-glucopyranosyl-nutanocumarin (Torrado Truiti *et al.*, 1998 and 2003).

## Properties Tested

Aqueous extracts have anti-inflammatory activity comparable to indomethacin in carrageenan-induced models in *in vivo* tests using female rats (Badilla *et al.*, 1999).

# Genus *Chiococca*

*Chiococca alba* (L.) C. L. Hitchc. (Rubiaceae), syn. *Ch. Brachiata* Ruiz and Pav., *Chiococca anguifuga*, Mart.

## Common Names

Nianca (chiriguano, Bolivia), bejuco de barraco (Dominican Republic), cainca, cipo cruz, raíz fedorenta, caninana (Brazil), ysypo-kurusu, (Gonzalez Torres, 1997) grass withers, ka'aysá, ka'a-ysa, ysypo-kurusu, (Gonzalez Torres, 1997); Toursarkissian, (1980) brings *Ch. angifuga*, isipo curuzu, bejuco de verraco, cipó-cruz, "yianoa" (chiriguano, Bolivia).

## Background

Effective against the venom of vipers (González Torres, 1997). In the Dominican Republic against snake bites and in Chaco (Argentina) the plant is known to be effective against poisons (De Luca and Zalles, 1992), indicated against cobra bite (Oliveira Simoes *et al.*, 1998).

## Chemical Constituents

Pentacyclic triterpenoids, free or as glycosides of α-amyrin, β-amyrin (Bhattacharyya and Cunha, 1992), and oleanolic acid and ursolic ketoalcohols, lignans and coumarins (Abd El-Hafiz *et al.*, 1991), which would form a complex and would be responsible for the neutralizing the effects of snakebite. Methanol extract of roots: one $C_{19}$ *nor*-seco-pimarane: merrilactone (Argaez Borges *et al.*, 2001).

## Properties Tested

Analgesic and anti-inflammatory properties were found using *in vitro* tests (Ruppelt *et al.*, 1990).

# Genus *Cissampelos*

*Cissampelos glaberrima* A. St.Hil. (Menispermaceae)

## Common Names

Southern Brazil: South ka'apeva, jarrinha (Gonzalez Torres, 1997).

## Chemical Constituents

Aporphynic alkaloids have been isolated from roots: cissaglaberrimine, trilobinine and a tricyclic lactone: eletefine (Barbosa-Filho, 1997).

*Cissampelos pareira* L. (Menispermaceae), syn. *C. pareira* L. var. *australis*; *Cissampelos pareira* L. var. *australis* (A. St.-Hil-) Diels, *C. pareira* L. var. *caapeba* var. Eichler.

Domínguez also brings *C. pareira* var. *australis*, (St. Hil.), as a variety of *C. pareira* L. (Dominguez, 1928).

## Common Names

In Brazil this plant is well known as abutua, and in Peru as mullein. References to abuta in herbal commerce today, however, may refer to either *Cissampelos pariera* or an entirely different plant, *Abuta grandiflora*, another tropical vine containing different products and used for different purposes in South American herbal medicine. The confusion is because in Peru, *A. grandiflora* is called chiric sanago as well as abuta. In Brazil: ysypo cobra, parreira brava, abutúa (Gonzalez Torres, 1997), caá pebá, pareira brava, (Toursarkissian, 1980; Ricciardi *et al.*, 1996), mil hombres, pareira falsa, (Amorín, 1988), hoja de mono, butua, (Manfred, 1977), zarza, caá-pebá (Gupta, 1995), (Dimitri and Parodi, 1988), (Marzocca, 1997). Perennial, climbing with a woody base and climbing found throughout north central Argentina. Abuta is found throughout the Amazon in Peru, Brazil, Ecuador, and Colombia, and is cultivated by many to beautify gardens.

## Background

Guatemala: Antidote, anti-rheumatic, cramps, diuretic, erysipelas, fever, menstrual disorders, snakebite, sweat promoter -Nicaragua: bites, fever, skin rash, sores, stings, venereal disease -Venzuela: bladder disorder, diuretic treatment, kidney stones, snakebite.

## Chemical Constituents

Contains tetrandrine a bis-benzylisoquinoline alkaloid, reported as having analgesic, anti-inflammatory, fever reducing and cardioactive properties as well as hypotensive effects and also action against cancer and leukemia cells. Other isoquinoline alkaloids present are berberine (hypotensive antimicrobial), cissampeline, a skeletal muscle relaxant drug in Ecuador.

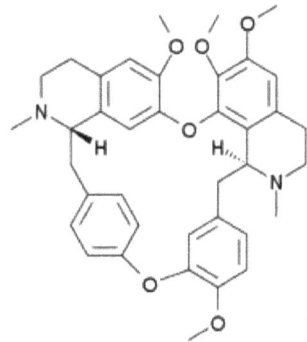

**tetrandrine**

In aerial parts: flavonoids (Ramirez *et al.*, 2003), leaves and roots contain alkaloids (Bhatnagar *et al.*, 1967a, Leger *et al.*, 2004; Basu, 1970); Bhatnagar *et al.*, 1967b; Kupchan *et al.*, 1965; Anwer *et al.*, 1968; Morita *et al.*, 1993; Morita *et al.*, 1995).

## Properties Tested

Tetrandrine is the compound studied in greatest detail. The alkaloid has analgesic, anti-inflammatory and antipyretic properties. It also exhibits pareirubrine A and B activities; however tetrandrine is too toxic for use in humans. Berberine is also hypotensive, cisampelina is a smooth muscle relaxant, marketed as such (Farnsworth *et al.*, 1989), and palmatine has hypotensive and sedative activity. The anti-inflammatory activity of *C. pareira* can probably be attributed to the suppression of nitric oxide mediated inflammation by bis benzylisoquinolenic alkaloids (Kondo,

1993). Paste obtained from roots is used topically for fistules, pruritus, skin disorders and snake venom poisoning (Amresh *et al.*, 2003). The entire plant may be considered to have very potent anti-inflammatory activity, with ethanol extract of roots being a convenient way for collection and storage (Amresh *et al.*, 2007). Alcohol extract of leaves of species in northeastern Argentina completely inhibits hemolytic activity of venom from *B. diporus* (Torres *et al.*, 2007). Incubation of *B. diporus* venom with alcohol extract of the entire plant results in the disappearance of all protein bands including PLA$_2$ (ratio 1: 7 (venom: dried extract) by SDS-PAGE. Aqueous and alcohol extracts, however, did not affect venom coagulant activity (Sosa de Torres *et al.*, 2004).

## Toxicity

*C. pareira* can be contraindicated for persons with low blood pressure. Tetrandine alkaloid present was documented to have various effects on heart function in animals and humans. Those with heart conditions or are taking heart medication should consult with a medical doctor before use of plant material.

# Genus *Clematis*

*Clematis bonariensis* Juss. ex DC. (Ranunculaceae), syn.: *C. maldonadenses.*

## Common Names

Larranaga, cabello de angel, barba de viejo (Lahitte *et al.*, 1998).

*Clematis montevidensis* Spreng. (Ranunculaceae), syn.: *Clematis hilarii* Spreng, *C. denticulata* Spreng.

## Common Names

Loconte, cabellos de angle, barba de viejo (Domínguez, 1928).

## Chemical Constituents

In roots: ranunculin (protoanemonin glucoside with irritating action on skin), stigmasterol, 3-β-*O*-D-glucoside of stigmasterol, sitosterol and campesterol, and in aerial parts: *p*-hydroxycinnamic, oleanolic acid, campesterol, stigmasterol, sitosterol (Pettenati *et al.*, 2005).

## Properties Tested

Infusion of the root and aerial parts of *C. montevidensis* showed a moderate diuretic activity in rats. This effect could be due, at least in part, to the levels of oleanolic acid present in this plant (Alvarez *et al.*, 2003).

# Genus *Cyperus*

*Cyperus obtusatus* (J. and K. Presl) Mattf. and Kükenth. (Cyperaceae), syn.: *Kyllinga vaginata, K. pungens* Link. (Gonzalez Torres, 1997).

## Background

Roots and leaves have aromatic calmative and antispasmodic activities, and that can neutralize properties of snake venom (Cabrera and Zardini, 1978), (Toursarkissian, 1980).

*Cyperus sesquiflorus* Mattf. and Kuek. (Cyperaceae) *Kyllinga odorata* Vahl syn.: C. *sesquiflorus* (Torr.) Mattf. and Kuk., C. *sesquiflorus* (Torr.) Mattf. and Kükenth. ex Kukenth., *K. sesquiflora* Torr. (Domínguez, 1928), (Matoso, 1983).

## Common Names

Junco, jahapé, iasapé, capi-catí, capii catí, capim cheiroso, in Brazil: capim cedreira, esquinanto (Gonzalez Torres, 1997), capií catí, capim cheiroso (Toursarkissian, 1980; Ricciardi *et al.*, 1996).

## Genus *Dorstenia*

*Dorstenia brasiliensis* Lam. (Moraceae), syn. *Dorstenia montevidensis* Miq.

## Common Names

Contrayerba, taropé (guaraní), kaapiá, higuerilla, tiú-tiú, carapiá, caiapiá taropé, tayapiá (Gonzalez Torres, 1997), contrayerba, caapiá-assú, eiga-eiga, chupa-chupa, carapiá (Gupta, 1995; Olivera Simoes *et al.*, 1998; Toursarkissian, 1980; Domínguez, 1928; Matoso, 1893; Sorarú and Bandoni, 1978; Amorín, 1988; Ratera and Ratera, 1980).

## Parts Used

Leaves in infusion, decoction of roots.

## Background

Digestive, emmenagogue, it has been widely used in the treatment of bites or stings from venomous animals or plant poisons. The roots of *Dorstenia brasiliensis*, have been used as a folk medicine for the treatment of digestive system disease and typhoid fever (Uchiyama *et al.*, 2002).

## Chemical Constituents

In roots: triterpenoids; coumarins and furocumarins 7-hydroxycumarin (Cussans and Huckerby, 1975); marmin: psoralen, bergaptene, (Elgamal *et al.*, 1979) 22 -(13-hydroxy-13-metylethyl)-psoralen (Quader *et al.*, 1992) and bergaptene, in rhizomes: furanocumarin monoterpenoid: 5-[3-(4,5-dyhydro-5,5-dimethyl-4-oxo-2-furanyl)butoxy]-7H-furo[3-2-g][1]benzopyran-7-one, 5-[[3-(4,5-dyhydro-5,5-dimethyl-4-oxo-2-furanyl-2-butenyl-oxy]-7H-furo[3-2-g][1]benzopyran-7-one (Kuster *et al.*, 1994); (22*S*, 32*R*)-32-hydroxymarmesin (Vilegas and Pozetti *et al.*, 1993); glycosides of hydroxymarmesin, (22*S*, 32 *R*) – hydroxymarmesin, 42-O-b-D-glucopy ranoside; glucosides of hydroxymarmesin (22*S*,32*R*)-32-hydroxymarmesin, 42-*O*-β-D-glucopyranoside; (Lemmich *et al.*, 1983); (22*S*,32*R*)-32-hydroxymarmesin 42-*O*-β-D-glucopyranoside; (22*S*)-marmesin 42-*O*-α-L- rhamnopyranosyl (1→6)-*O*-β-D-glucopyra noside (Srivastava and Srivastava, 1993); phenoline (probably responsible for the anti-venom activity); in roots: two 3,4-seco-adianane-type triterpenoids (Uchiyama *et al.*, 2002), sterols: sitosterol, stigmasterol, 3-*O*-β-glucosylsitosterol (Kuster *et al.*, 1994).

## Properties Tested

The furanocoumarins or psoralens are photosensitizers that can cause genetic damage and burns from exposure to UV radiation. *In vitro* testing has shown these compounds have analgesic and anti-inflammatory properties (Ruppelt *et al.*, 1991). Alcohol extracts of roots inhibit procoagulant activity of *B. diporus* venom (Torres *et al.*, 2008).

*Dorstenia tenuis* Bonpl. ex Bureau, (Moraceae), contrayerba, taropé. (Gonzalez Torres, 1997), (Sorarú and Bandoni, 1978; Matoso, 1893).

*Dorstenia tubicina* Ruiz and Pav. (Moraceae), syn.: *D. tubicina* f. *subexcentrica* Hassl.; *D. tubicina* var. *opifera* (Mart.) Hassl., caá-apiá, cayá-piá, carapiá (Brazil), contra-yerba (Peru).

*Echium vulgare* L. (Boraginaceae), snake tounge, (Manfred, 1977) Pyrrolizidine alkaloids have an accumulative toxicity in the liver.

# Genus *Eclipta*

*Eclipta prostrata* (L.) L. (Asteraceae), syn. *Eclipta alba* (L.) Hassk., *Cotula alba* L., *C. prostrata* L., *E. brachypoda* Michx., *E. erecta* L., *E. longifolia* DC., *E. parvifolia* DC., *E. procumbens* Michx., *E. prostrata* L., *E. thermalis* DC., *E. zipeliana* DC., *Micrelium asteroides* Forsk., *M. tolak* Forsk., *Verbesina alba* L., *V. conysoides* Trew., *V. prostrata* L.

## Common Names

Eclipta, hesillo, tangará-ka'á (Gonzalez Torres, 1997; Domínguez, 1928). It is an ascending, erect, ramifying, invasive annual grass in paddy fields, found throughout the northcentral part of the country.

## Background

This plant is used against snakebite in Paraguay by Amazonian Indians (Marzocca, 1997), who may also take infusions as a preventative mesure against possible bite by venomous snakes (Mors *et al.*, 1989).

## Chemical Constituents

Contains wedelolactone, dimethylwedelolactone; eclalbasaponins I-VI; taraxastone glucosides of: eclalbasaponins VII, VIII, IX and X (Yahara *et al.*, 1997) and a triterpenoid glucoside, ecliptasaponin C (28-*O*-β-D-glucopyranoside of the 3-β-*O*-β-D-glucopyranosyl-19-β-hydroxy olean-12-ene-28-oic acid) (Zhang and Chen, 1996; Zhang and Guo, 2001), daucosterol and 3-*O*-glucoside of the estigmasterol.

## Properties Tested

Wedelolactone, estigmasterol and sitosterol act synergistically *in vitro* to neutralize the myotoxicity and hemorrhagic effect of a number of Crotalid venoms including those of *B. jararaca*, *B. jararacussu* and *Lachesis muta*. The activities implicated are anti-proteolytic, anti-PLA$_2$ and therefore anti-inflammatory (Melo *et al.*, 1994). Methanol extract of aerial parts of *E. prostrata* neutralize the letal activity of South American rattlesnake venom (Mors *et al.*, 1989; Melo *et al.*, 1994; Martz, 1992; Houghton and Osibogun, 1993; Melo and Ownby, 1999). Butanol extract, the major component being dimethylwedelolactone, neutralize the lethal hemorrhagic, proteolytic and PLA$_2$ activity of the Malayan pit viper (*Calloselasma rhodostoma*) (Pithayanukulo *et al.*, 2004).

Hexane extract of aerial parts inhibits procoagulant and haemolytic activity of *B. diporus* venom (Torres *et al.*, 2007 and 2008).

## Genus *Erythrina*

*Erythrina crista-galli*, L. (Fabaceae-papilionoideae).

### Common Names

Ceibo, zuinandí, ivirá-iputezú.

### Chemical Constituents

Erythrina alkaloids: erisovine, erisopine, erisodine, erisovine, eritramine, eritraline and eritratine have strong anesthetic and neuromuscular blocking activities.

*Erythrina dominguezii* Hassl., Chacho seibo, seibo rosa, in northeast Argentina.

*Erythrina falcata* Benth., ceibo, suinana.

## Genus *Eupatorium*

*Eupatorium ayapana* Veuten (Asteraceae), syn. *E. oblongifolium* (Spreng.), Baker, *E. triplinerve*, *Conyza oblongifolia* Spreng.

### Common Names

In Brazil: yerba de serpiente, iapana, diapana, guaco (Di Stasi *et al.*, 1989; Manfred, 1977), ajapá, yerba de lagarto.

*Eupatorium oblongifolium* (Spreng.) Baker, (Asteraceae), syn. *Conyza oblongifolia* Spreng.

### Common Names

In Brazil: ajapa (Gonzalez Torres, 1997).

### Background

In Paraguay infusions of the plant are used at a dilution of 20-30/1000 for treatment of snakebite (González Torres, 1997).

Species of this genus have essential oils containing terpenes, sesquiterpenes (Maia *et al.*, 2002).

## Genus *Euphorbia*

*Euphorbia dichotoma* Forsk, (Euphorbiaceae).

### Common Names

Mbói-ka'á, yerba del colmillo, solimán de la tierra (Gonzalez Torres, 1997).

*Euphorbia hirta* L. (Euphorbiaceae) syn. *Chamaesyce hirta* (L.) Millsp.

### Common Names

Hierba de la víbora (Gonzalez Torres, 1997), ka'á-ysy-guasú, (Parodi and Dimitri, 1988), erva de sapo, tupasy camby (Marzocca, 1997; Manfred, 1977; Domínguez, 1928).

## Chemical Constituents

The ellagitannins, euphorbine A and B, are probably responsible of the anti-venom activity much like the triterpenoids, taraxerol and α- and β-amirin. Polyphenols that have been identified in leaves include: galic acid, quercitrin, myricetrin, 3,4-di-O-galloylquinic acid, 22,4,6-tri-O-galloyl-D-glucose, and 1,2,3,4,6-penta-O-galloyl-ß-D-glucose (Chen, 1991; Galvez *et al.*, 1993a; Galvez *et al.*, 1993b); in the milky sap there are diterpenoid esters that are irritating to eyes and skin. The flavonoid rutin (= 3-O-rutinoside of quercetin) is likely responsable for the anti-venom activity.

## Properties Tested

Extracts of leaves were found to increase excretion of urine and eletrolytes in rats (Johnson *et al.*, 1999).

*Euphorbia hypericifolia* L. (Euphorbiaceae).

## Common Names

hierba de la víbora (Gonzalez Torres, 1997), mbói-ka'á, yerba del colmillo de la víbora, solimán de la tierra (Gonzalez Torres, 1997).

*Euphorbia serpens* Kuntze (Euphorbiaceae), syn. *Euphorbia serpens* Kuntz. var. *microphylla* Müll. Arg. Cab. A perennial prostrate herb distributed widely throughout Argentina found in two varieties: *E. serpens* Kunth. var. *microphylla* Müll. Arg. and *E. serpens* Kunth. var. *serpens*.

## Common Names

Hierba de la víbora (Gonzalez Torres, 1997), yerba meona, yerba de la golondrina (Toursarkissian, 1980; Amorín, 1988), ka'á-kambý, caá ambuy, in Paraguay (Domínguez, 1928; Gupta, 1995; Marzocca, 1997).

## Background

The plant is known as yerba meona because of it diuretic activity.

## Chemical Constituents

Lanosterol, quercetin, salicylic acid and the alcohol cycloartenol:

## Genus *Fevillea*

*Fevillea trilobata* L. (Cucurbitaceae), syn. *F. cordifolia* Vell.

## Common Names

Andirova, andyrová, Cipó de jabotí, haba de San Ignacio (Domínguez, 1928; Manfred, 1977; Gonzalez Torres, 1997).

## Background

Scraped or ground seeds or oil of the seeds have anti-venom activity against snakebite (Rodriguez, 1980).

## Chemical Constituents

Methanol, aqueous or alcohol seeds extracts contains: fevicordin A glucoside, fevicordin F gentiobioside, cayaponoside B, cayaponoside D, norcucurbitacinic glycosides: andirobicine A glucoside, andirobicine A gentiobioside, andirobicine B glucoside, andirobicine C gentiobioside (Valente *et al.*, 1993; Valente *et al.*, 1994).

# Genus *Flaveria*

*Flaveria bidentis* (L.) Kuntze, (Asteraceae), syn. *F. bonariensis* D.C.; *F. contrayerba* (Cav.) Pers., *Ethulia bidentis* L., *Milleria chiloensis* Hort., *M. contrayerba* Cav.

## Common Names

Carapiá, ka'apiá, pique, chasca, (Gonzalez Torres, 1997; Toursarkissian, 1980), matagusanos, fique, flor amarilla (Amorín, 1988), balda, chasca, (Domínguez, 1928; Ratera and Ratera, 1980).

## Background

The plant constitutes a popular remedy for both internal and external use against snakebite (Gonzalez Torres, 1997; Sorarú and Bandoni, 1978; Lartigue, 1971).

## Chemical Constituents

This species is characterized by having flavonoid sulfates: 3,7-isorhamnetin disulfate (Cabrera and Juliani, 1976, 1977 and 1979) 3,7,32-quercetin trisulfate (Cabrera *et al.*, 1985); the 7,3',4'-trisulfate of 3-acetyl quercetin and the 3,7,3',4'-tetrasulfate of quercetin (Pereyra de Santiago and Juliani, 1972). These quercetin sulfates have been demonstrated *in vitro* to affect anti-thrombotic and anti-coagulant activity by increasing the partial time of thromboplastin activated (Guglielmone *et al.*, 2002); in roots and aerial parts: α-terthienyl and 5-(1-buten-1-inyl)-2,2'-bithienyl (Agnese *et al.*, 1999).

# Genus *Ipomoea*

*Ipomoea indica* (Burr. f.) Merr, (Convolvulaceae), syn. *I. acuminata* auct. non (Vahl.) Roem. and Schult.

## Common Names

Bejuco, suspiro, campanilla (Cabrera, 1993).

## Background

Anti-venom (Marzocca, 1997; Hieronymus, 1882), some species have alkaloid derivatives of lysergic acid and ergoline.

# Genus *Iresine*

*Iresine diffusa* Humb. and Bonpl. ex Willd. var. *diffusa*, (Amaranthaceae), syn. *I. celosioides* L., *Iresine celosioides* L. (Domínguez, 1928).

## Common Names

Mbói-ka'á, yerba del colmillo de la víbora, solimán de la tierra (Gonzalez Torres, 1997).

## Chemical Constituents

Compounds identified in aerial parts from this genus: drimenes: 3β,14-dihydroxy-δ(7,8)-drimen-11,12-acetonide, 3β,7β,14-trihydroxy-δ(8,9)-drimen-11, 12-olide and 3β,7α,14-trihydroxy-δ(8, 9)-drimen-11,12-olide (Rios and Berber, 2005).

### Properties Tested

Alcohol extract of aeral parts inhibits procoagulant activity of *B. diporus* venom (Torres *et al.*, 2008).

# Genus *Kyllinga*

*Kyllinga odorata* Vahl, (Cyperaceae); syn. *Cyperus sesquiflorus* Torr. (Mattf. and Kük.) ex and Kük.

## Common Names

Capií catí, esquinanto, esquinanto menor, jahapé, iasapé, cortadera, capim cedreira.

## Background

Roots and leaves are considered to have viper poison neutralizing properties (Gonzalez Torres, 1997).

## Chemical Constituents

Composition of the essential oil of roots depends on the plant's origin, and sesquiterpenoids have been identified for the most part, including: β-pinene, camphene, 1,8-cineole, *p*-cymene, limonene, δ-cadinene, cadalene, γ-calacorene, *trans*-calamenene, α-copaene, copadiene, cipera-2,4-diene, cipera-2,4(15)-diene, ciperorotundene, epoxyguaiene, γ-gurjunene, α-muurolene, γ-muurolene, nootkatene, α-selinene, *epi*-α-selinene, β-selinene, selinatriene, rotundene, (-) *nor*-rotundene, (-) *iso*rotundene, patchoulene, valencene, ylanga-2,4-diene, ciperol, isociperol, α-rotunol, β-rotunol, rotundenol, 4α,5α-osidoeudesm-11-en-3α-ol,α-ciperon, β-ciperon, ciperenon, ciperolon, ciperadion, ciperorotundon, kobudson, isokobuson, muskaton, patchoulenon, rotundon, (Sonwa and König, 2001), 4,5-secoeudesmanolid and epimer and cyclic acetal, (Ohira *et al.*, 1998), oleic, linoleic, linolenic, oleanolic and myristic acid.

# Genus *Laennecio*

*Laennecio sophiifolia* (Kunth.) G. L. Nesom, (Asteraceae), syn. *Conyza serpentaria* Griseb.

## Common Names

Hierba del zorro, hierba de la serpiente (Hieronymus, 1882).

## Chemical Constituents

In aerial parts compounds identified include: 2β- hydroxyhardwickiic acid (Jolad *et al.*, 1988), hautriwaic acid, (Hsü *et al.*, 1971), apigenine, β-sitosterol, 12-*epi*-bacchotricuneatin A (Simirgiotis *et al.*, 2000).

## Properties Tested

Components of aerial parts extracted with non-polar solvent were shown to have strong anti-inflammatory effects after injection into mice (Cifuente *et al.*, 2001).

# Genus *Macfadyena*

*Macfadyena unguis-cati* (L.) A. H. Gentry, (Bignoniaceae), syn.: *Bignonia unguis-cati* L.

## Common Names

U  a de gato, teyú isipó, mbaracayá poapé (Lahitte *et al.*, 1998).

## Background

The decoction of bark provides an antidote for snakebite (Lahitte *et al.*, 1998); the aerial parts are generally considered to have anti-inflammatory, anti-malarial and anti-venom properties (Duarte *et al.*, 2000).

## Chemical Constituents

Corimbside, vicenin-2, quercitrin, chlorogenic and isochlorogenic acids, lupeol, β-sitosterol, β-sitosterylglucoside, alantoin and lapachol (Duarte *et al.*, 2000).

# Genus *Mimosa*

*Mimosa pudica* L. (Fabaceae, Mimosoideae); syn. *Mimosa tetrandra* Humb. and Bonpl. ex Willd.

## Common Names

Mimosa, sensitiva, dormidera, malicia de mujer.

## Background

Infusions of roots are used for treatment of snakebite in Paraguay (Gonzalez Torres, 1997).

## Chemical Constituents

Identified in aerial parts: β-[$N$-(3-hydroxy-4-pyridone)]-α-amino propionic acid (Kleipool and Wibaut, 1950; Tiwari and Spencer, 1965; Tiwari *et al.*, 1967), crocetin, *p*-coumaric acid (*p*-hydroxycinnamic acid); 7,32 ,42-trihydroxy-3,8-dimethoxyflavone; 7,32,42-triacetoxy-3,8-dimethoxyflavone (Kirk *et al.*, 2003; Lobstein *et al.*, 2002). Two rare C-glycosylflavones (43-hydroxymaisin and cassiaoccidentalin B) were isolated from plant material from Africa (Englert *et al.*, 1994) that was not found in American material. Other researchers found sitosterol (Jiang *et al.*, 1990) and a bufadienolid (Yadava and Yadav, 2001).

## Properties Tested

Aqueous extract of roots, at specific doses, inhibits activity of hyaluronidase and protease in venom of Indian snakes (*Naja naja*, *Vipera russellii* and *Echis carinatus*) (Girish *et al.*, 2004). The aqueous extract of dried root also exhibits a significant effect on reversing lethality, myotoxicity and toxic enzyme activity in venom of cobra *N. kaouthia*, however this activity was absent in alcohol extracts (Mahanta and Mukherjee, 2001).

# Genus *Mikania*

*Mikania cordifolia* (L.f.) Willd. (Asteraceae), syn. *Cacalia cordifolia* L.f., *Willoughbya cordifolia* (L.f.) O. Kuntze.

## Common Names

Guaco rebalsero, guaco verde (Rodríguez, 1980; Martínez Crovetto, 1981; Iturralde, 1925).

## Chemical Constituents

Alcohol extract of leaves can be fractionated on a column using a mixture of solvents with increasing polarity (hexane, ethyl acetate, methanol), in which saponins elute first and were separated from alkaloids in subsequent fractions (Camargo *et al.*, 2010).

## Properties Tested

Butanol extracts of leaves of *M. cordifolia* demonstrated strong *in vitro* activity against *Trypanosoma cruzi* and *Trichomonas vaginalis* (Muelas Serrano *et al.*, 2000). In addition, a number of other pharmacological activities were found including: anti-fungal, anti-microbial, bronchodilatory, anti-allergic, and particularly snake venom specific anti-inflammatory activity.

Alcohol extract of aerial parts and roots and hexane extract of roots of *M. periplocifolia*, alcohol and hexane extracts of aerial parts as well as those of roots of *M. micrantha* and alcohol extracts of aerial parts and alcohol and hexane extracts of root of *M. coridifolia* showed inhibitory activity *in vitro* against the proteolytic activity of *B. diporus* venom (Torres *et al.*, 2010).

*Mikania micrantha* H.B.K. (Asteraceae).

## Common Names

Guaco (Toursarkissian, 1980).

*Mikania periplocifolia* **Hook.** *et* **Arn.** (Asteraceae), syn. *M. scandens* Willd. var. *periplocifolia* (Hook *et* Arn.) Baker.

## Common Names

Guaco (Toursarkissian, 1980; Amorín, 1988).

## Background

After cooking the plant, it is useful against the bites/stings of snakes and insects (Valeta, 1935).

## Chemical Constituents

The genus *Mikania* is characterized by the presence of alkaloids, lactones and sesquiterpenes like mikanolide, dihydromikanolide, deoxy-mikanolide, miscandenin, some with toxicity and being carcinogenic (Ricciardi, 2005).

Hexane extract of aerial parts and roots, alcohol extract of roots and essential oils of *M. coridifolia*; hexane extract of aerial parts and aqueous extract of roots of

*M. micrantha* and hexane extract of aerial parts of *M. periplocifolia* inhibit procoagulant activity of *B. diporus* venom (Torres *et al.,* 2008). Aqueous extract of aerial parts and roots of *M. micrantha* indirectly inhibits *B. diporus* venom hemolytic activity (Torres *et al.,* 2007).

# Genus *Morrenia*

*Morrenia odorata* (Hook. and Arn.) Lindl, (Asclepiadaceae), syn.: *Cynanchum odoratum* Hook. and Arn.

## Common Names

Tasi, isipó (Martínez Crovetto, 1981); tasi, doca, (Domínguez, 1928), leche-leche, uruma, tasi, doca, (Parodi and Dimitri, 1988), tasi fragante, guaicurú rembiú, isipó a, tasi, doca, (Marzocca, 1997).

## Background

The thin woody vines can be used as part of medical devices such as for tubal ligations (Martínez Crovetto, 1961).

# Genus *Nectandra*

*Nectandra angustifolia* (Schrad.) Nees and Mart. ex Nees. (Lauraceae), syn.: *N. angustiflora* var. *falcifolia* Nees.; *N. falcifolia* (Nees) J.A. Castigl. (Parodi and Dimitri, 1988) ex Mart. Crov. and Piccinini; *N. membranacea var. falcifolia* (Nees.) Hassl.; *Ocotea angustifolia* Schrad.

## Common Names

Ajuí, ajuý or laurel (Gonzalez Torres, 1997) laurel amarillo, aju'y hû, laurel-hu, laurel del río, laurel mini (Bertucci *et al.,* 2008), louro-branco (Marques, 2001). The seeds of *Nectandra sp.* acquired the name *Ishpingo* in colonial period of the 16th century, but are now referred as *hamalas* (Carod-Artal and Vazquez Cabrera, 2007).

## Background

Leaf infusión of *N. angustifolia* (Lauraceae) are used in folk medicine for digestion (Martinez Crovetto, 1961), for treatment of rheumatism, arthritis and pain (Bertucci *et al.,* 2008), for digestion and for anti-venom properties against sting/bite of snakes (Gonzalez Torres, 1997).

## Chemical Constituents

Essential oils of aromatic species were studied for the first time by Fester *et al.* (Fester *et al.,* 1961). Plant material for study was from Alto Verde and Puerto Ocampo (Santa Fe, Argentina). The terpenes identified were: α-pinene, 1,8-cineole and safrole; later Torres *et al.* (2005) determined the composition of essential oils in samples from San Isidro (Corrientes): α-pinene; β-pinene; β-myrceno; α-phellandrene; *p*-mentha-1(7),8-diene; limonene; (*E*) β-ocimene; γ-terpinene; α-terpinolene; δ-elemene; β-elemene; aromadendrene; germacrene D; bicyclogermacrene and α-chamigrene.

This species is stable with respect to expression of secondary metabolites, evaluated by production of volatiles, without significant variation in composition in different vegetative states in that essential oils in fall, spring and summer were consistently characterized by high levels of terpene hydrocarbons: 69.4 per cent, 72.2 per cent and 63.6 per cent, respectively. Variation in specific compounds was, respectively: $p$-1(7),8-menthadiene (25.4 per cent 25.2 per cent and 24.7 per cent), α-terpinolene (18.8 per cent, 20.9 per cent and 3.3 per cent), α-pinene (8.8 per cent, 10 per cent and 13.4 per cent) and β-pinene (2.8 per cent 3.7 per cent and 5.2 per cent). The very small oxygenated fraction contained: oxygenated monoterpenes (0.2 per cent, 0.4 per cent and 11.7 per cent) and oxygenated sesquiterpenes (4.3 per cent, 4.5 per cent and 8.4 per cent). Other compounds found in this fraction included ($E$)-asarone, which has proven analgesic activity. The sesquiterpene hydrocarbon fraction (21.1 per cent, 19.8 per cent and 12.8 per cent respectively) was primarily represented by germacrene D (6 per cent, 5.2 per cent and 1.3 per cent). Enantiomeric relations found were: α-pinene 83 per cent (1$R$,5$R$)-(+) and 17 per cent (1$S$,5$S$)-("); β-pinene 80 per cent (1$R$,5$R$)-(+), 20 per cent (1$S$,5$S$)-(-) and limoneno 44 per cent (4$R$)-(+) 56(4$S$)-(-) (Torres *et al.*, 2011; Torres, 2011).

Alcohol extracts (EtOH: $H_2O$ 70: 30) of leaves of *N. angustifolia* were found to have triterpenes/steroids, flavonoids and glycosides; in acetone extracts: triterpenes/steroids, low molecular weight terpenes and flavonoids and in chloroform extracts: triterpenes/steroids, low molecular weight terpenes and alkaloids (Bertucci *et al.*, 2008).

Separation by flash chromatography using solvent mixtures of increasing polarity (hexane, ethyl acetate, methanol) allowed for identification of flavonoids in fractions 2, 4 and 8; saponins in 2, 3, 4, 8 and 9; triterpenes/steroids in fraction 4, phenols in fractions 4 and 8 and alkaloids in 6, 7 and 8 (Torres *et al.*, 2010b).

## Properties Tested

Truiti (2004) showed that *N. angustifolia* leaf crude extract reduced capillary permeability inflammatory activity caused by inducing agents. In 2005, the same author demonstrated this activity in a carrageenan-induced pleurisy model in which leaf ethanol extract was administered orally (nasogastric tube). Anti-inflammatory activity of leaf crude extract was also examined by Oliveira de Melo *et al.* (2006). They showed that the extract caused a reduction in intensity of rat paw edema response 2 and 4 h after injection of carrageenan inducer. Nasogastic administration of crude extract to rats reduced levels of pleural inflammatory effusion induced by carrageenan and without altering the number of leukocytes. The total nitrates concentration ($NO_3^-$ + $NO_2^-$) in the induced pleural effusion was not different with *N. falcifolia* extract treatment.

Plant extracts were also effective in inhibiting trasudation of liquid and chemotactic influx of cells by *Croton* oil-induced edema on mice ears. This was also indicated by reduction in activity of the enzyme myeloperoxidase. Biological trials of ethanol extracts from leaves of *N. falcifolia* collected near the shores of the Upper Paraná River had antiprotozoal and molluscicidal activities (Truiti *et al.*, 2005). These extracts showed antileishmaniasic activity with an $LD_{50}$ of 138.5 μg/mL for promastigote form of *Leishmania* (Viannia) *braziliensis* and 65.6 ± 5.4 per cent growth inhibition at a higher dose of 320 µg/mL. The inhibition of the growth of epimastigote of *Trypanosoma cruzi* requires concentrations in the range of 1000 µg/mL.

Camargo *et al.* (2005) showed ethanol extracts of leaves to have concentration-dependent *in vitro* anti-hemolytic activity against venom of *B. neuwiedi diporus*.

Torres *et al.* (2011b and 2010) found that alcohol extracts of leaves had the highest *in vitro B. diporus* anti-venom activity. In addition, this activity had higher levels of anti-coagulant activity when isolated from plants in autumn and spring (Torres *et al.*, 2008), and highest hemolytic activity when isolated in spring (Torres *et al.*, 2007). There were also no differences in proteolytic activity isolated from plant material from different stages of vegetative growth (Torres *et al.*, 2010b). These results indicate that the plant must be harvested in springtime for preparation of alcohol leaf extracts. Alcohol and hexane extracts of leaves inhibit proteolytic acitivity in *B. diporus* venom (Torres *et al.*, 2010).

In *in vivo* tests, this active fraction was not toxic and provided protection ($ED_{50}$) from the lethal action ($4LD_{50}$) of venom injected intraperitoneally in experimental mice (Torres *et al.*, 2012b).

Flash chromatography was used to fractionate venom extract using (SDS-PAGE) to test for activity. Sub-fraction 5 was found to be the most effective in reducing intensity or completely eradicating bands present in venom after preincubation of venom and subfraction in a 1:2 ratio (venom-subfraction). Chemical analysis indicated presence of phenols in this active subfraction, and further separation of this fraction by HPLC-DAD showed the presence of five components, all with a UV absorption spectrum typical of flavonoids (Torres *et al.*, 2010b and 2011b).

Leaf essential oils in aqueous distillates of plant material collected in autumn or spring exhibited high levels of inhibition against coagulation activity in venom of *B. diporus*.This activity was greater than that from essential oil aqueous distillates obtained in summer. None of the essential oils had hemolytic activity in *in vitro* test. Inhibition of proteolytic action in *B. diporus* venom by both essential oils as well as extracts of aqueous distillates was found using a venom-mediated casein proteolysis assay, but large venom: extract ratios (>1:300) were necessary to measure this activity (Torres *et al.*, 2011c).

*Nectandra megapotamica* (Spreng.) Mez. syn.: *Nectandra membranacea* (Spreng.) Hassl., hom. illeg.; *N. membranacea var. saligna* (Nees and Mart. ex Nees) Hassl.; *N. membranacea var. saligna floribunda* Hassl.; *N. saligna* Nees and Mart. ex Nees; *N. saligna var. obscura* Meisn.; *N. tweediei* (Meissn.) Mez.; *Oreodaphne tweediei* Meisn.; *Tetranthera megapotamica* Spreng.

## Common Names

Canela preta (Ilza and Naría, 2000), laurel hu, canela imbuia, canela-fedorenta, ayuí-hú, laurel negro, laurel hú, laurel canela, laurel amarillo, canela negra, laurel, ayuy morotí, ayuy pará, laurel ayuy, laurel blanco, laurel overo (Richter and Dallwitz, 2008). Canela-preta, Canela-so-mato (Da Silva Filho *et al.*, 2004b), canelinha (Melo *et al.*, 2004a) canela-amarela (Marqués, 2001).

## Chemical Constituents

Psychoactive compounds were reported in seeds of *hamala* (Carod-Artal and Vazquez Cabrera, 2007). Da Silva Filho *et al.* (2004a) isolated three main components from hydroalcoholic crude extracts of *N. megapotamica*: α-asarone, galgravin and veraguensin. These compounds have analgesic and anti-inflammatory activities in mouse models with α-asarone being the main component responsible for the analgesic effect. The anti-coagulant activity in *Nectandra* sp. is due to papain content (Carod-Artal and Vazquez Cabrera, 2007). The tree is considered psychodelic and toxic because it contains NMT (*N*-methyl tryptamine) and β-carboline. Dos Santos Filho and Gilbert (1975) detected the presence of indolalquilamine, *N*-methyl tryptamine and 6-methoxy-*Nb*-methyl-1,2,3,4-tetrahydro-β-carboline.

Starting with crude ethanol extracts of leaves, Da Silva Filho *et al.* (2004b), isolated three tetrahydrofuran lignans: nectandrin C, D, and E; and eight described previously: machilin G, galgravin, nectandrin A, nectandrin B, calopiptin, veraguensin, aristolignin and ganschisandrin.

Melo *et al.* (2004), isolated spatulenol from leaves of *N. saligna* (only from leaves in spring), alismol (only in autumn), two phenylpropanoids: elemicin and asaron; two neolignans: veraguensin and calopiptin. Liriotulipiferin (the aporphine alkaloid) was isolated by standard methods starting with bark methanol extracts.

De Luca *et al.* (2004) isolated three tetrahydrofuran lignans with activity against *T. cruzi* from leaves of *N. megapotamica*.

Ilza and Naría (2000) found cyanogenic glycosides in leaves using picro sodium paper.

Regarding essential oils (yielding 0.11-0.18 per cent), Romoff *et al.* (2010), determined the chemical composition in plant samples collected in S o Paulo at different times of the year and hours of the day by GC-FID and GC-MS. They identified 19 compounds which were predominantly sesquiterpenes such as δ-elemeno (8.2-22.6 per cent) and α-bisabolol (62.3-69.4 per cent). They also found that there was a relatively large accumulation of oxygenated compounds during the month of February (70 per cent) compared with samples collected in August (63.9-65.1 most likely due to corresponding to decreases in sesquiterpene hydrocarbon levels. The structure of the major component, α-bisabolol, was later confirmed by NMR after purification. The monoterpenes identified: α-pinene, β-pinene, camphene, β-mircene and limonene, varied (1-9.3 per cent) depending on the season.

Torres *et al.* (2011c) characterized essential oils from *N. megapotamica* and found sesquiterpene hydrocarbon levels to be: 68.7 per cent in autumn, 58.4 per cent in spring and 62.1 per cent in summer. The major components found were: bicyclogermacrene (26.9 per cent, 24.6 per cent and 27.9 per cent), and germacrene D (17.8 per cent; 16.9 per cent and 18.5 per cent). The minor components found were: β-caryophyllene, δ-cadinene and δ-elemene. The monoterpene hydrocarbons were those in greatest abundance (24.7 per cent, 31.6 per cent and 21 per cent with α-pinene in highest amounts (9.8 per cent, 12.9 per cent and 9 per cent) followed by β-pinene (8.4 per cent, 10.6 per cent and 6.8 per cent) and limonene. Results obtained differed from those of Romoff *et al.* (2010), for the same class of natural products from the same plant species. These investigators in contrast found α-bisabolol to be the major component of the oxygenated sesquiterpene fraction, as noted above. These differences might be attributed to regional differences in the climate of S o Paulo (Brazil) versus Corrientes (Argentina) or a consequence of genetic difference within this species selected for different ecosystems to result in different ecotypes or chemotypes.

## Background

The essential oil has antimicrobial anti-inflammatory and antitumor properties (Apel *et al.*, 2006), and alcohol extracts have analgesic and anti-inflammatory activities (Da Silva Filho *et al.*, 2004a). The root extracts have been used to alleviate muscle pain because of the analgesic and anti-inflammatory activities (Melo *et al.*, 2004).

## Properties Tested

The analgesic and anti-inflammatory properties have been studied by Da Silva Filho *et al.* (2004a) using either crude alcohol extracts or isolated components (α-asarone, galgravin and veraguensin) at different doses. Bioassays used included testing abdominal contraction induced by acetic acid in mice, edema of paw induced by carrageenan in rats and the hot plate test with rats. All compounds showed activity in the abdominal contraction assay, but only α-asarone had activity in the hot plate test. Galgravin and veranguesin were active in the anti-inflammatory assay.

The same authors found that oral administration of 500 mg/kg of crude hydroalcoholic extract resulted in 82.2 per cent inhibition of contractions induced by acetic acid, and that 20 mg/kg dose of α-asarone, galgravin or veraguensin inhibited induced contractions 60.5 per cent, 71.3 per cent and 70.6 per cent respectively (2007a). Inhibition of carrageenan-induced paw edema was 41.2 per cent with veraguensin and 71.4 per cent with galgravin added at a dose of 20 mg/kg. Hot plate test activity was evident using either 20 mg/kg α-asarone or 300 mg/kg crude extract. In the *in vitro* assay COX-2 and NF-êB were inactive. The results suggest that veraguensin and galgravin have peripheral analgesic activity, while α-asarone is responsible for analgesic activities in the crude extract.

Apel *et al.* (2006) verified *in vitro* antimicrobial activity in essential oil which was able to inhibit growth of *S. aureus* 71 per cent and growth of *P. aeruginosa* 51 per cent. The material had no effect on growth of *E. coli* or *C. albicans*. Anti-inflammatory activity in oil was measured as inhibition of leukocyte migration (16.2 ± 3.8 mm in Borden boxes). The oil also has selective toxicity for prostate and multiple melanoma cancer cells. These results suggest that the essential oils can be used as a source of antimicrobial, anti-inflammatory and antitumor agents.

Trypanocidal activity in ethanol extracts and isolated components was investigated by Da Silva Filho *et al.* (2004b) using the trypomastigote form of the parasite. Crude extract used at 4 mg/mL was inactive perhaps due to antagonists or low lignan content, however, chlorofom extraction could be used to concentrate inhibitory activity such that this fraction at 2 mg/mL was 100 per cent active, with respect to purified components with highest activity, machilin G ($IC_{50}$ 2.2 mM) and calopiptin ($IC_{50}$ 4.4 mM).

Da Silva Filho *et al*. (2007b) studied the *in vitro* antileishmanial and antimalarial activities of seven tetrahydrofuran lignans isolated from *N. megapotamica*: machilin G and veraguensin had high activity, whereas galgravin, nectandrin A, nectandrin B, calopeptin and ganshisandrin were inactive against *L. donovani*. In assays against *Plasmodium falciparum*, it was found that calopeptin had moderate activity, but that machilin G, nectandrin B, veraguensin and ganshisandrin were inactive. Bioassays done with leaf ethanol extracts using *Artemia salina* or on microbial growth inhibition (*E. coli*, *S. aureus* y *P. aeruginosa*) indicated low toxicity.

With regard to *B. diporus* anti-venom activity, Torres (2011) found that essential oil of *N. megapotamica* had very low activity, in which large relative amounts of this were needed in the bioassays to obtain inhibition of *B. diporus* venom. Leaf essential oils from aqueous distillates obtained from material collected in autumn and spring were active in inhibiting venom coagulant activity, and leaf essential oils collected in autumn inhibited venom hemolysis and both preparations were active at a ratio of >1: 300 against venom-mediated proteolysis of casein.

Hexane extracts also had activity against venom-indiced serum clotting, but this activity was only present in spring and was only detected at very high extract: venom ratios (Torres *et al.*, 2008). Aqueous extracts obtained from plants in autumn and summer inhibited hemolytic activity (Torres *et al.*, 2007), whereas alcohol and hexane extracts showed proteolysis inhibitory activity in plant material collected year round (Torres *et al.*, 2010). A general conclusion is that *N. megapotamica* has very little anti-venom activity. This is reflected in the observations that anti-venom activities are not concentrated in a particular extract and that each specific activity also varies greatly with the stage of vegetative growth (Ricciardi *et al.*, 2008). While most people will tend to confuse the laurel amarillo (*N. angustifolia*) with the laurel negro (*N. megapotamica*), both species have anti-venom activity but clearly distinct activities. It is thus important to develop ways to easily distinguish these two species for use as well as collection and study (Torres, 2011).

# Genus *Paederia*

*Paederia brasiliensis* (Hook. f.) Puff. (Rubiaceae), syn.: *P. diffusa* (Brittson) Standl.

## Common Names

Bejuco blanco (Beni, Bolivia), bejuco hedion-do (Santa Cruz, Bolivia), boa, (Chacobo), yuraq waji, (quechua, Chapare), isopore (chiriguano), Janq'o waji (aymara, Yungas).

## Background

Leaves can be folded in the form of a compress for treatment of snakebite (De Lucca and Zalles, 1992).

# Genus *Peltodon*

*Peltodon radicans* Pohl. (Lamiaceae)

## Common Names

Mbói-ka'á, yerba del colmillo de la víbora, solimán de la tierra, (Gonzalez Torres, 1997). Hortela do mato, parakarí (Toursarkissian, 1980).

## Background

In Brazil properties promoting broncho-dilation have been attributed to this plant.

## Chemical Constituents

In the active extract: aliphatic hydrocarbons, 3β-OH,β–amirin, 3β-OH,α-amirin, β-sitosterol, stigmasterol, ursolic acid, 2α,3β,19α-trihydroxy-urs-12-en-28-oic acid (tormentic acid), methyl 3β-hydroxy,28-methyl-ursolate, sitosterol-3-*O*-β-D-glucopyranoside, and stigmasterol-3-*O*-β-D-glucopyranoside (Rocha da Costa *et al.*, 2008).

## Properties Tested

Neutralization of the main biological activities of venom from *C. durissus ruruima* and *B. atrox* (Cavalcanti-Neto *et al.*, 1996). The flower extracts had highest anti-edematogenic activity (Rocha da Costa *et al.*, 2008).

## Genus *Persea*

*Persea americana* Mill. (Lauraceae), syn.: *Laurus persea* L., *Persea americana* var. *angustifolia* Miranda, *P. americana* var. *drymifolia* (Schltd. and Cham.) S.F. Blake, *P. americana* Schltd. and Cham., *P. edulis* Raf., *P. drymifolia* Cham.; *P. gratissima* Gaertn., *P. persea* (L.) Cokerell.

## Common Names

In Peru: palta (avocado) (Mejía and Rengipo, 1995).

## Chemical Constituents

Persenone A and persenone B (Kim *et al.*, 2000), carotenoids: lutein, zeaxanthin, α-carotene, and β-carotene (Lu *et al.*, 2005); in seeds, phytohormones derived from abscisic acid: β-D-glucoside of the acid (1'$S$,6'$R$)-8'-hydroxyabscisic and β-D-glucoside (of this acid), (1'$R$,3'$R$,5'$R$,8'$S$)-*epi*-dihydrophaseic (del Refugio Ramos *et al.*, 2004).

The oil of the fruit is mainly monounsaturated fatty acids: oleic, linoleic, palmitic, palmitoleic, linolenic and stearic (in descending order of abundance) whose derivatives: (2$E$,5$E$,12$Z$,15$Z$)-1-hydroxyeicosa-2,5,12,15-tetraen-4-one, (2$E$,12$Z$,15$Z$)-1-hydroxy eicosa-2,12,15-trien-4-one, acetate of (5$E$,12$Z$)-2-hydroxy-4-oxoeicosa-5,12-dien-1-yl, and acetate of (2$R$)-(12$Z$,15$Z$)-2-hydroxy-4-oxoeicosa-12,15-dien-1-yl (Kawagishi *et al.*, 2001), in leaves there is a toxin, persin: ($Z$,$Z$)-1-(acetyloxy)-2-hydroxy-12,15-eicosadien-4-one (Oelrichs *et al.*, 1995).

## Properties Tested

Aqueous extracts of leaves were shown to have analgesic and anti-inflammatory activities in laboratory animals (Adeyemi *et al.*, 2002).

☆ The persenones were shown to strongly inhibit generation of superoxide and nitric oxide (NO) in cell culture systems, acting as antioxidants preferentially suppressing the generation of free radicals, and having anti-inflammatory effects (Kim *et al.*, 2000).

☆ Ester and ketone derivatives of fatty acids were very effective in counteracting liver damage (Kawagishi *et al.*, 2001).

☆ Aqueous, ethanol and ethyl acetate extracts injected into mice simultaneously with *B. asper* venom completely inhibited the hemorrhagic activity of the venom. This activity was attributed to the action of catechins, flavones, anthocyanins and condensed tannins that can chelate zinc required for venom metalloprotease activity (Castro *et al.*, 1999), (Borges *et al.*, 2001).

☆ Persin was shown to have necrotic activity using mammary gland and myocaridial epithelia. Feeding leaves of *P. americana* to livestock causes an intoxication effect (Oelrichs *et al.*, 1995).

## Genus *Petiveria*

*Petiveria alliacea* L. (Phytolaccaceae), syn.: *P. alliacea* L. var *alliacea*; *P. alliacea* var. *tetranda* (B.A. Gómez) Haumann.

### Common Names

Pipí, calauchín apacin, guiné, tipi is widely distributed in tropical Latin America.

### Chemical Constituents

Essential oil is found throughout the plant, but is in greater abundance in root: benzyl-2-hydroxyethyl-trisulphide (Szczepanski *et al.*, 1972), trithiolaniacine (*cis* 3,5-diphenyl-1,2,4,-tritiolan), benzaldehye, benzoic acid and *trans*-stilbene (Adegosan, 1974), dibenzyl trisulphide and *trans*-N-methyl-4-methoxy proline (Sousa *et al.*, 1990), isoarborinol, isoarborinol cinnamate; in roots: dipropyl disulfide, dibenzyl sulfide, dibenzyl disulfide, dibenzyl trisulfide, dibenzyl tetrasulfide, benzylhydroxymethyl sulphide and di(benzyltritio) methane (Coelho Benevides *et al.*, 2001), benzaldehyde, benzyl alcohol, benzyl benzoate, *cis* and *trans*-stilbene, diethyl sulphide, dibenzyl trisulfide (Ayedoun *et al.*, 1998), *S*-benzyl-*L*-cysteine sulfoxide (Kubec and Musah, 2001), (*R*)-*S*-(2-hydroxyethyl) cisteine sulfoxide, along with ($R_sR_C$)- and ($S_sR_C$)-*S*-(2-hydroxyethyl)cisteine, also along with vestiges of derivative of *S*-methyl-, *S*-ethyl- and *S*-propylcisteine (Kubec *et al.*, 2002). Also, according to Kubec *et al.* (2003), the lachrymatory principle from roots is (*Z*)-thiobenzaldehyde *S*-oxide.

## Genus *Pfaffia*

*Pfaffia tuberosa* (Spreng.) Hickey (Amaranthaceae), syn.: *Gomphrena tuberosa* Spreng.

### Common Names

Batatilla de don Antonio, caáparí mirí (Ricciardi *et al.*, 1996), in Paraguay caá pari.

## Genus *Philibertia*

*Philibertia gilliesii* Hook. and Arn. (Campanulaceae), syn: *Philibertia gilliesii* Hook. and Arn. var. *gracilis* (Don) T. Mey., *Sarcostemma gilliesii* (Hook. and Arn.) Decne

## Common Names

Aru-cumaé (chiriguano).

## Background

The decoction of leaves serves as an antidote against snakebite (De Lucca *et al.*, 1992).

## Chemical Constituents

Extracts have cysteine type proteases with specific proteolytic activity; the extracts are also anti-inflammatory (Sequeiros *et al.*, 2003).

# Genus *Pilocarpus*

*Pilocarpus pennatifolius* Lem. (Rutaceae), syn.: *P. pennatifolius* Lem. var. *selloanus* (Engl.) Hassl.(Domínguez, 1928); *P. selloanus* Engl.

## Common Names

Aguarandio mirí, jaborandí (tupí) (Ricciardi *et al.*, 1996), jaborandí, jaborandí del Paraguay (Amorín, 1988), ybirá-taí, yaguarandí, jaborandí (Domínguez, 1928), in Paraguay yvyrá-taí.

## Chemical Constituents

Alkaloids are in an approximate proportion of 1 per cent (Sawaya *et al.*, 2011), are derived from histidine (imidazole), pilocarpine, about 50 per cent of the total number of alkaloides, isopilocarpine, pilocarpidine, isopilocarpidine, pilosine, isopilosine, jaborine, jaborandine, jaboridine, carpiline (pilossine) as well as joboric and pilocarpic acids; terpenoids: α-pinene, limonene, myrcene; other constituents: 2-undecanone, sandaracopimaradiene, vinyl dodecanoate (Gaillard and Pepin, 1999).

# Genus *Rumex*

*Rumex brasiliensis* Link (Polygonaceae), syn.: *R. obtusifolius* L., *R. cuneifolius* Campdera.

## Common Names

Lengua de vaca (Toursarkissian, 1980).

*Rumex cuneifolius* Campd. (Polygonaceae).

## Common Names

Lengua de vaca, maquichi (Marzocca, 1997).

*Rumex crispus* L. (Polygonaceae).

## Common Names

Lengua de vaca (Gonzalez Torres, 1997), romaza (Domínguez, 1928), (Gupta, 1995), (Cabrera and Zardini, 1978).

## Background

Leaves macerated in cane brandy or spirits are applied directly to snake bites (González Torres, 1997).

## Chemical Constituents

From roots: 1,5-dihydroxyanthraquinones and anthrone (Gunaydin *et al.*, 2002), chrysophanol (chrysophanic acid), parietin (3-methyl ether of emodin), and nepodin (musizin, 1,8-dihydroxy-2-acetyl-3-methylnaphthalene) (Gyung Ja *et al.*, 2004); species of *Rumex* have between 6.6 and 11.1 per cent oxalic acid (plant dry weight) and ingestion can result in toxicity typical of oxalate-induced toxicosis (Panciera *et al.*, 1990).

Also identified: aesculetin, α-ionone, α-terpineol, benzylbenzoate, benzylsalicylate, β-bisabolene, β-cyclocitralbrunfelsene, β-damascenone, β-eudesmol, β-safranal, brunfelsene, brunfelsamidine, elemol, 2-ethylfuran, farnesol, farnesyl, geraniol, geranyl hopeanine, ionones, isobutylsalicylate, lavandulal, limonene, linalool, linoleic acid, linolenic acid, manaceine, manacine, mandragorine, methylfurans, methylanisoles, myrcene, myristic acid, *n*-decane, *n*-heneicosane, *n*-heptadecane, *n*-heptane, *n*-hexadecane, nerolidol, *n*-nonadecane, nonanes, *n*-octane, *n*-pentacosane, *n*-pentadecane, neophytadiene, *n*-tricosane, ocimene, pentadecanoic acid, palmitic acid, pinoresinols, salicylic acid esters, scopoletin, scopolin, and terpinolene (Gunaydin *et al.*, 2002).

*Rumex obtusifolius* L. (Polygonaceae), syn.: *Acetosa oblongifolia* (L.) A. and D. Löve, *Rumex obtusifolius* ssp. *agrestis* (Fries) Danser, *R. obtusifolius* ssp. *sylvestris* (Wallr.) Rech. f., *R. obtusifolius* var. *sylvestris* (Wallr.) Koch.

## Common Names

Lengua de vaca (Marzocca, 1997).

## Chemical Constituents

α-picolin (Wilkinson, 1958), anthraquinones: aloe-emodin, chrysophanol and emodin (Fairbairn and Muhtadi, 1972); there is a significant proportionproportion of flavonols: quercetin, kaempferol, miricetin, small amounts of isorhammnethin, as well as trace amounts of flavones: apigenin and luteolin (Trichopoulou *et al.*, 2000).

Polyphenols and flavonoids confer antioxidant properties (Trichopoulou *et al.*, 2000).

## Properties Tested

The latex or milk secreted from this plant has a high concentration of tannins and oxalic acid. A tincture of this plant can thus be useful for menopausal problems. The whole plant contains abundant iron (and is antianemic), phosphorus, tannins and glycosides (Ibá ez-Calero *et al.*, 2009). Compounds present may help cleansing of the digestive system and well as kidney (diuretic activities). Root components are considered safe for use as a laxative, and use is indicated in cases of constipation. Use is also indicated for treatment for eczema, a sluggish digestive system and in iron deficiency anemia. Applied externally, crushed leaves and roots have a healing effect on ulcers and skin sores (Cornara *et al.*, 2009).

# Genus *Sapindus*

*Sapindus saponaria* L. (Sapindaceae), syn.: *S. divaricatus* Cambess.; *S. inaequalis* DC.

## Chemical Constituents

In leaves, stems, fruits and seeds: saponins (3-β-$O$-[α-$L$-rhammnopyranosyl (1→3) β -$D$-glucopyranosyl] hederagenin) (Lemos *et al.*, 1992), carbohydrates and steroids; in stems and leaves: flavonoids; in new stems from stumps: tannins, essential oils and anthraquinones in seeds: β-sitosterol, α-amyrin and β-amyrin, and seeds and leaves: rutin, luteolin and 4-methoxyflavone (Wahab and Selim, 1985).

## Properties Tested

Complete inhibition of the hemorrhagic effect of *B. asper* (Costa Rica) was found in rat bioassys after intradermal injection of aqueous, ethanolic or ethyl acetate extracts of plant material (Castro *et al.*, 1999). Compounds identified in these extracts considered to be responsible for the inhibitory effects include catechins, flavones, anthocyanins and condensed tannins.

# Genus *Sapium*

*Sapium glandulosum* (L.) Morong (Euphorbiaceae), syn.: *Sapium aucuparium* Jacq.

## Common Names

Kurupika'ý (Gonzalez Torres, 1997).

## Background

According to González Torres (1997) infusions of leaves for drinking, and according to Parodi, decoction of wood and applicaiton of the sticky juice to wounds can be used to avoid potentially devastating results after being bitten by a coral snake or a rattlesnake.

*Sapium haematospermum* Müll. Arg. (Euphorbiaceae).

## Common Names

Kurupika'ý, lecherón, pega pega, (Gonzalez Torres, 1997), blanquillo, curupí, curupicaí, (Toursarkissian, 1980), árbol de leche, blanquillo, cambí, curupí, curupícaí, lecherón, pega-pega, lecherón, curupí-cay, curupí, pega-pega, (Domínguez, 1928), (Parodi and Dimitri, 1988).

## Background

Parodi: prepare a decoction of the wood and apply the viscous juice on coral and rattlesnake bites to avoid fatal results.

## Properties Tested

Alcohol extracts of stems and leaves of plants from Corrientes (Argentina) demonstrated *in vitro* activity inhibitory against blood coagulation caused by *B. diporus* venom (Torres *et al.*, 2008).

*Sapium longifolium* Huber (Euphorbiaceae), syn.: *S. longifolium* Müll. Arg. (Domínguez, 1928), (Parodi and Dimitri, 1988), *S. biglsadulosum* var. *lomgifolium* Müll. Arg.

## Common Names

Kurupika'ý, (Gonzalez Torres, 1997), kurupytá, kurupi, kurupika'y, curupica, lecherón, pega-pega.

## Background

In Paraguay: effective against snakebite (Gonzalez Torres, 1997).

# Genus *Sida*

*Sida rhombifolia* L. (Malvaceae).

## Common Names

Arrowleaf Sida; typychá (Gonzalez Torres, 1997), afata, escobadura, mata alfalfa, pichana (Toursarkissian, 1980; Gupta, 1995).

## Background

The leaves in a compress are effective against snakebite (Marzocca, 1997).

## Chemical Constituents

The leaves contain relatively higher amounts of nutrients including protein, carbohydrates, fiber, fat as well as ash. The root containes ephedrine and saponin, choline, pseudoephedrine, β-phenethylamine, vascin, hipaphorine and related indole alkaloids (Ahmad *et al.*, 1976).

*Sida spinosa* L. (Malvaceae); *Sida spinosa* L. var. *spinosa*, syn.: *S. angustifolia* Lam.; *S. spinosa* L. var. *angustifolia* (Lam.) Griseb., *S.angustifolia* Lam. (Gonzalez Torres, 1997), *S. spinosa* L. var. *riedelii* (K. Schum.) Rodrigo, syn. *S. riedelii* K. Schum., *S. spinosa* L. var. *angustifolia* Lam. (Griseb.) f. *riedelii*. (K. Schum.).

## Common Names

Típica, afata hembra.

## Chemical Constituents

Esters: 1-glyceryl eicosanoate; *p*-hydroxyphenylethyl *trans*-ferulate, ecdysteroids: 20-hydroxyecdysone, turkesterone; 20-hydroxyecdysone 20,22-monoacetonide, 20-hydroxy-24-hydroxymethyl ecdysone. (Darwish and Reinecke, 2003), also has unidentified alkaloids.

# Genus *Sidastrum*

*Sidastrum paniculatum* (L.) Fryxel (Malvaceae), syn.: *Sida paniculada* L. (Ricciardi *et al.*, 1996).

## Common Names

Malva-hú, makaguá-ka'á (Gonzalez Torres, 1997), cited by Martín Dobrizhoffer as macuanga caá (hierba del pato, duck grass).

## Background

P. de Montenegro (1711) *comidas sus ojas verdes como una cuarta de ellas luego que pica la víbora y así mismo mascada y aplicada a la mordedura, queda sin lesión y sin accidentes el herido. Si hubiere algunas horas que haiga mordido se toma una dragma de sus polvos, ó ojas machacadas en vino tibio y asimismo se aplica a la herida* (sic).

(Green leaves taken as food after the snakebite is recommended. Also, they should be chewed and applied directly to the bite. Powdered material suspended in warm wine and applied to the bite wound is also useful).

## Genus *Sinapis*

*Sinapis alba* L. (Brassicaceae), *syn.: B. alba* Rabenh. non L., *B. hirta* Moench; (Manfred, 1977), *B. alba* (L.) Boiss. (Parodi and Dimitri, 1988).

### Common Names

Mostaza blanca (Parodi and Dimitri, 1988) (Manfred, 1977).

### Chemical Constituents

Un roots and seeds: sinalbin and gluconasturtiin, and two aromatic glucosinolates (Kjaer, 1960).

## Genus *Tabernaemontana*

*Tabernaemontana catharinensis* A. DC. (Apocynaceae) syn.: *T. australis* Müll. Arg., *Peschiera australis* (Müll. Arg.) Myers, *P. catharinensis* A. DC. Myers.

### Common Names

Sapiranguí, jasmim, casca da cobra.

### Chemical Constituents

Indole alkaloids: 12-methoxy-4-methylvoachalotine (Batina *et al.,* 2000), coronaridine, voacangine, hydroxyindolenine voacangin; voacristin, hydroxylindolenin voacristin and 3-hydroxylcoronaridin (Pereira *et al.,* 2004).

### Properties Tested

The alkaloids have antibacterial activity (Guida *et al.,* 2003); extracts of root bark have been shown to have inhibitory activity against the myotoxic and lethal venom of rattlesnake, *C. durissus terrificus,* and this activity was correlated with the presence of quaternary alkaloid 12-methoxy-4-methylvoachalotin (Batina *et al.,* 2000). Aqueous plant extracts were fractionated on Sephadex G-10, and active fractions neutralized indole alkaloids 12-methoxy-4-methylvoachalotine (Batina *et al.,* 2000), coronaridine, voacangine, hydroxyindolenine voacangin; voacristin, hydroxylindolenin voacristin and 3-hydroxylcoronaridin (Pereira *et al.,* 2004). Activity of the venom after intramuscular injection neutralized $2LD_{50}$ doses of crotoxin (de Almeida *et al.,* 2004). Aqueous extracts also partially neutralized the myotoxic acitivity of the two *B. jararacussu* myotoxins bothropstoxin-I (BthTX-I) (catalytically inactivated) and BthTX-II on preparations of rat soleus muscle as well as live animals. Reduction of the tissue necrosis was observed without inhibition of $PLA_2$ activity for both myotoxins (Trombone *et al.*, 1999; Veronese *et al.*, 2005).

# Genus *Teucrium*

*Teucrium vesicarium* Mill. (Lamiaceae), syn.: *T. inflatum* Sw.

## Common Names

Makaguá-ka'á.

## Chemical Constituents

Aerial parts of species in this genus are rich in furoclerodane diterpenes (Piozzi *et al.*, 1998).

# Genus *Trixis*

*Trixis divaricata* (Kunth) Spreng. (Asteraceae), *Trixis divaricata* (Kunth). Spreng. subsp. *divaricata* (Ferrucci *et al.*, 2002).

## Common Names

Juan de la calle, malva del monte, contraveneno (Rodríguez, 1980).

## Properties Tested

Hexane extract of the aerial parts and essential oils and essential oil from aqueous distillates of plant material from Argentina inhibits procoagulant activity of *B. diporus* venom *in vitro* (Ricciardi *et al.*, 2008; Dellacassa *et al.*, 2008).

# Genus *Uncaria*

*Uncaria guianensis* (Aubl.) Gmel. (Rubiaceae).

## Common Names

In Peru: u a de gato (Mejía and Rengipo, 1995).

## Chemical Constituents

Contains 2.1 per cent alkaloids (bark and roots) and mainly oxindole acid glycosides (Sandoval *et al.*, 2002).

## Properties Tested

Activity promoting immune function: antileukemic, antitumor, antimutagenic and anti-inflammatory activity (Sandoval *et al.*, 2002).

# Conclusions

Disgarding what are called white bites or dry bites in which the snake's venom is not injected, it can be estimated that in 30 to 35 per cent of the cases there is the additional danger of microbial infection such as tetanus. It is evident that some constituents in various plant species interact with a range of different toxic materials that are associated with a snake bite. In effect it appears that plant secondary metabolites can provide a range of protective agents with effects on snakebite and the snake's venom being only one example. The basic underlying mechanisms for anti-venom compounds are to alter or reverse toxicity. In the case of anti-venom activities in plants we describe, characteristic common features for detoxification involve inducing analgesic or anti-inflammatory activities or both as in *Brunfelsia uniflora*, *Cynara scolymus*, *Dorstenia brasiliensis*, *Mikania glomerata* and *Trianosperma tayuya*. Some snake venoms cause the release of bradykinin which in turn causes a sharp drop in blood pressure, pain and contraction of smooth muscle in various organs. As such, most plants commonly used as alexiteric as having anti-venom activity in various products in effect act to reverse or inhibited the release of bradykinin (Biondo *et al.*, 2003). One well-described example would be the use of alcohol extract of Apocinacea *Mandevilla vellutina* root on rat uterus.

Anti-venom agents also might act by modifing primary, secondary or tertiary structure of enzymes and proteins of venom that mediate toxicity. The mechanisms may involve immobilization or some form of modification to inhibit or prevent activity. Many of the physiological responses originate by a stimulatory process involving a transmitter (endogenous or exogenous) that bind to a specific receptor in a "lock and key" manner to triggers some activity. A venom toxin may modify or provide an alternative key, *e.g.*, a toxin, or a product of toxic enzyme activity, to inactivate or disable normal cellular activity. Inhibition of ATPase, or as discussed phospholipase $PLA_2$ by ACA. Another possible mechanism in which specific plant compounds may counteract the effects of venom is by sequestration of metals needed as co-factors for enzymes mediating the toxic effects such as metalloproteinases.

Many mechanistic details remain to be explained as to why the mere ingestion or contact with the appropriate plant constituents may counteract the action of venom enzymes or toxins: cardiotoxins (cytotoxins), hemotoxins, hemolysins, hemorrhagins, myotoxins, necrotoxins, nephrotoxins, neurotoxins, proteases etc. The toxic effects of venom are varied and numerous in which nervous and muscle tissue can be severely affected, and with lethal consequences *in-situ* if transported through the booldstream. There is no doubt that harnessing the diverse protective effects naturally provided by plants to counteract the lethality and discomfort of a snake bite would be a useful additional approach to treatment beyond that of heterologous antivenom serum.

# References

Abd El-Hafiz, M.A., Weniger, B., Quirion J.C., and Anton, R. (1991). Ketoalcohols, lignans and coumarins from *Chiococca alba*. *Phytochemistry*, **30:** 2029-2031.

Abhijit, D., Jitendra, and Nath, D. (2012). Phytopharmacology of antiophidian botanicals: A review. *Int J Pharmacol.*, **8:** 62-79.

Adegosan, E.K. (1974). Trithiolaniacin, a novel trithiolan from *Petiveria alliacea*. *J Chem Soc Chem. Commun.*, **21:** 906-907.

Adeyemi, O.O., Okpo, S.O., and Ogunti, O.O. (2002). Analgesic and antiinflamatory effects of the aqueous extract of leaves of *Persea americana* Mill. (Lauraceae). *Fitoterapia*, **73:** 375-80.

Agnese, A.M., Nu ez Montoya, S., Ariza Espinar, L., and Cabrera, J.L. (1999). Chemotaxonomic features in Argentinian species of *Flaveria* (*Compositae*). *Biochem Syst Ecol.*, **27:** 739-742.

Ahmad, M.U., Husain, S.K., Ahmad, M., and Osman, S.M. (1976). Cyclopropenoid fatty acids in seed oils of *Sida acuta* and *Sida rhombifolia* (Malvaceae). *J Am Oil Chem Soc.*, **53:** 698-699.

Ahumada, Z. (1978). Aristoloquiáceas. *In:* Flora Ilustrada Catarinense, Itajaí, Santa Catarina, Brasil. ARIS.

Alvarez, L., Marquina, S., Villarreal, M.l., Alonso, D., Aranda, E., and Delgado, G. (1996). Bioactive polyacetylenes from *Bidens pilosa. Planta Med.*, **62:** 355-357.

Alvarez, M.E., Maria, A.O., Villegas, O., AND Saad, J.R. (2003). Evaluation of diuretic activity of the constituents of *Clematis montevidensis* Spreng. (Ranunculaceae) in rats. *Phytother Res.*, **17:** 958-960.

Amorín, J.L. (1988). Guía Taxonómica con Plantas de Interés Farmacéutico. Colegio Oficial de Farmacéuticos y Bioquímicos de la Capital Federal. Bs. As.

Amresh, G., Rao, Ch., Mehrotra, S., and Shirwaikar, A. (2003). Dissertation. Manipal, Academy of Higher Education, Karnataka, India, pp. 37-40.

Amresh, G., Reddy, G., Rao, Ch., and Singh, P. (2007). Evaluation of anti-inflammatory activity of *Cissampelos pareira* root in rats. *J Ethnopharmacol.*, **110:** 526-531.

Anwer, F., Popli, S.P., Srivastava, R.M., and Khare, M.P. (1968). Studies in medicinal plants. 3. Protoberberine alkaloids from the roots of *Cissampelos pareira* Linn. *Experientia*, **24:** 999.

Apel, M., Lima, M., Souza, A., Cordeiro, I., Young, M., Sobral, M., Suffredini, I., and Moreno, P. (2006). Screening of the biological activity from essential oils of native species from the Atlantic rain forest (Sao Paulo–Brazil). *Pharmacology online*, **3:** 376-383.

Ayedoun, M.A., Mondachirou, M., Sossou, P.V., Garneau, F.X., Gagnon, H., and Jean, F.L. (1998). Volatile constituents of the root oil of *Petiveria alliacea* L., from Benin. *J Essent Oil Res.*, **10:** 645-646.

Badilla, B., Mora, G., and Poveda, L.J. (1999). Anti-inflammatory activity of aqueous extracts of five Costa Rican medicinal plants in Sprague-Dawley rats. *Rev Biol Trop.*, **47:** 723-727.

Bandoni, A. (ed.) (2000). Los Recursos Vegetales Aromáticos en Latinoamérica. CYTED, Editorial de la Universidad Nacional de La Plata, Bs. As., Argentina, pp 367-368.

Barbosa-Filho, J.M., Da-Cunha, E.V.M., Lopes Cornélio, M., Da Silva Dias, C., and Gray, A.I. (1997). Cissaglaberrimine, an aporphine alkaloid from *Cissampelos glaberrima. Phytochemistry*, **44:** 959-961.

Baruah, R., Bohlmann, F., and King, R. (1985). Novel sesquiterpene lactones from *Anthemis cotula. Planta Med.*, **51:** 531-532.

Bassols, G., and Gurni, A. (1996). Especies de género *Lippia* Utilizadas en Medicina Popular Latinoamericana. *Dominguezia*, **13:** 14-25.

Basu, D.K. (1970). Studies on curariform activity of hayatinin methochloride, an alkaloid of *Cissampelos pareira. Jpa. J Pharmacol.*, **20:** 246-252.

Batina, M., Cintra, A.C., Veronese, E.L., Lavrador, M.A., Giglio, J.R., Pereira, P.S., Dias, D.A., Franca, S.C., and Sampaio, S.V. (2000). Inhibition of the lethal and myotoxic activities of *Crotalus durissus terrificus* venom by *Tabernaemontana catharinensis*: identification of one of the active components. *Planta Med.*, **66**: 424-428.

Bertucci, A., Haretche, F., Olivaro, C., and Vázquez, A. (2008). Prospección química del bosque de galería del río Uruguay. *Rev Bras Farmacogn.*, **19**: 21-25.

Bhatnagar, A.K., Bhattacharji, S., Roy, A.C., Popli, S.P., and Dhar, M.L. (1967a). Chemical examination of the roots of *Cissampelos pareira* Linn. IV. Structure and stereochemistry of hayatin. *J Org Chem.*, **32**: 819-820.

Bhatnagar, A.K., and Popli, S.P (1967b). Chemical examination of the roots of *Cissampelos pareira* Linn. V. Structure and stereochemistry of hayatidin. *Experientia*, **23**: 242-243.

Bhattacharyya, J., and Cunha, E.V.L. (1992). A triterpenoid from the root-bark of *Chiococca alba*. *Phytochemistry*, **31**: 2546-2547.

Biondo, R., Pereira, A.M., Marcussi, S., Pereira, P.S., Franca, S.C., and Soares, A.M. (2003). Inhibition of enzymatic and pharmacological activities of some snake venoms and toxins by *Mandevilla velutina* (Apocynaceae) aqueous extract. *Biochimie*, **85**: 1017-1025.

Bolzani, V. da S., Young, M.C., Furlan, M., Cavalheiro, A.J., Araujo, A.R., Silva, D.H., and Lopes, M.N. (1999). Search for antifungal and anticancer compounds from native plant species of Cerrado and Atlantic Forest. *An Acad Bras Cienc.*, **71**: 181-7.

Borges, M.H., Soares, A.M., Rodrigues, V.M., Andriao-Escarso, S.H., Diniz H., Hamaguchi, A., Quintero, A., Lizano, S., Gutierrez, J.M., Giglio, J.R., and Homsi-Brandeburgo, M.I. (2000). Effects of aqueous extract of *Casearia sylvestris* (Flacourtiaceae) on actions of snake and bee venoms and on activity of phospholipases A2. *Comp Biochem Physiol B.*, **127**: 21-30.

Borges, M.H., Soares, A.M., Rodrigues, V.M., Oliveira, F., Fransheschi, A.M., Rucavado, A., Giglio, J.R., and Homsi-Brandeburgo, M.I. (2001). Neutralization of proteases from *Bothrops* snake venoms by the aqueous extract from *Casearia sylvestris* (Flacourtiaceae). *Toxicon*, **39**: 1863-1869.

Borges-Argaez, R., Medina-Baizabal, L., May-Pat, F., Waterman, P.G., and Pena-Rodriguez, L.M. (2001). Merilactone, an unusual C19 metabolite from the root extract of *Chiococca alba*. *J Nat Prod.*, **64**: 228-231.

Brand o, M.G.L, Krettli, A.U., Soares, L.S.R., Nery, C.G.C, and Marinuzzi, H.C. (1997). Antimalarial activity of extracts and fractions from *Bidens pilosa* and other *Bidens* species (Asteraceae) correlated with the presence of acetylene and flavonoid compounds. *J Ethnopharmacol.*, **57**: 131-138.

Brand o, M.G.L., Nery, C.G.C. Mam o, M.A.S., and Krettli, A.U. (1998). Two methoxylated flavone glycosides from *Bidens pilosa*. *Phytochemistry*, **48**: 397-399.

Braz-Filho, R. (1994). Role of phyto-organics substances as therapeutic agents. *Rev Med da UFC.*, **34**: 1-2.

Burkart, A. (ed.). (1979). Flora Ilustrada de Entre Ríos. Colección Científica del INTA, Buenos Aires. V. Dicotiledoneas. I.N.T.A., Buenos Aires, Argentina.

Cabrera, A.L., and Zardini, E.M. (1978). Manual de la flora de los alrededores de Buenos Aires. 2ª ed. Ed. ACME S.A.C.I., Buenos Aires.

Cabrera, J.L., Juliani, H.R., and Gros, E.G. (1985). Quercetin 3,7,32-trisulphate from *Flaveria bidentis*. *Phytochemistry*, **24**: 1394-1395.

Cabrera, J.L., J and uliani, H.R. (1976). Quercetin/isol ethyl-7,3',4'-trisulphate from *Flaveria bidentis*. *Lloydia*, **39**: 253-254.

Cabrera, J.L., and Juliani, H.R. (1977). Isorhamnetin 3,7-disulphate from *Flaveria bidentis*. *Phytochemistry*, **16**: 400.

Cabrera, J.L., and Juliani H.R. (1979). Two new quercetin sulphates from leaves of *Flaveria bidentis*. *Phytochemistry*, **18**: 510-511.

Cabrera, A.L. (ed). (1993). Flora de la Provincia de Jujuy. Colección Científica XIII, parte IX: *Verbenáceas* a *Caliceraceas*. INTA. Bs.As. Argentina.

Cáceres, A. (1996). Plantas de Uso Medicinal en Guatemala. Editorial Universitaria, Universidad de San Carlos de Guatemala, pp. 208-210.

Cáceres Wenzel, M.I., Ricciardi, G.A., Torres, A.M.; Ricciardi, A.I.; and Dellacassa, E. (2012). Actividad antiveneno de los aceites y extractos de *Aloysia citriodora*, contra yarará chica. *Comunicaciones Científicas y Tecnológicas UNNE*. CE063.

Camargo, F.J., Torres, A.M., Ricciardi, G.A., Dellacassa, E.S., and Ricciardi, A.I.A. (2010). Fraccionamiento y caracterizacion fitiquimica del extracto alcohólico de partes aereas de *Mikania coridifolia* (L. f.) Willd de la Provincia de Corrientes. *Comunicaciones Científicas y Tecnológicas UNNE*.

Camargo, F., Torres, A., Tressens, S., Dellacassa, E., and Ricciardi, A. (2005). Inhibición de la actividad hemolítica del veneno de *Bothrops neuwiedi diporus* Cope (yarará chica) por extractos de plantas del nordeste argentino. *Comunicaciones Científicas y Tecnológicas UNNE*, E009.

Camargo, F., Torres, A., Ricciardi, G., Ricciardi, A., and Dellacassa, E. (2011). SDS-PAGE, una herramienta útil para la evaluación de actividad alexítera de extractos de plantas. *B.L.A.C.P.M.A.*, **10**: 440-445.

Carnat, A., Carnat, A.P., Fraisse, D., and Lamaison J.L. (1999). The aromatic and polyphenolic composition of lemon verbena tea. *Fitoterapia*, **70**: 44-49.

Carod-Artal, F., and Vázquez-Cabrera, C. (2007). Semillas psicoactivas sagradas y sacrificios rituales en la cultura Moche. *Rev Neurol.*, **44**: 43-50.

Carrasco, P., Mattoni, C., Leynaud, G., and Scrocchi, G. (2012). Morphology, phylogeny and taxonomy of South American bothropoid pitvipers (Serpentes, Viperidae). *Zool Scri.*, **41**: 109-124.

Castro, O., Gutierrez, J.M., Barrios, M., Castro, I., Romero, M., and Umana, E. (1999). Neutralización del efecto hemorrágico inducido por veneno de *Bothrops asper* (Serpentes: Viperidae) con extractos de plantas tropicales. *Rev Biol Trop.*, **47**: 605-616.

Cavalcanti-Neto, A.J., Borges, C.C., Boechat, A.L.R., Verçosa-Dias, A., Santos, W.C, Arruda, L.F.M.R., and dos-Santos, M.C. (1996). Verification of the efficacy of the plant *Peltodon radicans* in the neutralization of the main biological activities of *Crotalus durissus ruruima* and *Bothrops atrox* venoms. *Toxicon,* **34**: 16.

Chandra, V., Jasti, J., Kaur, P., Srinivasan, A., Betzel, Ch., and Singh, T.P. (2002). Structural basis of phospholipase A2 inhibition for the synthesis of prostaglandins by the plant alkaloid aristolochic acid from a 1.7 A crystal structure. *Biochemistry-US.*, **41**: 10914-10919.

Chen, L. (1991). Polyphenols from leaves of *Euphorbia hirta* L. *Zhongguo Zhong Yao Za Zhi,* **16**: 38-39.

Chippaux, J.P., Rakotonirina, V.S., Rakotonirina, A., and Dzikouk G. (1997). Drugs and plant substances which antagonise venom or potentiate antivenom. *Bull.Soc.Pathol.Exot.* **90(4)**: 282-285.

Cifuente, D.A., Simirgiotis, M.J., Favier, L.S., Rotelli, A.E., and Pelzer, L.E. (2001). Antiinflammatory activity from aerial parts of *Baccharis medullosa, Baccharis rufescens* and *Laennecia sophiifolia* in mice. *Phytother Res.*, **15**: 529-531.

Coelho Benevides, P.J., Young, M.C.M., Giesbrecht, A.M., Roque, N.F., and Bolzani, V. da S. (2001). Antifungal polysulphides from *Petiveria alliacea* L. *Phytochemistry,* **57**: 743-747.

Cornara, L., La Rocca, A., Marsili, S., and Mariotti, M.G. (2009) Traditional uses of plants in the Eastern Riviera (Liguria, Italy). *J Ethnopharmacol.,* **125**: 16-30.

Crohare, R., Priestap, H.A., Fari a, M., Cedola, M., and Rúveda, E.A. (1974). Aristololactams of *Aristolochia argentina. Phytochemistry,* **13**: 1957-1962.

Cussans, N.J., and Huckerby, T.N. (1975). Carbon-13 NMR spectroscopy of heterocyclic compounds–IV. *Tetrahedron,* **31**: 2719-2726.

Da Silva, G.L., de Abreu Matos, F.J., and Rocha Silveira, E. (1997). 42-dehydroxycabenegrin A-I from roots of *Harpalyce brasiliana. Phytochemistry,* **46**: 1059-1062.

Da Silva, G.L., Lacerda Machado, M.I., and de Abreu Matos, F.J., Braz-Filho, R. (1999). A new isoflavone isolated from *Harpalyce brasiliana. J Braz Chem Soc.,* **10**: 438-442.

Da Silva Filho, A., Andrade e Silva, M., Carvalho, J., and Bastos, J. (2004a). Evaluation of analgesic and anti-inflammatory activities of *Nectandra megapotamica* (Lauraceae) in mice and rats. *J Pharm Pharmacol.,* **56**: 1179-1184.

Da Silva Filho, A., Albuquerque, S., Silva M., Eberlin, M., Tomazela D., and Bastos, J. (2004b). Tetrahydrofuran lignans from *Nectandra megapotamica* with trypanocidal activity. *J Nat Prod.,* **67**: 42-45.

Da Silva Filho, A., Costa, E., Cunha, W., Silva M., Nanayakkara, N., and Bastos, J. (2007a). *In vitro* antileishmanial and antimalarial activities of tetrahydrofuran lignans isolated from Nectandra megapotamica (Lauraceae). *Phytother Res.,* **22:** 1307-1310.

Da Silva Filho, A., Costa, E., Rezende, K., Fukui, M., Cunha, W., e Silva, M., Nanayakkara, D., Khan, S., and Bastos, J. (2007b). Anti-inflammatory and analgesic activities of crude extract and isolated compounds from Nectandra megapotamica (Lauraceae). 1° BCNP Brazilian Conference on Natural Products S o Paulo, Brasil.

Dos Santos Filho, D., and Gilbert, B. (1975). The alkalois of *Nectandra megapotamica. Phytochemistry,* **14:** 821-822.

Darwish, F.M.M., and Reinecke, M.G. (2003). Ecdysteroids and other constituents from *Sida spinosa* L. *Phytochemistry,* **62:** 1179-1184.

De Almeida, L., Cintra, A.C.O., Veronese, E.L.G., Nomizo, A., Franco, J.J., Arantes, E.C., and Giglio, J.R., Vilela Sampaio, S. (2004). Anticrotalic and antitumoral activities of gel filtration fractions of aqueous extract from *Tabernaemontana catharinensis* (Apocynaceae) *Comp Biochem Phys Part C.,* **137:** 19-27.

De Carvalho, P.R., Furlan, M., Young, M.C., Kingston, D.G., and Bolzani, V.S. (1998). Acetylated DNA-damaging clerodane diterpenes from *Casearia sylvestris. Phytochemistry,* **49:** 1659-1662.

De Lucca, D.M., and Zalles, J.A. (1992). Flora Medicinal Boliviana. Diccionario Enciclopedico. Editorial Los Amigos del Libro – Werner Guttentag. La Paz-Cochabamba, Bolivia.

De Luca, A., Nunomura, S., and Yoshida, M. (2004). Phytochemical analices of Lauraceous Species. XIII *Congresso Italo-Latino Americano di Etnomedicina,* PO39.

Del Refugio Ramos, M., Jerz, G., Villanueva, S., Lopez-Dellamary, F., Waibel R., and Winterhalter, P. (2004). Two glucosylated abscisic acid derivates from avocado seeds (*Persea americana* Mill. Lauraceae cv. Hass). *Phytochemistry,* **65:** 955-62.

Dellacassa, E., Ricciardi, A.I.A., Torres, A.M., Camargo, F.J., and Ricciardi, G. (2008). Fitoquímica y actividad biológica de espécies vegetales aromáticas y medicinales del nordeste argentino. Actas de *II Congresso de Fitoterápico do MERCOSUR VI Reunido da Sociedade Latinoamericana de Fitoquímica.* Belo Horizonte, Brasil.

DeSmet, P.A.G.M. (1995). Health risk of herbal remedies. *Drug Safety,* **13:** 81-93.

Di Stasi, L.C., Guimar es Santos, E.M., Moreira Dos Santos, C.; and Akiko Hiruma, C. (1989). Plantas Medicinais na Amazonia. S o Paulo, Editora Universidade Estadual Paulista.

Domínguez, J.A. (1928). Contribuciones a la Materia Médica Argentina. Trabajos del Instituto de Botánica y Farmacología, N° 44, Facultad de Ciencias Médicas, Jacobo Peuser, Buenos Aires.

Duarte, D.S., Dolabela, M.F., Salas, C.E., Raslan, D.S., Oliveiras, A.B., Nenninger, A., Wiedemann, B., Wagner, H., Lombardi, J., and Lopes, M.T. (2000). Chemical characterization and biological activity of *Macfadyena unguis-cati* (Bignoniaceae). *J Pharm Pharmacol.*, **52:** 347-52.

Duke, J., Bogenschutz-Godwin, M.J., and Ottesen, A.R. (2009). Duke's Handbook of medicinal plants of Latin America. CRC 28-30.

Elgamal, M.H.A., Elewa, N.H., Elkhrisy, E.A.M., and Duddcck, H. (1979). $^{13}$C-NMR chemical shifts and carbon-proton coupling constants of some furocoumarins and furochromones. *Phytochemistry*, **18:** 139-143.

El-Sebakhy, N., Richomme, P., Taaima, S., and Shamma, M. (1989). (-)-Temuconine, a new bisbenzylisoquinoline alkaloid from *Aristolochia elegans. J Nat Prod.*, **52:** 1374-1375.

El-Sebakhy, N., and Waterman, P.G. (1994). (")-( *R, R*)-72-*O*-methylcuspidaline from the leaves of *Aristolochia elegans. Phytochemistry*, **23:** 2706-2707.

Englert, J., Jiang, Y., Cabalion, P., Oulad-Ali, A., and Anton, R. (1994). *Planta Med.*, **60:** 194.

Esteso, S. (1985). Ofidismo en la República Argentina. Ed. Arpón, Córdoba, Argentina.

Fairbairn, J.W., and Muhtadi, F.J. (1972). The biosynthesis and metabolism of anthraquinones in *Rumex obtusifolius. Phytochemistry*, **11:** 215-219.

Farnsworth, N.R. (1988). Screening plants for new medicines. Wilson E.O. (ed.) IX. Biodiversity, National Academic Press, Washington DC, pp. 83-97.

Farnsworth, N.R., Akerlele, O., Bingel, A.S., Guo, Z.G., and Soejarto, D.D. (1989). Plantas medicinales. *Bol Ofic Sanit Panam.*, **107:** 314- 329.

Fenwick, A., Gutberlet, R., Evans, J., and Parkinson, C. (2009). Morphological and molecular evidence for phylogeny and classification of South American pitvipers, genera *Bothrops, Bothriopsis*, and *Bothrocophias* (Serpentes: Viperidae). *Zool J Linn Soc.*, **156:** 617-640.

Ferreira, L.A.F., Henriques, O.B., Andreoni, A.A.S., Vital, G.R.F., Campos, M.M.C., Habermehl, G.G., and de Moraes, V.L. (1992). Antivenom and biological effects of ar-turmerone isolated from *Curcuma longa* (Zingiberaceae). *Toxicon*, **30:** 1211-1218.

Ferrucci, M.S., Cáceres Moral, S.A., Galbany Casals, M., and Martínez, W.J. (2002). Las plantas trepadoras del macrosistema Iberá. *13$^{ra}$ Reunión de Comunicaciones Científicas y Técnicas, Facultad de Ciencias Agrarias, UNNE*, 06- B003.

Fester, G.A., Martinuzzi, E A., Retamar, J.A., and Ricciardi, A.I. (1961). Aceites Esenciales de la República Argentina. Academia Nacional de Ciencias Córdoba.

Gaillard, Y., and Pepin, G. (1999) Poisoning by plant material: review of human cases and analytical determination of main toxins by high-performance liquid chromatography–(tandem) mass spectrometry. *J Chromatogr B.*, **733:** 181-229.

Galvez, J., Crespo, M.E., Jimenez, J., Suarez, A., and Zarzuelo, A. (1993a) Antidiarrhoeic activity of quercitrin in mice and rats. *J Pharm Pharmacol.*, **45:** 157-159.

Galvez, J., Zarzuelo, A., Crespo, M.E., Lorente, M.D., Ocete, M.A., and Jimenez, J. (1993b). Antidiarrhoeic activity of *Euphorbia hirta* extract and isolation of an active flavonoid constituent. *Planta Med.*, **59:** 333-336.

Geissberger, P., and Séquin, U. (1991). Constituents of *Bidens pilosa* L.: do the components found so far explain the use of this plant in traditional medicine?. *Acta Trop.*, **48:** 251-261.

Girish, K.S., Mohanakumari, H.P., Nagaraju, S., Vishwanath, B.S., and Kemparaju, K. (2004). Hyaluronidase and protease activities from Indian snake venoms: neutralization by *Mimosa pudica* root extracts. *Fitoterapia*, **75:** 378-380.

González Torrres, D.M. (1997). Catálogo de Plantas Medicinales (y alimenticias y útiles) Usadas en Paraguay. Asunción.

Guglielmone, H.A., Agnese, A.M., Nu ez Montoya, S.C., and Cabrera, J.L. (2002). Anticoagulant effect and action mechanism of sulphated flavonoids from *Flaveria bidentis*. *Thromb Res.*, **105:** 183-188.

Guida, A., De Battista, G., and Bagardi, S. (2003). Actividad antibacteriana de alcaloides de *Tabernaemontana catharinensis* A.D.C. *Ars Pharmaceutica*, **44:** 167-173.

Gunaydin, K., Topcu, G., and Ion, R.M. (2002). 1,5-dihydroxyanthraquinones and an anthrone from roots of *Rumex crispus*. *Nat Prod Lett.*, **16:** 65-70.

Gupta, M. (ed.). (1995). 270 Plantas Medicinales Iberoamericanas. Programa Iberoamericano de Ciencia y Tecnología para el Desarrollo, Subprograma de Química Fina Farmacéutica; Santafe de Bogotá, Colombia, pp. 553–554.

Gyung Ja, Choi, Seon-Woo, Lee, Kyoung Soo, Jang, Jin-Seog, Kim, Kwang Yun, Cho, and Jin-Cheol, K. (2004). Effects of chrysophanol, parietin, and nepodin of *Rumex crispus* on barley and cucumber powdery mildews. *Crop Prot.*, **23:** 1215-1221.

Habib, A.A., and El-Sebakhy, N.A. (1981). Ent-kaurane-16 α,17-diol and (-)-Cubebin as natural products from *Aristolochia elegans*. *Pharmazie*, **36:** 291-294.

Hieronymus, J. (1882). Plantae diaphoricae florae argentinae. *Bol Acad Nac Ciencias*, **4:** 3-4.

Houghton, P.J., and Osibogun, I.M. (1993). Flowering plants used against snakebite. *J Ethnopharmacol.*, **39:** 1-29.

Hsü, H., Pan, Chen, Y., and Kakisawa, H. (1971). Structure if hautriwaic acid *Phytochemistry*, **10:** 2813-2814.

Hutt, M.J., and Houghton, P.J. (1998). A survey from the literature of plants used to treat scorpion stings. *J Ethnopharmacol.*, **60:** 97–100.

IARC. (2002). *Aristolochia* species and aristolochic acids herbal remedies containing plant species of the genus *Aristolochia* (Group 1). *IARC Monog Eval Carc.*, **82:** 69-128.

Ibá ez-Calero, S.L., Jullian, V., and Sauvain, M. (2009) A new anthraquinone isolated from *Rumex obtusifolius*. *Rev Bol Quim.*, **26:** 49-56.

Ioset, J.R., Raoelison, G.E., and Hostettmann K. (2002). An LC/DAD-UV/MS method for the rapid detection of aristolochic acid in plant preparations. *Planta Med.*, **68:** 856-858.

Itokawa, H., Totsuka, N., Morita, H., Takeya, K., Iitaka, Y., Schenkel, E.P., and Motidome, M. (1990). New antitumor principles, casearins A-F, for *Casearia sylvestris* Sw. (Flacourtiaceae). *Chem Pharm Bull.*, **38:** 3384-88.

Itokawa, H., Totsuka, N., Takeya, K., Watanabe, K., and Obata, E. (1988). Antitumor principles from *Casearia sylvestris* Sw. (Flacourtiaceae), structure elucidation of new clerodane diterpenes by 2-D NMR spectroscopy. *Chem Pharm Bull.*, **36:** 1585-88.

Iturralde, P. (1925). Erbe Medicinali del Chaco. Pubblicazioni dell' Istituto Cristóforo Colombo, Roma.

lza, A., and Naría, H. (2000). Cyanogenic glycosides in plants. *Braz Arch Biol Tech.*, **43:** 487-492.

Izidoro, L.F., Rodrigues, V.M., Rodrigues, R.S., Ferro, E.V., Hamaguchi, A., Giglio, J.R., and Homsi-Brandeburgo, M.I. (2003). Neutralization of some hematological and hemostatic alterations induced by neuwiedase, a metalloproteinase isolated from *Bothrops neuwiedi pauloensis* snake venom, by the aqueous extract from *Casearia mariquitensis* (Flacourtiaceae). *Biochimie.*, **85:** 669-75.

Januário, A.H., Santos, S.L, Marcussia S., Mazzib, M.V., Pietro, R.C.L.R., Sato, D.N., Ellena, J., Sampaio, S.V., França, S.C., and Soares, A.M. (2004). Neo-clerodane diterpenoid, a new metalloprotease snake venom inhibitor from *Baccharis trimera* (Asteraceae): anti-proteolytic and anti-hemorrhagic properties. *Chem-Biol Inter.*, **150:** 243-251.

Jiang, Y., Haag, M., Quirion, J.C., Cabalion, P., Kuballa, B., Anton, R. (1990). Structure of a new saponin from the seed of *Mimosa pudica. Planta Med.*, **56:** 555.

Jiménez-Ferrer, J.E, Pérez-Terán, Y.Y., Román-Ramos, R., and Tortoriello, J. (2005). Antitoxin activity of plants used in Mexican traditional medicine against scorpion poisoning. *Phytomedicine,* **12:** 116-122.

Johnson, P.B., Abdurahman, E.M., Tiam, E.A., Abdu-Aguye, I., and Hussaini, I.M. (1999). *Euphorbia hirta* leaf extracts increase urine output and electrolytes in rats. *J Ethnopharmacol.*, **65:** 63-9.

Jolad, S.F., Timmermann, B.N., Hoffmann, J.J., Bates, R.B., Camou, F.A. (1988). Diterpenoids of *Conyza coulteri. Phytochemistry*, **27:** 1211-1212.

Jolís, J. (1972). Ensayo sobre la historia natural del Gran Chaco. Faenza 1789, UNNE.

Jozamí, J.M., and Mu oz, J. de D. (1982). Árboles y arbustos indígenas de la Provincia de Entre Ríos. IPNAYS. Santa Fe, Argentina.

Kawagishi, H., Fukumoto, Y., Hatakeyama, M., He, P., Arimoto, H., Matsuzawa, T., Arimoto, Y., Suganuma, H., Inakuma, T., and Sugiyama, K. (2001). Liver injury suppressing compounds from avocado (*Persea americana*). *J Agric Food Chem.*, **49:** 2215-2221.

Kim, O.K., Murakami, A., Nakamura, Y., Takeda, N., Yoshizumi, H., and Ohigashi, H. (2000). Novel nitric oxide and superoxide generation inhibitors, persenone A and B, from avocado fruit. *J Agric Food Chem.*, **48:** 1557-1563.

Kirk, L.F., Mrller, M.V., Christensen, J., Stærk, D., Ekpe, P., and Jaroszewski, J.W. (2003). A 5-deoxyflavonol derivative in *Mimosa pudica. Biochem Syst Ecol.*, **31:** 103-105.

Kjaer, A. (1960). Fortschitte der Chemische Organischer Naturstoffe 18, 122.

Kleipool, R.J.C., and Wibaut, J.P. (1950). *Mimosine (leucaenine)*. 80th Communication on pyridine derivatides. *Rec Trav Chim Pays Bas.*, **69:** 37-44.

Kondo, Y. (1993). Inhibitory effect of bisbenzylisoquinoline alkaloids on nitric oxide production in activated macrophages. *Biochem Pharmacol.*, **46:** 1887-1892.

Kubec, R., Kim, S., and Musah, R.A. (2002). S´Substituted cysteine derivatives and thiosulfinate fromation in *Petiveria alliacea*-part II. *Phytochemistry*, **61:** 675-680.

Kubec, R., Kim, S., and Musah, R.A. (2003). The lachrymatory principle of *Petiveria alliacea*. *Phytochemistry*, **63:** 37-40.

Kubec, R., and Musah, R.A. (2001). Cysteine sulfoxide derivatives in *Petiveria alliacea*. *Phytochemistry*, **58:** 981-985.

Kupchan, S.M., Patel, A.C., and Fujita, E. (1965). Tumor inhibitors. VI. Cissampareine, new cytotoxic alkaloid from *Cissampelos pareira*. Cytotoxicity of bisbenzylisoquinoline alkaloids. *J Pharm Sci.*, **54:** 580-83.

Kuster, R.M., Bernardo, R.R., da Silva, A.J.R., Parente, J.P., and Mors, W.B. (1994). Furocoumarins from the rizomes of *Dorstenia brasiliensis. Phytochemistry*, **36:** 221-223.

Lahitte, H.B., Hurrell, J.A., Belgrano, M.J., Jankowski, L., Haloua, P., and Mehltreter, K. (1998). Plantas medicinales rioplatenses. Buenos Aires, Argentina.

Lartigue, J. (1971). Historia de la medicina argentina. *Rassegna*, **4:** 52-56.

Leger, J.M, Guillon, J., Massip, S., and Saxena, A.K. (2004). Crystal structure of daijisong. *Anal Sci.*, **20:** 105-106.

Lemmich, J., Havelund, S., and Thastrup, O. (1983). Dihydrofurocoumarin glucosides from *Angelica archangelica* and *Angelica silvestris. Phytochemistry*, **22:** 553–555.

Lemos de Arruda Camargo, M.T. (1999). Herbario etnobotanico. Humanitas, S o Paulo.

Lemos, T.L.G, Mendes, A.L., Souza, M.P, and Braz-Filho, R. (1992). New saponin from *Sapindus saponaria* L. *Fitoterapia*, **63:** 515–517.

Lobstein, A., Weniger, B., Um, B.H.M., Steinmetz, L., Declercq, R., and Anton, R. (2002). 43-Hydroxymaysin and cassiaoccidentalin B, two unusual C-glycosylflavones from *Mimosa pudica* (Mimosaceae). *Bioch Syst Ecol.*, **30:** 375-377.

López Sáez, J., and Pérez Soto, J. (2009). Plantas alexitéricas: antídotos vegetales contra las picaduras de serpientes venenosas. *Med Natur.*, **3:** 17-24.

Lu, Q.Y., Arteaga, J.R., Zhang, Q., Huerta, S., Go, V.L., and Heber, D. (2005). Inhibition of prostate cancer cell growth by an avocado extract: role of lipid-soluble bioactive substances. *J Nutr Biochem.*, **16:** 23-30.

Mahanta, M., and Mukherjee, A.K. (2001). Neutralisation of lethality, myotoxicity and toxic enzymes of *Naja kaouthia* venom by *Mimosa pudica* root extracts. *J Ethnopharmacol.*, **75:** 55-60.

Maia, J.G.S., Zoghbi, M. das G.B., Andrade, E.H.A., da Silva, M.H.L., Luz, A.I.R., and da Silva, J.D. (2002). Essential oils composition of *Eupatorium* species growing wild in the Amazon. *Bioch Syst Ecol.*, **30:** 1071-1077.

Maestri, D.M., and Guzmán, C.A. (1995). Fatty acid composition of *Brunfelsia uniflora* (Solanaceae) seed oil. *Grasas Aceites.*, **46:** 96-97.

Makhija, I., and Khamar, D. (2010). Anti-snake venom properties of medicinal plants. *Der Pharm Lett.*, **2:** 399-411.

Manfred, L. (1977). 7000 Recetas botánicas a base de plantas medicinales. 11 ª ed. Ed. Kier, Buenos Aires, Argentina.

Marqués, C. (2001). Importancia econômica da familia Lauraceae Lindl. *Floresta e Ambiente*, **8:** 195-206.

Martínez Crovetto, R. (1961). Plantas Utilizadas en Medicina en el NO de Corrientes. Fundación Miguel Lillo, Tucumán.

Martz, W. (1992). Plants with a reputation against snakebite. *Toxicon,* **30:** 1131-1142.

Marzocca, A. (1997). Vademecum de Malezas Medicinales de la Argentina, Indígenas y Exóticas. Orientación gráfica, Buenos ires, Argentina.

Matoso, E. (1893). Cien industrias. Notas sobre Plantas útiles escogidas de la flora correntina. 3ª parte. Corrientes, Argentina.

Mejia, C.K., and Rengipo, S.E. (1995). Plantas medicinales de uso popular en la Amazonia Peruana. AECI y IIAP: Lima, Perú.

Melo, P.A. (1997). SubstancesThat Antagonize The Myotoxic Effect Of Crotalid Snake Venoms. *J Venom Anim Toxins,* **3:** 355-355.

Melo, P.A., do Nascimento, M.C., Mors, W.B., and Suarez-Kurtz, G. (1994). Inhibition of the myotoxic and hemorrhagic activities of crotalid venoms by *Eclipta prostrata* (Asteraceae) extracts and constituents.*Toxicon,* **32:** 595-602.

Melo, P.A., Mors, W.B., Nascimento, M.C., and Suarez-Kurtz, G. (1993). Antiproteolytic and antiphospholipasic activity of wedelolactone. *An Acad Bras Ci.,* **65:** 331-332.

Melo, P.A., and Ownby, C.L. (1999). Ability of wedelolactone, heparin, and para-bromophenacyl bromide to antagonize the myotoxic effects of two crotaline venoms and their PLA2 myotoxins. *Toxicon,* **37:** 199-215.

Melo, N., Collantes, I., and Yacida, M. (2004). Chemical constituents of *Nectandra saligna* Nees (*Lauraceae*). *XIII Congresso Italo-Latino Americano di Etnomedicina,* PO133.

Mongelli, E., Pampuro, S., Coussio, J., Salomon, H., and Ciccia, G. (2000). Cytotoxic and DNA interaction activities of extracts from medicinal plants used in Argentina. *J Ethnopharmacol.*, **71**: 145-151.

Moreno, J.J. (1993). Effect of aristolochic acid on arachidonic acid cascade and *in vivo* models of inflammation. *Immunopharmacology*, **26**: 1-9.

Morita, H., Matsumoto, K., Takeya, K., Itokawa, H., and Iitaka, Y. (1993). Structures and solid state tautomeric forms of two novel antileukemic tropoloisoquinoline alkaloids, pareirubrines A and B, from *Cissampelos pareira. Chem Pharm Bull.*, **41**: 1418-22.

Morita, H., Nakayama, M., Kojima, H., Takeya, K., Itokawa, H., Schenkel, E.P., and Motidome, M. (1991). Structures and cytotoxic activity relationship of casearins, new clerodane diterpenes from *Casearia sylvestris* Sw. *Chem Pharm Bull.*, **39**: 693-697.

Morita, H., Takeya, K., and Itokawa, H. (1995). A novel condensed tropone-isoquinoline alkaloid, pareitropone, from *Cissampelos pareira. Bioorg Med Chem Lett.*, **5**: 597-598.

Mors, W.B., do Nascimento, M.C., Parente, J.P., Silva, M.H., Melo, P.A., and Suarez-Kurtz, G. (1989). Neutralization of lethal and myotoxic activities of South American rattlesnake venom by extracts and constituents of the plant *Eclipta prostrata* (Asteraceae).*Toxicon*, **27**: 1003-1009.

Mors, W.B., Nascimento, M.C., Pereira, B.M., and Pereira, N.A. (2000). Plant natural products active against snake bite, the molecular approach. *Phytochemistry*, **55**: 627-42.

Muelas Serrano, S., Nogal, J., Martínez Diaz, R., Escario, J., Martínez Fernández, A., and Gomez Barrios, A. (2000). *In vitro* screening of american plant extracts of *Trypanosoma cruzi* and *Trichomonas vaginalis*. *J Ethnopharmacol.*, **71**: 101-107.

Nakagawa, M., Nakanishi, K., Darko, L.L., and Vick, J.A. (1982). Structures of cabenegrins A-I and A-II, Potent Anti-Snake Venom. *Tetrahedron Lett.*, **23**: 3855-3858.

Oberlies, N.H., Burgess, J.P., Navarro, H.A., Pinos, R.E., Fairchild, C.R., Peterson, R.W., Soejarto, D.D., Farnsworth, N.R., Kinghorn, A.D., Wani, M.C., and Wall, M.E. (2002). Novel bioactive clerodane diterpenoids from the leaves and twigs of *Casearia sylvestris. J Nat Prod.*, **65**: 95-99.

Oelrichs, P.B., Ng, J.C., Seawright, A.A., Ward, A., Schaffeler, L., and MacLeod, J.K. (1995). Isolation and identification of a compound from avocado (*Persea americana*) leaves which causes necrosis of the acinar epithelium of the lactating mammary gland and the myocardium. *Nat Toxins*, **3**: 344-349.

Ogawa, K, and Sashida, Y. (1992). Caffeoyl derivatives of a sugar lactone and its hydroxy acid from the leaves of *Bidens pilosa. Phytochemistry*, **31**: 3657-3658.

Ohira, S., Hasegawa, T., Hayashi, K., Hoshino, T., Takaoka, D., Nozaki, H. (1998). Sesquiterpenoids from *Cyperus rotundus. Phytochemistry*, **47**: 1577-1581.

Oksay, M, Usame Tamer, A, Ay, G, Sari, D, and Aktas, K. (2005). Antimicrobial activity of the leaves of *Lippia triphylla* (L Her) O. Kuntze (Verbenaceae) Against on Bacteria and Yeasts. *J Biol Sci.*, **5**: 620-622.

Oliva, M. de las M., Beltramino, E., Dallucci, N., Cacero, C., Zygadlo J., and Demo, M. (2010). Antimicrobial activity of essential oils of *Aloysia triphylla* (L'Her) Briton from different region of Argentine. *BLACPMA.*, **9**: 29-37.

Oliveira de Melo, J., Truiti, M., Muscará, M., Bolonheis, S., Dantas, J., Caparroz-Assef, S., Cuman, R., and Bersani-Amado, C. (2006). Anti-inflammatory activity of crude extract and fractions of *Nectandra falcifolia* leaves. *Biol Pharm Bull.*, **29**: 2241-2245.

Oliveira, F.Q, Andrade-Neto, V., Krettli, A.U., and Brand o, M.G.L. (2004). New evidences of antimalarial activity of *Bidens pilosa* roots extract correlated with polyacetylene and flavonoids. *J Ethnopharmacol.*, **93**: 39-42.

Oliveira Sim es, C.M., Mentz, L.A., Schenkel, E.P., Irgang, B.E., and Stehmann, J.R. (1998). Plantas da Medicina Popular no Rio Grande do Sul. 5ª ed. Editora da Universida-de/UFRGS. Porto Alegre, Brasil.

Otero, R., Fonnegra, R., Jiménez, S.L., Nú ez, V., Evans, N., Alzate, S.P., Garcí-a, M.E., Saldarriaga, M., Del Valle, G., Osorio, R.G., Dí-az, A., Valderrama, R, Duque, A., and Vélez, H.N. (2000a). Snakebites and ethnobotany in the northwest region of Colombia: Part I: traditional use of plants. *J Ethnopharmacol.*, **71**: 493-504.

Otero, R., Nú ez, V., Barona, J., Fonnegra, R., Jiménez, S.L., Osorio, R.G., Saldarriaga, M., and Díaz, A. (2000c). Snakebites and ethnobotany in the northwest region of Colombia. Part III: Neutralization of the haemorrhagic effect of *Bothrops atrox* venom. *J Ethnopharmacol.*, **73**: 233-241.

Otero, R., Nú ez, V., Jiménez, S.L., Fonnegra, R., Osorio, R.G., García, M.E., and Díaz, A. (2000b). Snakebites and ethnobotany in the northwest region of Colombia. Part II: Neutralization of lethal and enzymatic effects of *Bothrops atrox* venom. *J Ethnopharmacol.*, **71**: 505-511.

Panciera, R.J., Martin, T., Burrows, G.E., Taylor, D.S., and Rice, L.E. (1990). Acute oxalate poisoning attributable to ingestion of curly dock (*Rumex crispus*) in sheep. *J Am Vet Med Assoc.*, **196**: 1981-1984.

Parodi, L.R., and Dimitri, M.J. (1988). Enciclopedia Argentina de Agricultura y Jardinería. 3ª ed. Ed. ACME S.A.C.I., Buenos Aires, Argentina.

Pereira, B.M.R., Gonçalves, L.C., and Pereirra, N A. (1992). Abordagem farmacológica de plantas recomendadas pela medicina folclórica como antiofídicas III-Atividade antiedematogênica. *Rev Bras Farm.*, **73**: 85-86.

Pereira, C.G., Marques, M.O.M., Barreto, A.S., Siani, A.C., Fernandes, E.C., and Meireles, M.A.M. (2004). Extraction of indole alkaloids from *Tabernaemontana catharinensis* using supercritical $CO_2$+ethanol: an evaluation of the process variables and the raw material origin. *J Supercrit Fluid.*, **30**: 51-61.

Pereira, N.A., Ruppelt Pereira, B.M., do Nascimento, M.C., Parente, J.P., and Mors, W.B. (1994). Pharmacological screening of plants recommended by folk medicine as snake venom antidotes, IV. Protection against jararaca venom by isolated constituents. *Planta Med.*, **60:** 99-100.

Pereira, R.L., Ibrahim, T., Lucchetti, L., da Silva, A.I., and Gonçalves de Moraes, V.L. (1999). Immunosuppresive and anti inflammatory effects of methanolic extracts and the polyacetylene isolated from *Bidens pilosa* L., *Immunopharmacology,* **43:** 31-37.

Pereyra de Santiago, O.J., and Juliani, H.R. (1972). Isolation of quercetin 3,7,3',4'-tetrasulphate from *Flaveria bidentis* L. Otto Kuntze. *Experientia,* **28:** 380-381.

Petenatti, M.E., Alvarez, M.E., Petenatti, E.M., Del Vitto, L.A., Saad, J.R., Tévez, M.R., and Giordano, O.S. (2005). Medicamentos de Herbarios en el Centro-oeste Argentino, V. *Clematis montevidensis* va. *montevidensis*, caracterización de la droga. *Lat Am J Pharm.*, **24:** 190-196.

Petricic, J. (1966). Desgluco-uzarin, the principal cardanolide glycoside of *Asclepias mellodora* St. Hil. *Naturwissenschaften,* **53:** 332.

Piozzi, F., Bruno, M., and Roselli, S. (1998). Further furoclerodanes from *Teucrium* genus. *Heterocycles,* **48:** 2185-2203.

Pithayanukul, P., Laovachirasuwan, S., Bavovada, R., Pakmanee, N., and Suttisri, R. (2004). Anti-venom potential of butanolic extract of *Eclipta prostrata* against Malayan pit viper venom. *J Ethnopharmacol.*, **90:** 347-52.

Posey, D.A. (2002). Commodification of the sacred through intellectual property rights. *J Ethnopharmacol.*, **83:** 3-12.

Priestap, H.A. (1982). Phenanthrene derivatives from *Aristolochia argentina. Phytochemistry,* **21:** 2755-2756.

Priestap, H.A. (1985a). Seven aristololactams from *Aristolochia argentina. Phytochemistry,* **24:** 849-852.

Priestap, H.A. (1985b). Two carboxy- and two hydroxymethyl-substituted aristololactams from *Aristolochia argentina. Phytochemistry,* **24:** 3035-3039.

Priestap, H.A. (1987). Minor aristolochic acids from *Aristolochia argentina* and mass spectral analysis of aristolochic acids. *Phytochemistry,* **26:** 518-529.

Priestap, H.A., Bonafede, J.D., and Rúveda, E.A. (1977). Argentilactone, a novel 5-hydroxyacid lactone from *Aristolochia argentina. Phytochemistry,* **16:** 1579-1582.

Priestap, H.A., Iglesias, S.L., Desimone, M.F., and Díaz, L.E. (2003). Determination of aristolochic acids by capillary electrophoresis. *J Capillary Electrophor.*, **8:** 39-43.

Priestap, H.A., van Baren, C.M., Di Leo Lira, P., Coussio, J.D., and Bandoni, A. (2003b). Volatile constituents of *Aristolochia argentina* L. *Phytochemistry,* **63:** 221-225.

Quader, M.A., El-Turbi, J.A., Armstrong, J.A., Gray, A.I., and Waterman P.G. (1992). Coumarins and their taxonomic value in the *Genus phebalium. Phytochemistry,* **31:** 3083–3089.

Ramirez, I., Carabot, A., Melendez, P., Carmona, J., Jimenez, M., Patel, A.V., Crabb, T.A., Blunden, G., Cary, P.D., Croft, S.L., and Costa, M. (2003). Cissampeloflavone, a chalcone-flavone dimer from *Cissampelos pareira*. *Phytochemistry*, **64**: 645-647.

Raslan, D.S., Jamal, C.M., Duarte, D.S., Borges, M.H., De Lima, M.E. (2002). Anti-PLA2 action test of *Casearia sylvestris* Sw. *Bull Chim Farm.*, **141**: 457-60.

Rastrelli, L., Capasso, A., Pizza, C., De Tommasi, N., and Sorrentino, L. (1997). New protopine and benzyltetrahydroprotoberberine alkaloids from *Aristolochia constricta* and their activity on isolated guinea-pig ileum. *J Nat Prod.*, **60**: 1065-1069.

Ratera, E.L., and Ratera, M.O. (1980). Plantas de la Flora Argentina Empleadas en Medicina Popular. Ed. Hemisferio Sur S.A., Buenos Aires, Argentina.

Reyes Chilpa, R., and Jimenez Estrada, M. (1995). Química de las plantas alexíteras. *Interciencia*, **20**: 257-263.

Reyes-Chilpa, R., Gomez-Garibay, F., Quijano, L., Magos-Guerrero, GA, and Rios, T. (1994). Preliminary results on the protective effect of (-)-edunol, a pterocarpan from *Brongniartia podalyrioides* (Leguminosae), against *Bothrops atrox* venom in mice. *J Ethnopharmcacol.*, **42**: 199-203.

Ricciardi, A. (2005). Plantas con tradición de uso como alexíteras en la medicina popular. Conferencia del Curso de Actualización y Perfeccionamiento para graduados *Vegetales de importancia médica y Toxicológica. Control, Legislación y Fiscalización*. Confederación Farmacéutica Argentina y Asociación Amigos del Museo de Farmacobotánica, Buenos Aires, Argentina.

Ricciardi, A.I.A., Caballero, N.E., and Chifa, C. (1996). Identificación botánica de plantas descriptas en Materia Médica Misionera usadas en accidentes ofídicos. *Rojasiana*, **3**: 239-245.

Ricciardi, G., van Baren, C., Di Leo Lira, P., Ricciardi, A., Lorenzo, D., Dellacassa, E., and Bandoni, A. (2006). Volatile constituents from aerial parts of *Aloysia gratissima* (Gillies and Hook.) Tronc. var. gratissima growing in Corrientes, Argentina. *Flav Frag J.*, **21**: 698-703.

Ricciardi, G., Torres, A., Camargo, F., Agrelo de Nassiff, A., Dellacassa, E., and Ricciardi, A. (2008). Caracterización quimiotaxonómica y evaluacion de la actividad contra veneno de yarará de *Nectandra megapotamica* (Spreng.) Mez. Congreso Ítalo- Latinoamericano de Etnomedicina (SILAE) Palermo, Italia.

Ricciardi, G., Torres, A.M., Camargo, F.J., Ricciardi, A.I.A., and Dellacassa, E. (2008). Quimiotaxonomía de *Trixis divaricata* Spreng. especie utilizada contra venenos de víboras en el nordeste argentino. *Comunicaciones Científicas y Tecnológicas UNNE*, E098.

Ricciardi, G., Torres, A., Bubenik, A., Ricciardi, A., Lorenzo, D., and Dellacassa, E. (2011). Environmental effect on essential oil composition of *Aloysia citriodora* from Corrientes (Argentina). *Nat Prod Comm.*, **6**: 1711-1714.

Ricciardi Verrastro, B., Torres, A., Camargo, F., Ricciardi, G., Ricciardi, A., and Dellacassa, E. (2012). Actividad antiveneno de especies del género *Asclepia* contra yarará chica (*Bothrops diporus*). *XIV Jornadas de la Sociedad Uruguaya de Biociencias*. Piriàpolis, Maldonado. Uruguay.

Richter, H., and Dallwitz, M. (2008). Maderas comerciales. http://delta-intkey.com/wood/es/www/launemeg.htm.

Rios, M.Y., and Berber, L.A. (2005). (1)H and (13)C assignments of three new drimenes from *Iresine diffusa* Humb. and Bonpl. ex Willd. *Magn Reson Chem.*, **43:** 339-342.

Rocha da Costa, H.N., dos Santos, M.C., de Carvalho Alcântara, A.F., Conceiç o Silva, M., Cabral França, R., and Piló-Veloso, D. (2008). Chemical constituents and antiedematogenic activity of *Peltodon radicans* (*Lamiaceae*). *Quim Nova.*, **31:** 744-750.

Rodriguez, P.M. (1980). Plantas de la Medicina Popular Venezolana de venta en herbolarios. *Publ Soc Venezolana Cienc Nat.*, 168.

Romoff, P., Ferreira, M., Padilla, R., Toyama, D., Fávero, O., and Lago, J. (2010). Chemical composition of volatile oils from leaves of *Nectandra megapotamica* Spreng. (Lauraceae). *Quim Nova.*, **33:** 119-1121.

Ruppelt, B.M., Gonçalves, L.C., and Pereira, N.A. (1990). Abordagem farmacológicas de plantas recomendadas pela medicina folclórica como antiofídicas. II. Bloqueio da atividade de permeabilidade capilar e na letalidade do veneno de jararaca (*Bothrops jararaca*). *Rev Bras Farm.*, **71:** 57-58.

Ruppelt, B.M., Pereira, E.F., Gonçalves, L.C., and Pereira, N.A. (1990). Abordagem farmacológicas de plantas recomendadas pela medicina folclórica como antiofídicas. I. Atividades analgésica e antinflamatória. *Rev Bras Farm.*, **71:** 54-56.

Ruppelt, B.M., Pereira, E.F., Gonçalves, L.C., and Pereira, NA. (1991). Pharmacological screening of plants recommended by folk medicine as anti-snake venom-I. Analgesic and anti-inflammatory activities. *Mem Inst Oswaldo Cruz*, **86:** 203-205.

Ruppelt, B.M., Pereira, E.F., Gonçalves, L.C., and Pereira, N.A. (1992). Abordagem farmacológica de plantas recomendadas pela medicina folclórica como antiofídicas III-Atividade antiedematogênica. *Rev Bras Farm.*, **73:** 85-86.

Sandoval, M., Okuhama, N.N., Zhang, X.J., Condezo, L.A., Lao, J., Angeles, F.M., Musah, R.A., Bobrowski, P., and Miller, M.J.S. (2002) Anti-inflammatory and Antioxidant activities of cat's claw (*Uncaria tomentosa* and *Uncaria guianensis*) are independent of their alkaloid content. *Phytomedicine*, **9:** 325-337.

Sarg, T.M., Ateya, A.M., Farraq, N.M., and Abbas, F.A. (1991). Constituents and biological activity of *Bidens pilosa* L. grown in Egypt. *Acta Pharm Hungarica*, **61:** 317-323.

Sarker, S.D., Bartholomew, B., Nash, R.J., and Robinson, N. (2000). 5-*O*-methylhoslundin: An unusual flavonoid from *Bidens pilosa* (Asteraceae). *Bioch Syst Ecol.*, **28:** 591-593.

Sartoratto, A., Machado, A.L., Delarmelina, C., Figueira, G.M., Duarte, M.C., and Rehder, V.L. (2004). Composition and antimicrobial activity of essential oils from aromatic plants used in Brazil. *Braz J Microbiol.*, **35:** 275-280.

Sawaya, A.C.H.F., Vaz, B.G., Eberlin, M.N., and Mazzafera, P. (2011) Screening species of *Pilocarpus* (Rutaceae) as source of pilocarpine and other imidazole alkaloids. *Genet Resour Crop Ev.*, **58:** 471-480.

Stashenko, E., Jaramillo, B., and Martínez, J.R. (2003). Comparación de la Composición Química y de la Actividad Antioxidante *in vitro* de los metabolitos secundarios volátiles de las plantas de la família Verbenaceae. *Rev Acad Colomb Cienc.*, **27:** 579-597.

Sequeiros, C., López, L.M.I., Caffini, N.O., and Natalucci, C.L. (2003). Proteolytic activity in some Patagonian plants from Argentina. *Fitoterapia*, **74(6):** 570-577.

Shi, L.S., Kuo, P.C., Tsai, Y.L., Damu, A.G., and Wu, T.S. (2004). The alkaloids and other constituents from the root and stem of *Aristolochia elegans. Bioorg Med Chem.*, **12:** 439-446.

Simirgiotis, M.J., Favier, L.S., Rossomandoi, P.S., Giordano, O.S., Tonn, C.E., Padrón, J.I., and Vázquez, J.T. (2000). Diterpenes from *Laemnecia sophiifolia. Phytochemistry*, **55:** 721-726.

Sonwa, M.M., and König, W.A. (2001). Chemical study of the essential oil of *Cyperus rotundus. Phytochemistry*, **58:** 799-810.

Sorarú, S.B., and Bandoni, A. (1978). Plantas de la Medicina Popular Argentina. Ed. Albatros, Buenos Aires, Argentina.

Sousa, J.R., Demuner, A.J., Pinheiro, J.A. Breitmaier, E., and Cassels, B.K. (1990). Dibenzyl trisulfide and trans-N-methyl-4-methoxy proline from *Petiveria alliacea. Phytochemistry*, **29:** 3653-3655.

Srivastava, S., and Srivastava, S. (1993). New constituents and biological activity of the roots of *Murraya koenigii. J Indian Chem Soc.*, **70:** 655–659.

Staneva, J.D., Todorova, M.N., Evstatieva, L.N. (2005). New linear sesquiterpene lactones from *Anthemis cotula* L. *Bioch Syst Ecol.*, **33:** 97-102.

Strack Dieter, W.V., Metzger, J., and Grosse, W. (1992). Two anthocyanins acylated with gallic acid from the leaves of *Victoria amazonica. Phytochemistry*, **31:** 989-991.

Szczepanski, C., Zgorzelak, P., and Hoyer, G.A. (1972). Isolierung, Strukturaufklärung und Synthese einer antimikrobiell Wirksamen Substanz aus Petiveria alliacea L. *Arzneim.-Forsch.*, **22:** 1975-1976.

Tiwari, H.P., Penrose, W.R., and Spenser, I.D. (1967). Biosynthesis of mimosine: Incorporation of serine and of α-aminoadipic acid. *Phytochemistry*, **6:** 1245-1248.

Tiwari, H.P., and Spenser, I.D. (1965). Precursors of mimosine in *Mimosa pudica. Can J Biochem.*, **43:** 1687-1691.

Torrado Truiti, M. da C., and Sarragiotto, M.H. (1998). Three 5-methylcoumarins from *Chaptalia nutans*. *Phytochemistry*, **47:** 97-99.

Torrado Truiti, M. da C., Sarragiotto, M.H., de Abreu Filho, B.A., Nakamura, C.V., and Dias Filho, B.P. (2003). *In vitro* antibacterial activity of a 7-O-beta-D-glucopyranosyl-nutanocoumarin from *Chaptalia nutans* (Asteraceae). *Mem Inst Oswaldo Cruz.*, **98:** 283-286.

Torres de Sosa, A.M., Camargo, F. J., Avanza de Temporetti, M.V., Tressens, S. G., and Ricciardi, A.I.A. (2004). Interacción entre extractos de órganos de plantas y veneno de *Bothrops neuwiedi diporus* Cope (yarará chica). *VIII Simposio Argentino y XI Simposio Latinoamericanoo de Farmacobotánica*, 84.

Torres, A.M., Ricciardi, G.A., Agrelo de Nassiff, A., Ricciardi, A.I.A., Lorenzo, D., and Dellacassa, E. (2005). Examen del aceite esencial de *Nectandra angustifolia* (Schrad.) Nees and Mart. ex Nees. *Comunicaciones Científicas y Tecnológicas UNNE*, E013.

Torres, A., Camargo, F., Dellacassa, E., and Ricciardi, A. (2007). Acción antihemolítica *in vitro* de extractos de plantas del nordeste argentino sobre el veneno de *Bothrops neuwiedi diporus* Cope (yarará chica). *Congreso Ítalo- Latinoamericano de Etnomedicina* (SILAE), La Plata Argentina.

Torres, A., Camargo, F., Ricciardi, G., and Dellacassa, E., Ricciardi, A. (2008). Inhibición de la actividad procoagulante del veneno de yarará chica (*Bothrops neuwiedi diporus*), por extractos de plantas del NE argentino. *Congreso Ítalo-Latinoamericano de Etnomedicina* (SILAE) Palermo, Italia, P178.

Torres, A.M., Camargo, F.J., Ricciardi, G.A., Ricciardi, A.I.A., and Dellacassa, E. (2010). Inhibición de la actividad proteolítica del veneno de *Bothrops diporus* por extractos de plantas del nordeste argentino. *Comunicaciones Científicas y Tecnológicas UNNE*, Corrientes, Argentina.

Torres, A.M., Camargo, F.J., Ricciardi, G.A., Ricciardi, A.I.A., and Dellacassa, E. (2010b). Fraccionamiento y caracterización fitoquímica del extracto alcohólico de hojas del laurel amarillo *Nectandra angustifolia* (Schrad.) Nees and Mart. ex Nees. *Comunicaciones Cinetíficas y Tecnológicas UNNE*. Corrientes, Argentina.

Torres, A., Camargo, F., Ricciardi, G., Dellacassa, E., and Ricciardi, A. (2011). Neutralizing effects of *Nectandra angustifolia* extracts against *Bothorps neuwiedi* snake venom. *Nat Prod Comm.*, **6:** 1393-1396.

Torres, A. (2011). Caracterización quimiotaxonómica de especies pertenecientes al género *Nectandra* en el Nordeste Argentino. Evaluación de su potencial actividad antiveneno en accidentes provocados por *Bothrops neuwiedi diporus* (COPE) "yarará chica". Universidad Nacional del Nordeste. PhD Thesis.

Torres, A., Camargo, F., Ricciardi, G., Ricciardi, A., Guerra, M., and Dellacassa, E. (2011b). Fitoquímica de una fracción del extracto alcohólico de *Nectandra angustifolia*, activa contra el veneno de *Bothrops diporus*. *Comunicaciones Cinetíficas y Tecnológicas UNNE*, Chaco, Argentina.

Torres, A.M., Camargo, F.J., Ricciardi, G.A., Ricciardi, A.I., Martínez, N., Lorenzo, D., and Dellacassa, E. (2011c). Neutralizing effects of *Nectandra angustifolia* essential oil against *Bothrops diporus* snake venom. *VI SBOE* Simpósio Brasileiro de Óleos Essenciais. Campinas, Brasil.

Torres, A., Camargo, F., Ricciardi, G., Ricciardi, A., and Dellacassa, E. (2012a). Actividad antiveneno de dos especies del género *Aristolochia* contra veneno de *Bothrops diporus*. *Rev Fitoter.*, **12**: 123.

Torres, A.M., Camargo, F.J., Ricciardi, G.A., Ricciardi, A.I., Dellacassa, E., Lozina, L., and Maru ak, S., Acosta de Perez, O. (2012b). Inhibición de la actividad letal del veneno de Bothrops diporus por extractos de *Nectandra angustifolia*. *Rev Fitoter.*, **12**: 124.

Toursarkissian, M. (1980). Plantas Medicinales de la Argentina. Ed. Hemisferio Sur S.A., Buenos Aires, Argentina.

Trichopoulou, A., Vasilopoulou, E., Hollman, P., Chamalides, Ch., Foufa, E., Kaloudis, Tr., Kromhout, D., Miskaki, Ph., Petrochilou, I., Poulima, E., Stafilakis, K., and Theophilou, D. (2000). Nutritional composition and flavonoid content of edible wild greens and green pies: a potential rich source of antioxidant nutrients in the Mediterranean diet. *Food Chem.*, **70**: 319-323.

Trombone, A.P.F., Veronese, E.L.G., Marques, L.B., Cintra, A.C.O., and Sampaio, S.V. (1999). Avaliaç o da aç o do extrato de *Tabernaemontana catharinensis* sobre a miotoxicidade induzida pelo veneno de *B. jararacussu* e BthTX-I. *VII Simpósio de Iniciaç o Científica da Universidade de S o Paulo*. Ribeir o Preto, Brasil.

Truiti, M. (2004). Chemical study and evaluation of biological activities of naturally occurring species in the Brazilian stretch of the Upper Paraná River. *Ph.D. Thesis*, State University of Maringá, Maringá, Brazil.

Truiti, M., Ferreira, I., Zamuner, M., Nakamura, C., Sarragiotto, M., and Souza, M. (2005). Antiprotozoal and molluscicidal activities of five Brazilian plants. *Braz J Med Biol Res.*, **38:** 1873-1878.

Ubillas, R.P., Mendez, C.D., Jolad, S.D., Luo, J., King, S.R., Carlson, T.J., and Fort, D.M. (2000). Antihyperglycemic acetylenic glucosides from *Bidens pilosa*. *Planta Med.*, **66:** 82-83.

Uchiyama, T., Hara, S., Makino, M., and Fujimoto, Y. (2002). *Seco*-Adianane-type triterpenoids from *Dorstenia brasiliensis* (Moraceae). *Phytochemistry*, **60:** 761-764.

Valente, L.M, Gunatilaka, A.A., Glass, T.E., Kingston, D.G., Pinto, A.C. (1993). New norcucurbitacin and heptanorcucurbitacin glucosides from *Fevillea trilobata*. *J Nat Prod.*, **56:** 1772-1778.

Valente, L.M., Gunatilaka, A.A, Kingston, D.G, and Pinto, A.C. (1994). Norcucurbitacin gentiobiosides from *Fevillea trilobata*. *J Nat Prod.*, **57:** 1560-1563.

Valeta, A. (1935). Botánica Práctica. Plantas medicinales. Ed Higiene y Salud, 6a ed., Montevideo, Uruguay.

Van Klink, J., Becker, H., Andersson, S., and Boland, W. (2003). Biosynthesis of anthecotuloide, an irregular sesquiterpene lactone from *Anthemis cotula* L. (Asteraceae) via a non-farnesyl diphosphate route. *Org Biomol Chem.*, **1:** 1503-1508.

Velazquez, E., Tournier, H.A., Mordujovich de Buschiazzo, P., Saavedra, G., and Schinella G.R. (2003). Antioxidant activity of Paraguayan plant extracts. *Fitoterapia*, **74:** 91-97.

Veronese, E.L., Esmeraldino, L.E., Trombone, A.P.F., Santana, A.E., Bechara, G.H., Kettelhut, I., Cintra, A.C.O., Giglio, J.R., and Sampaio, S.V. (2005). Inhibition of the myotoxic activity of *Bothrops jararacussu* venom and its two major myotoxins, BthTX-I and BthTX-II, by the aqueous extract of *Tabernaemontana catharinensis* A. DC. (Apocynaceae). *Phytomedicine*, **12:** 123-130.

Vila, R., Mundina, M., Muschietti, L., Priestap, H.A., Bandoni, A.L., Adzet, T., and Ca igueral, S. (1997). Volatile constituents of leaves, roots and stems from *Aristolochia elegans*. *Phytochemistry*, **46:** 1127-1129.

Vilegas, W., and Pozetti, G.L. (1993). Coumarins from *Brosimum gaudichaudii*. *J Nat Prod.*, . **56:** 416-417.

Vishwanath, B.S., Appu, Rao, A.G., and Gowda, T.V. (1987a). Interaction of phospholipase A from *Vipera russelli* with aristolochic acid: a circular dicroism study. *Toxicon*, **25:** 939-946.

Vishwanath, B.S., and Gowda, T.V. (1987b). Interaction of aristolochic acid with *Vipera russelli* phospholipase $A_2$: Its effect on enzymatic and pathological activities. *Toxicon*, **25:** 929-937.

Vishwanath, B.S, Kini, R.M., and Gowda, T.V. (1987c). Characterization of three edema-inducing phospholipase $A_2$ enzymes from habu (*Trimeresurus flavoviridis*) venom and their interaction with the alkaloid aristolochic acid. *Toxicon*, **25:** 501-515.

Vishwanath, B.S., Kini, R.M., and Gowda, T.V. (1985). Purification of an edema inducing phospholipase $A_2$ from *Vipera russelli* venom and its interaction with aristolochic acid. *Toxicon*, **23:** 617.

Wahab, A.S.M., and Selim, M.A. (1985). Lipids and flavonoids of *Sapindus saponaria* L. *Fitoterapia*, **61:** 167-168.

Wang, J., Yang, H., Lin, Z-W., Sun, H.D. (1997). Flavonoids from *Bidens pilosa* var. *radiate. Phytochemistry*, **46:** 1275-1278.

Wat, C.K., Biswas, R.K., Graham, E.A., Bohm, L., Towers, G.H., and Waygood, E.R. (1979). Ultraviolet mediated cytotoxic activity of phenylheptatriyne from *Bidens pilosa*. *J Nat Prod.*, **42:** 103-111.

Wilkinson, S. (1958). Alpha-Picoline from *Rumex obtusifolius* L. *Nature*, **181(4609):** 636-7.

Wu, T.S., Tsai, Y.L., Damu, A.G., Kuo, P.C., and Wu, P.L. (2002). Constituents from the root and stem of *Aristolochia elegans*. *J Nat Prod.*, **65:** 1522-1525.

Wu, T.S., Tsai, Y.L., Wu, P.L., Lin, F.W., and Lin, JK. (2000). Constituents from the leaves of *Aristolochia elegans*. *J Nat Prod.*, **63:** 692-693.

Wüster, W., Salomo, M., Quijada-Mascare as, J., and Thorpe, R., BBBSP (2002). Origins and evolution of the South American pitvipers fauna: evidence from mitochondrial DNA sequence analysis. *In:* Biology of the Vipers. Eagle Mountain, UT. 111-129.

Yadava, R.N., and Yadav, S. (2001). A novel buffadienolide from the seeds of *Mimosa pudica* Linn. *Asian J Chem.*, **13:** 1157-1160.

Zhang, J.S., and Guo, Q.M. (2001), Studies on the chemical constituents of *Eclipta prostrata* (L). *Yao Xue Xue Bao.*, **36:** 34-37.

Zhang, M, and Chen, Y. (1996). Chemical constituents of *Eclipta alba* (L.) Hassk. *Zhongguo Zhong Yao Za Zhi.*, **21:** 480-481.

Zygadlo, J.A., Lamarque, A.L., Guzmán, C.A., and Grosso, NR. (1995). Composition of the Flower Oils of Some *Lippia* and *Aloysia* Species from Argentina; *J. Essent. Oil Res.*, **7:** 593-595.

Zuloaga, F.O., and Morrone, O. (eds.) (1996). Catálogo de las plantas vasculares de la República Argentina I: Pteridophyta. Gimnospermae y Angiospermae (*Monocotyledoneae*), Vol. 60 Missouri Botanical Garden Press, E.E.U.U., 323 págs.

Zuloaga, F.O., and Morrone, O. (eds.). (1999). Catálogo de las plantas vasculares de la República Argentina II. 74. Missouri Botanical Garden Press. E.E. U.U.

Zulueta, M.C.A., Tada, M., and Ragasa, C.Y. (1995). A diterpene from *Bidens pilosa*. *Phytochemistry*, **38:** 1449-1450.

Utilisation and Management of Medicinal Plants Vol. 2 (2014)    *Pages* **63–91**
*Editor-in-Chief:* **V.K. Gupta**
*Published by:* **DAYA PUBLISHING HOUSE, NEW DELHI**

# 2

# Phytotherapeutic Potentialities and Efficacy of *Cymbopogon citrates* in the Traditional Systems of Medicine

Kaliyaperumal Karunamoorthi[1, 2]*

## ABSTRACT

The utilization and management of traditional folk medicinal plants (TFMPs) is as old as human civilization/history. It is estimated that nearly 25 per cent of all conventional medicines are directly or indirectly derived from TFMPs. Globally, *Cymbopogon citratus* or lemongrass is well-known due to its omnipotent nature and an infinite number of phytotherapeutic applications and potentialities. The present scrutiny results suggesting that the tea, infusion and extracts of *C. citratus*, which are prepared with fresh or dry leaves, are often used in the popular medicine as a restorative, digestive, anti-tussis and an effective drug against colds. Worldwide, numerous scientific studies show that the lemongrass essential oil (LGEO) has an analgesic, antithermic, anti-inflammatory of urinary ducts, diuretic, antibacterial, antifungal, antimalarial, antileashmanial, antiacaricidal, antifilarial, and antiallergic effect. In addition, there are studies reporting that *C. citratus* has mosquito larvicidal and repellent activities too. Besides the lemon grass extracts and their essential oils are also used in food (flavoring), perfume, soap and cosmetics industries.

Therefore, presently there is a renewed interest on *C. citratus* as a source of developing new natural sustainable and affordable drugs and for perfumery and

1    Unit of Medical Entomology and Vector Control, Department of Environmental Health Science and Technology, College of Public Health and Medical Sciences, Jimma University, Jimma, Ethiopia.

2    Research and Development Centre, Bharathiar University, Coimbatore, Tamil Nadu, India.

*    *Corresponding author*: E-mail: k_karunamoorthi@yahoo.com

cosmetics due to its user-friendly nature. It calls for the continuous efforts in order to identify and validate the traditional usage of *C. citratus* by identifying the bioactive molecules and secondary metabolites. Since, we are living in the era of synthetic chemistry and biotechnology techniques, we can employ these most innovative techniques to develop the plant-derived pharmaceuticals and to obtain enviable results. In addition, there is a necessity to adopt a holistic interdisciplinary approach too. As per my knowledge, understanding and observation many prefer the traditional system of medicine. Nevertheless, the major barrier is the lack of phytochemical standardization and therefore there is a pressing need for more clinical assay to validate ethnomedical usage of *C. citratus* in the near future.

*Keywords*: *Cymbopogon citratus*, Herb, Traditional phytotherapy, Traditional system of medicine.

# Introduction

Traditional system of medicine (TSM) is one of the centuries-old practice and long-serving companion to the human kind to fight against disease and to lead a healthy life. Indigenous people have been using their unique approach of TMS practice where among, the Chinese, Indian, African TMSs are world-wide renowned. India has a unique Indian System of Medicines (ISM) consisting of Ayurveda, Siddha, Unani, Naturopathy and Homoeopathy (Karunamoorthi *et al.*, 2012).

In the past, the induction of modern health care services has posed immense threat to indigenous health practices due to their potential speedy therapeutic effect and the traditional medicinal systems are disappearing, displaced, and undervalued by the people (Karunamoorthi *et al.*, 2012). Despite globalization and modernization, still a sizeable fraction of the rural poor purely relies on the traditional medicines as their primary health care modality. Over the past three decades, the worldwide scientific community is extremely inclined to explore the plant-based products for the betterment of human from our long–standing traditional knowledge and medicinal systems (Karunamoorthi, 2012a).

Plants are factual expressions of the eternal kindness of the nature on which all living beings including man are dependent to fulfil their multifarious requirements. Man is a powerful and thoughtful creature on Earth who takes advantage of the nature in all possible ways. Man's interest in plants began for his requirement of food and shelter. Next he sought remedies for injuries through plants, during his nomadic life. Rig Veda says that man learnt to distinguish edible plants from the poisonous ones by observing the way animals used those (Manilal, 1989).

# Traditional Folk Medicinal Plants: A Source of Phytotherapeutic Agent

At present, more than 80 per cent of the world's population relies on traditional healing modalities and herbals for primary health care and wellness (WHO, 2002). In fact, the global market for traditional medicines was estimated to be US$ 83 billion annually in 2008, with a rate of increase that has been exponential (Robinson and

Zhang, 2011). The population rise, inadequate supply of drugs, prohibitive cost of treatments, side effects of several allopathic drugs and development of resistance to currently used drugs for infectious diseases have led to the increased emphasis on the use of plant materials as a source of medicine for a wide variety of human ailments. Global estimates indicate that 80 per cent of about 4 billion population cannot afford the products of the Western Pharmaceutical Industry and have to rely upon the use of traditional medicines which are mainly derived from plant material (Joy *et al.*, 1998).

Natural products from plants offer vast and valuable sources of bioactive compounds and have been traditionally used in folk medicine to treat various illnesses from headaches to parasite infections (Anthony *et al.*, 2005; Halberstein, 2005). Folk medicines, mainly based on plants, enjoy a respectable position today, especially in the developing countries, where modern health service is limited. Safe, effective and inexpensive indigenous remedies are gaining popularity among the people of both urban and rural areas, especially in India and China (Katewa *et al.*, 2004).

Use of medicinal plants as a source of relief and cure from various illness is as old as humankind itself. Even today, medicinal plants provide a cheap source of drugs for majority of world's population. Plants have provided and will continue to provide not only directly usable drugs, but also a great variety of chemical compounds that can be used as starting points for the synthesis of new drugs with improved pharmacological properties (Potterat and Hostettmann, 1995).

## Plant Species with Therapeutic Value Under Different Plant Groups

Of the 2,50,000 higher plant species on earth, more than 80,000 species are reported to have at least some medicinal value. There are about 400 families in the world of flowering plants. Among them, at least 315 are represented in India. According to WHO, around 21,000 plant species have the potential for being used as medicinal plants (Joy *et al.*, 1998). According to Jiaxiang (1997) 4,877 plant species belonging to different plant groups, have potential therapeutic value (Table 2.1).

**Table 2.1**: Plant species with therapeutic value under different plant groups.

| Sl.No. | Name of the Plant Groups | Number of Species |
|--------|--------------------------|-------------------|
| 1. | Thalophytes | 230 |
| 2. | Bryophytes | 39 |
| 3. | Pteridophytes | 382 |
| 4. | Gymnospermae | 55 |
| 5. | Angiospermae | |
| | a) Monocotyledones | 676 |
| | b) Dicotyledones | 3495 |
| | Total | 4877 |

Today's use of medicinal plants and bioactive phytocompounds worldwide and our scientific knowledge of them comprise the modern field of the "phytosciences". The phytosciences have been created from the integration of disciplines that have never been linked before, combining diverse areas of economic, social, and political fields with chemistry, biochemistry, physiology, microbiology, medicine, and agriculture. At the moment, considerable research has been carried out on phyto-chemistry and bioactivity of indigenous flora worldwide.

In this context, the present chapter becomes more significant and pertains, as we describe about the *C. citratus* distribution, habitat, physical characteristic features and bioactive molecules. It also deals about *C. citratus* phytotherapeutic potentialities and their efficacy in traditional systems of medicine. Besides, it also explains the existing daunting challenges and issues in the utilization and management of medicinal plants, more particularly, *C. citratus* and the opportunities to make traditional medicinal plants as a curative agent in the developing countries owing to fragile health care system and management.

## *Cymbopogon citratus*: A Source of Ethnomedicines

The *Cymbopogon citratus* (DC) Stapf, is an important species of Poaceae family, which has about 660 genera and 9,000 species (Clayton, 1968). *C. citratus* has been commonly used in the folk world-wide with infinite number of applications. The Cymbopogon genus presents more than 100 species found in tropical countries (Lorenzi and Matos, 2003), and from those, 56 are aromatic and rich in essential oils (concentrated hydrophobic liquid, which carries volatile and aromatic compounds) (Silva *et al.*, 2010).

### Scientific Classification of *C. citratus*

**Table 2.2**: Taxonomic position of *Cymbopogon citratus* (DC.) Stapf.

| | |
|---|---|
| **Kingdom** | Plantae–Plants |
| **Subkingdom** | Tracheobionta–Vascular plants |
| **Super division** | Spermatophyta–Seed plants |
| **Division** | Magnoliophyta–Flowering plants |
| **Class** | Liliopsida–Monocotyledons |
| **Subclass** | Commelinidae |
| Order | Cyperales |
| **Family** | Poaceae–Grass family |
| **Genus** | Cymbopogon Spreng.–Lemon grass |
| *Species* | Cymbopogon citratus *(DC.) Stapf*–Lemon grass |

### Etymology

The botanical genus name Cymbo-pogon is derived from Greek kymbe [κυμβη] [boat] and pogon [Πωγων] [beard]; it refers to the boat-shaped spathes and the many-eared inflores-cences which remind about a beard. The species name citratus obviously relates to the prominent lemon fragrance of the plant.

**Table 2.3**: The vernacular name of the *C. citratus* worldwide, in their respective language.

| Sl.No. | Language | Vernacular Name(s) |
|---|---|---|
| 1. | Amharic | Lomi-saar |
| 2. | Arabic | limun-Hashisha al حَشِيشَة اللَّيْمُون |
| 3. | Bengali | লেমন গ্রাস, গন্ধবেনা (Leman gras, Gandhabena) |
| 4. | Burmese | Zabalin, Sabalin |
| 5. | Chinese (Cantonese) | 草薑 [chóu gèung], 風茅 [fùng màauh], 檸檬草 [nìhng mùng chóu], 檸檬香茅 [nìhng mùng hèung màauh], 香巴茅 [hèung bā màauh], 香茅屬 [hèung màauh sūk] |
| 6. | Chinese (Mandarin) | 草薑 [cǎo jiāng], 風茅 [fēng máo], 檸檬草 [níng méng cǎo], 檸檬香茅 [níng méng xiāng máo], 香巴茅 [xiāng bā máo], 香茅屬 [xiāng máo shǔ], 柠檬香茅 [níng méng xiāng máo] |
| 7. | English | Lemon grass, Citronella, Squinant |
| 8. | French | Verveine des Indes |
| 9. | German | Zitronengras, Citronella, Lemongras |
| 10. | Greek | Λεμονόχορτο, Κιτρονέλλα (Cymbopogon nardus); Lemonochorto; Kitronella (Cymbopogon nardus) |
| 11. | Hebrew | Limonit rehanit, Limon gras, Essef limon לימון, עשב לימון, לימונית ריחנית, לימון גראס |
| 12. | Hindi | सेरा, गंधत्रिण (Sera, Gandhatrina) |
| 13. | Indonesian | Sereh, Serai |
| 14. | Italian | Cimbopogone |
| 15. | Japanese | レモングラス, レモンソウ (Remon-sō, Remonso, Remon-guraso) |
| 16. | Kannada | ಮಜ್ಜಿಗೆ ಹುಲ್ಲು (Majjige hullu) |
| 17. | Malay | Serai, Serai dapur |
| 18. | Malayalam | ഇഞ്ചിപ്പുല്ല്, തെരവ പുല്ല് (Enchipullu, Inchi-pullu, Terava-pullu, Vasana-pullu) |
| 19. | Manipuri (Meitei-Lon) | হাওনা (Haona) |
| 20. | Nepali | पिर्हे घाँस, काँश (Pirhe ghans, Kansh) |
| 21. | Romanian | Citronella, Ierba de Lămâie |
| 22. | Russian | Лимонное сорго, Лимонная трава (Limonnoe sorgo, Limmonaya trava) |
| 23. | Spanish | Zacate de limón, Te de limón, Caña de Limón, Citronella, Hierba de Limón, Malojillo |
| 24. | Swedish | Citrongräs |
| 25. | Tamil | கர்ப்பூரப்புல், போதைப்புல் (Karppurappul, Potaippul) |
| 36. | Telugu | కామంచి కసు Kamanchi kasu, Nimmagaddi |
| 27. | Thai | ตะไคร้, ตะไคร้หอม, จะไคร, จะไคร้, ไคร (Takrai, Takhrai hom, Chakrai, Krai, Soet kroei) |
| 28. | Urdu | Agan ghas اگن گھاس |
| 29. | Vietnamese | Sả chanh, Xả (Sa chanh, Xa) |

# Origin of *C. citratus*

The genus *Cymbopogon* has about 56 important species, indigenous in tropical and semi-tropical areas of Asia. However, they are cultivated in South and Central America, Africa and other tropical countries too. The so-called East-Indian lemon grass (*Cymbopogon flexuosus* [Nees ex Steudel] J.F. Watson) is native to India, Sri Lanka, Burma and Thailand. West-Indian lemon grass C. citratus is native to India and the nearby island of Sri Lanka. It is found growing naturally in tropical grasslands and also extensively cultivated throughout tropical Asia. Although the two species can almost be used interchangeably, *C. citratus* is more relevant for cooking. In India, it is cultivated as a medical herb and for perfumes, but not used as a spice; in the rest of tropical Asia (Sri Lanka and even more South East Asia), it is an important culinary herb and spice.

## Physical Characteristics

Lemongrass (LG) is a frost-tender clumping perennial grass and is widely cultivated in Southeast Asia. It is also an outstanding ornamental grass that lends great beauty to garden areas regardless of whether its culinary uses are to be tapped. This tropical grass grows in dense clumps that can grow upto 6 ft (1.8 m) in height, about 4 ft (1.2 m) in width and with a short underground rhizome (Reitz, 1982).

The straps like leaves are simple, alternate, linear, 5.0-7.0 cm. long, 0.5-1.5 cm. wide, sheathed, apex acute, parallel venation, central vein appear more in lower

**Figure 2.1**: *Cymbopogon citratus* plant.

epidermis, and have gracefully drooping tips. The evergreen leaves are bright bluish-green and release a citrus aroma when crushed. It is the leaves that are used as flavoring and in medicine. They are steam distilled to extract lemongrass oil, an old standby in the perfumer's palette of scents. Inflorescences are 3.0 to 6.0 cm long and nodding and the partial inflorescences are paired racemes of spikelets subtended by spathes (Ross, 1999).

The lemon grass plants that you are likely to encounter are cultivars and do not typically produce flowers but flowering panicles are rarely formed. Extracted plant oils have been used for many years in herbal medicines and perfumes. Lemongrass is equally versatile in the garden. All parts of the grass are lemon flavored. Leaves are leathery-textured sheaths (Karunamoorthi *et al.*, 2010).

## Ecology of *C. citratus*

*C. citratus* grows best under sunny, warm, and humid conditions. It performs best below 500 m altitude, but is grown up to 750 m. Short periods with a daily maximum temperature over 30°C do not harm growth but severely reduce oil content. Hot dry winds may not only desiccate the crop but can also evaporate the oil. Water requirement is very high; the highest yields are obtained where the average annual rainfall is 2,500–3,000 mm. Long periods of dry sunny weather strongly reduce herbage yield and oil yield, although oil content is generally higher during the dry season (Plant Resource, 1998).

## Cultivation

If possible plant lemongrass in fertile loam–but it will tolerate many other types of soils, including sand, if given some care.

- ☆ **Light:** Bright sun preferred but will grow in light shade.
- ☆ **Moisture:** Likes moisture but can survive some drought although its appearance will suffer.
- ☆ **Hardiness:** This is a tender plant that suffers leaf damage from frosts and is killed back to the roots by hard freezes.
- ☆ **Propagation:** By division of old clumps in the spring and summer. Also by seed which is not readily available.

## Chemical Composition of *C. citratus*

As a consequence of innumerous applications of *C. citratus*, several studies have been done aiming at enlarging the knowledge about the chemical composition of lemon grass leaves, which are the parts used for medicinal purposes. These studies have revealed that, although the chemical composition of the essential oil and aqueous extracts of *C. citratus* varies according to the geographical origin, the isolated and identified substances from the leaves are mainly alkaloids, saponin, asitosterol, terpenes, alcohols, ketone, flavonoids, chlorogenic acid, caffeic acid, p-coumaric acid and sugars (Matouschek and Stahl, 1991; Chisowa *et al.*, 1998; Negrelle and Gomes, 2007).

## Main Constit-uents of Lemon Grass Essential Oil (LGEO)

Essential oils (EOs) constitute a relatively common group of natural products derived from aromatic medicinal plants. They are volatile liquids usually with pleasant and sometimes intensive odors (aroma). They also are referred to as volatile oils, ethereal oils, or essences of many plants. The chemical composition is quite different from one plant to another and the main chemical component determines the aroma and its biological activities. Therapeutically, they exert a wide spectrum of biological activities such as: antiseptic, stimulant, carminative, diuretic, anthelmintic, analgesic and many others according to the chemical composition (Ikam, 1969). In the world-wide, flavour and fragrance market and essential oils constitute about 17 per cent. The estimate of world production of essential oils varies from 40,000 to 60,000 tonnes per annum (Joy *et al.*, 1998).

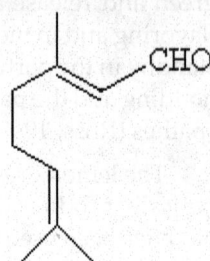

Citral

Citral    Citronellal

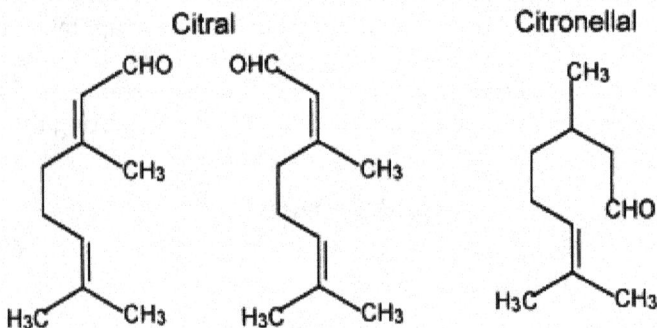

Geranial (Citral a)    Neral (Citral b)

EOs are a rich source of biologically active compounds. There has been an increased interest in looking at antimicrobial properties of extracts from aromatic plants particularly essential oils (Milhau *et al.*, 1997). LG is one of the few herbs from the grass family (Paranagama *et al.*, 2003). The compounds identified in C. citratus are mainly terpenes, alcohols, ketones, aldehyde and esters. Some of the reported phyto-constituents are EOs that contain Citral α, Citral β, Geraniol (Citral a), Nerol (Citral b), Citronellal, Terpinolene, Geranyl acetate, Myrecene and Terpinol Methylheptenone. The plant also contains reported phyto-constituents such as

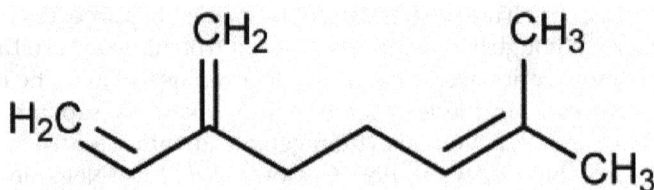

Myrecene

flavonoids and phenolic compounds, which consist of luteolin, isoorientin 2'-O-rhamnoside, quercetin, kaempferol and apiginin (Shah *et al.*, 2011).

The mean essential oil (EO) of *C. citratus* contents from the blade and sheath were 0.42 and 0.13 per cent respectively and citral contents of the blade and sheath essential oils were 87.28 and 82.39 per cent respectively (Ming *et al.*, 1996).

Citral is a mixture of two stereo-isomeric mono-terpene alde-hydes; in lemon grass oil, the trans-isomer geranial (40 to 62 per cent) domi-nates over the cisisomer neral (25 to 38 per cent). The EO of C. citratus contains Citral α (40 per cent), Citral β (32 per cent), nerol (4.18 per cent), geranicol (3.04 per cent), citronellal (2.10 per cent), terpinolene (1.23 per cent), geranyl acetate (0.83 per cent) etc. and all are used as important raw material in the pharmaceutical, perfumery and cosmetics industries, especially for the synthesis of Vitamin A and ionones (Ravinder *et al.*, 2010).

## Biological Activities of *C. citratus*

In recent times, there has been renewed interest on plants as sources of antimicrobial agents due to their ethnomedicinal uses and the fact that a good portion of the world's population, particularly in developing countries, rely on plants for the treatment of infectious and non-infectious diseases. The plant-derived drugs have been reported to be safe and without side-effects and the antimicrobial properties of plant volatile oils have been recognized since antiquity (Cowan, 1999).

LG is an important source of ethnomedicines as well as citral (3,7dimethyl-2,6-octadienal). It is the major constituent of *C. citratus*, possesses various biological activities i.e neurobehavioral effect, larvicidal activity, hypoglycemic and hypolipidemic effects, hypocholesterolaemic effect, free radical scavengers and antioxidants effect, ascaricidal activity, antiprotozoan activity, antinociceptive effect, antimycobacterial activity, antimalarial activity, anti-inflammatory activity, antifungal activity, antifilarial activity, antidiarrhoeal activity, anti bacterial activity and anti amoebic effect (Ravinder *et al.*, 2010). Besides, the EO of *C. citratus* has significant public health importance not only as a phytotherapeutic agent. It also serves as an imperative tool in the vector control programmes in terms of mosquito larvicidal and repellent.

## Antibacterial Activity

The *C. citratus* oil possesses gram positive and gram negative antibacterial activities. The essential oil of *C. citratus* leaves exhibits activity against *Staphylococcus aureus, Bacillus subtilis, Escherichia coli* and *Pseudomonas aeruginosa* (El Kamali *et al.*, 1998). S. aureus: the neat had an effect of 87 per cent with respect to the standard Gentamicin and the dilution 1/32 had the best effect with 48 per cent with respect to the standard. *E. coli* the neat had an effect of 45 per cent in respect to the standard Gentamicin and the dilutions were all more active up to 1/16 with 55 per cent of the activity (Armando and Rahma, 2009). The essential oils of the plant are reported to have bactericidal effect against *Helicobacter pylori* without the development of acquired resistance (Ohno *et al.*, 2003).

## Antifungal and Antiyeast Activity

Candida albicans is the most common species associated with candidiasis and is the most frequently recovered species from hospitalized patients. It encompasses infections that range from superficial, such as oral thrush (Fidel, 2002) and vaginitis, to systemic and potentially life-threatening diseases. Use of essential oils for controlling *C. albicans* growth has gained significance due to the resistance acquired by pathogens towards a number of widely-used drugs. LG essential oil exhibited the strongest antifungal effect followed by mentha (*Mentha piperita*) and eucalyptus (*Eucalyptus globulus*) essential oil. The Minimum Inhibitory concentration (MIC) of LG essential oil in liquid phase (288 mg/l) was significantly higher than that in the vapour phase (32.7 mg/l) and a 4 h exposure was sufficient to cause 100 per cent loss in viability of *C. albicans* cells (Tyagi and Malik, 2010).

In particular, oils of *C. citratus* and *Syzygium aromaticum* have long been used in traditional practices by many ancient cultures. These essential oils have been recommended as home remedies for treatment of oral and vaginal fungal infections by numerous books and articles and products containing these essential oils for treatment of such infections are being used in many parts of the world (Barnes, 1989; Newall *et al.*, 1996). LG oil is active against such dermatophytes as *Trichophyton mentagrophytes, T. rubrum, Epidermophyton floccosum, Microsporum gypseum* (Wannissorn *et al.*, 1996) and is one among the most active against human dermatophyte. Other studies reported that LG oil has actions against keratinophilic fungi, 32 ringworm fungi (Kishore *et al.*, 1993; Abe *et al.*, 2003) and food storage fungi (Mishra and Dubey, 1994).

LG oil assays were conducted against *C. albicans* and were observed to hinder the growth. It is vivid that the concentration of the vapor inside the Petri dish is very active and avoids the growth of the *C. albicans* strain (Armando and Rahma, 2009). By using disk diffusion assay, a study was conducted to evaluate the antifungal activity of LG oil and citral against yeasts of Candida species (*C. albicans, C. glabrata, C. krusei, C. parapsilosis* and *C. tropicalis*) and found that the LG oil and citral have a potent *in vitro* activity against *Candida* spp. (Silva *et al.*, 2008). Efficient anti-biofilm activity of *C. citratus* and *Syzygium aromaticum* against *C. albicans* suggests that these oils are effective against adaptive mechanisms of resistance exhibited by *C. albicans* biofilms against amphotericin B and fluconazole (Khan and Ahmad, 2012).

A study aimed to verify the effectivenesses of *C. citratus* essential oil to inhibit the growth/survival of some fungi (*Alternaria alternata, Aspergillus niger, Fusarium oxysporum,* and *Penicillium roquefortii*) and yeasts (*Candida albicans, Candida oleophila, Hansenula anomala, Saccharomyces cerevisiae, Schizosaccharomyces pombe, Saccharomyces uvarum,* and *Metschnikowia fructicola*). *C. citratus* essential oil showed effectiveness in inhibiting the growth of all fungi by disc diffusion and broth dilution bioassay (Irkin and Korukluoglu, 2009).

Matasyoh *et al.* (2011) evaluated the antifungal activity of essential oil of *C. citratus* against five mycotoxigenic species of the genus *Aspergillus* and the test results showed that the oil was active against all the five *Aspergillus* species. The mycelial growth of *Aspergillus flavus* Link was completely inhibited using 1.5 microl/

ml or 2.0 microl/ml of *C. citratus* essential oil applied by fumigation or contact method in Czapek's liquid medium, respectively. This oil was found also to be fungicidal at the same concentrations (Helal *et al.*, 2007).

## Antimalarial Activity

The EOs of *C. citratus* were found to produce 86.6 per cent suppression in growth of *Plasmodium berghei* when compared to chloroquine (Tchoumbougnang *et al.*, 2005). A study investigated the *in vivo* antiplasmodial activity of *C. citratus* and *Vernonia amygdalina* leaves extracts when mice treated with a combination of *C. citratus* and *V. amygdalina* extracts at a dose of 600 mg/kg each and this showed no parasitemia after treatment using the 4-day suppressive test. This was not the case when the extracts were administered singly (Melariri *et al.*, 2011). The number of existing antimalarials is still inadequate especially in the face of parasite resistance. The results of these studies could help encourage more identification and validation of natural products which has shown antiparasitic properties thus facilitating the development of a new generation of antimalarials (Melariri *et al.*, 2011).

## Antileishmanial Activity

A recently conducted study results showed that the antiproliferative activity of *C. citratus* EO on promastigotes and axenic amastigotes, and intracellular amastigote forms of *Leishmania amazonensis* was significantly better than citral, and indicated a dosedependent effect. The promastigote forms of *L. amazonensis* underwent remarkable morphological and ultrastructural alterations. This study revealed that citral-rich essential oil from *C. citratus* has promising antileishmanial properties, and is a good candidate for further research to develop a new antiprotozoan drug (Santin *et al.*, 2009). Luize *et al.* (2005) reported that a crude hydroalcoholic extract of *C. citratus* is active against promastigote and amastigote forms of *Leishmania amazonensis* and *Trypanosoma cruzi*, with inhibition rates over 90 per cent at 100 µg/ml.

Similarly, a study demonstrated the antileishmanial effect of EOs from *C. citratus*, *Lippia sidoides*, and *Ocimum gratissimum* on growth and ultrastructure of *Leishmania chagasi* promastigote forms. All essential oils showed *in vitro* inhibitory action on *L. chagasi* promastigotes growth in a dose-dependent way, with $IC_{50}/72$ h of 45, 89, and 75 µg/mL for *C. citratus*, *L. sidoides*, and *O. gratissimum*, respectively (Oliveira *et al.*, 2009).

The most prominent ultrastructural effect observed in promastigotes treated with the EO from *C. citratus* were the appearance of aberrant-shaped cells, some of them presenting abnormal mitotic processes, besides an increase in both number and electron density of lipid droplets. Treated parasites also presented cytoplasmic vacuolation and swelling of cell body, flagellar pocket, mitochondria, and acidocalcisomes. Although further studies are necessary to examine each oil component separately and in combination to ascertain whether they act alone or synergistically and to evaluate their potential toxicity *in vivo* for clinical use as a potential drug against *L. chagasi* (Oliveira *et al.*, 2009).

## Ascaricidal Activity

The fresh leaf essential oil is observed to have ascaricidal activity (Chungsamarnvart *et al.*, 1992). Hanifah *et al.* (2011) demonstrated that *C. citratus* extract has more acaricidal activity against *Der-matophagoides farina* and *D. pteronyssinus* than *Azadirachta indica* at 50 per cent concentra-tion. A study was conducted to determine the effect of lemongrass alcoholic extracts on the control of *Boophilus microplus*. Twelve animals were allocated in three groups of four animals. Group 1 was treated with amitraz at 0.025 per cent, Group 2 was treated with lemongrass extracts at 1.36 per cent and Group 3 with the same product at 2.72 per cent of the plant. Engorged ticks were evaluated on animals with length superior to 4.0 mm, before (mean of days -3, -2, -1) and at 1, 2, 3, 4, 5, 6, 7 and 14 days after treatment. The mean efficacy of amitraz was 97.93 per cent. Lemongrass extract at 2.72 per cent reduced the tick infestation by 40.3, 46.6 and 41.5 per cent on day 3, 7 and 14 post-treatment, respectively (Heimerdinger *et al.*, 2006).

## *C. citratus* and HIV/AIDS

A study was to investigate the safety and efficacy of lemon juice and lemon grass in the treatment of oral thrush in HIV/AIDS patients when compared with the control group using gentian violet aqueous solution 0.5 per cent. Oral thrush is a frequent complication of HIV infection. In the Moretele Hospice, due to financial constraints, the treatment routinely given to patients with oral thrush is either lemon juice directly into the mouth or a lemon grass infusion made from *C. citratus* grown and dried at the hospice. These two remedies have been found to be very efficacious and are used extensively (Wright *et al.*, 2009).

## Antinociceptive Effect

EO of *C. citratus* possesses a significant antinociceptive activity. The EO from leaves of *C. citratus* increased the reaction time to thermal stimuli both after oral (25 mg/kg) and intraperitoneal (25–100 mg/kg) administration. EO (50–200 mg/kg, p.o. or i.p.) strongly inhibited the acetic acid-induced writhings in mice. In the formalin test, EO (50 and 200 mg/kg, i.p.) inhibited preferentially the second phase of the response, causing inhibitions of 100 and 48 per cent at 200 mg/kg, i.p. and 100 mg/kg, p.o., respectively. On the other hand, the opioid antagonist naloxone blocked the central antinociceptive effect of EO, suggesting that EO acts both at peripheral and central levels (Viana *et al.*, 2000).

## Antioxidant Activity

In a comparison of 11 essential oils isolated from various plant sources, LG oil ranked highly in terms of antioxidant activity – radical scavenging and protection of lipids from peroxidation – as well as antimicrobial activity against a range of food spoilage yeasts (Sacchetti *et al.*, 2005). With a slightly different focus, Nakamura *et al.* (2003) reported that citral isolated from LG induced the activity of the phase 2 enzyme, glutathione-S-transferase (GST), which plays important detoxification and anti-cancer roles in the body (see Hedges and Lister 2006 for further information on phase 2 enzymes.) Using an animal skin cancer model, this study also showed that topically applied citral had an antioxidant effect in mouse skin according to biomarkers of oxidative damage.

Five C-glycosylflavonoids namely, orientin, isoorientin, isoscoparin, swertiajaponin and isoorientin 2½-O-rhamnoside, isolated from *C. citratus*, have shown potent antioxidant activity by significantly inhibiting lipid peroxidation in erythrocytic membranes (Orrego *et al.*, 2009). It is also reported to exhibit anti-oxidant activity by scavenging of peroxide anion and inhibition of the enzyme xanthine oxidase and lipid peroxidation in human erythrocytes (Cheel *et al.*, 2005).

## Anti-inammatory Effect

Several studies have reported the traditional use of *C. citratus*, but little is known about their inuence on the immune system. A previous work reported that the treatment of mice with LG water extract inhibited macrophages to produce IL1β and IL-6 production, suggesting the anti-inammatory action of this spice *in vivo* (Sforcin *et al.*, 2009). Figueirinha *et al.* (2010) demonstrated that *C. citratus* leaf infusion signicantly inhibited other inammatory parameters like nitric oxide (NO) production and inducible NO synthase expression by LPS-stimulated mouse skin dendritic cells, suggesting its anti-inammatory activity. The potential of both LG and citral as immunomodulatory and anti-inammatory agents are required to be further explored. The analgesic activities produced by crude extracts of both *Anacardium occidentale* and *C. citratus*, demonstrate their potential analgesic properties and also support their use in traditional medical practice as analgesics (Sha'a *et al.*, 2011).

## Neurobehavioral Effect

The EO of *C. citratus* was effective in increasing the sleeping time, the percentage of entries and time spent in the open arms of the elevated plus maze as well as the time spent in the light compartment of the light/dark box. In addition, the EO delayed clonic seizures induced by pentylenetetrazole and blocked the tonic extensions induced by maximal electroshock, indicating the elevation of the seizure threshold and/or blockage of the seizure spread (Blanco *et al.*, 2009).

*C. citratus* and *C. winterianus* are popularly used in Brazil, mainly due to their actions on the central nervous system (CNS). Both the *C. winterianus* and *C. citratus* leaf EOs have a CNS depressant activity and anticonvulsant properties, justifying the use of the plant infusion by traditional medicine practitioners in the treatment of epilepsy. Results suggest that the EO from *C. citratus* has the potential to alter the course of convulsive episodes, interfering in the seizure threshold and/or blocking the seizure propagation (Silva *et al.*, 2010). *C. citratus*, commonly known as lemongrass, is widely used in traditional medicine as an infusion or decoction for treating nervous disturbances (Carlini *et al.*, 1986). In Mexico, *C. citratus* is used as a sedative (Tortoriello and Romero, 1992), and in Brazil, an infusion or the cold juice of the leaves has been employed as a sedative and analgesic (Hiruma-Lima *et al.*, 2002).

## Anti-Mutagenic and Anti-Carcinogenic

The 80 per cent ethanol extract of the plant leaves has also been documented to be anti-mutagenic and anti-carcinogenic (Suaeyun *et al.*, 1997). The plant is used in antitumor formation, tranquilizer, pebbles and kidney diseases (ElGhazali *et al.*, 1997). LG extract showed inhibitory effects on the early phase of induced liver cancer (Puatanachokchai *et al.*, 2002) and on the formation of induced DNA adducts (abnormal pieces of DNA, bonded to a cancer-causing chemical) and aberrant crypt foci (precursors of colon cancer) in the rat colon (Suaeyun *et al.*, 1997).

## Hypocholesterolaemic Effect

High cholesterol in the blood (hypercholesterolemia) is associated with an increased risk of various disorders, such as coronary heart disease and stroke (Hornstra *et al.*, 1988). These disorders are caused by blood vessels becoming narrowed with fatty deposits leading to reduced blood flow to vital organs, like brain. Arthrosclerosis is caused by hardening and narrowing of arteries (Sundram *et al.*, 1995). The elevated cholesterol concentration was significantly lowered in the animals given the plant extract. This reduction was found to be dose dependent. This result shows that the extract possesses hypocholestecolaemic potential. Therefore, ethanolic extracts of fresh leaves of *C. citratus* may explain its use in management of heart disease in ethnomedicine (Agbafor and Akubugwo, 2007).

## Hypoglycemic and Hypolipidemic Effects

Fresh leaf aqueous extract of *C. citrates*, administered in normal rats, lowered fasting plasma glucose and total cholesterol, triglycerides, low-density lipoproteins, very low-density lipoprotein dose dependently while raising the plasma high-density lipoprotein level in same dose-related fashion but with no effect on plasma triglycerides level (Adeneye and Agbaje, 2007).

## Public Health Importance of *C. citratus*

Vector-borne diseases are major causes of morbidity and mortality in many tropical and subtropical countries. Principally the devastating nature of malaria in sub-Saharan Africa particularly in country like Ethiopia is indubitably intolerable (Karunamoorthi and Ilango, 2010). Plants have been used since ancient times to repel/kill blood-sucking insects in the human history and even now, in many parts of the world people are practicing plants substances to repel/kill the mosquitoes and other blood-sucking insects (Karunamoorthi *et al.*, 2008). The explosive global economic development, people movement, water projects, climate change and increased urbanization have substantially altered the disease transmission dynamics and pattern too and this in turn requires implementing need-based and community-oriented malaria control strategies (Karunamoorthi, 2012b).

Essential oils have received attention as potentially controlling vectors of mosquito-borne disease (Sutthanont *et al.*, 2010). Natural products used as insecticides may have less of environmental impact due to shorter latency, possibly resulting in reduced resistance (Hardin and Jackson, 2009). Several studies have also focused on lemongrass oil for controlling mosquitoes as a larvicide and a repellent with varied results. *C. citratus* is also well known for its medicinal and insects/mosquitoes repellent properties among the rural residents of Ethiopia (Karunamoorthi and Ilango, 2010).

## C. *citratus* as a Mosquito Larvicidal Agent

Indeed, source reduction is one of the key components in the malaria vector control programme since the target is exceptionally specific unlike adult control. However, application of conventional insecticides into the mosquitoes breeding sites may lead to adverse side effects in the aquatic ecosystem. In this context, innovative vector control strategy like use of phytochemicals as alternative sources of insecticidal/larvicidal agents in the fight against the vector-borne diseases has become inevitable. In the recent epoch, around the globe, phytochemicals has gained massive attention by various researchers because of their bio-degradable and eco-friendly values (Karunamoorthi and Ilango, 2010).

Methanol leaf extracts of Ethiopian traditional medicinal plant *viz.*, Lomisar [vernacular name (local native language, Amharic); *C. citratus*] was screened for larvicidal activity against late third instar larvae of Anopheles arabiensis Patton, a potent malaria vector in Ethiopia. The $LC_{50}$ and $LC_{90}$ values of *C. citratus* were 74.02 and 158.20 ppm, respectively. It has been observed to exhibit potent larvicidal activity at lower concentrations. The investigation establishes that *C. citratus* extracts could serve as potent mosquito larvicidal agent against *An. arabiensis* (Karunamoorthi and Ilango, 2010).

Sukumar *et al.* (1991) reported that *C. citratus* causes significant growth inhibition and mortality in later developmental stages of *Aedes aegypti*. The essential oils obtained by hydrodistillation of dry leaves from *C. citratus* was analyzed for their larvicidal activity against fourth instar larvae of *An. gambiae*. The essential oil has remarkable larvicidal properties and could induce 100 per cent mortality in the larvae of *An. gambiae* at the concentration of 100 ppm for *C. citratus*. The essential oil of *C. citratus* was found to be the most efficient, with respective values of: $LC_{50} = 18$ ppm and $LC_{80} = 25$ ppm (Tchoumbougnang *et al.*, 2009).

The essential oils of *Cananga odorata* (ylang ylang), *Citrus sinensis* (orange), *Cymbopogon citratus* (lemongrass), *Cymbopogon nardus* (citronella grass), *Eucalyptus citriodora* (eucalyptus), *Ocimum basilicum* (sweet basil) and *Syzygium aromaticum* (clove), were tested for their insecticide activity against *Aedes aegypti*, *Culex quinquefasciatus* and *Anopheles dirus* using the WHO standard susceptibility test. *C. citratus* oil is found to have high insecticidal activity against *Ae. aegypti*, *Cx. quinquefasciatus* and *An. dirus*, with $LC_{50}$ values of <0.1, 2.22 and <0.1 per cent, respectively (Phasomkusolsil and Soonwera, 2011).

*C. citratus* essential oil when studied, has shown toxicity against *Cx. quinquefasciatus* larvae with a $LC_{50}$ value of 24 mg/l (Nazar *et al.*, 2009). *C. citratus* and *Lippia sidoides* had larvicidal activity against *Ae. aegypti* causing 100 per cent mortality at a concentra-tion of 100 ppm (Cavalcanti *et al.*, 2004). However, the mode of action and larvicidal efficiency under the field conditions should be scrutinized and determined in the near future (Karunamoorthi and Ilango, 2010). LG is also effective as a potential herbicide and an insecticide, because of the antimicrobial effects.

## C. *citratus* as an Insect Repellent

Insect repellents may be chemical or plant-based substances, preventing from arthropod bites, which eventually help to achieve the reduction of man-vector contact.

Repellents are typically applied to exposed skin or can be applied to clothing or other surfaces to discourage arthropods from landing or climbing onto treated surfaces (Dolan and Panella, 2011). Generally, synthetic repellents have several limitations, including reduced efficacy owing to sweating, unpleasant odor, expensive and can cause allergic reactions (Karunamoorthi, 2011). Besides their toxicity and adverse side effects, a few of them require electricity for their usage too (Karunamoorthi *et al.*, 2008a).

In the human history since time immemorial, plants have been used as insect repellent in order to avoid insects' annoyance. Even today, in many parts of the world, people use various plants as repellents to prevent disease transmission (Karunamoorthi *et al.*, 2008). However, since their efficiency is uncertain, conducting intensive research on the identification and development of novel potent and low-cost plant-based repellents is quite essential and inevitable (Karunamoorthi, 2012c).

Pushpanathan *et al.* (2006) showed that the skin repellent test at 1.0, 2.5, and 5.0 mg/cm$^2$ concentration of *C. citratus* essential oil against the filarial mosquito *Cx. quinquefasciatus* gave 100 per cent protection up to 3.00, 4.00, and 5.00 h, respectively. A laboratory study was carried out to evaluate the repellent efficacy of a methanol-leaf extract of *C. citratus* against *An. arabiensis*.

Several studies evidently suggest that the traditionally used plant–based insect repellents are promising and could potentially contain vectors of disease (Karunamoorthi *et al.*, 2008; 2008a). Numerous widely-known repellent plants are in use by the indigenous rural people in the SSA countries though they are quite unaware of the complete elucidation of the mechanism of repellency of those plants. Furthermore analysis and scientific experiments are required to identify whether the bioactive molecules of the repellent plants have greater effect on the vectors rather than the intrinsic nature of smoke and also to evaluate the rate of effectiveness of the repellent plants and their insecticidal activities. Besides these, testing for their mammalian toxicity is also required to be warranted (Karunamoorthi, 2012).

In Ethiopia, Karunamoorthi and Husen (2012) have reported that the local residents burn *C. citratus* dried leaves in order to drive-away mosquitoes, house flies and fleas by smoking and surfacing the leaves over the floor of the house. Ethnobotanical surveys result reveal that *C. citratus* plant has been one of the most common and widely used insect/mosquito repellent against various blood-sucking insects in Addis Zemen town, North-Western Ethiopia (Karunamoorthi *et al.*, 2009a). Similarly another ethnobotanical survey conducted among the Oromo ethnic group of Ethiopia results have revealed that the majority of them use *C. citratus* as an insect repellent (Karunamoorthi *et al.*, 2009b). These results are indicating that *C. citratus* is one of the most popular and widely used medicinal plant as an insect repellent among the majority of the Ethiopians. The dried leaves are also available as insect repellent in majority of the Ethiopian town markets.

Globally, *C. citratus* is renowned for its therapeutic values. Above and beyond, due to its user- as well as environmental-friendly nature, it should be promoted among the marginalized populations in order to reduce man-vector contact. In addition, this appropriate strategy affords the opportunity to minimize chemical repellent usage and the risks associated with adverse side effects (Karunamoorthi *et al.*, 2010).

## *C. citratus* against Other Public Health Important Insects

Samarasekera *et al.* (2006) found that *C. citratus* oil has good knockdown and mortality activity against adult *Musca domes-tica* at $LD_{50}$ of 1.71 µg in Sri Lanka. Purwal *et al.* (2010) evaluated the activity of *C. citratus* and *Mentha piperita* essential oils in combination against *Pe-diculus humanus capitis* and found a mean time to death of 60 min. *C. citratus* may be included in disease vector control programs, since it can be obtained easily and at low cost. *C. citratus* oil may be considered an effective insecticide and repellent against mosquitoes (Phasomkusolsil and Soonwera, 2011) and other public health important insects could be used instead of chemical insecticides.

## Traditional Phytotherapeutic Usage Custom

Among various species with medicinal properties, *C. citratus*, shows an infinite number of applications and is popularly used by people in many countries. In Brazil, for example, the tea, infusion and extracts of *C. citratus*, which are prepared with fresh or dry leaves, are often used in the popular medicine as a restorative, digestive, anti-tussis, effective drug against colds, with an analgesic, anti-hermetic,

**Figure 2.2**: The author discusses about the traditional phytotherapuetic potentialities of *Cymbopogon citratus* with an Ethiopian key informant.

anticardiopatic, antithermic, anti-inflammatory of urinary ducts, diuretic, antispasmodic, diaphoretic and antiallergic effect (Negrelle and Gomes, 2007). *C. citratus* is a plant used in traditional folk medicine in Brazil for the treatment of nervous and gastrointestinal disturbances, and in various other countries to treat fevers (Melo *et al.*, 2001).

C. *citratus* is used in different parts of the world in the treatment of digestive disorders, fevers, menstrual disorder, rheumatism and other joint pains (Simon *et al.*, 1984). The essential oil of the plant is used for the treatment of skin diseases (Balch and Phyllis, 1990). Traditional medicine practitioners in the Eastern part of Nigeria use the plant's preparation in the form of tincture (solution of plant extract in local alcohol) for treating coronary heart disease and related conditions, such as cardiovascular disorders (Agbafor and Akubugwo, 2007). Its sedative and anticonvulsant properties as well as its use as an anxiolytic agent has been documented (Blanco *et al.*, 2009). It is also used in the treatment of fever, jaundice, hypertension and as analgesic (Onabanjo *et al.*, 1993).

LG is used for symptomatic relief of asthma and it is one of the more popular traditional indigenous West Indian medicinal plants used for this purpose (Clement *et al.*, 2005). The fresh leaves of *C. citratus* are a good source of an essential oil (EO) rich in citral, and its tea is largely used in the Brazilian folk medicine as a sedative (Silva *et al.*, 2010).

## Culinary Uses of *C. citratus*

The lemon and zesty taste that LG provides is ideal for many spicy dishes. In Thai cooking, lemongrass provides a fresh citric flavor and aroma to many of its dishes. Lemon juice (or lime) may be substituted for LG but citrus fruits will not be able to fully replicate its complex and sophisticated flavour profile. LG is also thought to have numerous health benefits, especially when used in combination with other spices such as garlic, fresh chillies, and coriander. In fact, scientists are now studying Thailand's favorite soup: Tom Yum Kung, which is thought to be capable of combating colds, flus, and even some cancers (Joshi, 2012). Besides, the LG essential oils are also used in the food (flavoring), perfume and cosmetics industries, this use being of reasonable economical importance in various countries (Oliveira *et al.*, 1997). The volatile oil obtained from fresh leaves of this plant is widely used by the perfume and cosmetics industries. It has also been used in chemical synthesis, because of its high citral content (Rauber *et al.*, 2005).

## Other Uses

In Nigeria, lemon grass leaves are commonly used for making soup and beverage. Hot decoction of the plant leaves, prepared by boiling in hot water for about 1 h, is often prescribed as 50–70 ml/kg of body weight twice daily for several weeks to their suspected diabetic and obese patients (Adeneye and Agbaje, 2007). Lemon grass oil is an essential oil used in deodorants, skin care products, fragrances, insect repellents, and for aromatherapy (Cheel *et al.*, 2005). *C. citratus* is used in soap making and in insect repellants (Onabanjo *et al.*, 1993).

Traditionally, communities in south East Asia and in particular India have used Lemongrass for the following purposes (Joshi, 2012);

☆ Detoxifies the digestive organs like pancreas, liver, kidney and bladder.

☆ Stimulates digestion and blood circulation in the body and hence, keeps gastric and indigestion problems at bay.

☆ The sweet smelling oil obtained from lemongrass is of significant use in aromatherapy, as it helps in relieving stress and tension from the body.

☆ A few drops of lemongrass oil can be used to combat greasy hair and as a deodorant to curb perspiration.

☆ Can be used to combat cramps and muscle pain.

Although plant extracts of lemon grass have been extensively used in the folk medicine, scientific research has found some potentially toxic substance in this species. Hepatotoxic and nephrotoxic effects in mice treated with fluid extracts of *C. citratus* (30 per cent and 80 per cent) were observed (Guerra *et al.*, 2000), to indicate the necessity of more detailed research on its cytotoxicity (Negrelle and Gomes, 2007).

## Utilization and Management of Medicinal Plants: Major Challenges and Thorny Issues

The demand for medicinal compounds has increased in recent years due to the limitation of processes for obtaining products based on chemical synthesis. Additionally, the cost of the biopharmaceutical product limits its accessibility in a wide sector of the market, and does not satisfy the need of the world's population due to increased demand for natural compounds from botanical origin that improve the health or reduce the incidence of illnesses (Misawa, 1994). However, the indiscriminate medicinal use of plants, usually toxic ones, may entail risks to health, because, similar to the allopathic drugs, there is a threshold dosage for each phytotherapeutic agent. Thus, after an inadequate use, several disorders may occur, from intoxications to mutation events in somatic and germinative tissue, and it can lead to the development of somatic diseases, teratogenic effects and inherited genetic damages (Celik and Aslantürk, 2007; Pugliesi *et al.*, 2007, Lubini *et al.*, 2008).

In recent years, folk medicine is no more an attraction to the younger generation and many young people migrate to urban areas for education and job opportunities. This rich medicinal plant knowledge is, however, seriously threatened due to deforestation, environmental degradation and acculturation, currently taking place in the country. As a consequence, only the elder people possess the knowledge of herbs and it is estimated that only a handful of people are able to use the traditional remedies to treat illness. Thus, the traditional knowledge is rapidly eroding. Therefore, the assessment and documentation of ancestral knowledge of indigenous people on traditional plant medicines would fill the gap associated knowledge between the elders and young people on medicinal plants (Tangjang *et al.*, 2011).

## Future Perspective and Opportunities

Ethnic groups are the repositories of the knowledge on herbal remedy which need to be documented and tapped properly. The use of traditional antimalarial plants is one of the prehistoric age-old customs and it is largely conceived by long-term observation and trial and error mode (Karunamoorthi and Tsehaye, 2012). This expertise has been passed down to many generations mostly through word of mouth. This mode of information conveyance may result in distortion or loss of indigenous knowledge and usage custom of plants. Therefore, right now documenting and safeguarding these practices have become crucial core issues (Karunamoorthi *et al.*, 2009a; 2009b), due to the rapid degradation of natural habitats and ecosystems leading to loss of plant and cultural diversity become crucial core issues.

The disappearance of traditional cultural and natural resources due to population growth, urbanization, and the erosion of botanical knowledge in developing countries suggests that the unrecorded knowledge (information) may be lost forever (Dawn *et al.*, 2008), requiring to document traditional knowledge especially about medicinal plant diversity (Mshana *et al.*, 2001; Van Wyk *et al.*, 2002; Van Wyk and Wink, 2004). This documentation is of high priority in order to support the discoveries of drugs beneting mankind in treating diseases such as malaria (Pieroni, 2000). Similarly, there are many questions about the standardization techniques for the production and exchange of phytotherapeutic agents (Negrelle and Gomes, 2007).

Conserving our traditional knowledge and long-standing age old custom on traditional insect repellent plants (TIRPs) is extremely imperative for the betterment of humankind in order to minimize the cause of many arthropod-borne diseases in particular malaria. Besides, at the moment generating the regional and international data-base on TIRPs could pave the way to isolate/identify the untapped pool of bio-active molecules accountable for the repellent efficacy. Furthermore, conducting more laboratory and field based studies to evaluate their repellent efficiency and human safety is extremely inevitable (Karunamoorthi and Husen, 2012).

## Conclusion

Human civilization is closely associated and inter-related with utilization and management of medicinal plants. In the past decades, it has been considered that although traditional system of medicine (TSM) is an age-old practice it may not be suitable or worthwhile for the current scenario. However, in the recent past the people's perception about TSM has changed a lot and the importance has been greatly recognized and acknowledged. It is purely because of the lack of affordability, inaccessibility and adverse side-effect of conventional modern drugs. Besides, the emergence of drug resistance is also one of the major phenomenons.

Earlier, people are extremely concerned about speedy recovery from the ailments and least bothered or oblivious about the adverse side-effects. But now, the acuity has changed and people are more interested on plant-based drugs owing to safety. It has led to revitalization of traditional knowledge on herbs. Besides, we are living in the era of synthetic chemistry and biotechnology techniques and we must use this prospect and employ these techniques to develop the plant-derived pharmaceuticals and to deliver the desirable results. As per my knowledge and observation everybody prefer the TSM but the major barrier is the lack of phytochemical standardization techniques therefore there is a pressing need for more clinical assay to validate ethnomedical usage of *C. citratus* in the near future.

# Acknowledgements

I would like to thank Mrs. Melita Prakash for her sincere assistance in editing the manuscript. My last but not least heartfelt thanks go to my colleagues from the Department of Environmental Health Science, College of Public Health and Medical Sciences, Jimma University, Ethiopia, for their kind support and cooperation.

# References

Abe, S., Sato, Y., Inoue, S., Ishibashi, H., Maruyama, N., Takizawa, T., Oshima, H., and Yamaguchi, H., (2003). Anti-candida albicans activity of essential oils including lemongrass (*Cymbopogon citratus*) oil and its component, citral. *Nippon Ishinkin Gakkai Zasshi*, **44(4)**: 285-291.

Adeneye, A.A., and Agbaje, E.O. (2007). Hypoglycemic and hypolipidemic effects of fresh leaf aqueous extract of *Cymbopogon citratus* Stapf. in rats. *Journal of Ethnopharmacology*, **112**; 440-444.

Agbafor, K.N., and Akubugwo, E.I. (2007). Hypocholesterolaemic effect of ethanolic extract of fresh leaves of *Cymbopogon citratus* (lemongrass). *African Journal of Biotechnology*, **6**: 596-598.

Anthony, J.P., Fyfe, L., and Smith, H. (2005). Plant active components—a resource for antiparasitic agents? *Trends in Parasitology*, **21**: 462–8.

Armando, C.C., and Rahma, H.Y. (2009). Evaluation of the yield and the antimicrobial activity of the essential oils from: *Eucalyptus globulus*, *C. citratus* and *Rosmarinus officinalis* in Mbarara district (Uganda). *Revista Colombiana de Ciencia Animal*, **1(2)**: 240-249.

Balch, J.F., and Phyllis, A.B. (1990). Prescription for nutritional healing. 2nd edition, Avery publishing group, New York, pp. 869-873.

Barnes, B. (1989). The development of topical applications containing tea tree oil for vaginal conditions, in; Modern phytotherapy-the clinical significance of tea tree oil and other essential oils. In: Proceedings of a conference in Sydney, September 17 1989, Vol. 1. Australian Tea Tree Industry Association, Coraki, Australia, pp. 27–35.

Blanco, M.M., Costa, C.A.R.A., Freire, A.O., Santos Jr, J.G., and Costa, M. (2009). Neurobehavioral effect of essential oil of *Cymbopogon citratus* in mice. *Phytomedicine*, **16**: 265-270.

Carlini, E.A., Contar, J.D.P., Silva-Filho, A.R., Silveira-Filho, N.G., Frochtengarten, M.L., and Bueno, O.F.A. (1986). Pharmacology of lemongrass (*Cymbopogon citratus* Stapf). I. Effects of teas prepared from the leaves on laboratory animals. *Journal of Ethnopharmacology*, **17**: 37-64.

Cavalcanti, E.S., Morais, S.M., Lima, M.A., and Santana, E.W. (2004). Larvicidal activity of essential oils from Brazilian plants against *Aedes aegypti* L. *Mem Inst Oswaldo Cruz.*, **99**: 541-544.

Celik, T.A., and Aslantürk, O.S. (2007). Cytotoxic and genotoxic effects of *Lavandula stoechas* aqueous extracts. *Biologia*, **62**: 292-296.

Cheel, J., Theoduloz, C., Rodriguez, J., *et al.* (2005). Free radical scavengers and anti-oxidants from lemongrass (*Cymbopogon citratus* DC strapf). *Journal of Agricultural and Food Chemistry*, **53**: 2511-2517.

Chisowa, E.H., Hall, D.R., and Farman, D.I. (1998). Volatile constituents of the essential oil of *Cymbopogon citratus* Stapf grown in Zambia. *Flavour* and *Fragrance Journal*, **13**: 29–30.

Chungsamarnvart, N., and Jiwajinda, S. (1992). Acaricidal activity of volatile oil from lemon and citronella grasses on tropical cattle ticks. *Kasetsart Journal: Natural Science*, **26**:46-51.

Clayton, W.D. (1968). Gramineae. In Flora of West Africa: *Tropical Africa*, **3(2)**:349-512.

Clement, Y.N., Williams, A.F., Aranda, D., Chase, R., Watson, N., Mohammed, R., Stubbs, O., and Williamson, D. (2005). Medicinal herb use among asthmatic patients attending a specialty care facility in Trinidad. *BMC Complementary and Alternative Medicine*, **5**: 3.

Cowan, M.M. (1999). Plant products as antimicrobial agents. *Clinical Microbiology Reviews*, **12**: 564–582.

Dawn, T.A., Jialin, W., Zhihong, J., Hubiao, C., Guanghua, L., and Zhongzhen, Z. (2008). Ethnobotanical study of medicinal plants used by Hakka in Guangdong, China. *Journal of Ethnopharmacology*, **117**: 41-50.

Dolan, M.C., and Panella, N.A. (2011). Recent Developments in Invertebrate Repellents. Editor(s): Gretchen E. Paluch, Joel R. Coats. Volume 1090. Chapter 1, pp. 1-19.

El Kamali, H.H., Ahmed, A.H., Mohammed, A.S., Yahia, A.A.M., El Tayeb, I., and Ali, A.A., (1998). Antibacterial properties of essential oil from Nigella sativa seeds, *Cymbopogon citrates* leaves and *Pulicaria undulata* aerial parts. *Fitoterapia*, **69(1)**: 77-78.

ElGhazali, G.E.B., El tohami, M. S., El Egami, A. A. B.., Abdalla W.S., and Mohamed, M.G. (1997). Medicinal plants of the Sudan. Medicinal plants of Northern Kordofan, mdurman Islamic University Press, Omdurman.

Fidel, P.L. (2002). Immunity to Candida. *Oral Diseases*, **8**: 69-75.

Figueirinha, A., Cruz, M.T., Francisco, V., Lopes, M.C., and Batista, M.T. (2010). Anti-in?ammatory activity of *Cymbopogon citratus* leaf infusion in lipopolysaccharide-stimulated dendritic cells: contribution of the polyphenols. *Journal of Medicinal Food*, **13**: 681-690.

Guerra, M.J.M., Badell, J.B., Albajés, A.R.R., Pérez, H.B., Valencia, R.M., and Azcuy, A.L. (2000). Toxicologic acute evaluation of the fluid extracts 30 and 80 porciento of *Cymbopogon citratus* (D.C.) Stapf (lemon grass). *Revista Cubana* de *Plantas Medicinales*, **5**: 97-101.

Halberstein, R.A. (2005). Medicinal plants: historical and cross-cultural usage patterns. *Annals of Epidemiology*, **15**: 686–699.

Hanifah, A.L., Awang, S.H., Ming, H.T., Abidin, S.Z., and Omar, M.H. (2011). Acaricidal activity of *Cymbopogon citratus* and *Azadirachta indica* against house dust mites. *Asian Pacific Journal of Tropical Biomedicine*, **1**: 365-369.

Hardin, J.A., and Jackson, F.L.C. (2009). Applications of natural products in the control of mosquito-transmitted diseases. *African Journal of Biotechnology*, **8**: 7373-7378.

Heimerdinger, A., Olivo, C.J., Molento, M.B., Agnolin, C.A., Ziech, M.F., Scaravelli, L.F., Skonieski, F.R., Both, J.F., and Char o PS. (2006). Alcoholic extract of lemongrass (*Cymbopogon citratus*) on the control of *Boophilus microplus* in cattle. *Revista Brasileira de Parasitologia Veterinária*, **15(1)**: 37-39.

Helal, G.A., Sarhan, M.M., Abu Shahla, A.N.K., and Abou El-Khair, E.K. (2007). Effects of *Cymbopogon citratus* L. essential oil on the growth, morphogenesis and aflatoxin production of *Aspergillus flavus* ML2-strain. *Journal of Basic Microbiology*, **47(1)**: 5-15.

Hiruma-Lima, C.A., Guimar es, E.M., Santos, C.M., and Di Stasi, L.C. (2002). Commelinidae medicinais. In: Di Stasi, L.C., Hiruma-Lima, C.A. (Eds.), Plantas Medicinais na Amazônia e na Mata Atlântica. Editora UNESP, S o Paulo, pp. 41-50.

Hornstra, G. (1988). Dietary lipids and cardiovascular diseases. *Oleagineux*, **43**: 75-81.

Ikam, R. (1969). Natural Products: a laboratory guide. Chapter V. Elsevier Publication, London.

Irkin, R., and Korukluoglu, M. (2009). Effectiveness of *Cymbopogon citratus* L. essential oil to inhibit the growth of some filamentous fungi and yeasts. *Journal of Medicinal Food*, **12(1)**: 193-197.

Jiaxiang, S. (1997). Introduction to the Chinese Materia Medica. In UNDP, 1997.

Joshi K. (2012). Social: Lemongrass Health Benefits and Uses. 5th April 2012. Available at: http://www.ecademy.com/node.php?id=176167 (accessed on 5th May 2012).

Joy, P.P., Thomas, J., Mathew, S., and Skaria, B.P. (1988). Medicinal Plants. Kerala Agricultural University, Aromatic and Medicinal Plants Research Station, Odakkali, Kerala, India.1998. Available at: http://amprs.kau.edu/

Karunamoorthi, K. (2011). Vector Control: A cornerstone in the malaria elimination campaign. *Clinical Microbiology and Infection*, **17(11)**: 1608-1616. DOI: 10.1111/j.1469-0691.2011.03664.x.

Karunamoorthi, K. (2012c). Plant-Based Insect Repellents: Is That a Sustainable Option to Curb the Malaria Burden in Africa? *Medicinal and Aromatic Plants*, **1**:e106. doi:10.4172/map.1000e106. Available at: http://omicsgroup.org/journals/MAP/MAP-1-e106.php?aid=4081 (accessed on 5th June 2012).

Karunamoorthi, K. (2012b). Global Malaria Burden: Socialomics Implications. *Journal of Socialomics*, **1**:e108. doi:10.4172/jsc.1000e108. Available at: http://omicsgroup.org/journals/JSC/JSC-1-e108.php?aid=4298 (accessed on 5th June 2012).

Karunamoorthi, K., and Husen, E. (2012). Knowledge and self-reported practice of the local inhabitants on traditional insect repellent plants in Western Hararghe zone, Ethiopia. *Journal of Ethnopharmacology*, **141(1)**: 212-219.

Karunamoorthi, K., Jegajeevanram, K., Jeromem, X., Vijayalakshmi, J., and Melita, L. (2012). Tamil Traditional Medicinal System – Siddha: An Indigenous Health Practice in the International Perspectives. *International Journal Genuine Traditional Medicine*, **2**:2.e Available at: http://www.e-tang.org (accessed on 5th June 2012).

Karunamoorthi, K., Mulelam, A., and Wassie, F. (2008b). Laboratory evaluation of traditional insect/mosquito repellent plants against *Anopheles arabiensis*, the predominant malaria vector in Ethiopia. *Parasitology Research*, **103(3)**: 529-534.

Karunamoorthi, K., Ramanujam, S., and Rathinasamy, R. (2008a). Evaluation of leaf extracts of *Vitex negundo* L. (Family: Verbenaceae) against larvae of *Culex tritaeniorhynchus* and repellent activity on adult vector mosquitoes. *Parasitology Research*, **103(3)**: 545-550.

Karunamoorthi, K., and Tsehaye, E. (2012). Ethnomedicinal knowledge, belief and self-reported practice of local inhabitants on traditional antimalarial plants and phytotherapy. *Journal of Ethnopharmacology*, **141(1)**:143-150.

Karunamoorthi, K. (2012a). "Neem Oil: biological activities and usage". "Recent Progress in Medicinal Plants" (RPMP)" Vol. 33: Fixed Oils and Fats of Pharmaceutical Importance. J.N. Govil (Eds). Studium Press LLC., P.O. Box-722200, Houston, Texas-77072, United States of America. ISBN-13: 9781933699233.

Karunamoorthi, K., and Ilango, K. (2010). Larvicidal activity of *Cymbopogon citratus* (DC) Stapf. and *Croton macrostachyus* Del. against *Anopheles arabiensis* Patton (Diptera: Culicidae), the principal malaria vector. *European Review of Medical and Pharmacological Science*, **14(1)**: 57-62.

Karunamoorthi, K., Ilango, K., and Endale, A. (2009b). Ethnobotanical survey of knowledge and usage custom of traditional insect/mosquito repellent plants among the Ethiopian Oromo ethnic group. *Journal of Ethnopharmacology*, **125(2)**: 224-229.

Karunamoorthi, K., Ilango, K., and Murugan, K. (2010). Laboratory evaluation of traditionally used plant-based insect repellent against the malaria vector *Anopheles arabiensis* Patton (Diptera: Culicidae). *Parasitology Research*, **106(5)**: 1217-1223.

Karunamoorthi, K., Mulelam, A., and Wassie, F. (2009a). Assessment of knowledge and usage custom of traditional insect/mosquito repellent plants in Addis Zemen Town, South Gonder, North Western Ethiopia. *Journal of Ethnopharmacology*, **121(1)**: 49-53.

Katewa, S.S., Chaudhary, B.L., and Jain, A. (2004). Folk herbal medicines from tribal area of Rajasthan, India. *Journal of Ethnopharmacology*, **92**: 41-46.

Khan, M.S.A., and Ahmad, I. (2012). Biofilm inhibition by *Cymbopogon citratus* and *Syzygium aromaticum* essential oils in the strains of *Candida albicans*. *Journal of Ethnopharmacology*, **140**: 416-423.

Kishore, N., Mishra, A.K., and Chansouria, J.P.N. (1993). Fungitoxicity of essential oils against dermatophytes. *Mycoses*, **36**: 211-215.

Lorenzi, H., and Matos, F.J.A. (2003). Plantas medicinais no Brasil: nativas e exóticas. Plantarum, Nova Odessa, S o Paulo.

Lubini, G., Fachinetto, J.M., Laughinghouse. H.D., Paranhos, J.T., Silva, A.C.F., and Tedesco, S.B. (2008). Extracts affecting mitotic division in root-tip meristematic cells. *Biologia*, **63**: 647-651.

Luize, P.S., Tiuman, T.S., Morello, L.G., Maza, P.K., Ueda-Nakamura, T., Dias Filho, B.P., Cortez, D.A.G., Mello, J.C.P., and Nakamura, C.V. (2005). Effects of medicinal plants extracts of *Leishmania* (L.) *amazonensis* and *Trypanosoma cruzi*. *Brazilian Journal of Pharmaceutical Sciences*, **41**: 85-94.

Manilal, K.S. (1989). Linkage of Ethnobotany with other sciences and disciplines. *Ethnobotany*, **1**: 15-24.

Matasyoh, J.C., Wagara, I.N., Nakavuma, J.L., and Kiburai, A.M. (2011). Chemical composition of *Cymbopogon citratus* essential oil and its effect on mycotoxigenic *Aspergillus* species. *African Journal of Food Science*, **5(3)**: 138-142.

Matouschek, B.K., and Stahl, B.E. (1991). Phytochemical study of non volatile substance from *Cymbopongon citratus* (D.C) Stapf (Poaceae). *Pharmaceutica Acta Helvetiae*, **66**: 242-245.

Melariri, P., Campbell, W., Etusim, P., and Smith, P. (2011). *In vitro* and *in vivo* antiplasmodial activities of extracts of *Cymbopogon citratus* Staph and *Vernonia amygdalina* Delile leaves. *Journal of Natural Products*, **4**: 164-172

Melo, S.F., Soares, S.F., Costa, R.F., Silva, C.R., Oliveira, M.B.N., Bezerra, R.J.A.C., Caldeira-de-Araújo, A., and Bernardo-Filho, M. (2001). Effect of the *Cymbopogon citratus*, *Maytenus ilicifolia* and *Baccharis genistelloides* extracts against the stannous chloride oxidative damage in *Escherichia coli*. *Mutation Research*, **496**: 33-38.

Milhau, G., Valentin, A., Benoit, F., Mallie, M., Bastide, J., Pelissier, Y., and Bessiere, J: (1997). *In vitro* antimicrobial activity of eight essential oils. *Journal of Essential Oil Research*, **9**:329-333.

Ming, L.C., Figueirdo, R.O., Machado, S.R., Andrade, R.M.C., Craker, L.E., Nolan, L., and Shelty, K. (1996). Yield of essential oil and citral content in different parts of lemongrass leaves *(Cymbopogon citratus* D.C.Stapf.) Poaceae. International symposium USA, 27-30. *Acta Horticulturae*, **426**: 555-559.

Misawa, M. (1994). Plant tissue culture: an alternative for production of useful metabolites. FAO *Agricultural Services Bulletin*, FAO, Rome, pp. 57.

Mishra, A.K., and Dubey, N.K. (1994). Evaluation of some essential oils for their toxicity against fungi causing deterioration of stored food commodities. *Applied and Environmental Microbiology*, **60**: 1101-1105.

Mshana, R.N., Abbiw, D.K., Addae-Mensah, I., Adjanouhoun, E., Ahyi, M.R.A., Ekpere, J.A., Enow-Rock, E.G., Gbile, Z.O., Noamesi, G.K., Odei, M.A., Odunlami, H., OtengYeboah, A.A., Sarpong, K., Sofowora, A., and Tackie, A.N. (2001). Traditional Medicine and Pharmacopoeia: Contribution to the Revision of Ethnobotanical and Floristic Studies in Ghana. Institute for Scienti?c and Technological Information, Accra, pp. 919.

Nakamura, Y., Miyamoto, M., Murakami, A., Ohigashi, H., Osawa, T., and Uchida. K. (2003). A phase II detoxification enzyme inducer from lemongrass: identification of citral and involvement of electrophilic reaction in the enzyme induction. *Biochemical and Biophysical Research Communications*, **302(3)**: 593-600.

Nazar, S., Ravikumar, S., Prakash, W.G., Syed, A.M., and Suganthi, P. (2009). Screening of Indian coastal plant extracts for larvicidal activity of *Culex quinquefasciatus*. *Indian Journal of Science and Technology*, **2**: 24-7.

Negrelle, R.R.P., and Gomes, E.C. (2007). *Cymbopogon citratus* (D.C) Stapf: chemical composition and biological activities. *Revista Brasileira de Plantas Medicinais*, **9**: 80-92.

Newall, C.A., Anderson, L.A., and Phillipson, J.D. (1996). Herbal Medicines: A Guide for Health Care Professionals. The Pharmaceutical Press, London.

Ohno, T., Kita, M., and Yamaoka, Y., *et al.* (2003). Antimicrobial activity of essential oils against *Helicobacter pylori*. *Helicobacter*, **8**: 207-215.

Oliveira, A.C.X., Ribeiro, P.L.F, and Paumgarttem, F.J.R. (1997). *In vitro* inhibition of CYP2B1 monooxygenase by amyrcene and other monotherpenoide compounds. *Toxicology Letters*, **92**: 39-46.

Oliveira, V.C., Moura, D.M.S., Lopes, J.A.D., de Andrade, P.P., da Silva, N.H., and Figueiredo, R.C.B.Q. (2003). .Effects of essential oils from *Cymbopogon citratus* (DC) Stapf., *Lippia sidoides* Cham., and *Ocimum gratissimum* L. on growth and ultrastructure of Leishmania chagasi promastigotes. *Parasitology Research*, **104**: 1053-1059.

Onabanjo, A.O., Agbaje, E.O., and Odusote, O.O. (1993). Effects of aqueous extracts of *Cymbopogon citratus* in Malaria. *Journal of Proteome Research*, **3**: 40-45.

Orrego, R., Leiva, E., and Cheel, J. (2009). Inhibitory effect of three C-glycosyflavonoids from *Cymbopogon citratus* (Lemon grass) on human low density lipoprotein oxidation. *Molecules*, **14**: 3906-3913.

Paranagama, P.A., Abeysekera, K.H.T., Abeywickrama, K., and Nugaliyadde, L. (2003). Fungicidal and anti-aflatoxigenic effects of the essential oil of *Cymbopogon citratus* (DC.) Stapf. (lemongrass) against *Aspergillus flavus* Link. Isolated from stored rice. *Letters in Applied Microbiology*, **37(1)**: 86-90.

Phasomkusolsil, S., and Soonwera, M. (2011). Efficacy of herbal essential oils as insecticide against A*edes aegypti* (linn.), C*ulex quinquefasciatus* (say) and A*nopheles dirus* (peyton and harrison). *Southeast Asian Journal of Tropical Medicine and Public Health*, **42(5)**:1083-1092.

Pieroni, A. (2000). Medicinal plants and food medicines in the folk traditions of the Upper Lucca Province, Italy. *Journal of Ethnopharmacology*, **70**: 235-273.

Plant Resources of South-East Asia No 19. 1998.

Potterat, O., and Hostettmann, K. (1995). Plant source of natural drugs and compounds. In: *Encyclopedia of Environmental Biology*, **3**: 139-152.

Puatanachokkchai, R., Kishida, H., Denda, A., Murata, N., Konishi, Y., Vinitketkumnuen, U., and Nakae, D. (2002). Inhibitory effects of lemon grass (*Cymbopogon citratus*, Stapf) extract on the early phase of hepatocarcinogenesis after initiation with diethylnitrosamine in male Fischer 344 rats. *Cancer Letters*, **183(1)**: 9-15.

Pugliesi, G.C., Andrade, S.F., Bastos, J.K., and Maistro, E.L. (2007). *In vivo* clastogenicity assessment of the *Austroplenckia populnea* (Celastraceae) leaves extract using micronucleus and chromosomes aberration assay. *Cytologia*, **72**: 1-6.

Purwal, L., Shrivastava, V., and Jain, U. (2010). Assessment of pediculicidal potential of formulation containing essential oils of *Mentha piperita* and *Cymbopogon citratus*. *Research Journal of Pharmaceutical, Biological and Chemical Sciences*, **1**: 366-372.

Pushpanathan, T., Jebanesan, A., and Govindarajan, M. (2006). Larvicidal, ovicidal and repellent activities of *Cymbopogan citratus* Stapf (Graminae) essential oil against the filarial mosquito *Culex quinquefasciatus* (Say) (Diptera: Culicidae). *Tropical Biomedicine*, **23(2)**: 208-212.

Rauber, C.S., Guterres, S., and Schapoval, E.E.S. (2005). LC determination of citral in *Cymbopogon citratus* volatile oil. *Journal of Pharmaceutical and Biochemical Analysis*, **37**: 597-601.

Ravinder, K., Pawan, K., Gaurav, S., Paramjot, K., Gagan, S., and Appramdeep, K. (2010). Pharmacognostical investigation of *Cymbopogon citratus* (DC) Stapf. *Der Pharmacia Lettre*, **2(2)**: 181-189.

Reitz, R. (1982). Flora illustrates catarinense. 1309-1314.

Robinson, M.M., and Zhang, X. (2011). The World Medicines Situation 2011. Traditional medicines: global situation, issues and challenges. Available at: http://apps.who.int/medicinedocs/documents/s18063en/s18063en.pdf (accessed on 30th May 2012).

Ross, I.A. (1999). Medicinal plants of the world, Part 1: Chemical constituents, Traditional and Modern medicinal uses. Humana Press Inc. New Jersery. 119-125.

Sacchetti, G., Maietti, S., Muzzoli, M., Scaglianti, M., Manfredini, S., Radice, M., and Bruni, R. (2005). Comparative evaluation of 11 essential oils of different origin as functional antioxidants, antiradicals and antimicrobials in foods. *Food Chemistry*, **91(4)**: 621-632.

Samarasekera, R., Kalhari, K.S., and Weerasinghe, I.S. (2006). Insecticidal activity of essential oils of *Ceylon cinnamomum* and *Cymbopogon* species against *Musca domestica*. *Journal of Essential Oil Research*, **18**: 352-354.

Santin, M.R., dos Santos, A.O., Nakamura, C.V., Dias Filho, B.P., Ferreira, I.C.P., and Ueda-Nakamura, T. (2009). *In vitro* activity of the essential oil of *Cymbopogon citratus* and its major component (citral) on *Leishmania amazonensis*. *Parasitology Research*, **105**: 1489-1496.

Sforcin, J.M., Amaral, J.T., Fernandes Jr., A., Sousa, J.P.B., and Bastos, J.K. (2009). LG effects on IL-1 and IL-6 production by macrophages. *Natural Product Research*, **23**: 1151-1159.

Sha'a, K.K., Oguche, S., and Ajayi, J.A. (2011). The *in vivo* analgesic activity of aqueous and ethanolic extracts of *Anacardium Occidentale Occidentale* Linn and *Cymbopogon citrates* DC. *Journal of Medicine in the Tropics*, **13(2)**: 115-118.

Shah, G., Shri, R., Panchal, V., Sharma, N., Singh, B., and Mann, A.S. (2011). Scientific basis for the therapeutic use of *Cymbopogon citratus*. *Journal* of *Advanced Pharmaceutical Technology* & *Research*, **2(1)**: 3-8.

Silva, C.B., Guterres, S.S., Weisheimer, V., and Schapoval, E.E.S. (2008). Antifungal activity of the lemongrass oil and citral against *Candida* spp. *Brazilian Journal of Infectious Diseases*, **12(1)**: 63-66.

Silva, M.R., Ximenes, R.M., Martins da Costa, J.G., Leal, L.K.A.M., de Lopes, A.A., and de Barros Viana, G.S. (2010). Comparative anticonvulsant activities of the essential oils (EOs) from *Cymbopogon winterianus* Jowitt and *Cymbopogon citratus* in mice. *Naunyn*-Schmiedeberg's *Archives* of *Pharmacology*, **381**: 415-426.

Simon, J.E., Chadwick, A.F., and Cracker, L.E. (1984). The scientific literature on selected herbs and medicinal plants of the remperate zone, 2nd edition, Archon books New York. U.S National Cholesterol Education Program Expert Panel (1988). Detection, evaluation and treatment of high blood cholesterol in adults, *The Archives of Internal Medicine*, **148**: 36-39.

Suaeyun, R., Kinouchi, T., Arimochi, H., Vinitketkumnuen, U., and Ohnishi, Y. (1997). Inhibitory effects of lemon grass (*Cymbopogon citratus* Stapf) on formation of azoxymethane-induced DNA adducts and aberrant crypt foci in the rat colon. *Carcinogenesis*, **18**: 949-955.

Sukumar, K., Perich, M.J., and Boobar, L.R. (1991). Botanical derivatives in mosquito control: a review. *Journal of American Mosquito Control Association*, **7**: 210-237.

Sundram, K., Hayes, K.C., and Siru, O.H. (1995). Dietary 18:2 and 16: 0 may be required to improve the serum LDL/HDL cholesterol ratio in noncholesterolemic men. *Journal of Nutritional Biochemistry*, **6**: 179-187.

Sutthanont, N., Choochote, W., and Tuetun, B., *et al.* (2010). Chemical composition and larvicidal activity of edible plant-derived essential oils against the pyrethroid-susceptible and -resistant strains of *Aedes aegypti* (Diptera: Culicidae). *Journal of Vector Ecology*, **35**: 106-115.

Tangjang, S., Namsa, N.D., Aran, C., and Litin, A. (2011). An ethnobotanical survey of medicinal plants in the Eastern Himalayan zone of Arunachal Pradesh, India. *Journal of Ethnopharmacology*, **134**: 18-25.

Tchoumbougnang, F., Dongmo, P.M.J., Sameza, M.L., Mbanjo, E.G.N., Fotso, G.B.T., Zollo, A., and Menut, C. (2009). Larvicidal activity against *Anopheles gambiae* Giles and chemical composition of essential oils from four plants cultivated in Cameroon. *Biotechnology, Agronomy, Society and Environment*, **13**: 77-84.

Tchoumbougnang, F., Zollo, P.H.A., Dagne, E., and Mekonnen, Y. (2005). *In vivo* antimalarial activity of essential oils from *Cymbopogon citratus* and *Ocimum gratissimum on* mice infected with *Plasmodium berghei*. *Planta Medica*, **71**: 20-23.

Tortoriello, J., and Romero, O. (1992). Plants used by Mexican traditional medicine with presumable sedative properties: an ethnobotanical approach. *Archives of Medical Research*, **2**: 111-116.

Tyagi, A.K., and Malik, A. (2010). Liquid and vapour-phase antifungal activities of selected essential oils against *Candida albicans*: microscopic observations and chemical characterization of *Cymbopogon citratus*. *BMC Complementary and Alternative Medicine*, **10**:65.

Van Wyk, B.E., and Wink, M. (2004). Medicinal Plants of the World: An Illustrated Scienti?c Guide to Important Medicinal Plants and their Uses. Timber Press, Portland, Oregon, USA, pp. 480.

Van Wyk, B.E., Van Oudshoorne, B., and Gericke, N. (2002). Medicinal Plants of South Africa. Briza Publications, Pretoria, South Africa, pp. 336.

Viana, G.S.B., Vale T.G., Pinho R.S.N., and Matos F.J.A. (2000). Anti-nociceptive effect of the essential oil from *Cymbopogon citratus* in mice. *Journal of Ethnopharmacology*, **70**: 323-327.

Wannissorn, B., Jarikasem, S., and Soontorntanasart, T. (1996). Antifungal activity of lemon grass and lemon grass oil cream. *Phytotherapy Research*, **10**: 551-554.

World Health Organization: WHO traditional medicine strategy 2002–2005. WHO, Geneva, Available at: http://whqlibdoc.who.int/hq/2002/who_edm_trm_2002.1.pdf (accessed on 6th June 2012).

Wright, S.C., Maree, J.E., and Sibanyoni, M. (2009). Treatment of oral thrush in HIV/AIDS patients with lemon juice and lemon grass (*Cymbopogon citratus*) and gentian violet. *Phytomedicine*, **16(2-3)**:118-124.

Utilisation and Management of Medicinal Plants Vol. 2 (2014)    *Pages* **93–113**
*Editor-in-Chief:* **V.K. Gupta**
*Published by:* **DAYA PUBLISHING HOUSE, NEW DELHI**

# 3

# The Antimalarial Potential of Medicinal Plants Used for the Treatment of Malaria in Kenya

Charles Mutai[1]* and Lucia Keter[1]

## ABSTRACT

*Malaria remains one of the leading public health problems in Kenya like in many Sub-Saharan Africa countries. In the past decades, this situation has been aggravated by the increasing spread of drug-resistant Plasmodium falciparum strains. New anti-malarial drug leads are therefore urgently needed. Traditional healers have long used plants to prevent or cure infections. This article reviews the current status of botanical screening efforts in Kenya as well as experimental studies done on medicinal plants used by traditional health practitioners to treat malaria. Data was collected from 54 references from various research groups in the literature up to June 2007 shows that 217 different species have been cited for their use as anti-malarials in folk medicine in Kenya. About a hundred phytochemicals have been isolated from 26 species some among which are potential leads for development of new anti-malarials. Crude extracts and or essential oils prepared from 54 other species showed a wide range of activity on Plasmodium spp. The present study shows that Cameroonian flora represents a high potential for new anti-malarial compounds. Further ethnobotanical surveys and laboratory investigations are needed to fully exploit the potential of the identified species in the control of malaria.*

***Keywords***: Antimalarial potential, Medicinal plants, Phytochemicals, *Plasmodium falciparum*, Folk medicine.

---

1    Kenya Medical Research Institute, Centre for Traditional Medicine and Drug Research, PO Box 54840, Nairobi, Kenya.

*    *Corresponding author*: E-mail: cmutai@kemri.org

# Introduction

Malaria is a major public health problem general in the world and particularly in Sub-Saharan countries. According to the World Health Organization (WHO) report, more than 500 million infections occur per year with approximately 2.7 million deaths. It is becoming more difficult to prevent and to treat malaria due to the increasing resistance of the transmitting mosquito to the insecticides and of the malaria parasite to drugs such as chloroquine, quinine and more lately sulphadoxine/pyrimethamine based combination that have been commonly used. Artemisinin and derivatives have recently been introduced to combat drug resistant *Plasmodium falciparum* but it is estimated that the parasite will soon develop resistance to artemisinin-based drugs. An alternative to conventional drugs is the use of traditional medicines for the treatment of malaria and in the last decade, there has been increasing interest in potential of plants to provide novel molecules for the development of anti-malarial therapy. With the exception of the antifolate anti-malarial drugs, virtually all the other commonly used anti-malarial molecules are based upon plant-derived lead compounds a strong indication that ethno-botany is a rich source of new anti-malarial compounds or anti-malarial lead compounds.

Resistance of *Plasmodium falciparum* to commonly used anti-malarial drugs is increasing in Kenya as in other parts of Africa (Ref). This has resulted in resurgence in transmission and an increase in adverse outcomes due to therapy failure. Hence, new highly efficacious anti-malarial agents are urgently needed. For thousands of years, plants have constituted the basis of traditional medicine systems and recently, natural products have been a good source of lead compounds for drug development. A good example against malaria is quinine, isolated from *Cinchona* bark, which was used as a template for the synthesis of chloroquine and mefloquine. More recently, artemisinin isolated from the Chinese plant *Artemisia annua*, has been used successfully against chloroquine-resistant *P. falciparum* strains (Schwikkard and Van Heerden, 2006).

In Kenya, a large number of plant species have been identified as antimalarial medicinal plants. Pure products have been isolated from some of these plants amongst which are those whose anti-malarial activities are comparable to or more active than chloroquine on sensitive and resistant stains of *P. falciparum* (CTMDR ref). It is therefore imperative that anti-malarial drug development has to be pursued further. In the present review, we report on the plants, which have been identified as anti-malarial plants and the work done so far in evaluating their anti-malarial potential. The review is structured according to plant families and the extent of investigations carried out in the specific plant families to date.

# Methodology

The data on the medicinal plants were collected through a review of unpublished documents in the Library of the Centre for Traditional Medicine and Drug Research, Kenya Medical Research Institute (KEMRI), books on traditional medicine and through internet search in www.google.com, hinari and www.pubmed.gov. We also collected information on antimalarial medicinal plants from members of research groups in the Laboratory of Phytochemistry and Medicinal Plants Studies.

## Use of Traditional Remedies for the Treatment of Malaria and other Fevers in Kenya

The proportion of the populations using traditional remedies to treat malaria varies widely. Mostly in the rural areas, the use of plant medicines plays an important role in primary health care. Traditional medicines are even preferred to modern medicines in treatment of chronic diseases such as hypertension, diabetes mellitus and cancers. Traditional medicines are commonly sold in markets and public places or administered by Traditional Health Practitioners in their homes or clinics. Whole plants or parts of them are prepared and administered as oral decoction, steam baths, infusion or enema. Most remedies are a concoction of two or more plant species and solvents used include water (Kokwaro, 1993).

## An Overview of Studies on Kenyan Medicinal Plants Used as Anti-malarial

A wide variety of Kenyan plants have been identified through ethnobotanical surveys and ethnopharmacological studies as anti-malarial medicinal plants. Botanists have identified these plants and vouchers are found at the East African Herbarium, National Museums of Kenya, the Herbarium, Botany Department, University of Nairobi and the herbarium of the Faculty of Pharmacy, University of Nairobi, Kenya. Some of these plants have undergone various degrees of scientific investigation by various researchers mentioned in this paper. Following botanical identification and depending on the part used by the THPs, various plant parts such as the leaves, roots, fruits, stem, stem-barks or whole plant was collected. The plant part was air-dried under shade and then powdered using a laboratory mill or mortar and pestle (root bark and stem bark) and a kitchen blender (leaves). Ground plant material was extracted with distilled water and various organic solvents. The water extracts were lyophilized in a freeze-dryer while the organic extracts were concentrated using a rotary evaporator. The filtrate is concentrated by rotary evaporation to obtain the crude extract. The extract is then tested in various systems mainly *in vitro* incubation with several *Plasmodium falciparum* strains or *in vivo* in a malaria animal model infected with *plasmodium bergei*. The *in vitro* activity was assessed by the parasite lactate dehydrogenase (pLDH) assay method (Delhaes *et al.*, 1999) or the $^3$H-hypoxanthine radioisotope uptake method (Desjardins *et al.*, 1979; Le Bras and Deloron, 1983). The *in vivo* assay protocol used was based on Peter's 4–day suppressive (Peters *et al.*, 1982). Extracts with significant antiplasmodial activity are fractionated using various techniques and the fractions tested to identify their biological activity. Biologically active fractions then undergo purification to isolate the bioactive natural product(s). The pure product is further tested for anti-malarial activity *in vitro* and *in vivo* as mentioned above. The *in vitro* activity was categorized as high when IC50 was <5 µg/mL, moderate when IC50 was 5–20 µg/mL and weak/low when IC50 was 20–100 µg/mL while above 100 µg/mL was considered inactive. The *in vivo* activity was described as mild or moderate were parasitaemia suppression of 30 – 49 per cent was reported while 50 per cent and above was considered high or remarkable anti-malarial activity. In the following sections we assess some families, which have been investigated so far in Kenya.

## Acanthaceae

*Justicia betonica* L. green leaves infusion is taken for snake bite while the roots are roasted and chewed for cough (Kokwaro, 1993). The aerial part methanolic extracts have been reported to exhibit mild *in vitro* antiplasmodial activity (Muregi *et al.*, 2003).

## Aloeaceae

*Aloe kedongensis* Reynolds leaves and roots infusion is used to treat malaria, typhoid, skin diseases, colds, ear problems and wounds (Jeruto *et al.*, 2008).

## Amaranthaceae

*Cyathula schimperiana* Igifashi root decoction is taken for fever, malaria and stomach problems (Jeruto *et al.*, 2008; Gathirwa *et al.*, 2007; Kokwaro, 1993). Methanol root extracts reported to have moderate *in vitro* and *in vivo* anti-malarial activity against various Plasmodial strains (Gathirwa *et al.*, 2007).

*Cyathula cylindrica* root bark decoction is drunk as a remedy for malaria and leprosy (Jeruto *et al.*, 2008; Kokwaro, 1993).

*Celosia schweinfurthiana* Schinz. is used to treat malaria and bilharzias (Orwa *et al.*, 2007; Kokwaro, 1993).

## Anacardiaceae

*Rhus natalensis* Bernh.is used traditionally to treat malaria and the aqueous and methanol stem bark extracts reported to exhibit weak *in vitro* antiplasmodial activity but high *in vivo* antimalarial activity with the aqueous extract exhibiting chemo-suppression of 83.15 per cent against *P. berghei* (Gathirwa *et al.*, 2007).

*Lennea schweinfurthii* is used against malaria and fever. The aqueous and methanol stem bark extracts reported to exhibit both *in vivo* and *in vitro* antimalarial activity (*Lennea schweinfurthii*).

*Sclerocarya birrea* is used in forkmedicine to treat malaria and fever. The stem bark extracts have been reported to exhibit both *in vitro* and *in vivo* antimalarial activity with the methanol extract reporting high *in vitro* activity (IC50 5.91 µg/ml) (Gathirwa *et al.*, 2008).

## Annonaceae

*Uvaria scheffleri* Diels root bark decoction is used to cure malaria (Muthaura *et al.*, 2007a; Beentje, 1994; Kokwaro, 1993).

## Apocynaceae

The root decoction of *Carissa edulis* (Forssk.) Vahl. is used in fork medicine for malaria, indigestion and chest pains ((Jeruto *et al.*, 2008; Muthaura *et al.*, 2007a; Kokwaro, 1993). The aqueous and methanol roots extracts have been reported to exhibit no *in vitro* antiplasmodial activity (Kirira *et al.*, 2006).

*Catharanthus roseus* (L.) G. Don whole plant decoction taken for addominal pains and malaria (Orwa *et al.*, 2007; Gathirwa *et al.*, 2007; Kokwaro, 1993). Various leaf extracts have been reported to have both *in vivo* and *in vitro* antimalarial activity with the leaf methanol extract exhibiting high antiplasmodial activity, IC 50 of 4.65 and 5.34 µg/ml against chloroquine sensitive and resistant strains respectively (Gathirwa *et al.*, 2007).

### Asclepiadaceae

*Centella asiatica* (L.) Urban whole plant decoction is used to treat malaria and syphilis (Muthaura *et al.*, 2007a; Kokwaro, 1993).

*Curroria volubilis* (Schlecht.) Bullock bark decoction used to treat malaria (Jeruto *et al.*, 2008).

### Asteraceae

*Artemisia afra* Jacq. is traditionally used to treat malaria (Gathirwa *et al.*, 2007; Kuria *et al.*, 2001). High *in vitro* and *in vivo* antimalarial activity reported with the methanol leave extracts exhibiting IC50 of 3.98 µg/ml against the chloroquine resistant *P. falciparum* W2 strains and chemo-suppression of 77.45 per cent (Gathirwa *et al.*, 2007).

### Bignoniaceae

*Markhamia lutea* (Benth.) K. Schum. is used to treat malaria (Orwa *et al.*, 2007) while the leaves and root infusion of *Ehretia cymosa* Thonn. is used to treat wounds, pneumonia and malaria among many other conditions (Jeruto *et al.*, 2008).

### Caesalpinaceae

*Senna didymobotrya* (Fresen.) Irwin and Barne by leaf and root infusion is used as an emetic against malaria (Jeruto *et al.*, 2008; Beentje, 1994). The leaf extracts was reported to be in-active *in vitro* (Muregi *et al.*, 2004). The bitter tasting leaves of *Caesalpinia volkensii* Harms are used mainly for malaria while roots are used for gonorrhea and bilharzia (Kuria *et al.*, 2001; Gachathi, 1993). The petroleum ether extracts of *C. volkensii* has been shown to exhibited anti-plamodial activity against the chloroquine resistant *P. falciparum* W2 strains (Kuria *et al.*, 2001).

*Tamarindus indica* L. is used in folk medicine to treat fever and malaria (Orwa *et al.*, 2007; Kokwaro, 1993).

*Senna occidentalis* (L.) Link is also used to treat malaria (Orwa *et al.*, 2007).

### Capparidaceae

Traditionally the bark of *Boscia angustifolia* A. Rich. is boiled in water and the decoction drunk to cure malaria (Kokwaro, 1993). Moderate to high *in vitro* and *in vivo* anti-malarial activity reported with various stem bark extracts with the aqueous extracts $IC_{50}$ of ≤ 5 µg/ml (Muthaura *et al.*, 2007b).

*Boscia salicifolia* Oliv exhibits limited distribution in Kenya. Its extracts are traditionally used to treat backache (Gathirwa *et al.*, 2007; Beentje 1994). It has been reported to have high *in vitro* and *in vivo* anti-malarial activity with the methanol extracts of the stem bark exhibiting IC50 1.04 µg/ml and chemo-suppression of 86.5 per cent (Gathirwa *et al.*, 2007).

*Cleome gynandra* L. leaves and roots decoction is used for malaria and stomach problems (Jeruto *et al.*, 2008).

## Canellaceae

*Warburgia stuhlmannii* Engl. hot water decoction of stem bark and leaves used to treat malaria. Stem bark used for toothache and rheumatism (Muthaura *et al.*, 2007a; Kokwaro, 1993). The stem bark or leave decoctions of *Warburgia ugandensis* Sprague is taken as a cure for malaria (Kuria *et al.*, 2001; Kokwaro, 1993). Dried bark is a traditional remedy for several conditions such as stomachache, constipation, coughs, fever, toothache, muscle pains, weak joints and general body pains (Kokwaro, 1993).

## Celastraceae

*Maytenus senegalensis* (Lam.) Exell is widely used traditionally for fever, diarrhea, snake bite, rheumatism and eye infections. The root bark is used to treat malaria (Orwa *et al.*, 2007; Muthaura *et al.*, 2007a; Kokwaro, 1993). Anti-malarial activity of the extracts of *M. senegalensis* has been demonstrated (El-Tahir, 2001; Gessler, 1995). A group of quinone methide triterpenes isolated from this plant have been found to have anti-malarial properties (Gessler, 1995). Leaves and root bark aqueous extracts reported to be inactive *in vivo* (Muregi *et al.*, 2007).

*Maytenus putterlickioides* (Loes.) Exell and Mendonca root decoction is used to treat malaria, as aphrodisiac and leaves for hookworm (Muthaura *et al.*, 2007a; Kokwaro, 1993).

The roots decoction of *Maytenus undata* (Thunb.) Blakelock is traditionally used to treat malaria, syphilis and other diseases of urethra (Muthaura *et al.*, 2007a; Kokwaro, 1993).

*Maytenus heterophylla* (Eckl. and Zeyh.) Robson roots boiled in water is taken as an anthelmintic, cure syphilis and hernia (Kokwaro, 1993). The root bark water extracts reported to have mild *in vivo* anti-malarial activity (Muregi *et al.*, 2007).

## Chrysobalanaceae

The aerial parts of *Parinari curatellifolia* Benth. has been reported to have no antiplasmodial activity (Muregi *et al.*, 2004).

## Combretaceae

*Terminalia spinosa* Engl. stem bark infusion used to treat jaundice and malaria (Muthaura *et al.*, 2007a; Beentje, 1994).

## Commelinaceae

*Commelina forskalaei* Vahl. is used to treat malaria and the plant infusion is used as a wash to reduce fever (Orwa *et al.*, 2007; Kokwaro, 1993).

## Compositae

*Vernonia lasiopus* O.Hoffm. is traditionally used to treat malaria, scabies, stomachache, indigestion, venereal disease, as a purgative, as stimulant and pounded leaves are applied to sores to kill maggots (Beentje, 1994; Kokwaro, 1993). Root bark water extracts from this plant reported to have *in vivo* anti-malarial activity (Muregi *et al.*, 2007 and 2006). The leaf extracts reported to exhibit good antiplasmodial activity with IC50 values ≤ 5 µg/ml against the multi-drug resistant *P. falciparum* isolate, V1/S (Muregi *et al.*, 2003).

*Vernonia brachycalyx* O.Hoffm. leave infusion is used against malaria and the roots used to cure stomachache and as a purgative (Beentje, 1994; Kokwaro, 1993). Extracts of the leaves have been reported to show *in vitro* activity against *Plasmodium falciparum* (Oketch-Rabah *et al.*, 1998). The germacrane dilactone 16, 17-dihydrobrachycalyxolide and two isomeric 5-methylcoumarins (2'-epicycloisobrachycoumarinone epoxide and cycloisobrachycoumarinone epoxide) have been isolated from this plant and reported strong anti-plasmodial activity (Oketch-Rabah *et al.*, 1998 and 1997).

*Sphaeranthus suaveolens* (Forsk.) DC. is used to treat colds and rubbed on to the body of a person suffering from malaria (Kokwaro, 1993). *Schkuhria pinnata* (Lam) O. Ktze plant is used by herbalists to treat chest, liver and stomach pains (Kokwaro, 1993). Various whole plants extracts of these plants have shown moderate *in vitro* and *in vivo* antimalarial activity (Muthaura *et al.*, 2007b).

*Vernonia auriculifera* (Welw.) Hiern is taken for stomach troubles, fever and the leaves are used in the treatment of cows and sheep with diarrhoea (Kokwaro, 1993; Gathathi, 1989). Various leaf extracts demonstrated mild antiplasmodial activity (Muregi *et al.*, 2003).

*Artemisia afra* Willd. is used traditionally for malaria, sore throat, indigestion, intestinal worms, as an emetic and to cure fever (Kuria *et al.*, 2001; Kokwaro, 1993).

*Aspilia pluriseta* Schweinf. is used by herbalist to treat skin diseases and cut wounds (Kokwaro, 1993). The leaves hexane and methanol extracts have exhibited mild *in vitro* activity (Muregi *et al.*, 2003).

*Microglossa pyrifolia* (Lam.) O. Kuntze infusion of the leaves is taken as a remedy for malaria and for the treatment of lime fractures. Root infusion taken for the treatment of headache and colds (Kokwaro, 1993). Various leaf extracts demonstrated mild antiplasmodial activity (Muregi *et al.*, 2003).

*Bidens pilosa* L. is used by herbalists to cure malaria and to treat conjunctivitis, stomach-ache, intestinal worms, constipation (Kuria *et al.*, 2001; Kokwaro, 1993).

*Tridax procumbens* L. leaves are chewed as a remedy for malaria and for stomachache (Kokwaro, 1993) and cold water infusion of the whole plant is used to treat malaria (Muthaura *et al.*, 2007a).

### Cucurbitaceae

Traditionally *Cucumis figarei* Naud. is used for indigestion and constipation (Kokwaro, 1993). The whole plant extracts showed no antiplasmodial activity (Muregi *et al.*, 2004).

### Enenaceae

Traditionally, *Euclea divinorum* Hiern. is used to treat malaria (Orwa *et al.*, 2007).

## Euphorbiaceae

*Flueggea virosa* (Willd.) Voigt is a shrub widely distribute in all regions of Kenya. The root decoction is used for chest pains and malaria (Muthaura *et al.*, 2007a; Beentje, 1994).

*Euphorbia inaequilatera* Sond. is used to treat wounds, sores and to relieve labour pain. Leaves and stem chewed to treat gonorrhoea (Kokwaro, 1993). Whole plant extracts showed no *in vitro* antimalarial activity (Muregi *et al.*, 2003)

*Clutia abyssinica* Jaub and Spach cure headache, malaria, influenza, indigestion, liver pains and stomach ache. Both leaves and roots are used to treat malaria (Kokwaro, 1993). Organic and aqueous leaf extracts reported to exhibit moderate antimalarial activity (Muthaura *et al.*, 2007b; Kraft *et al.*, 2003).

The aqueous and the methanol leaf extracts of *Clutia robusta* Pax, E, FZ. is reported to exhibit both *in vitro* and *in vivo* antimalarial activity with the methanol extracts exhibiting high *in vitro* activity (IC50 3.41 g/ml) (Gathirwa *et al.*, 2007).

*Suregada zanzibarensis* Baill root bark and leaves used to treat malaria while the roots infusion is used for snake bite (Muthaura *et al.*, 2007a; Kokwaro, 1993).

*Neoboutonia macrocalyx* aqueous and methanol stem bark extracts have been reported to have low *in vitro* antiplasmodial activity (Kirira *et al.*, 2006).

## Flacourtiaceae

*Trimeria grandifolia* (Hochst) Warb is traditionally used to treat malaria (Orwa *et al.*, 2007).

## Guttiferae

A decoction from the root bark and stem bark of *Harungana madagascariensis* Poir is drunk as a remedy for malaria (Muthaura *et al.*, 2007a; Kokwaro, 1993).

## Labiatae

The leaves of *Ajuga remota* Benth. are known to relieve tooth ache, while a decoction or infusion from leaves is prescribed by Kenyan herbalists for severe stomachache, treatments of malaria and oedema associated with protein-calorie malnutrition disorders in infants when breast-feeding is terminated. The plant is also used for the treatment of pneumonia and liver problems (Kuria *et al.*, 2001; Gachathi, 1993; Kokwaro, 1993). Mild *in vitro* antiplasmodial activity with the methanolic leaves extracts have been reported (IC50 21.6 µg/ml) by Muregi *et al.* (2004). The antiplasmodial activity has been shown to be due to ergosterol-5, 8-peroxide (Kuria *et al.*, 2002).

*Leonotis mollissima* Gürke. is used traditionally against stomachache and diarrhea, to treat wounds, dysentery, intestinal worms, venereal disease and festering sores. Also use to treat conjunctivitis, stomach cramps, fever and oedema (Kokwaro, 1993). Extract reported to exhibit *in vitro* antiplasmodial activity (Muregi *et al.*, 2004).

*Leucas calostachys* Oliv. leaves chewed and juice swallowed as a cure for pneumonia and for serious stomachache (Kokwaro, 1993). Also used in treatment of dysmenorrheal, dyspepsia, stomachaches, gastric ulcer, colic cholitis, gastritis, expectorant, and stomach ulcers among the Marakwets of Kenya (Lindsay, 1978). Anti-plasmodial activity has been demonstrated (Muregi, 2004).

*Leonotis nepetifolia* R. Br. used to treat stomach troubles and malaria (Orwa *et al.,* 2007; Beentje, 1994; Kokwaro, 1993).

**Lauraceae**

The root infusion of *Ocotea usambarensis* Engl. is taken for backache, malaria and stomach pains (Kokwaro, 1993). Moderate antimalarial activity has been reported with the stem bark extracts (Muthaura *et al.,* 2007b).

**Leguminosae**

*Cassia abbreviata* Oliv. roots decoction is drunk to cure fever or malaria, stomach problems and for uterus complaints, relieve gonorrhoea, pneumonia, chest complaints and for suspected syphilis (Muthaura *et al.,* 2007a; Kokwaro, 1993).

*Acacia hockii* De Wild. is used for abdominal pains and abscess (Kokwaro, 1993). The aqueous root bark extracts reported to show mild anti-plasmodial activity (Muregi *et al.,* 2004).

*Acacia nilotica* (L.) Del. is used to treat fever, stomach trouble, chest pains, pneumonia, indigestion, gonorrhoea, throat and coughs and as an aphrodisiac sore while the root bark decoction is used to treat malaria (Muthaura *et al.,* 2007a; Kokwaro, 1993). The methanol stem bark extracts have been reported to have low *in vitro* antiplasmodial activity while the aqueous extract exhibit no *in vitro* antiplasmodial activity (Kirira *et al.,* 2006).

The bark or roots decoction of *Albizia anthelmintica* Brongn is used traditionally to treat malaria, fever, gonorrhoea and tapeworm (Muthaura *et al.,* 2007a; Kokwaro, 1993).

*Cassia occidentalis L.* is used in fork medicine for severe stomach-ache, malaria, fever, snake bites and kidney problems (Kuria *et al.,* 2001; Kokwaro, 1993).

**Lamiaceae**

*Fuerstia africana* T.C.E. Fries leaves are grounded boiled until liquid turns yellow; juice is squeezed and drunk to treat malaria. The young parts and leaves of the plant are pounded and extract used for treatment of stomach ulcers, pneumonia, urinary problems, and tongue infection, as a purgative and an anthelmintic (Gachathi, 1989; Kokwaro, 1993). *In vitro* and *in vivo* antimalarial activity have been demonstrated with the methanol of the whole plant exhibiting IC50 of 0.98 and 2.4 µg/ml against D6 and W2 *Plasmodium falciparum*, respectively. A compound named ferruginol with anti-malarial activity isolated (Muthaura *et al.,* 2007b; Koch *et al.,* 2006, 2005).

**Loganiaceae**

*Strychnos henningsii* Gilg is used to treat malaria, chest pains, rheumatism, arthritis, internal injuries and snake bite (Kuria *et al.,* 2001; Bentje, 1994; Kokwaro, 1993). The aqueous stem bark extracts have been reported to have weak *in vitro* antiplasmodial activity while the methanol extract exhibit no *in vitro* antiplasmodial activity (Kirira *et al.*, 2006).

## Meliaceae

*Azadirachta indica* A. Juss is commonly referred to as Mwarubaini or Neem Tree. It is native of India but it has been naturalized in tropical countries. It is described as good for treatment of malaria, skin diseases, fungal and bacterial infections, diabetes, and mild hypertension among many others (Muthaura *et al.*, 2007a; Kuria *et al.*, 2001). However, *in vivo* anti-malarial studies suggest that the extracts are more effective for prophylaxis than curative (Isah *et al.*, 2003; Agomo *et al.*, 1992; Obih and Makinde, 1985). Various extracts of the leaves reported to have no *in vitro* antiplasmodial activity (Kirira *et al.*, 2006; Muregi *et al.*, 2004).

*Melia azedarach* Linn. is well known for its medicinal uses. Its various parts have antihelmintic, antimalarial (Orwa *et al.*, 2007), cathartic and emetic properties. Leaves extracts reported to have no antiplasmodial activity (Muregi *et al.*, 2004).

*Ekebergia capensis* Sparrm roots is use by herbalists to treat diarrhoea (Gachathi, 1989). Total extracts obtained form this plant have been shown to possess antiplasmodial activity (Muregi *et al.*, 2007, 2004 and 2003). Various organic extracts from the stem bark have been shown to have high antiplasmodial activity with IC50 values of less than 5µg/ml (Muregi *et al.*, 2004).

*Turraea robusta* is used in fork medicine against malaria and febrifuge and both the aqueous and methanol root bark extracts have been reported to have *in vitro* anti-plasmodial and high *in vivo* antimalarial activity with the methanol extract exhibiting high activity against D6 (IC50 2.09 µg/ml) (Gathirwa *et al.*, 2008).

## Menispermaceae

Various total extracts from *Stephania abyssinica* (Dill. and A. Rich.) Walp. have been shown to exhibit antiplasmodial activity with the aqueous extracts giving an IC50 22.9 µg/ml (Muregi *et al.*, 2004).

*Cissampelos mucronata* A. Rich. roots are traditionally used for malaria and abdominal pains and as an antidote for snake bite (Muthaura *et al.*, 2007a; Kokwaro, 1993). Anti-plasmodial activity has been demonstrated (Gessler, 1994).

## Mimosoideae

*Albizia coriaria* Oliv. is used to treat malaria and venereal diseases (Orwa *et al.*, 2007; Kokwaro, 1993).

## Moraceae

*Ficus sur* Forssk. is used traditionally against stomachache, diarrhea and as cough remedy (Beentje, 1994). The chloroform and hexane extracts of the stem bark reported to exhibit good *in vitro* anti-malarial activity (IC 50 9.0 µg/ml and 19.2 µg/ml, respectively) (Muregi *et al.*, 2003).

## Myricaeae

*Myrica salicifolia* aqueous and methanol stem bark extracts have been reported to have low *in vitro* antiplasmodial activity (Kirira *et al.*, 2006).

## Olacaceae

*Ximenia Americana* L. traditionally used to treat malaria (Orwa *et al.*, 2007). Anti-*Plasmodium falciparum* activity has been reported (Benoit *et al.*, 1996). The methanol root bark extract reported to have moderate to low *in vitro* and *in vivo* antimalarial activity (Gathirwa *et al.*, 2007) supporting the earlier finding by Benoit *et al.*

## Onagraceae

*Ludwigia erecta* (L.) Hara whole plant is boiled and a bath is taken in the liquid against malaria (Kokwaro, 1993). Various whole plant extracts reported to exhibit *in vitro* and *in vivo* anti-malarial activity with the aqueous extract giving an IC50 of ≤ 2 µg/ml (Muthaura *et al.*, 2007b).

## Pittosporaceae

The bark of *Pittosporum viridiflorum* Sims is traditionally applied as an emetic, for chest complication, malaria and other fevers (Bentje, 1994; Kokwaro, 1993). Moderate *in vitro* and high *in vivo* anti-malarial activity reported with the leaves extracts. Chemosuppression of 89.76 per cent reported with the methanol leaf extract (Muthaura *et al.*, 2007b).

## Rhamnaceae

*Rhamnus prinoides* L'He´rit. leaves are used to treat malaria and a decoction of the root is used to treat gonorrhea, indigestion and rheumatism (Kuria *et al.*, 2001; Bentje, 1994; Kokwaro, 1993).Various extracts of the root bark reported to exhibit moderate *in vitro* antiplasmodial activity (methanol extract IC50 value 15.1 µg/ml) Muregi *et al.*, 2003). The water extracts of the root bark reported to exhibit remarkable *in vivo* antimalarial activity while the leaves water extract exhibited mild activity (Muregi *et al.*, 2007).

*Rhamnus staddo* A.Rich. roots and stem are used as a cure for malaria and venereal diseases (Kuria *et al.*, 2001; Kokwaro, 1993). Various extracts of the root bark reported to exhibit mild *in vitro* antiplasmodial activity (methanol extract IC50 value 15.1 µg/ml) Muregi *et al.*, 2003). The water extracts of the leaves and the root bark reported to have moderate *in vivo* antimalarial activity (Muregi *et al.*, 2007).

## Rubiaceae

A decoction of *Pentas bussei* Krause roots is taken as a remedy for malaria, venereal diseases and dysentery (Muthaura *et al.*, 2007a; Bentje, 1994; Kokwaro, 1993).

*Pentas agathisanthemum* KI. whole plant decoction is used to treat malaria (Muthaura *et al.*, 2007a).

*Pentas longiflora* Oliv. roots decoction is mixed with milk and taken as a cure for malaria. Roots also used as a cure for tapeworm, itchy rashes and for pimples (Muthaura *et al.*, 2007a; Kokwaro, 1993).

The leaf decoction of *Spermacoce princeae* (K.Schum.) Verdc. is taken for hepatic disease while the whole plant is used for malaria, wounds and general skin diseases (Muregi *et al.*, 2003; Kokwaro, 1993). Extracts from the whole plant showed no *in vitro* antiplasmodial activity (Muregi *et al.*, 2003).

The infusion made from *Vangueria acutiloba* Robyns pounded bark is drunk for treatment of malaria (Kokwaro, 1993). Various extracts of the stem bark have been reported to exhibit moderate *in vitro* and *in vivo* anti-malarial activity (Muthaura *et al.*, 2007b).

*Vangueria volkensii* K. Schum is used traditionally to treat malaria (Orwa *et al.*, 2007).

## Rutaceae

*Toddalia asiatica* is well distributed in most provinces in Kenya. The roots are used to treat malaria, fevers, indigestion and influenza and the fruits are used as a cough remedy (Muthaura *et al.*, 2007a; Orwa *et al.*, 2007; Kuria *et al.*, 2001; Kokwaro, 1993). Aqueous root bark extract reported to have potential antimalarial activity (Muregi *et al.*, 2007). An alkaloid nitidine isolated from *T. asiatica* root bark has anti-plasmodial activity (Gakunju *et al.*, 1995).

*Zanthoxylum chalybeum* Engl. roots or stem bark decoction is used as an emetic, against malaria, coughs, dizziness, colds and sore throat (Muthaura *et al.*, 2007a; Beentje, 1994). The extracts from the bark have been shown to have *in vitro* anti-plasmodial activity (Gessler, 1994).

*Zanthoxylum usambarense* (Engl.) stem bark and roots are traditionally used for treating cough, malaria and rheumatism (Kuria *et al.*, 2001; Kokwaro, 1993). The aqueous and methanol stem bark extracts *of Zanthoxylum usambarense* have been shown to have high anti-plasmodial activity against both chloroquine-resistant (ENT 30) (IC50 14.33 and 5.54 µg/ml) and chloroquine-sensitive (NF 54) (IC50 6.13 and 4.68 µg/ml) isolates (Kirira *et al.*, 2006), respectively. Nitidine, the most common anti-malarial benzophenanthridine alkaloid in *Zanthoxylum* species (Gakunju *et al.*, 1995), has been previously isolated from *Z. usambarense* (Kato *et al.*, 1996).

*Fagaropsis angolensis* (Engl.) Dale is used for treatment of malaria (Khalid and Waterman, 1985). The aqueous and methanol stem bark extracts of *Fagaropsis angolensis* reported to show high anti-plasmodial activity against both chloroquine-resistant (ENT 30) (IC50 10.65 and 5.04 µg/ml) and chloroquine-sensitive (NF 54) (IC50 5.25 and 3.20 µg/ml) isolates (Kirira *et al.*, 2006), respectively. Nitidine, the most common anti-malarial benzophenanthridine alkaloid in *Zanthoxylum* species (Gakunju *et al.*, 1995), has been previously isolated from *F. angolensis* (Khalid and Waterman, 1985). This may explain the observed anti-plasmodial activity.

*Teclea nobilis* Del. leaves and roots are both used in fork medicine. The leaf or root decoction used against pneumonia colds and chest problems, rheumatism and as an anthelminthic. Stem bark used to treat malaria while vapour from steamed leaves inhaled to cure fever (Kuria *et al.*, 2001; Kokwaro, 1993).

## Simaroubaceae

A decoction from the boiled roots of *Harrisonia abyssinica* Oliv. is taken for fevers, malaria, nausea, vomiting, tuberculosis and venereal diseases (Muthaura *et al.*, 2007a; Orwa *et al.*, 2007; Beentje, 1994; Kokwaro, 1993). The aqueous and methanol stem bark extracts have been reported to have weak *in vitro* antiplasmodial activity (Kirira *et al.*, 2006).

## Solanaceae

*Withania somnifera* (L.) Dunal root decoction is used for treatment of malaria, gonorrhea, stomachace and gastric ulcer and as a tonic for general illnesses (Muregi *et al.*, 2007; Bentje, 1994; Kokwaro, 1993). The aqueous and methanol roots extracts have been reported to exhibit no *in vitro* antiplasmodial activity (Kirira *et al.*, 2006). Root bark aqueous extract reported to be in active against plasmodial parasites *in vivo* (Muregi *et al.*, 2007).

## Tiliaceae

*Grewia bicolor* Juss. used to treat malaria (Orwa *et al.*, 2007).

## Urticaceae

*Urtica massaica* Mildbr. macerated roots and leaves used to treat hepatic diseases (Kokwaro, 1993). Extracts from the aerial parts showed no *in vitro* antiplasmodial activity (Muregi *et al.*, 2003).

## Verbenaceae

*Clerodendrum myricoides* (Hochst.) root decoction is used treatment of malaria, chest pain, indigestion, sore throat, tonsillitis, gonorrhea, rheumatism, amoebic dysentery, venereal diseases and as a purgative (Muthaura *et al.*, 2007a; Orwa *et al.*, 2007; Bentje, 1994; Kokwaro, 1993). Various extracts of this plant have been reported to have good anti-plasmodial activity with IC50 as low as 3.96 for the methanolic root bark extracts (Muthaura *et al.*, 2007b; Muregi *et al.*, 2004).

Extracts of pounded leaves of *Clerodendrum eriophyllum* Guerke is used for treating malaria and root decoction drunk for intestinal disorders (Kokwaro, 1993).

The leaves of *Lantana camara* L. is chewed for the treatment of toothache and the ash of the burnt leaves together with a little salt acts as a good remedy for coughs, sore throat and conjunctivitis. Leaves are inhaled for the treatment of headache and colds (Kokwaro, 1993). Also used to treat malaria (Kuria *et al.*, 2001).

# Kenyan Medicinal Plants with Antimalarial Activity

Table 3.1 shows a list of medicinal plants used in Kenyan fork medicine that have been reported to have anti-malarial activity. It is necessary to carry out detailed biological and phytochemical studies to identify the active constituents in these plants and used to develop cheap anti-malarial drugs.

The crude extracts of *V. lasiopus* exhibited good antiplasmodial activity comparable to those of Cinchona (0.5 mg/ml). It would be interesting to investigate this plant for novel antiplasmodial compounds.

The lack of antiplasmodial activity in these plants may not necessarily imply the same *in vivo* since compounds may either act as prodrugs (which must undergo metabolic changes to achieve the required activity), febrifuges (fever is one of the symptoms associated with uncomplicated severe *P. falciparum* malaria) or immuno-modulators. Besides the presence of bioactive compounds depends on many factors such as the season, age, intra-species variation, part collected, soil and climate. Therefore, lack of *in vitro* activity in this case does not disqualify the use of these

**Table 3.1**: List of Kenyan medicinal plants reported to have anti-malarial activity.

| Family | Species | Part(s) Used | Reference |
|---|---|---|---|
| Acanthaceae | *Justicia betonica* L. | Aerial part | Muregi *et al.*, 2003 |
| Aloeaceae | *Aloe kedongensis*Reynolds | Leaves, roots | Jeruto *et al.*, 2008 |
| Amaranthaceae | *Cyathula schimperiana* Igifashi | Roots | Gathirwa *et al.*, 2007 |
| | *Cyathula cylindrica* | Root bark | |
| | *Celosia schweinfurthiana* Schinz. | | |
| Anacardiaceae | *Rhus natalensis* Bernh. | Stem bark | Gathirwa *et al.*, 2007 |
| | *Lennea schweinfurthii* | Stem bark | Gathirwa *et al.*, 2008 |
| | *Sclerocarya birrea* | Stem bark | Gathirwa *et al.*, 2008 |
| Annonaceae | *Uvaria scheffleri* Diels | Root bark | |
| Apocynaceae | *Carissa edulis* (Forssk.) Vahl. | | |
| | *Catharanthus roseus* (L.) G. Don | Whole plant | |
| Asclepiadaceae | *Centella asiatica* (L.) Urban | Whole plant | Muthaura *et al.*, 2007a; Kokwaro, 1993 |
| | *Curroria volubilis* (Schlecht.) Bullock | Bark | |
| Asteraceae | *Artemisia afra* Jacq. | | |
| Bignoniaceae | *Markhamia lutea* (Benth.) K. Schum. | | Orwa *et al.*, 2007 |
| Caesalpinaceae | *Senna didymobotrya* (Fresen.) Irwin and Barneby | Roots, leaves | Jeruto *et al.*, 2008; Beentje, 1994 |
| | *Caesalpinia volkensii* Harms | Leaves | Kuria *et al.*, 2001; Gachathi, 1993 |
| | *Tamarindus indica* L. | | Orwa *et al.*, 2007; Kokwaro, 1993 |
| | *Senna occidentalis* (L.) Link | | Orwa *et al.*, 2007 |
| Canellaceae | *Warburgia stuhlmannii* Engl. | Stem bark, Leaves | |
| | *Warburgia ugandensis* Sprague | Stem bark, Leaves | Kuria *et al.*, 2001; Kokwaro, 1993 |
| Capparaceae | *Boscia angustifolia* A. Rich. | Bark | Kokwaro, 1993 |
| | *Boscia salicifolia* Oliv | | Gathirwa *et al.*, 2007; Beentje 1994 |
| | *Cleome gynandra* L. | Leaves, Roots | Jeruto *et al.*, 2008 |
| Celastraceae | *Maytenus senegalensis* (Lam.) Exell | Root bark | Orwa *et al.*, 2007; Muthaura *et al.*, 2007a; Kokwaro, 1993 |
| | *Maytenus putterlickioides* (Loes.) Exell and Mendonca | Root | Muthaura *et al.*, 2007a; Kokwaro, 1993 |
| | *Maytenus undata* (Thunb.) Blakelock | Root | Muthaura *et al.*, 2007a; Kokwaro, 1993 |
| | *Maytenus heterophylla* (Eckl. and Zeyh.) Robson | Roots | Kokwaro, 1993 |
| | *Maytenus acuminata* (L.f.) Loes | | |

*Contd...*

**Table 3.1**–*Contd...*

| Family | Species | Part(s) Used | Reference |
|---|---|---|---|
| Chrysobalanaceae | *Parinari curatellifolia* Benth. | Aerial parts | Muregi *et al.*, 2004 |
| Combretaceae | *Terminalia spinosa* Engl. | Stem bark | Muthaura *et al.*, 2007a; Beentje, 1994 |
| Commelinaceae Kokwaro, 1993 | *Commelina forskalaei* Vahl. | | Orwa *et al.*, 2007; |
| Compositae | *Vernonia lasiopus* O.Hoffm. | | Beentje, 1994; Kokwaro, 1993 |
| | *Vernonia brachycalyx* O.Hoffm. | leaves | Beentje, 1994; Kokwaro, 1993 |
| | *Sphaeranthus suaveolens* (Forsk.) DC. | | Kokwaro, 1993 |
| | *Schkuhria pinnata* (Lam) O. Ktze | | Kokwaro, 1993 |
| | *Vernonia auriculifera* (Welw.) Hiern | | Kokwaro, 1993; Gathathi, 1989 |
| | *Artemisia afra* Willd. | | Kuria *et al.*, 2001; Kokwaro, 1993 |
| | *Aspilia pluriseta* Schweinf | | |
| | *Microglossa pyrifolia* (Lam.) O.Kuntze | Leaves | Kokwaro, 1993 |
| | *Bidens pilosa* L. | | Kuria *et al.*, 2001; Kokwaro, 1993 |
| | *Tridax procumbens* L. | Leaves, Whole plant | Kokwaro, 1993; Muthaura *et al.*, 2007a |
| Cucurbitaceae | *Cucumis figarei* Naud. | | |
| Enenaceae | *Euclea divinorum* Hiern. | | Orwa *et al.*, 2007 |
| Euphorbiaceae | *Flueggea virosa* (Willd.) Voigt | Roots | Muthaura *et al.*, 2007a; Beentje, 1994 |
| | *Euphorbia inaequilatera* Sond. | | |
| | *Clutia abyssinica* Jaub and Spach | Leaves, Roots | Kokwaro, 1993 |
| | *Clutia robusta* Pax, E, FZ. | | Gathirwa *et al.*, 2007 |
| | *Suregada zanzibarensis* Baill | Root bark, Leaves | Muthaura *et al.*, 2007a; Kokwaro, 1993 |
| | *Neoboutonia macrocalyx* | Stem bark | |
| Flacourtiaceae | *Trimeria grandifolia* (Hochst) Warb | | Orwa *et al.*, 2007 |
| Guttiferae | *Harungana madagascariensis* Poir | Root bark, stem bark | Muthaura *et al.*, 2007a; Kokwaro, 1993 |
| Labiatae | *Ajuga remota* Benth. | Leaves | Kuria *et al.*, 2001; Gachathi, 1993; Kokwaro, 1993 |
| | *Leonotis mollissima* Gürke. | | Kokwaro, 1993 |
| | *Leucas calostachys* Oliv. | Leaves | Kokwaro, 1993 |

*Contd...*

**Table 3.1**–*Contd...*

| Family | Species | Part(s) Used | Reference |
|---|---|---|---|
| | *Leonotis nepetifolia* R. Br. | | Orwa *et al.*, 2007; Beentje, 1994; Kokwaro, 1993 |
| Lauraceae | *Ocotea usambarensis* Engl. | Roots | Kokwaro, 1993 |
| Leguminosae | *Cassia abbreviata* Oliv. | Roots | Muthaura *et al.*, 2007a; Kokwaro, 1993 |
| | *Acacia hockii* De Wild. | Root bark | Muregi *et al.*, 2004 |
| | *Acacia nilotica* (L.) Del. | Root bark | Muthaura *et al.*, 2007a; Kokwaro, 1993 |
| | *Albizia anthelmintica* Brongn | Bark, Roots | Muthaura *et al.*, 2007a; Kokwaro, 1993 |
| | *Cassia occidentalis* L. | | Kuria *et al.*, 2001; Kokwaro, 1993 |
| Lamiaceae | *Fuerstia africana* T.C.E. Fries | Leaves | Gachathi, 1989; Kokwaro, 1993 |
| Loganiaceae | *Strychnos henningsii* Gilg | | Kuria *et al.*, 2001; Bentje, 1994; Kokwaro, 1993 |
| Meliaceae | *Azadirachta indica* A. Juss | | Muthaura *et al.*, 2007a; Kuria *et al.*, 2001 |
| | *Melia azedarach* Linn. | | Orwa *et al.*, 2007 |
| | *Ekebergia capensis* Sparrm | | Muregi, 2003, 2006, 2004 |
| | *Turraea robusta* | Root bark | Gathirwa *et al.*, 2008 |
| Menispermaceae | *Stephania abyssinica* (Dill. and A. Rich.) Walp. | Roots | Muregi *et al.*, 2004 |
| | *Cissampelos mucronata* A. Rich. | Roots | Muthaura *et al.*, 2007a; Kokwaro, 1993 |
| Mimosoideae | *Albizia coriaria* Oliv. | | Orwa *et al.*, 2007; Kokwaro, 1993 |
| Moraceae | *Ficus sur* Forssk. | Stem bark | Muregi *et al.*, 2003 |
| Olacaceae | *Ximenia americana* L. | | Orwa *et al.*, 2007; Gathirwa *et al.*, 2007 |
| Onagraceae | *Ludwigia erecta* (L.) Hara | Whole plant | Kokwaro, 1993 |
| Pittosporaceae | *Pittosporum viridiflorum* Sims | Bark | Bentje, 1994; Kokwaro, 1993 |
| Rhamnaceae | *Rhamnus prinoides* L'Hé´rit. | Leaves | Kuria *et al.*, 2001; Bentje, 1994; Kokwaro, 1993 |
| | *Rhamnus staddo* A.Rich. | Roots, Stem | Kuria *et al.*, 2001; Kokwaro, 1993 |

*Contd...*

**Table 3.1**–*Contd...*

| Family | Species | Part(s) Used | Reference |
|---|---|---|---|
| Rubiaceae | *Pentas bussei* Krause | Roots | Muthaura *et al.*, 2007a; Bentje, 1994; Kokwaro, 1993 |
| | *Pentas agathisanthemum* Kl. | Whole plant | Muthaura *et al.*, 2007a |
| | *Pentas longiflora* Oliv. | Roots | Muthaura *et al.*, 2007a; Kokwaro, 1993 |
| | *Spermacoce princeae* (K.Schum.) Verdc. | Whole plant | Muregi *et al.*, 2003 |
| | *Vangueria acutiloba* Robyns | Stem bark | Kokwaro, 1993 |
| | *Vangueria volkensii* K. Schum | | Orwa *et al.*, 2007 |
| Rutaceae | *Toddalia asiatica* | Roots | Muthaura *et al.*, 2007a; Orwa *et al.*, 2007; Kuria *et al.*, 2001; Kokwaro, 1993 |
| | *Zanthoxylum chalybeum* Engl. | Roots, Stem bark | Muthaura *et al.*, 2007a; Beentje, 1994 |
| | *Zanthoxylum usambarense* (Engl.) | Stem bark, Roots | Kuria *et al.*, 2001; Kokwaro, 1993 |
| | *Fagaropsis angolensis* (Engl.) Dale | | Khalid and Waterman, 1985 |
| | *Teclea nobilis* Del. | Stem bark | Kuria *et al.*, 2001; Kokwaro, 1993 |
| Simaroubaceae | *Harrisonia abyssinica* Oliv. | Roots | Muthaura *et al.*, 2007a; Orwa *et al.*, 2007; Beentje, 1994; Kokwaro, 1993 |
| Solanaceae | *Withania somnifera* (L.) Dunal | Roots | Muregi *et al.*, 2007 |
| Tiliaceae | *Grewia bicolor* Juss. | | Orwa *et al.*, 2007 |
| Urticaceae | *Urtica massaica* Mildbr. | | Muregi *et al.*, 2003 |
| Verbenaceae | *Clerodendrum myricoides* (Hochst.) | Roots | Muthaura *et al.*, 2007a; Orwa *et al.*, 2007; Bentje, 1994; Kokwaro, 1993 |
| | *Clerodendrum eriophyllum* Guerke | Leaves | Kokwaro, 1993 |
| | *Lantana camara* L. | | Kuria *et al.*, 2001 |

plants as traditional antimalarials. While plant extracts may not display *in vitro* activity they may display *in vivo* activity (Gessler *et al.*, 1995) or vice versa. It is, therefore, necessary to undertake *in vivo* investigation of these plants before any conclusion on their efficacy as antimalarials could be drawn.

# Conclusion

Many communities around the world use these plants as traditional anti-malarials. However, activity depends on many factors such as the season in which the plant is collected, the age of the plants, intraspecies variation, part collected and the environmental conditions among others. Therefore, lack of *in vitro* anti-plasmodial activity in this case does not disqualify the use of these plants as traditional anti-malarial. After detailed *in vivo* antimalarial evaluation and thorough toxicological studies, some of these plants may be recommended as antimalarials in known dosages especially in rural communities where the conventional drugs are unaffordable or unavailable and the health facilities inaccessible.

# References

Agnew, A.D.Q., and Agnew, S. (1994). A flora of the ferns and herbaceous flowering plantsof upland Kenya. East Africa natural history society.

Agomo, P.U., Idigo, J.C., and Afolabi, B.M. (1992) "Antimalarial" medicinal plants and their impact on cell populations in various organs of mice. *Afri. J. Med. Med. Sci.*, **21** (2): 39-46.

Beentje, H.J. (1994). Kenya trees, shrubs and lianas. National Museums of Kenya.

Benoit, F., Valentin, A., Pelissier, Y., *et al.* (1996). *In vitro* antimalarial activity of vegetal extract used in West African traditional medicine. *Amer. J. Trop. Med. Hyg.*, **54**: 67 – 71.

Delhaes, L., Lazaro, J.E., Gay, F., Thellier, M., and Danis, M. (1999). The microculture tetrazolium assay (MTA): another colorimetric method of testing *Plasmodium falciparum* chemosensitivity. *Annals of Tropical Medicine and Parasitology*, **93(1)**: 31-40.

Desjardins, R.E. Canfield, R.E., Hayness, C.J., and Chuby, J.D. (1979). Quantitative assessment of antimalarial activity *in vitro* by automated dilution technique. *Antimicrobial agent chemotherapy*, **16**: 710-718.

El-Tahir, A., Satti, G.M.H., and Khalid, S.A. (2001). A novel anti-plasmodial activity of pristmerin isolated from *Maytenus senegalensis* (Lam.) Excell. *Journal of Saudi chemical society*, **5 (2):** 157-163.

Gachathi, F.N. (1989). Kikuyu Botanical Dictionary of Plant Names and Uses. FN, Gachathi, Nairobi.

Gachathi, F.N. (1993). Kikuyu Botanical Dictionary of Plant Names and Uses. AMREF, Nairobi, Kenya.

Gakunju, D. M. N., Mberu, E. K., Dossaji, S. F., Gray, A. I., Waigh, R. D., Waterman, P. G., and Watkins, W. M. (1995). Potent antimalarial activity of the alkaloid nitidine, isolated from a Kenyan herbal remedy. *Antimicrobial Agents and Chemotherapy*, **39**: 2606–2609.

Gathirwa, J.W., Rukunga, G.M., Njagi, E.N. M., Omar, S.A., Guantai, A.N., Muthaura Charles, N., Mwitari Peter, G., Kimani, C.W., Kirira, P.G., Tolo, F.M., Ndunda, T. N., and Ndiege, I.O. (2008). The *in vitro* and *in vivo* anti-malarial efficacy of combinations of some medicinal plants used traditionally for treatment of malaria by the Meru community in Kenya. *Journal of Ethnopharmacology*, **115**: 223–231.

Gathirwa, J.W., Rukunga, G.M., Njagi, E.N. M., Omar, S.A., Guantai, A.N., Muthaura Charles, N., Mwitari Peter, G., Kimani, C.W., Kirira, P.G., Tolo, F.M., Ndunda, T. N., and Ndiege, I.O. (2007). *In vitro* anti-plasmodial and *in vivo* anti-malarial activity of some plants traditionally used for the treatment of malaria by the Meru community in Kenya. *J Nat Med.*, **61**: 261–268.

Gessler, M.C., Nkunya, M.H., Mwasumbi, L.B., Heinrich, M., and Tanner, M. (1994). Screening Tanzanian medicinal plants for antimalarial activity. *Acta Trop.*, **56(1)**: 65-77.

Gessler, M.C., Tanner, M., Chollet, J., Nkunya, M.H.H., and Heinrich, M. (1995). Tanzanian medicinal plants used traditionally for the treatment of malaria: *in vivo* antimalarial and *in vitro* cytotoxicity activities. *Phytotherapy research*, **9**: 504–508.

Isah, A.B., Ibrahim, Y.K., and Iwalewa, E.O. (2003). Evaluation of the antimalarial properties and standardization of tablets of *Azadirachta indica* (Meliaceae) in mice. *Phytotherapy Research*, **17(7)**: 807-10.

Jeruto, P., Lukhoba, C., Ouma, G., Otieno, D., and Mutai, C. (2008). An ethnobotanical study of medicinal plants used by the Nandi people in Kenya. *Journal of Ethnopharmacology*, **116**: 370–376,

Kato, A., Moriyasu, M., Ichimaru, M., and Nishiyama, Y. (1996). Isolation of alkaloidal constituents of *Zanthoxylum usambarense* and *Zanthoxylum chalybeum* using ion-pair HPLC. *Journal of Natural Products*, **59**: 316–318.

Khalid, S.A., and Waterman, P.G. (1985). 6-Hydroxymethyldihydronitidine from *Fagaropsis angolensis. Journal of Natural Products*, **48**: 118–119.

Kirira, P.G., Rukunga, G.M., Wanyonyi, A.W., Muregi, F.M., Gathirwa, J.W., Muthaura, C.N., Omar, S.A., Tolo, F., Mungai, G.M., and Ndiege, I.O. (2006). Anti-plasmodial activity and toxicity of extracts of plants used in traditional malaria therapy in Meru and Kilifi Districts of Kenya. *J Ethnopharmacology*, **106(3)**: 403-407.

Koch, A., Orjala, J., Mutiso, P.C., and Soejarto, D.D. (2006). An antimalarial abietane diterpene from *Fuerstia africana* T.C.E Fries. *Biochemical Systems and Ecology*, **34**: 270-272.

Koch, A., Tamez, P., Pezzuto, J., and Soejarto, D. (2005). Evaluation of plants used for antimalarial treatment by Maasai of Kenya. *Journal of Ethnophamacology*, **101**: 95-99.

Kokwaro, J.O. (1993). Medicinal plants of East Africa, 2$^{nd}$ edition. East African Literature Bureau

Kraft, C.K., Jenett-Siems, K., Siems, K., *et al.* (2003). Herbal remedies traditionally used against malaria. *In vitro* antiplasmodial evaluation of medicinal plants from Zimbabwe. *Phytother Res.*, **17**: 123–128.

Kuria, K.A.M., Chepkwony, H., Govaerts, C., *et al.* (2002). The antiplasmodial activity of isolates from *Ajuga remota*. *J Nat. Prod.*, **65**: 789–793.

Kuria, K. A. M., De Coster, S., Muriuki, G., Masengo, W., Kibwage, I., Hoogmartens, J., and Laekeman, G. M. (2001). Antimalarial activity of *Ajuga remota* Benth (Labiatae) and *Caesalpinia volkensii* Harms (Caesalpiniaceae): *in vitro* confirmation of ethnopharmacological use. *Journal of Ethnopharmacology*, **74:** 141-148.

Le Bras, J., and Deloron, P. (1983). *In vitro* study of drug sensitivity of *Plasmodium falciparum*: an evaluation of a new semi-micro-test. *American Journal of Tropical Medicine and Hygiene*, **32:** 447-51.

Lindsay, R.S., and F.N. Hepper. (1978). Medicinal plants of Marakwet, Kenya. Kew, Royal Botanic Gardens, United Kingdom, 49 p.

Muregi, F. W., Chhabra, S. C., Njagi, E. N. M., Lang'at-Thoruwa, C. C., . Njue, W. M,. Orago, A. S. S,. Omar, S. A., and Ndiege, I. O. (2004). Anti-plasmodial activity of some Kenyan medicinal plant extracts singly and in combination with chloroquine. *Phytotherapy Research*, **18(5)**: 379–384.

Muregi, F.W., Chhabra, S.C., Njagi, E.N., Lang'at-Thoruwa, C.C., Njue, W.M., Orago, A.S., Omar, S.A., and Ndiege, I.O. (2003). *In vitro* antiplasmodial activity of some plants used in Kisii, Kenya against malaria and their chloroquine potentiation effects. *J Ethnopharmacol.*, **84(2-3):** 235-239.

Muregi, F.W., Ishih, A., Miyase, T., Suzuki, T., Kino, H., Amano, T., Mkoji, G.M. and Terada, M. (2007). Antimalarial activity of methanolic extracts from plants used in Kenyan ethnomedicine and their interactions with chloroquine (CQ) against a CQ-tolerant rodent parasite, in mice. *J Ethnopharmacol.*, 111(1): 190-195.

Muregi, F.W., Ishih, A., Miyase, T., Suzuki, T., Kino, H., Amano, T., Mkoji, G.M., Miyase, T., and Terada, M. (2007). *In vivo* antimalarial activity of aqueous extracts from Kenyan medicinal plants and their chloroquine (CQ) potentiation effects against a blood-induced CQ-resistant rodent parasite in mice. *Phytother. Res.*, **21**: 337–343.

Muthaura, C. N., Rukunga, G. M., Chhabra, S. C., Mungai, G. M., and Njagi, E. N. M. 2007(a). Traditional antimalarial phytotherapy remedies used by the Kwale community of the Kenyan Coast. *Journal of Ethnopharmacology*, **114**: 377–386.

Muthaura, C. N., Rukunga, G. M., Chhabra, S. C., Omar, S. A., Guantai, A. N., Gathirwa, J. W., Tolo, F. M., Mwitari, P. G., Keter, L. K., Kirira, P. G., Kimani, C. W., Mungai, G. M., and Njagi, E. N. M. (2007(b). Antimalarial activity of some plants traditionally used in Meru district of Kenya. *Phytotherapy Research*, **21**: 860–867.

Obih, P.O., and Makinde, J.M. (1985). Effect of *Azadirachta indica* on *Plasmodium berghei berghei* in mice. *Afri. J. Med. Med. Sci.*, **14(1-2):** 51-54.

Oketch-Rabah, H.A., Brogger Christensen, S., Frydenvang, K., Dossaji, S.F., Theander, T.G., Cornett, C., Watkins, W.M., Kharazmi, A., and Lemmich, E. (1998).

Antiprotozoal properties of 16,17-dihydrobrachycalyxolide from *Vernonia brachycalyx*. *Planta Med.*, **64(6):** 559-62.

Oketch-Rabah, H.A., Lemmich, E., Dossaji, S.F., Theander, T.G., Olsen, C.E., Cornett, C., Kharazmi, A., and Christensen, S.B. (1997). Two new antiprotozoal 5-methylcoumarins from *Vernonia brachycalyx*. *J Nat Prod.*, **60(5):** 458-461.

Orwa, J.A., Mwitari, P.G., Matu, E.N., and Rukunga, G.M. (2007). Traditional healers and the management of malaria in Kisumu District, Kenya. *East African Medical Journal*, **84 (2):** 51-55.

Peters, W. (1982). Anti-malarial drug resistance. *British medical Bulletin*, **32**: 187-192.

Schwikkard, S., and Van Heerden, F.R. (2006). Antimalrial activity of plant metabolites. *Nat. Prod. Rep.*, **19**: 675-692.

Tits, M., Damas, J., Quetin-Leclercq, J., Angenot, L. (1991). From ethnobotanical uses of *Strychnos henningsii* to antiinflammatories, analgesics and antispasmodics. *J Ethnopharmacol.*, **34(2-3):** 261-267.

Utilisation and Management of Medicinal Plants Vol. 2 (2014)  *Pages* 115–152
*Editor-in-Chief:* V.K. Gupta
*Published by:* DAYA PUBLISHING HOUSE, NEW DELHI

# 4

# Histochemical Studies on Selected Medicinally Important Primitive Ferns of Western Ghats, South India

V. Irudayaraj[1], M. Johnson[1]*, A. Kala[1], I. Revathy[1],
N. Janakiraman[1] and A. Sivaraman[1]

## ABSTRACT

*In the present study, anatomical and histochemical studies have been carried out on the following 26 ferns and fern allies belonging to 23 genera under 15 families. The matured stipes of different species was collected from Upper Kothayar, Western Ghats, South India. They were used for anatomical and histochemical studies by taking free hand sections. For anatomical studies, the fresh free hand sections were stained using safranin. Histochemical tests were made on the fresh sections of the stipes treated with the following reagents to identify the presence or absence of chemicals like lipids, polyphenols, lignin and tannins. The anatomical characters such as shape, presence or absence of hairs, scales, groove and wing on the stipe, number of hypodermal layers, nature of ground tissue, number and shape of vascular bundles etc. have been studied for all the 26 species. Presence of lipids has been observed in almost all the parts of the stipe in all the species studied. Polyphenols are mildly present in the xylem and phloem of 12 species and in ground tissue and endodermis of 18 species. High concentration of polyphenols has been observed in the epidermis of 11 species and hypodermis of 12 species. Lignin is present at high concentration in the epidermis of 15 species and hypodermis of 16 species. It is averagely present in the ground tissue of 11 species. Tannin is present in generally in lesser amount in lesser number of species. It is present in high concentration only in the epidermis of 4 species and hypodermis of 5 species. It is present*

---

1   1   Centre for Plant Biotechnology, Department of Plant Biology and Plant Biotechnology, St. Xavier's College (Autonomous), Palayamkottai – 627 002, Tamil Nadu, India.

*   *Corresponding author*: E-mail: ptcjohnson@gmail.com

mildly in the epidermis of 7 species and ground tissue of 7 species. It is absent in the endodermis, xylem and phloem of all the 26 species. The present histochemical studies on lipids, polyphenols, lignins and tannins show the evolutionary progress of such chemicals from the primitive ferns towards the advance ferns.

**Keywords**: Anatomy, Primitive, Polyphenols, Epidermis, Evolution.

## Introduction

The most numerous Pteridophyta are ferns, with 12,000 living species, followed by *Selaginella* with about 600. At the other extreme are the Psilotales and Equisetales, each with about 20 species. The ferns have probably attracted attention longer and more extensively than other Pteridophyta. They were clearly recognized by the early Greek naturalists as unusual land plants, quite different from other herbaceous forms (Hort, 1916). Pteridophyta survive in the field as parts of much wider communities. Throughout their life-cycles they are undoubtedly influenced by the physical and biotic factors which arise in those communities, whilst these factors may themselves be modified as a result of the species present. In order to escape from the destruction by microbes, herbivorous insects and grazing animals, several metabolic pathways have evolved in this group of plants to synthesize several defense chemicals like tannin, phenols and phyto-ecydysteroids. The knowledge about the occurrence and distribution of such chemicals will throw much information about the adaptive potential of this successful group of plants. Moreover, several species of pteridophytes are ethnomedicinally important and several species have already been proved of having higher degree of antimicrobial activity.

Identification and localization of the active compounds will be useful in the field of pharmacy. Today pharmaceutical industries are deeply interested in large variety of chemical substances. Though these substances are generally extracted from plant parts, the plant tissue culture technique has widened the scope and opened new vistas for the production of secondary metabolites. To achieve this goal, it is essential to identify the part and tissue of the plant with high amount of the required chemicals. This can be successfully done by various simple histochemical tests. In the meantime morpho-anatomical, phytochemical and histochemical criteria are also very useful in the field of pharmacognosy of medicinal plants. Several ferns such as *Asplenium* sp. *Adiantum* sp. and *Polypodiaceous* sp. are medicinally important. Taxonomy is a science without data of its own, utilizing a wide range of information from diverse sources *viz.*, chemical, genetical, anatomical, and cytological and other investigations, in order to obtain the best sort of natural classification. Any data which show differences from species to species are of taxonomic significance, and thus constitute part of the information or evidence which may be used by taxonomists. The modern approach has given more reliable or fundamental solutions for the taxonomical problems than old or classical ones. In this way cytological occurrence of secondary metabolites (alkaloids, flavonoids, phenols, saponins, tannins etc.), protein, amino acid sequences and DNA analysis and so on have all at various times.

The anatomical and histochemical observations are used for clear determination of the species level identification or classification. There has been a remarkable revolution in the past 50 years in the investigation of vascular plant anatomy and its use in classification (Srivastava, 2008). It is now generally realized that anatomical and histochemical characters are just as valuable as morphological ones, and must not be neglected. By considering the above points, the present study was aimed to make anatomical and histochemical studies on some ferns and fern allies commonly available on upper Kothayar which is the nearest hilly area with rich of ferns and fern allies.

## Materials and Methods

Stipes of different species *viz.*, *Lycopodiella cernua* (L.) Pic. Ser. (Lycopodiaceae), *Selaginella involvens* (Sw.) Spring. *Selaginella inaequalifolia* (Hook. *et* Grev.) Spring and *Selaginella tenera* (Hook. *et* Grev.) Spring belonging to the family Selaginellaceae, *Angiopteris evecta* (Forst.) Hoffm. (Angiopteridaceae), *Marattia fraxinea* Sm. (Marattiaceae), *Lygodium micropyllum* (Cav.) R. Br. (Schizaeaceae), *Pteris argyraea* T. Moore and *Pteris confusa* T. G. Walker belonging to the family Pteridaceae, *Doryopteris concolor* (Lagsd. *et* Fisch.) Kuhn, *Cheilanthes viridis* (Forssk.) Swartz, and *Pellaea boivini* Hook. belonging to the family Sinopteridaceae, *Hemionitis arifolia* (Burm.) Moore and *Pityrogramma calomelanos* (L.) Link var. *calomelanos* belonging to the family Hemionitidaceae, *Adiantum raddianum* Presl (Adiantaceae), *Pteridium aquilinum* (L.) Kuhn v. Deck, *Histiopteris incisa* (Thunb.) J. Sm., *Hypolepis glandulifera* Brownsey *et* Chinnock and *Microlepia speluncae* (L.) Moore belonging to the family Dennstaedtiaceae, *Lindsaea ensifolia* Sw. and *Odontosoria chinensis* (L.) J. Sm. belonging to the family Lindsaeaceae, *Araiostegia hymenophylloides* (Bl.) Copel. (Davalliaceae), *Nephrolepis multiflora* (Roxb.) Jarret (Oleandraceae), *Trichomanes obscurum* Bl (Hymenophyllaceae), *Dicranopteris linearis* (Burm.f.) Underwood var. *sebastiana* (Gleicheniaceae) and *Cyathea nilgirensis* Holttum (Cyatheaceae) were collected from Upper Kothayar, Tirunelveli hills, Western Ghats, South India. The selected specimens were identified based on "Pteridophyte Flora of the Western Ghats, South India" (Manickam and Irudayaraj, 1992).

The fresh materials were used for anatomical and histochemical studies by taking free hand sections. For anatomical studies, the fresh free hand sections were stained using safranin. Histochemical tests were made on the fresh sections of the stipes treated with the following reagents to identify the presence or absence of chemicals like lipids, polyphenols, lignin and tannins. Sudan III is used to detect lipids (Ruthmann, 1970), Fast Blue BB salts for polyphenols (Gahan, 1927), Lugol's Iodine is used to detect lignin and tannin (Chamberlain, 1924; Haridass and Suresh Kumar, 1985). The stained sections were observed on Motic trinocular microscope (Japan). They were photographed at different magnifications and at different views. Based on the photographs taken, anatomical description and localization of tested chemicals were done.

## Results and Discussion

In the present study, anatomical and histochemical studies have been carried out on the following 26 ferns and fern allies belonging to 23 genera under 15 families.

The anatomical features and the results of histochemical analysis on the stipe of each and every species have been given in the Tables 4.1–4.3.

## Anatomy

The anatomical characters such as shape, presence or absence of hairs, scales, groove and wing on the stipe, number of hypodermal layers, nature of ground tissue, number and shape of vascular bundles etc. have been studied for all the 26 species. A comparative account on stipe anatomy of 26 species is given below.

## Shape

In transverse section, the outline of the stipe is circular or semicircular in almost all the species except in *L. ensifolia* in which the outline is tetragonal with slightly widened adaxial side [Plate 4.5(1a), Plates 4.9(1a), 4.12(1a) and 4.14(1a)]. Generally the adaxial side of the stipe is either flattened or grooved except in few cases like *L. cernua*, *Selaginella* spp., *D. concolor*, *P. boivini* and *T. obscurum*. The number and depth of groove may vary from position to position within a stipe. For example, in *N. multiflora*, the adaxial surface is either with single shallow groove [Plate 4.12(4a)] or with two shallow grooves with a raised middle ridge [Plate 4.17(a)].

## Epidermis

In all the species studied, the epidermis is single layered with thick or thin cuticle on the outer side. Usually it is smooth without bearing any epidermal appendages. Rarely the epidermis bears unicellular or multicellular hairs or rarely scales. It also depends upon the position of the stipe. Usually the appendages are commonly seen towards the base of the stipe since the rhizome in all the species bears scales or hairs. The epidermal cells are generally rectangular or barrel-shaped and colourless. The epidermis is without any stomata. Generally in land plants, the epidermis, particularly the cuticle plays an important role in the protection of the plants against pathogens. Usually the cuticle is made up of several waxy compounds which play an important role in defense mechanisms.

## Hypodermis

Sclerenchymatous hypodermis is present in all the 26 species studied and the number of layers varies from species to species. Usually it ranges from 4-8. In all the species, the sclerenchymatous hypodermis is present as a continuous ring next to the epidermis. The stipe of *L. micropyllum* is exceptional in having fully sclerenchymatous layers between the epidermis and endodermis without differentiated into hypodermis and ground tissue [Plate 4.2(3a,b) and Plate 4.6(7a,b)]. It is also to be noted that the hypodermis in *L. cernua* is chlorenchymatous followed by few layers of sclerenchyma [Plate 4.19(1a,c)].

The sclerenchymatous hypodermis gives mechanical support to the stipe. Depends upon the number of hypodermal layers the rigidity of the stipe varies. Thus the stem of *S. inaequalifolia* is soft with the presence of 2 or 3 sclerenchymatous layers [Plate 4.19(2a,b)]. This species is a long, horizontal trailer and so it does not require much of mechanical support. In contrast, *L. micropyllum* is a vertical climber. Thus, in order to stand erect, the stipe and rachis of species is with multilayered

**Table 4.1:** Results of histochemical analysis on ferns and fern allies.

| Name of the Species | Epidermis | | | | Hypodermis | | | | Ground Tissue | | | | Endodermis | | | | Xylem | | | | Phloem | | | |
|---|---|---|---|---|---|---|---|---|---|---|---|---|---|---|---|---|---|---|---|---|---|---|---|---|
| | Li | P | Lig | T | Li | P | Lig | T | Li | P | Lig | T | Li | P | Lig | T | Li | P | Lig | T | Li | P | Lig | T |
| S. inaequalifolia | 3+ | 3+ | 3+ | – | 2+ | 2+ | 3+ | – | 2+ | – | 3+ | + | 2+ | – | 3+ | – | – | – | 3+ | – | + | – | 3+ | – |
| S. involvens | – | 3+ | – | 3+ | 2+ | 2+ | – | 3+ | + | – | – | + | – | – | – | – | – | * | – | – | – | * | – | – |
| S. tenera | 3+ | 3+ | 3+ | 3+ | 2+ | 3+ | 3+ | 3+ | 2+ | – | 2+ | + | 2+ | – | – | – | 2+ | * | 3+ | – | + | * | 2+ | – |
| L. cernua | 2+ | + | + | + | 3+ | 2+ | + | + | + | + | + | – | – | 2+ | + | – | – | 2+ | + | – | – | 2+ | – | – |
| A. evecta | 2+ | * | 2+ | – | 2+ | 3+ | 2+ | – | 2+ | * | 2+ | – | – | * | 2+ | – | 2+ | * | 2+ | – | 2+ | * | 2+ | – |
| M. fraxinea | 2+ | 3+ | 2+ | 2+ | – | + | 3+ | 2+ | 2+ | * | 2+ | – | – | * | 2+ | – | + | + | + | – | + | + | + | – |
| L. microphyllum | 2+ | + | + | | – | + | + | – | – | * | 3+ | – | + | * | – | – | + | + | – | – | + | + | – | – |
| P. confusa | – | + | 2+ | | 2+ | 2+ | 2+ | – | – | + | 2+ | – | – | + | 2+ | – | – | + | 2+ | – | – | + | 2+ | – |
| P. argyraea | 3+ | – | – | 3+ | 3+ | 2+ | – | – | 3+ | + | – | – | – | + | 2+ | – | + | – | – | – | 3+ | – | – | – |
| C. viridis | 3+ | – | 3+ | 3+ | 3+ | – | 3+ | 3+ | 3+ | + | 3+ | + | 3+ | + | 3+ | – | 3+ | – | 3+ | – | – | – | 3+ | – |
| P. boivini | 3+ | – | 3+ | + | 3+ | – | 3+ | 3+ | 2+ | + | 3+ | + | – | + | 3+ | – | – | – | – | – | – | – | – | – |
| D. concolor | 3+ | – | 3+ | + | 3+ | + | 3+ | 2+ | 2+ | + | 2+ | + | + | + | 2+ | – | 3+ | – | – | – | 2+ | – | – | – |
| H. arifolia | 2+ | – | 3+ | + | – | * | 3+ | – | – | + | 3+ | – | + | + | 2+ | – | 3+ | – | 2+ | – | – | – | – | – |
| P. calomelanos | 3+ | + | 3+ | – | 3+ | + | 3+ | + | 3+ | + | 2+ | – | 3+ | + | * | – | – | – | 2+ | – | 3+ | – | 2+ | – |
| A. raddianum | 3+ | + | – | – | 2+ | + | – | – | + | + | – | – | – | + | – | – | – | * | – | – | 3+ | * | – | – |
| H. incisa | 3+ | + | – | – | 3+ | – | – | – | 3+ | + | – | – | – | + | 2+ | – | – | + | 3+ | – | – | + | 3+ | – |
| P. aquilinum | 3+ | + | 2+ | – | 2+ | 3+ | 2+ | – | 2+ | + | 2+ | – | 2+ | + | 3+ | – | 2+ | + | 2+ | – | 2+ | + | 2+ | – |
| M. speluncae | 3+ | 3+ | – | – | 3+ | 3+ | – | – | 2+ | + | 2+ | – | 3+ | + | 2+ | – | 2+ | + | 2+ | – | – | + | 2+ | – |
| H. glandulifera | 3+ | 3+ | 3+ | – | 3+ | 3+ | 3+ | – | 3+ | + | 3+ | – | 3+ | + | – | | + | + | 3+ | | 2+ | + | 3+ | – |
| L. ensifolia | 3+ | 3+ | 3+ | 2+ | 2+ | 3+ | 3+ | – | 3+ | + | – | – | – | + | 3+ | – | + | + | 3+ | – | – | + | – | – |
| O. chinensis | – | 3+ | 3+ | – | – | 3+ | 3+ | – | – | + | 3+ | – | – | + | 3+ | – | + | + | 3+ | – | – | + | 3+ | – |
| A. hymenophylloides | 3+ | 3+ | 3+ | – | 3+ | 3+ | 3+ | 3+ | 2+ | + | + | – | 2+ | + | 3+ | – | 2+ | + | 3+ | – | + | + | + | – |
| N. multiflora | 3+ | 3+ | 3+ | + | 2+ | 3+ | 3+ | – | 2+ | + | 2+ | – | – | + | 3+ | | – | ‡ | + | | + | ‡ | + | – |
| T. obscurum | 2+ | 3+ | 3+ | – | + | 3+ | 3+ | – | – | + | 2+ | – | – | + | 3+ | – | + | + | – | – | – | + | – | |
| D. linearis | 3+ | + | 3+ | + | + | 3+ | 3+ | – | – | + | 3+ | – | – | + | 3+ | – | + | + | * | – | – | + | * | – |
| C. nilgirensis | * | * | 3+ | + | * | 3+ | 3+ | 2+ | + | – | 2+ | – | * | – | – | – | * | * | 3+ | – | – | * | + | – |

+: Mild; 2+: Average; 3+: High; *: Data not available.

**Table 4.2**: Frequency distribution of different chemicals in different parts of the stipe.

| Chemicals | Different Parts in the Stipe | | | | | | | | | | | |
|---|---|---|---|---|---|---|---|---|---|---|---|---|
| | Epidermis | | Hypodermis | | Ground Tissue | | Endodermis | | Xylem | | Phloem | |
| Lipids | M | – | M | 2 | M | 4 | M | 3 | M | 4 | M | 5 |
| | A | 6 | A | 9 | A | 10 | A | 4 | A | 5 | A | 4 |
| | H | 14 | H | 8 | H | 5 | H | 4 | H | 3 | H | 2 |
| Polyphenol | M | 7 | M | 5 | M | 12 | M | 18 | M | 12 | M | 12 |
| | A | – | A | 5 | A | – | A | 1 | A | 2 | A | 2 |
| | H | 11 | H | 12 | H | – | H | – | H | – | H | – |
| Lignin | M | 2 | M | 2 | M | 2 | M | 1 | M | 3 | M | 3 |
| | A | 4 | A | 3 | A | 11 | A | 8 | A | 6 | A | 6 |
| | H | 15 | H | 16 | H | 7 | H | 7 | H | 7 | H | 4 |
| Tannin | M | 7 | M | 2 | M | 7 | M | – | M | – | M | – |
| | A | 2 | A | 3 | A | – | A | – | A | – | A | – |
| | H | 4 | H | 5 | H | – | H | – | H | – | H | – |

M: Mild; A: Average; H: High.

**Table 4.3**: Number of scores with high concentration in each species.

| Name of the Species | Number of Scores with High Concentration | Name of the Species | Number of Scores with High Concentration |
|---|---|---|---|
| S. inaequalifolia | 8 | S. involvens | 3 |
| S. tenera | 8 | L. cernua | 1 |
| A. evecta | 1 | M. fraxinea | 2 |
| L. microphyllum | 1 | P. confusa | – |
| P. argyraea | 4 | C. viridis | 13 |
| P. boivini | 7 | D. concolor | 5 |
| H. arifolia | 4 | P. calomelanos | 7 |
| A. raddianum | 2 | H. incisa | 5 |
| P. aquilinum | 3 | M. speluncae | 5 |
| H. glandulifera | 11 | | |
| L. ensifolia | 8 | O. chinensis | 8 |
| A. hymenophylloides | 8 | N. multiflora | 6 |
| T. obscurum | 5 | D. linearis | 5 |
| C. nilgirensis | 4 | | |

sclerenchymatous hypodermis without parenchymatous ground tissue [Plate 4.2(3a,b); Plate 4.6(7a,b)]. So it is very strong and highly flexible although the diameter of the stipe and rachis is below 1 mm in diameter. Due to this reason the stipe and

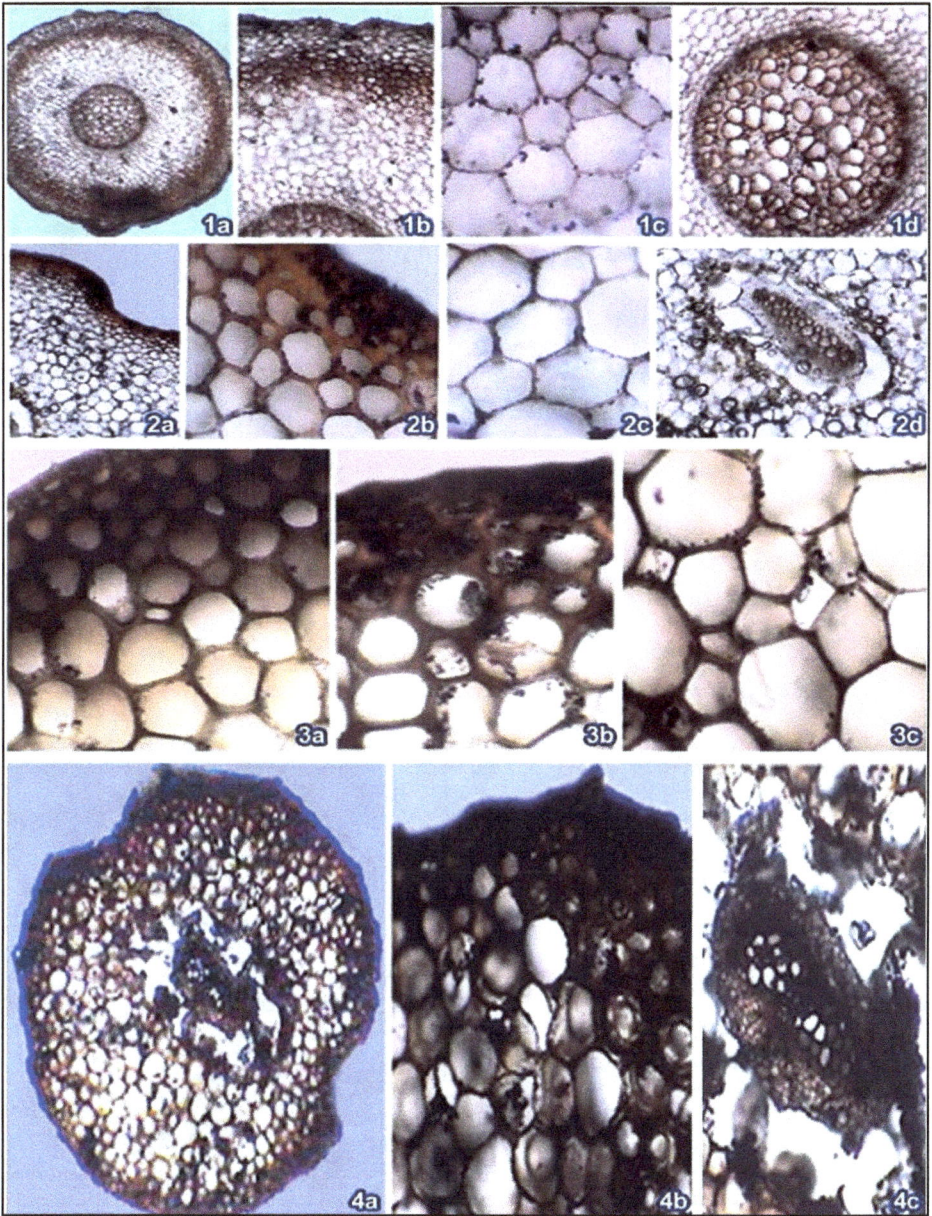

**Plate 4.1**: Results of histochemical analysis on lipids.

1: *Lycopodiella cernua* (L.) Pic. Ser.—1a: T.S. of stem–Entire view; 1b: T.S. of stem–Portion enlarged; 1c: Ground tissue; 1d: Vascular bundle; 2: *Selaginella inaequalifolia* (Hook. and Grev.) Spring—2a: Portion of Transverse Section of stem; 2b: Hypodermis; 2c: Ground tissue; 2d: Vascular bundle; 3: *Selaginella* species—3a-b: A portion of T.S. of stem showing epidermis, hypodermis and ground tissue; 3c: Ground tissue; 4: *Selaginella tenera* (Hook.& Grev.) Spring—4a: T.S. of stem. Entire view; 4b: T. S. of stem- A portion enlarged; 4c: Vascular bundle.

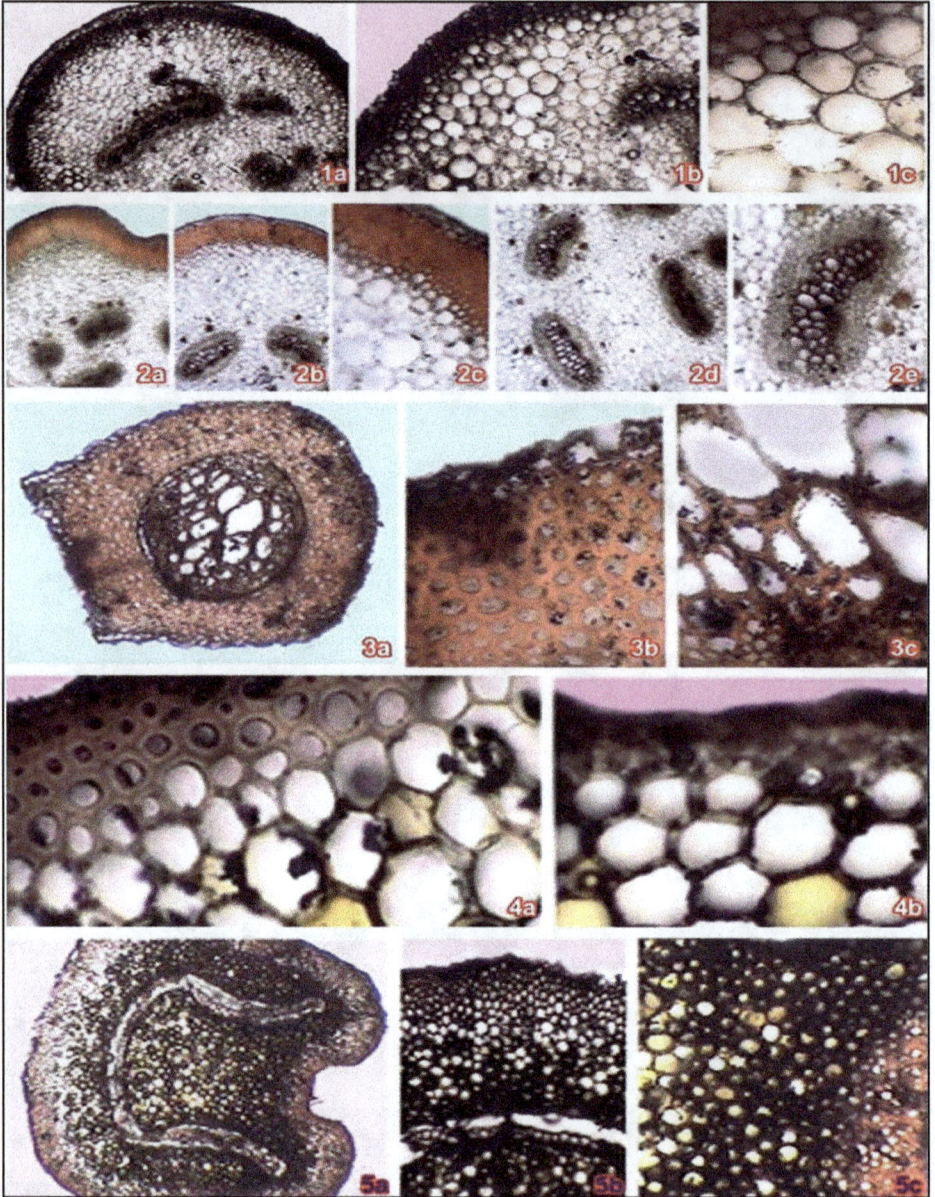

**Plate 4.2**: Results of histochemical analysis on lipids.

1: *Angiopteris evecta* (Forst.) Hoffm.—1a: T.S. of stipe – A portion; 1b: T.S. of stipe – A Portion enlarged; 1c: Ground tissue; 2: *Marattia fraxinea* Sm.—2a,b: T.S. of stipe – Different portions; 2c: T.S. of stipe-A Portion enlarged with hypodermis and ground tissue; 2d: Vascular bundles; 2e: A vascular bundle enlarged; 3: *Lygodium micropyllum* (Cav.) R.Br.— 3a: Transverse Section of stipe – Entire view; 3c: A portion enlarged showing hypodermis; 3c: Xylem enlarged; 4: *Pteris argyraea* T. Moore—4a,b: T.S. of stipe – A portion enlarged with hypodermis and ground tissue; 5: *Pteris confusa* T. G. Walker—5a: T.S. of stipe – Entire view; 5b: A portion enlarged; 5c: A portion enlarged showing ground tissue and hypodermis.

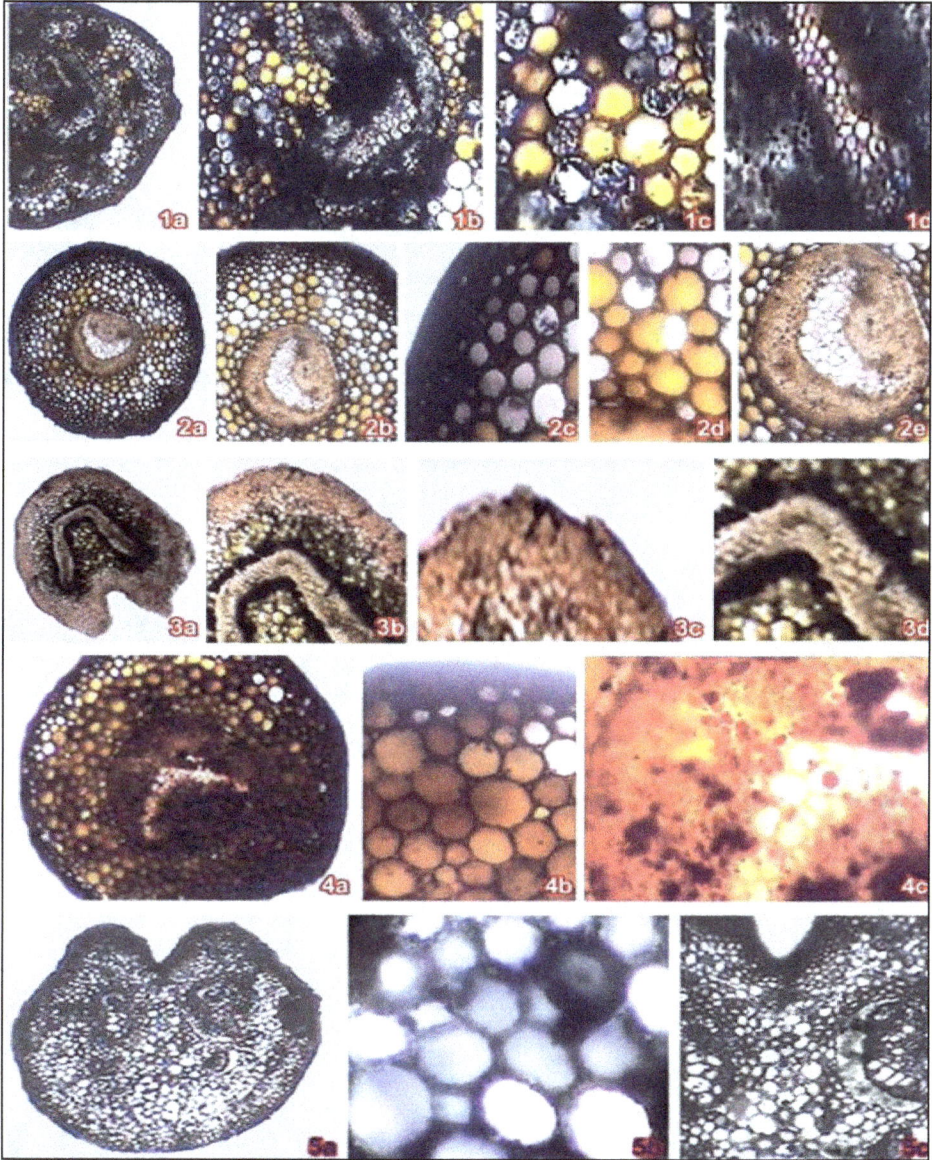

**Plate 4.3**: Results of histochemical analysis on lipids.

1: *Cheilanthes viridis* (Forssk.) Swartz—1a: T.S. of stipe–Half view; 1b: Vascular bundle; 1c: Ground tissue; 1d: Xylem and Phloem; 2: *Pellaea boivini* Hook.—2a: Transverse Section of stipe – Entire view; 2b: A portion enlarged with vascular bundle; 2c: Hypodermis; 2d: Ground tissue; 2e: Vascular bundle; 3: *Hemionitis arifolia* (Burm.) Moore—3a: T.S. of stipe–Entire view; 3b-c: A portion enlarged; 3d: A portion of vascular bundle with Xylem and Phloem; 4: *Doryopteris concolor* (Lagsd. *et* Fisch.) Kuhn—4a: T.S. of stipe - Entire view; 4b: A portion enlarged; 4c: Xylem and Phloem enlarged; 5: *Pityrogramma calomelanos* (L.) Link var. *calomelanos*—5a: T.S. of stipe - Entire view; 5b: A ground tissue; 5c: A portion enlarged of 4a.

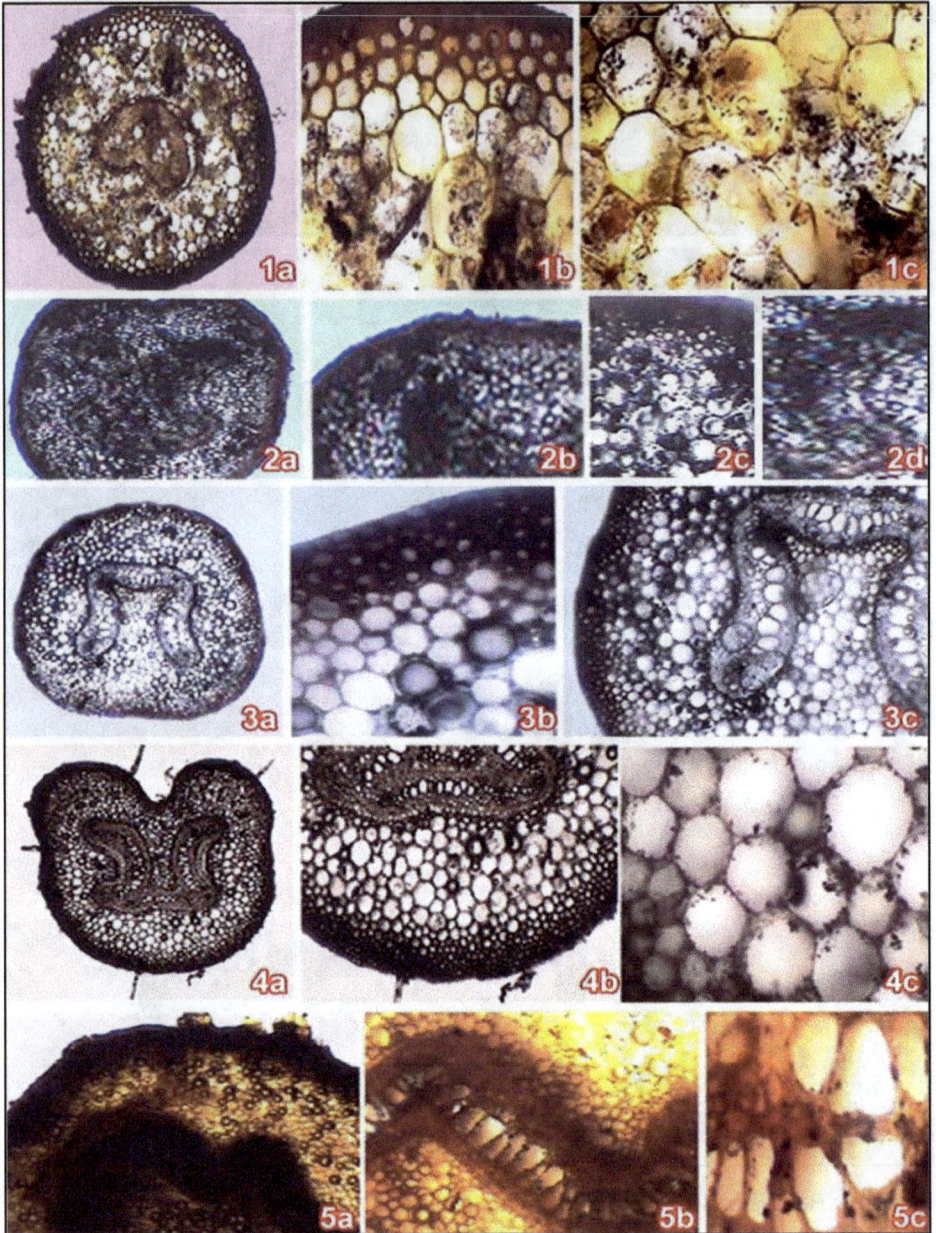

**Plate 4.4**: Results of histochemical analysis on lipids.

1: *Adiantum raddianum* Presl—1a: T.S. of stipe–Entire view; 1b: A portion of 1a -Enlarged; 1c: Ground tissue; 2: *Hypolepis glandulifera* Brownsey *et* Chinnock—2a: Transverse Section of stipe – Entire view; 2b-c: A portion enlarged; 2d: Vascular bundle; 3: *Microlepia speluncae* (L.) Moore—3a: T.S. of stipe–Entire view; 3b: A portion enlarged; 3c: Vascular bundle; 4: *Pteridium aquilinum* (L.) Kuhn v.Deck—4a: T.S. of stipe - Entire view; 4b: A portion enlarged; 4c: Ground tissue; 5: *Histiopteris incisa* (Thunb.) J. Sm.—5a: T.S. of stipe- A portion enlarged; 5b: A portion of vascular bundle enlarged; 5c: Xylem enlarged.

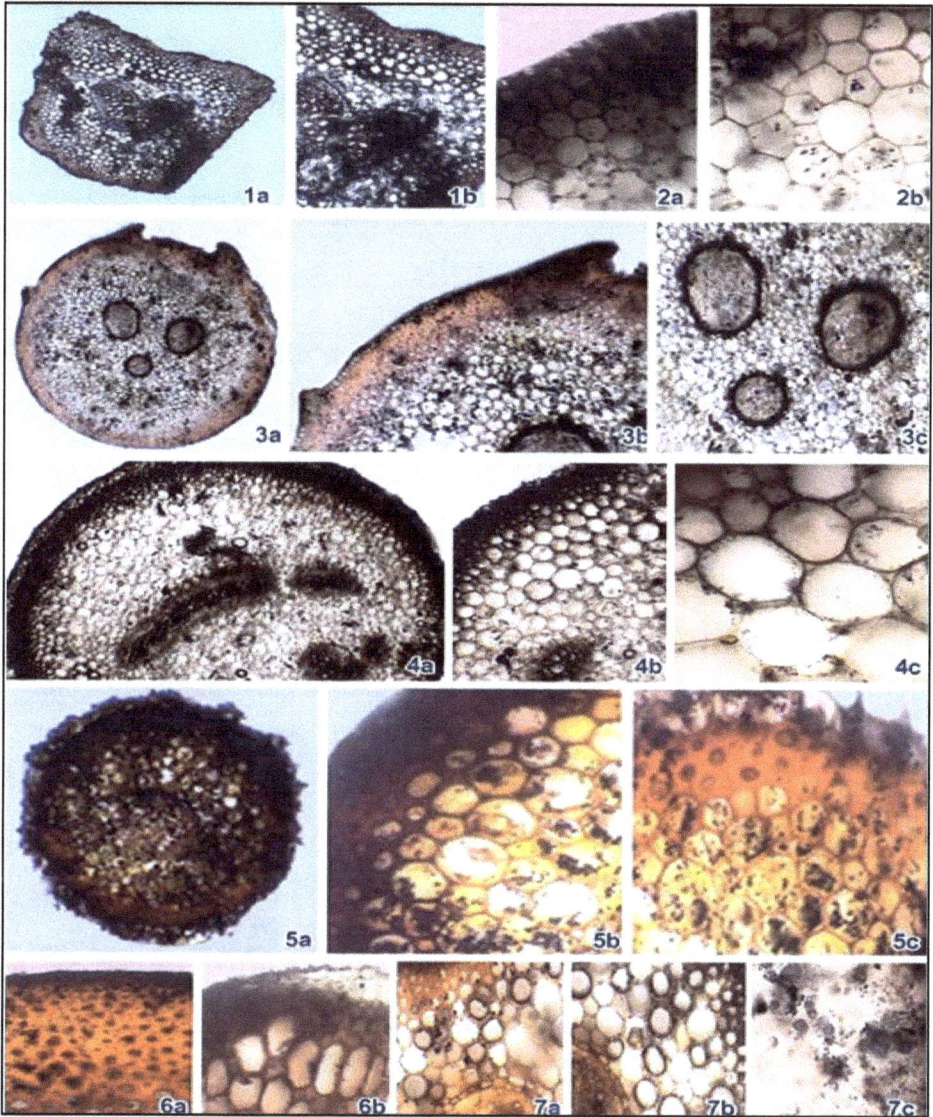

**Plate 4.5**: Results of histochemical analysis on lipids.

1: *Lindsaea ensifolia* Sw—1a: T.S. of stipe–Entire view; 1b: A portion of Enlarged; 2: *Odontosoria chinensis* (L.) J. Sm.—1a: T.S. of stipe – A portion with epidermis and Hypodermis; 1b: Ground tissue; 3. *Araiostegia hymenophylloides* (Bl.) Copel.—3a: T. S. of stipe – Entire view; 3b: A portion enlarged; 3c: Vascular bundles; 4: *Nephrolepis multiflora* (Roxb.) Jarret—4a: T.S. of stipe–Half view; 4b: A portion enlarged with hypodermis and ground tissue; 4c: Ground tissue enlarged; 5: *Trichomanes obscurum* Bl.—5a: T.S. of stipe-Entire view; 5b: A portion enlarged showing ground tissue; 5c: A portion enlarged showing hypodermis; 6: *Dicranopteris linearis* (Burm.f.) Underwood var. *sebastiana*—6a: T.S. of stipe- A portion enlarged showing hypodermis; 6b: Xylem enlarged; 7. *Cyathea nilgirensis* Holttum—7a. T.S. of stipe – A portion enlarged with hypodermis and ground tissue; 7b: Ground tissue–Enlarged; 7c: Phloem.

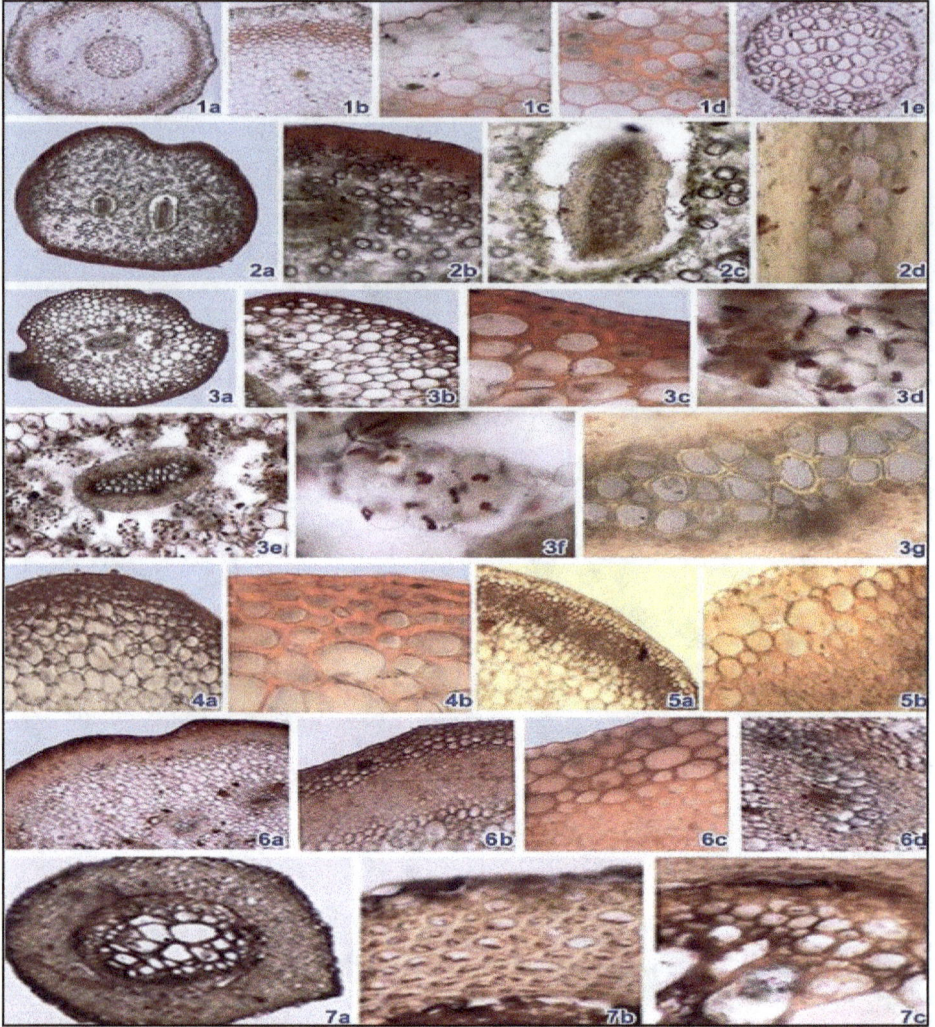

**Plate 4.6**: Results of histochemical analysis on Polphenols.

1: *Lycopodiella cernua* (L.) Pic. Ser.—1a: T.S. of stem–Entire view; 1b,c: T.S. of stem–Portions enlarged; 1d: Sclerenchymatous portion of cortex enlarged; 1e: Vascular bundle; 2: *Selaginella inaequalifolia* (Hook. and Grev.) Spring—2a: Transverse Section of stem – Entire view; 2b: A portion enlarged; 2c: Vascular bundle; 2d: Xylem and phloem enlarged; 3: *Selaginella tenera* (Hook.& Grev.) Spring—3a: T.S. of stem-Entire view; 3b: T.S. of stem-A portion enlargedl; 3c: Epidrmis and hypodermis enlarged; 3d: Ground tissue enlarged; 3e: Vascular bundle; 3f: Trabeculae; 3g: Xylem enlarged; 4: *Selaginella* species— 4a-b: T.S. of stem. Different portions enlarged showing epidermis, hypodermis and ground tissue; 5: *Angiopteris evecta* (Forst.) Hoffm.; 5a-b: T. S. of stipe. Different portions enlarged showing epidermis, hypodermis and ground tissue; 6: *Marattia fraxinea* Sm.—6a-c: T.S. of stipe. Different portions enlarged showing epidermis, hypodermis and ground tissue; 6d: Vascular bundle enlarged; 7: *Lygodium micropyllum* (Cav.) R.Br.—7a: T.S. of stipe – Entire view; 7b: A portion enlarged showing epidermis and hypodermis; 7c: Portion of vascular bundle enlarged showing xylem and phloem.

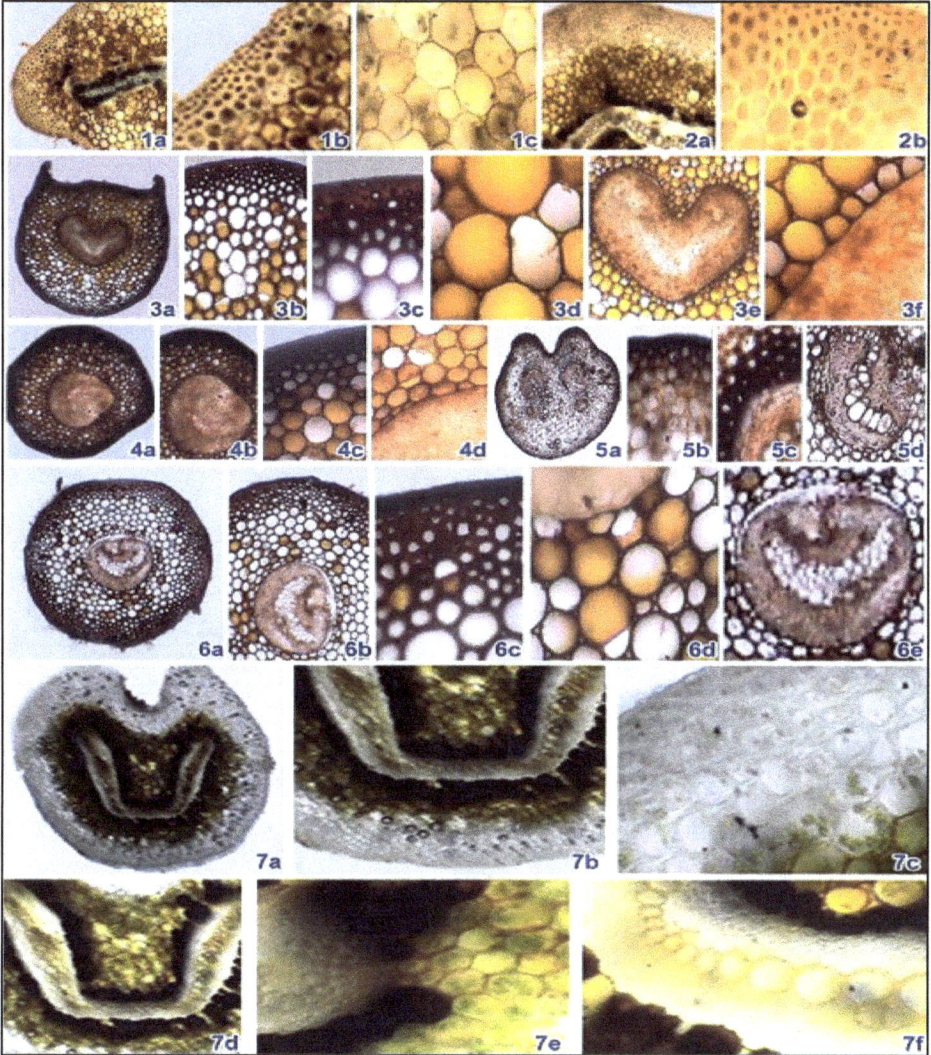

**Plate 4.7**: Results of histochemical analysis on Polphenols.

1: *Pteris argyraea* T. Moore—1a,b: T.S. of stipe – Different portions enlarged; 1c: Ground tissue enlarge; 2. *Pteris confusa* T. G. Walker—2a: T.S. of stipe-A portion; 2b: Hypodermis enlarged; 3. *Cheilanthes viridis* (Forssk.) Swartz—3a: T.S. of stipe. Entire view; 3b,c: A portion enlarged; 3d: Ground tissue enlarged; 3e: Vascular bundle; 3f: Endodermis and phloem; 4. *Doryopteris concolor* (Lagsd: *et* Fisch.) Kuhn—4a: T.S. of stipe – Entire view; 4b,c: Different portions enlarged; 4d: Portion enlarged showing endodermis and phloem; 5. *Pityrogramma calomelanos* (L.) Link var. *calomelanos*—5a: T.S. of stipe, Entire view; 5b,c: Outer (b) and inner (c) portions enlarged; 5d: Vascular bundle enlarged; 6. *Pellaea boivini* Hook—6a: T.S. of stipe, Entire view; 6b,c: Different portions enlarged; 6d: Ground tissue enlarged; 6e: Vascular bundle enlarged; 7. *Hemionitis arifolia* (Burm.) Moore—7a: T.S. of stipe, Entire view; 7b: A portion enlarged; 7c: Epidermis and hypodermis enlarged; 7d: Vascular bundle enlarged; 7e: Endodermis enlarged; 7f: Portion of vascular bundle enlarged showing xylem and phloem.

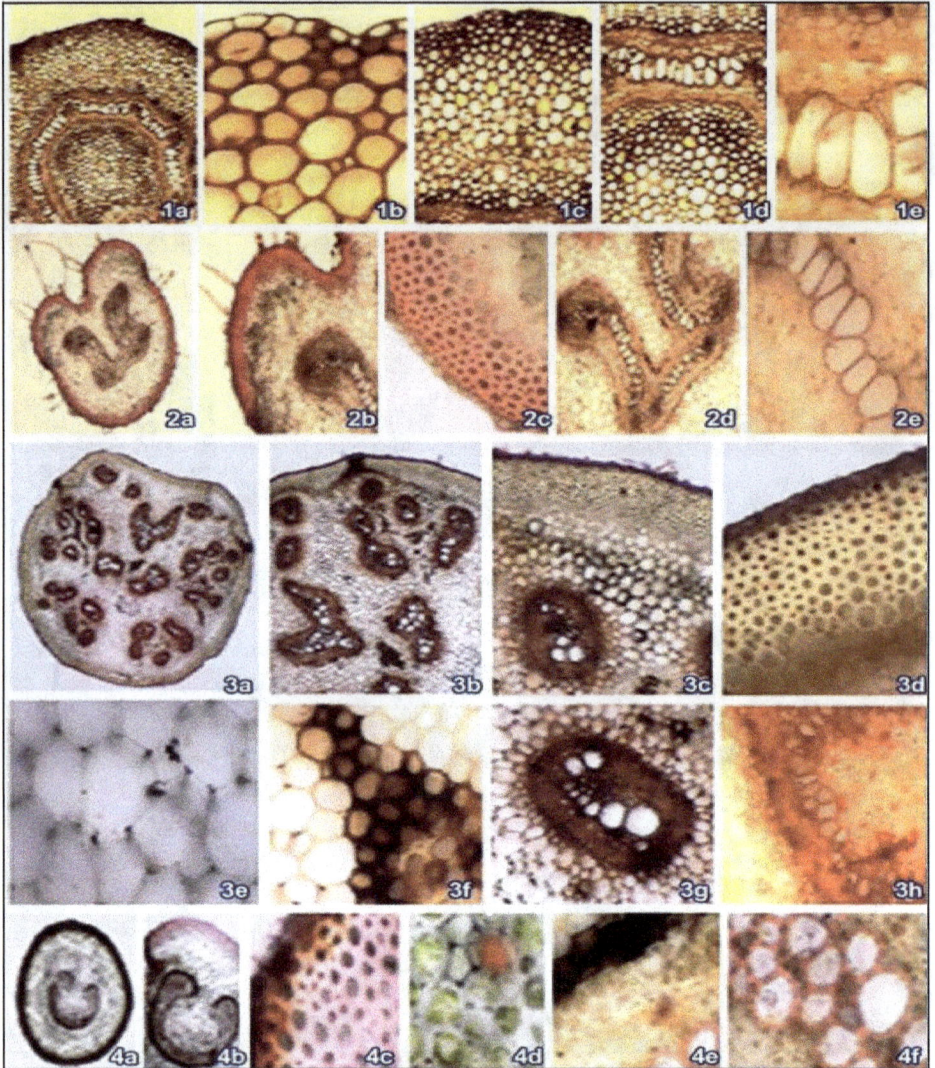

**Plate 4.8**: Results of histochemical analysis on Polphenols.

1: *Histiopteris incisa* (Thunb.) J. Sm—1a: T.S. of stipe- A portion enlarged; 1b: A portion enlarged showing epidermis and hypodermis; 1c: A portion enlarged showing epidermis, hypodermis and ground tissue; 1d: A portion of vascular bundle enlarged; 1e: Xylem enlarged; 2. *Hypolepis glandulifera* Brownsey *et* Chinnock—2a: Transverse Section of stipe – Entire view; 2b: A portion enlarged; 2c: A portion enlarged with epidermis and hypodermis; 2d: Vascular bundle; 2e: Xylem and phloem enlarged; 3. *Pteridium aquilinum* (L.) Kuhn v.Deck—3a: T.S. of stipe- Entire view; 3b,c: Different portions enlarged; 3c: A portion enlarged with epidermis and hypodermis; 3d: Ground tissue; 3e: Lignified tissue intermingled with ground tissue; 3g,h: Vascular bundle (g) and Xylem and phloem (h); 4: *Microlepia speluncae* (L.) Moore—4a: T.S. of stipe–Entire view; 4b: A portion enlarged; 4c,d: A portion enlarged showing hypodermis (c) and ground tissue (d); 4e: Endodermis and phloem enlarged; 4f: Xylem enlarged.

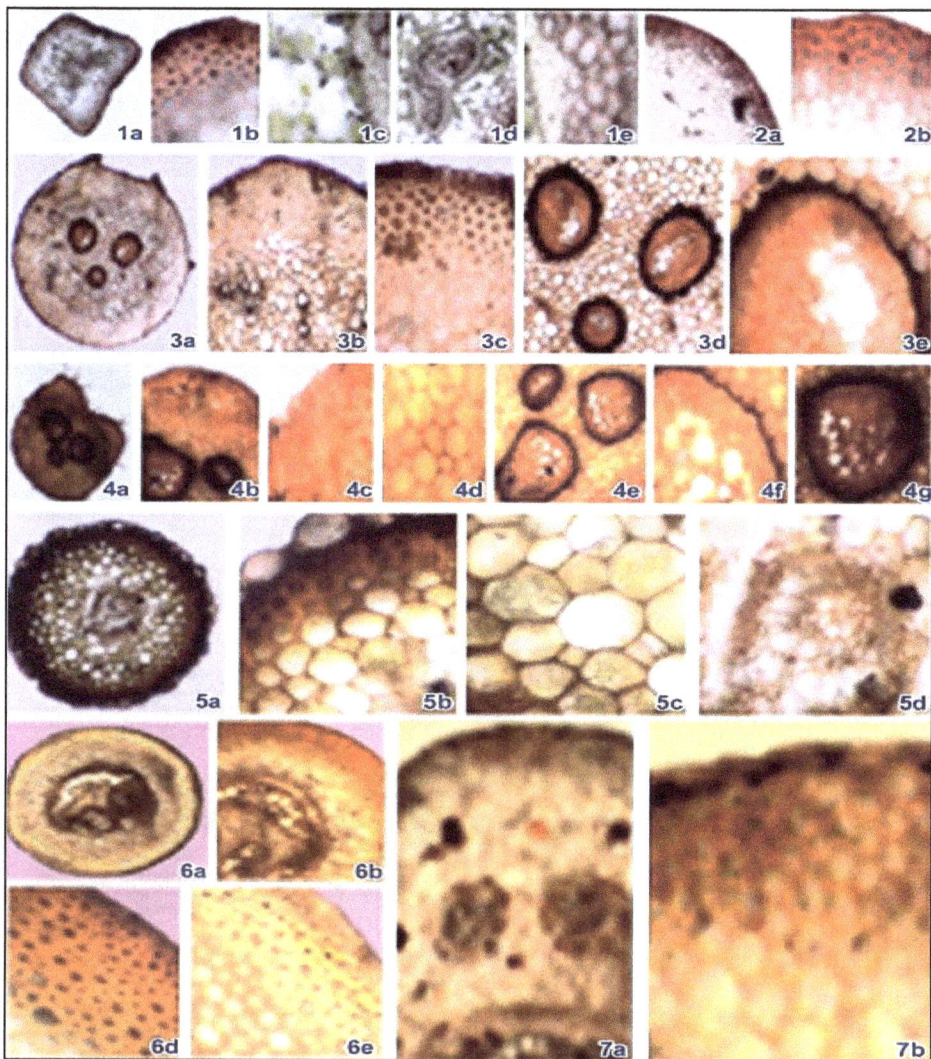

**Plate 4.9**: Results of histochemical analysis on Polphenols.

1: *Lindsaea ensifolia* Sw—1a: T.S. of stipe–Entire view; 1b: A portion enlarged; 1c: Ground tissue; 1d: Portion of vascular bundle; 1e: Xylem enlarged; 2. *Odontosoria chinensis* (L.) J. Sm—2a,b: T.S. of stipe – Different portions with epidermis and Hypodermis; 3: *Araiostegia hymenophylloides* (Bl.) Copel—3a: T.S. of stipe – Entire view; 3b,c: Different portions enlarged; 3c: Vascular bundles; 3d: Vascular bundle enlarged; 4: *Nephrolepis multiflora* (Roxb.) Jarret—4a: T.S. of stipe–Entire view, 4b,c: Different portions enlarged; 4d.Ground tissue enlarged, 4e: Vascular bundles 4f: Endodermis enlarged; 4g. Vascular bundle enlarged; 5. *Trichomanes obscurum* Bl; 5a: T.S. of stipe- Entire view; 5b: A portion enlarged showing epidermis and hypodermis; 5c: A portion enlarged showing ground tissue; 5d: Vascular bundle enlarged; 6. *Dicranopteris linearis* (Burm.f.) Underwood var. *sebastiana*—6a: T.S. of stipe- Entire; 6b: A portion enlarged; 6c,d: Different portions enlarged showing epidermis and hypodermis; 7: *Cyathea nilgirensis* Holttum—7a: T.S. of stipe – A portion enlarged; 7b: A portion enlarged with epidermis, hypodermis and ground tissue.

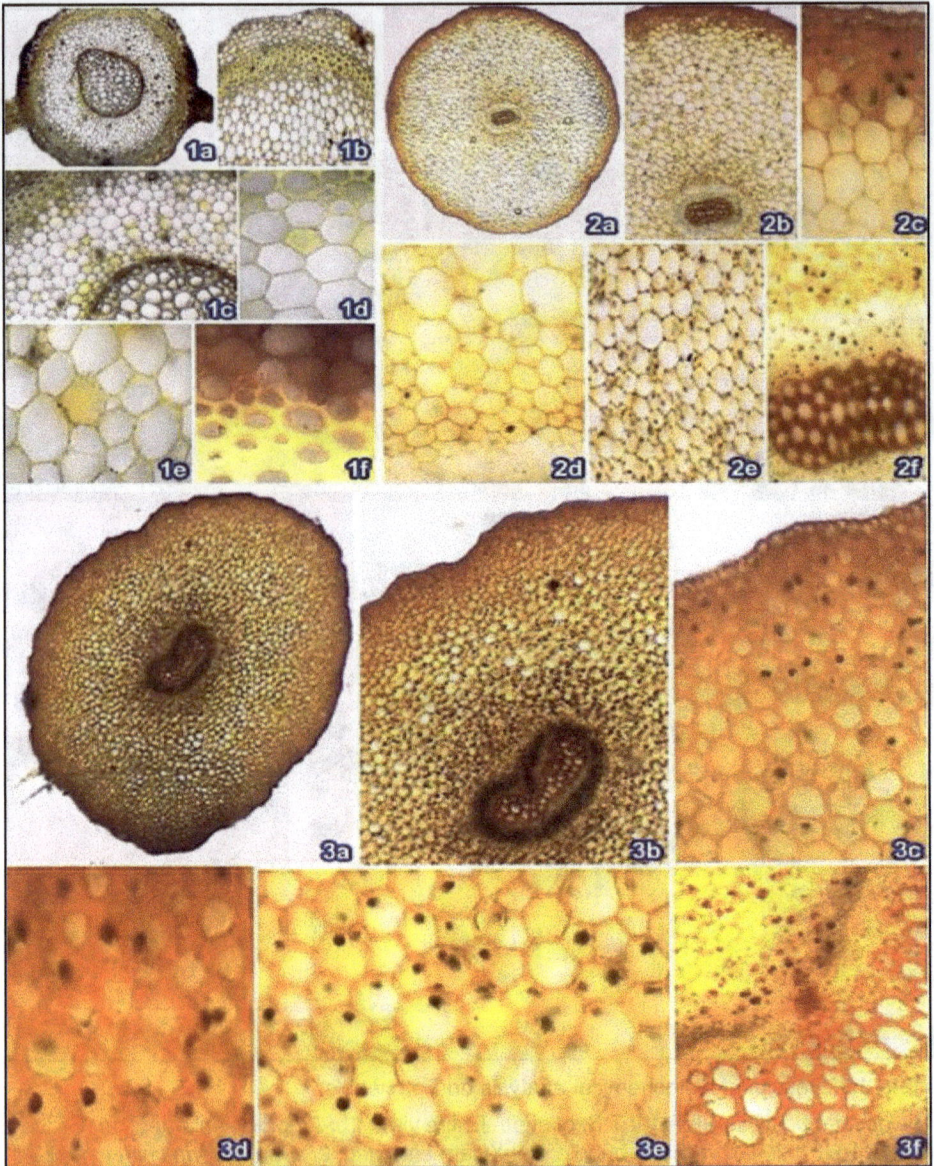

**Plate 4.10**: Results of histochemical analysis on Lignin.

1: *Lycopodiella cernua* (L.) Pic: Ser—1a: T.S. of stem–Entire view; 1b: T.S. of stem–Portion enlarged; 1c: Ground tissue and portion of vascular bundle enlarged; 1d: Ground tissue enlarged; 2. *Selaginella tenera* (Hook.& Grev.) Spring—2a: T.S. of stem. Entire view; 2b: T. S. of stem-A portion enlarged; 2c: Epidermis and hypodermis enlarged; 2d, e: Outer (e) and inner (d) ground tissue; 2f: Vascular bundle enlarged; 3. *Selaginella inaequalifolia* (Hook. and Grev.) Spring—3a: Transverse Section of stem – Entire view; 3b: A portion enlarged; 3c: A portion enlarged showing Epidermis, hypodermis and Ground tissue; 3d: Sclerenchymatous hypodermis enlarged; 3e: Parenchymatous ground tissue enlarged; 3f: Portion of vascular bundle enlarged showing xylem and phloem.

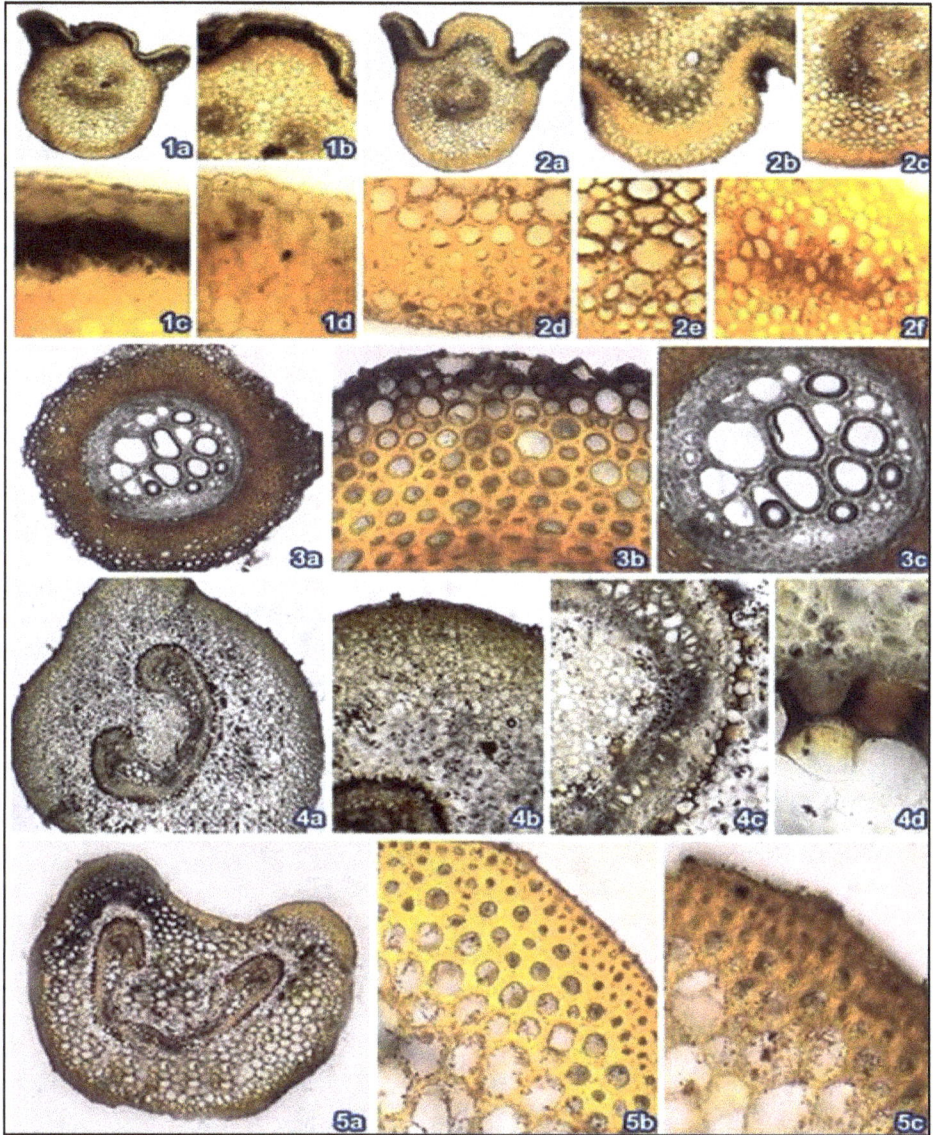

**Plate 4.11**: Results of histochemical analysis on Lignin.

1: *Angiopteris evecta* (Forst.) Hoffm—1a: T.S. of stipe – Entire view; 1b: T.S. of stipe – A Portion enlarged; 1c,d: Different portions enlarged towards periphery; 2: *Marattia fraxinea* Sm; 2a: T.S. of stipe–Entire view; 2b-e: T.S. of stipe – Different portions enlarged with hypodermis and ground tissue; 2f: A vascular bundle enlarged; 3. *Lygodium micropyllum* (Cav.) R.Br.—3a: Transverse Section of stipe – Entire view; 3c: A portion enlarged showing epidermis and hypodermis; 3c: Vascular enlarged; 4. *Pteris argyraea* T. Moore—4a: T.S. of stipe – Entire view; 4a: A portion enlarged with epidermis hypodermis and ground tissue; 4c: Portion of vascular bundle enlarged; 4d: Endodermis enlarged; 5. *Pteris confusa* T. G. Walker—5a: T.S. of stipe – Entire view; 5b,c: Different portions enlarged showing epidermis and hypodermis.

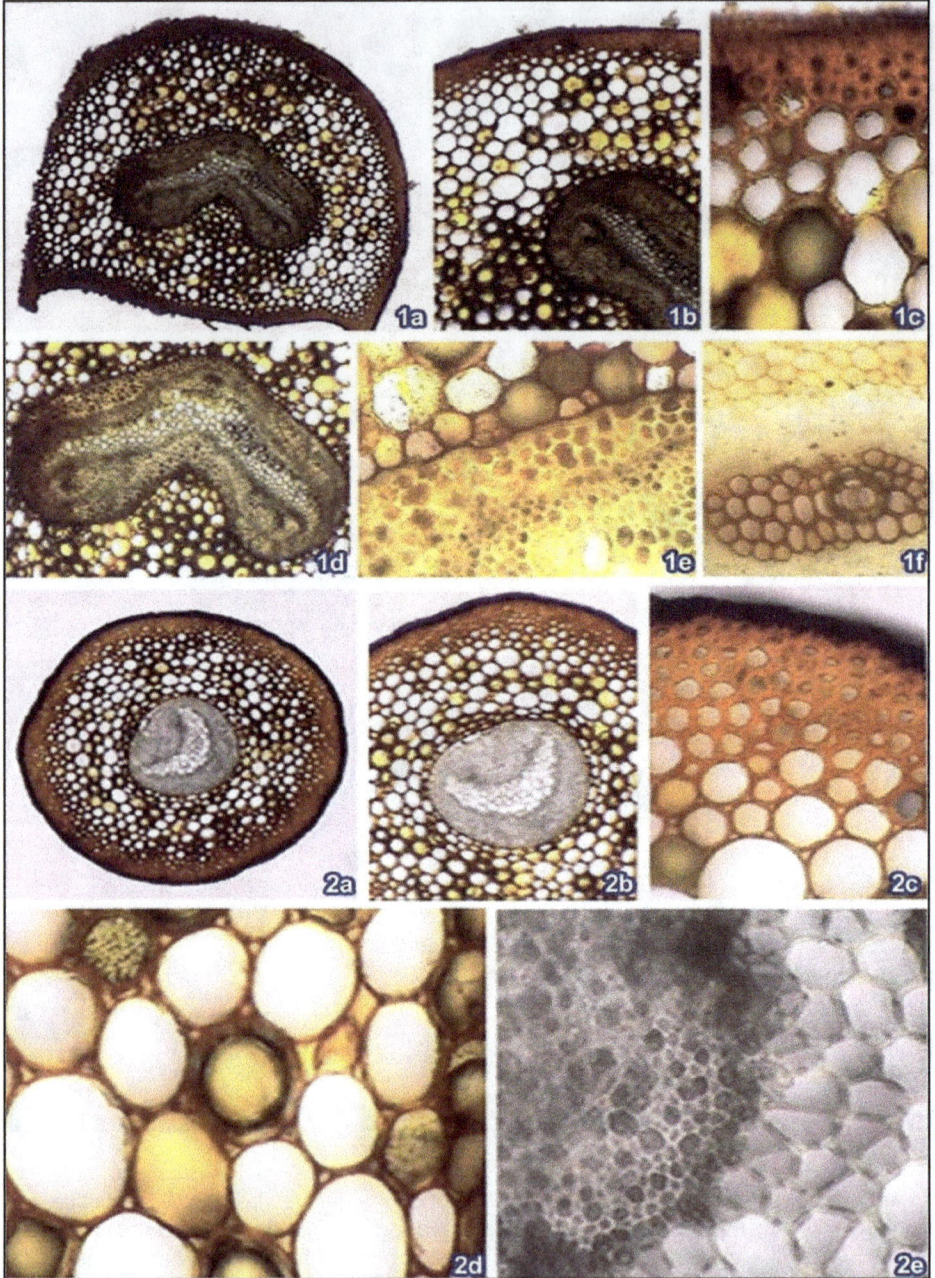

**Plate 4.12**: Results of histochemical analysis on Lignin.

1: *Cheilanthes viridis* (Forssk.) Swartz—1a: T.S. of stipe–Entire view; 1b,c: Different portions enlarged; 1d: Vascular bundle; 1e: Endodermis and pericycle enlarged; 1f: Xylem and Phloem enlarged; 2. *Pellaea boivini* Hook—2a: Transverse Section of stipe – Entire view; 2b: A portion enlarged with vascular bundle; 2c: A portion enlarged with Hypodermis and ground tissue; 2d: Ground tissue enlarged; 2e: Portion of vascular bundle enlarged.

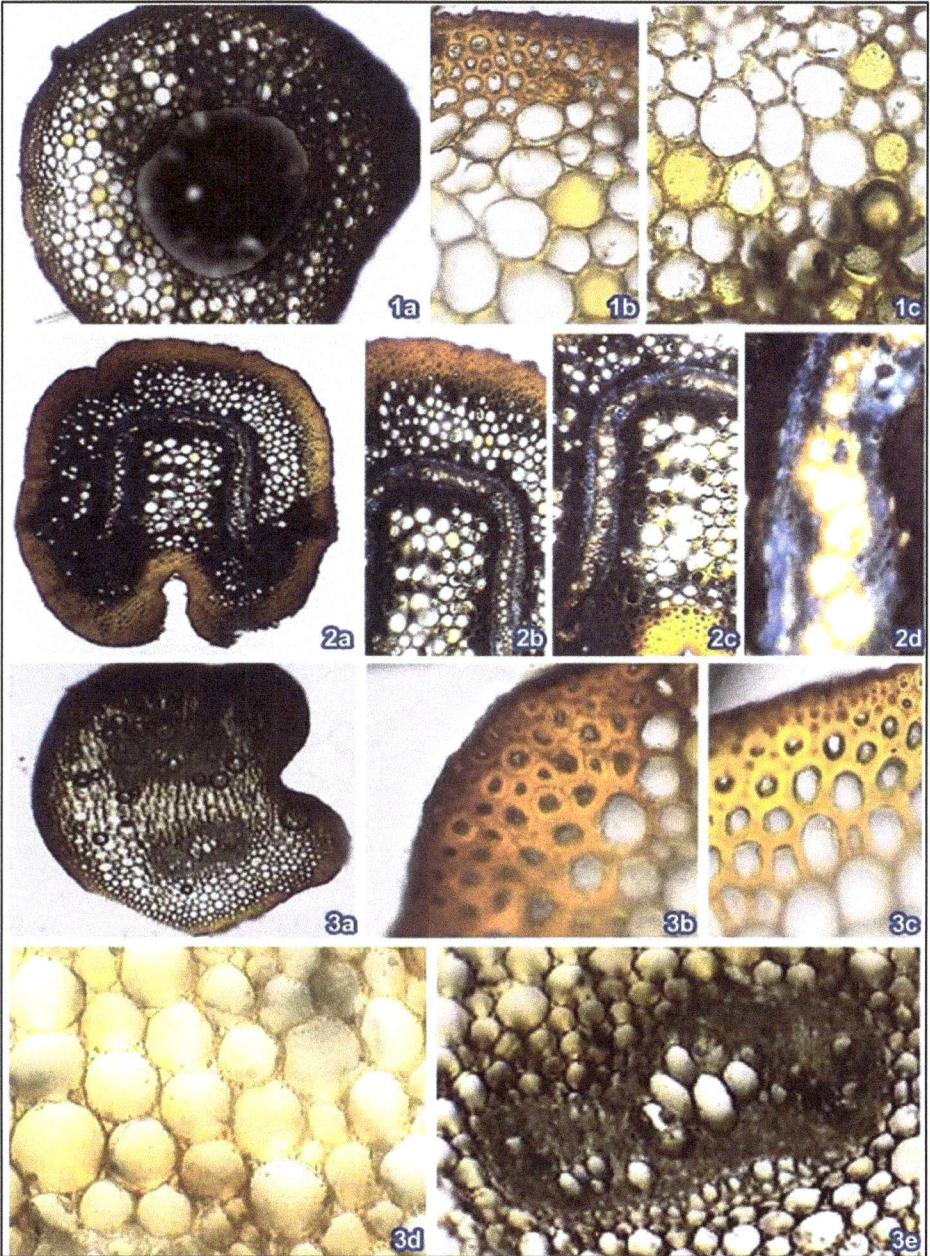

**Plate 4.13**: Results of histochemical analysis on Lignin.

1: *Doryopteris concolor* (Lagsd: *et* Fisch.) Kuhn—1a: T.S. of stipe- Entire view; 1b: A portion enlarged; 1c: Ground tissue enlarged; 2. *Hemionitis arifolia* (Burm.) Moore—2a: T.S. of stipe–Entire view; 2b: A portion enlarged; 2c,d: Different portions of vascular bundle enlarged showing showing Xylem and Phloem; 3: *Pityrogramma calomelanos* (L.) Link var. *calomelanos*—3a: T.S. of stipe- Entire view; 3b,c: Different portions enlarged showing epidermis and hypodermis; 3d: A ground tissue enlarged; 3e: Vascular bundle enlarged.

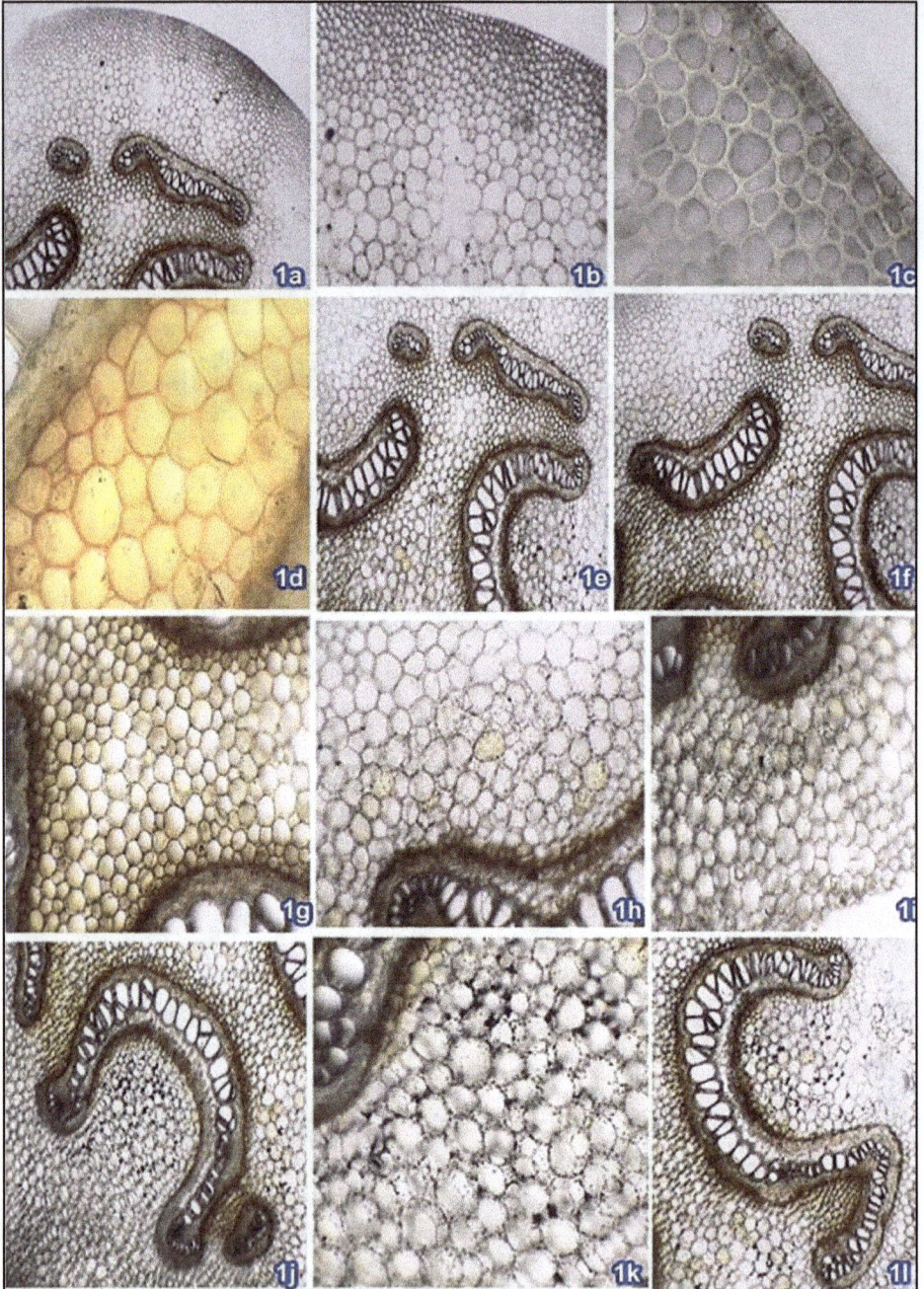

**Plate 4.14**: Results of histochemical analysis on Lignin.

1: *Pteridium aquilinum* (L.) Kuhn v.Deck—1a: T.S. of stipe - A portion enlarged; 1b: A portion enlarged showing epidermis, hypodermis and ground tissue; 1c, d: Different portions enlarged showing epidermis and hypodermis; 1g, h, i.: Ground tissue from different portions; 1e, f, j, l: Vascular bundles at different magnifications.

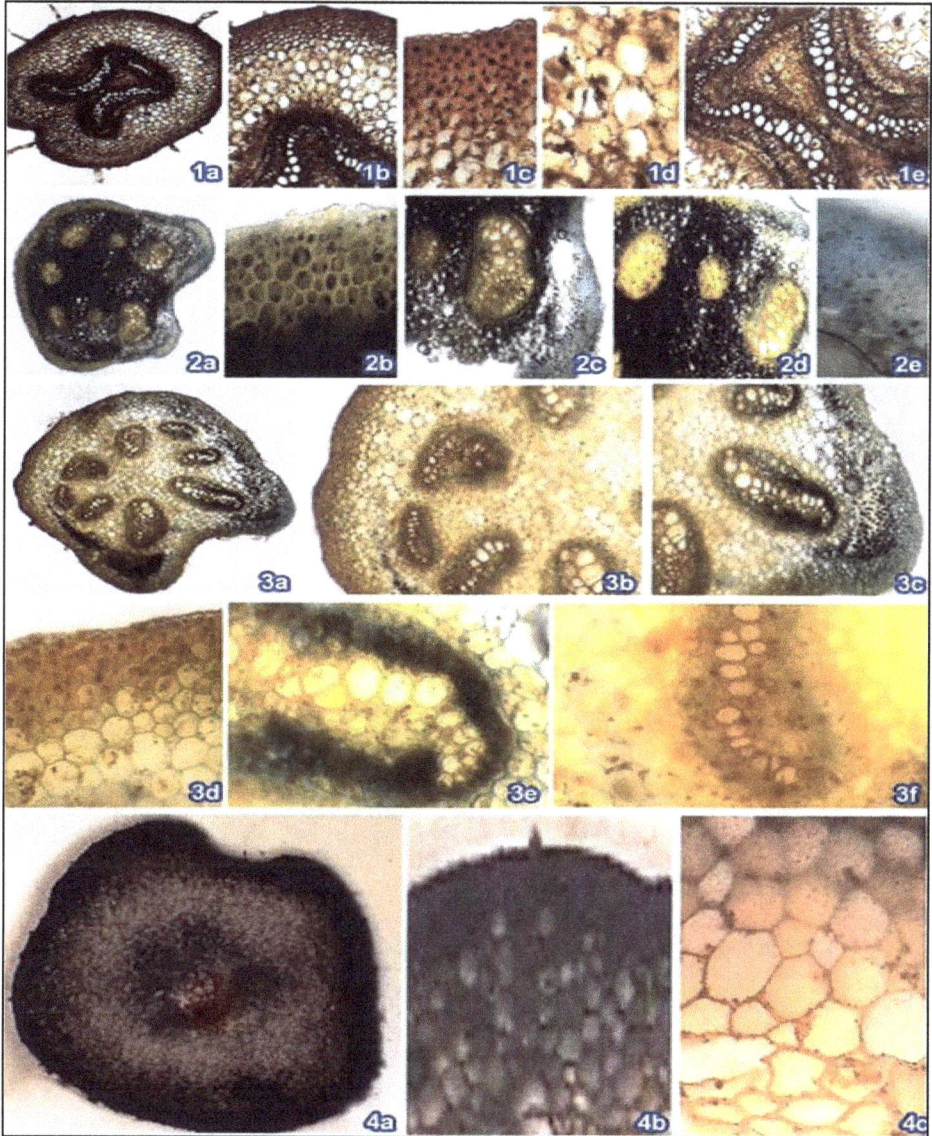

**Plate 4.15**: Results of histochemical analysis on Lignin.

1: *Hypolepis glandulifera* Brownsey *et* Chinnock—1a: Transverse Section of stipe – Entire view; 1b: A portion enlarged with epidermis, hypodermis, ground tissue and portion of vascular bundle; 1c: A portion enlarged showing epidermis and hypodermis; 1d: Ground tissue enlarged; 1d: Vascular bundle enlarged; 2. *Microlepia speluncae* (L.) Moore— 2a: T.S. of stipe–Entire view; 2b,e: Different portions enlarged showing epidermis and hypodermis; 2c,d: Vascular bundles; 3. *Pteridium aquilinum* (L.) Kuhn v.Deck—3a: T.S. of stipe- Entire view; 3b,c: Different portions enlarged; 3d: A portion enlarged with epidermis and hypodermis and ground tissue; 3e: Vascular bundle enlarged, 3f: Xylem and phloem enlarged; 4. *Odontosoria chinensis* (L.) J. Sm.—4a: T.S. of stipe – Entire view; 4b: A portion enlarged showing epidermis, hypodermis and ground tissue; 4c: Ground tissue enlarged.

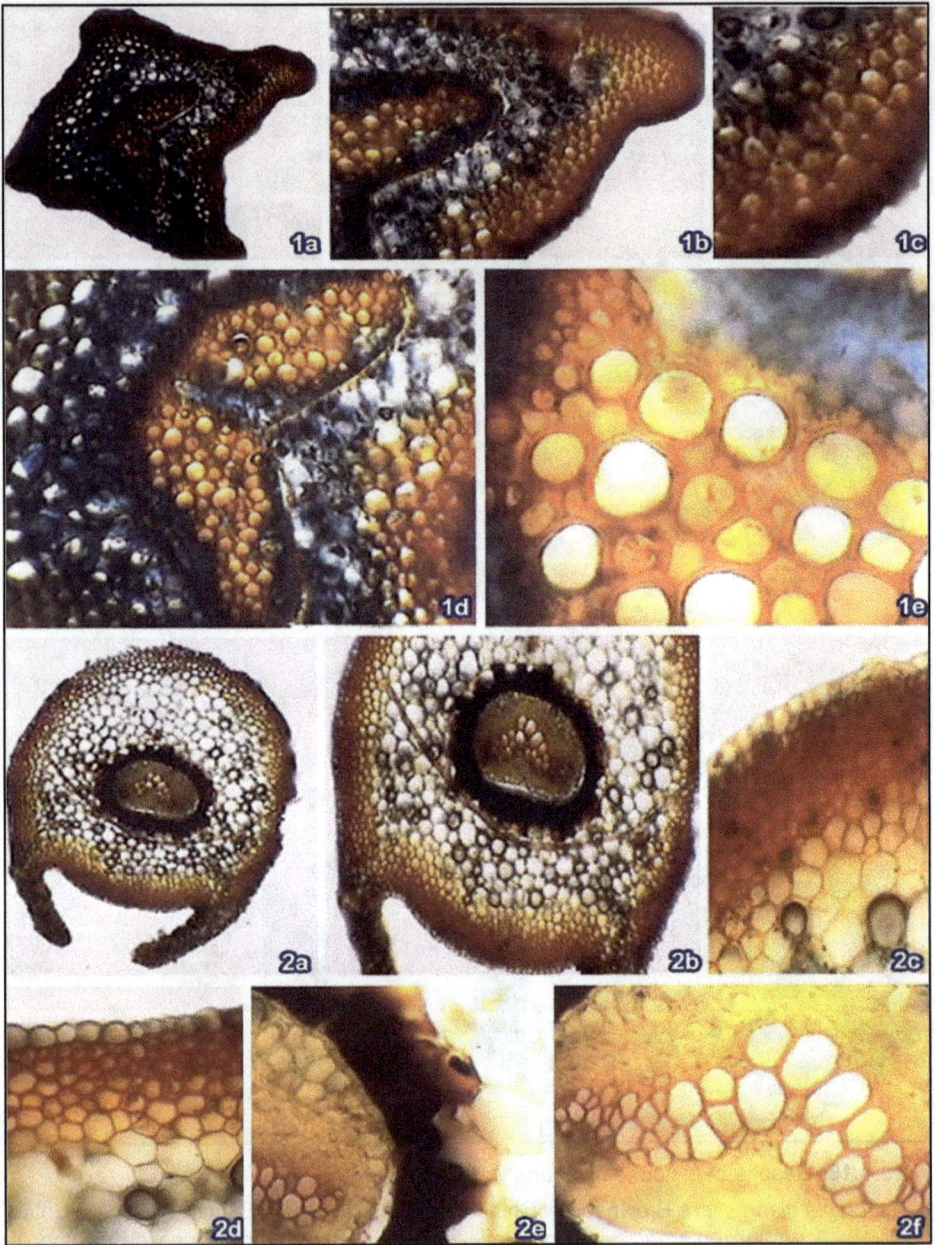

**Plate 4.16**: Results of histochemical analysis on Lignin.

1: *Lindsaea ensifolia* Sw.—1a: T.S. of stipe–Entire view; 1b: A portion of Enlarged; 1c: A portion enlarged showing epidermis and hypodermis; 1d: Vascular bundle enlarged; 1e: Xylem and phloem enlarged; 2. *Araiostegia hymenophylloides* (Bl.) Copel.—2a: T.S. of stipe – Entire view; 2b: A portion enlarged; 2c,d: Different portions enlarged showing epidermis and hypodermis; 2e: Endodermis and pericycle enlarged; 2f: Xylem and phloem enlarged.

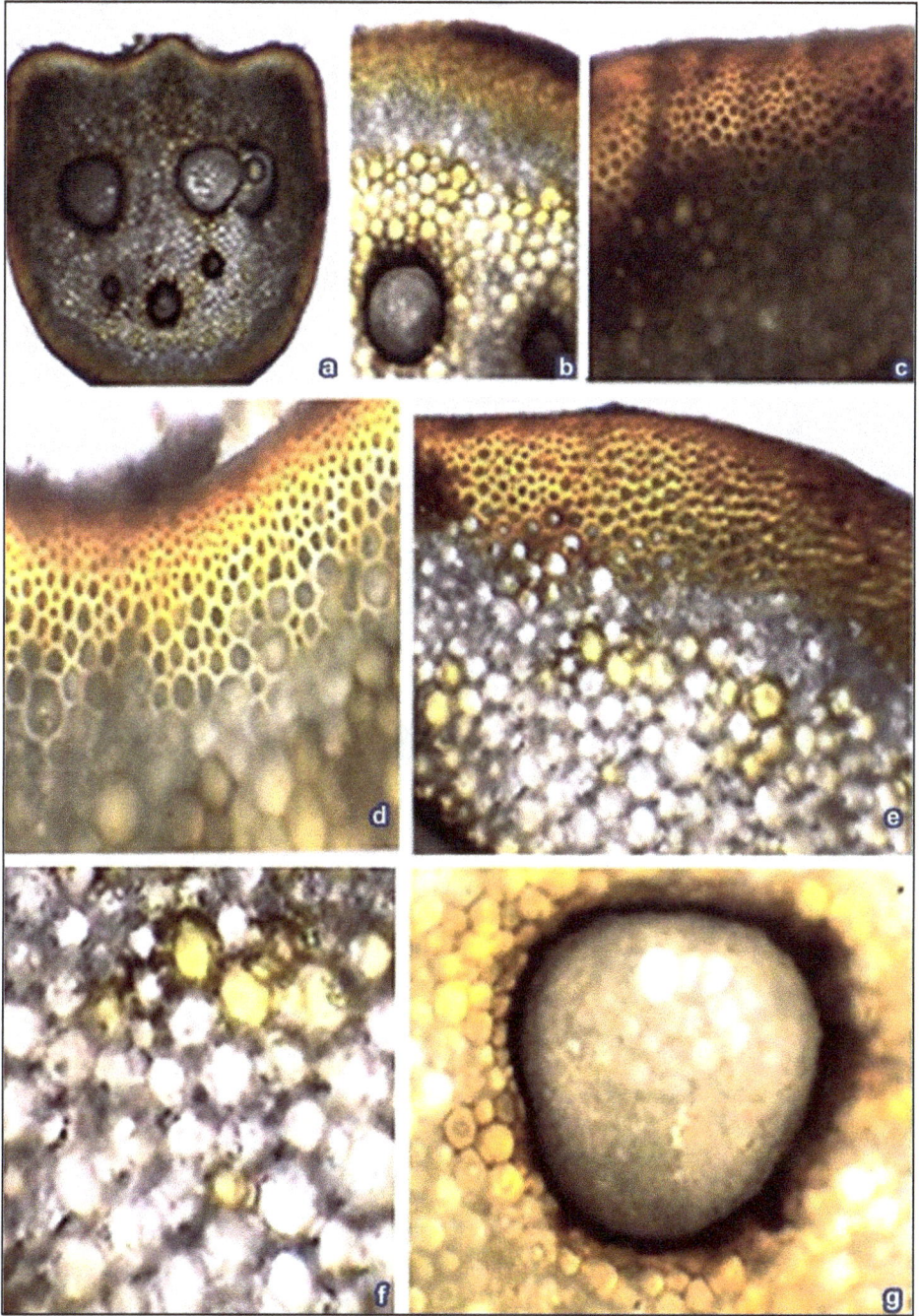

**Plate 4.17**: Results of histochemical analysis on Lignin.

a–g: *Nephrolepis multiflora* (Roxb.) Jarret—a: T.S. of stipe – Entire view; b–e: A portion enlarged with epidermis, hypodermis and ground tissue; f: Ground tissue enlarged; g. Vascular bundle enlarged.

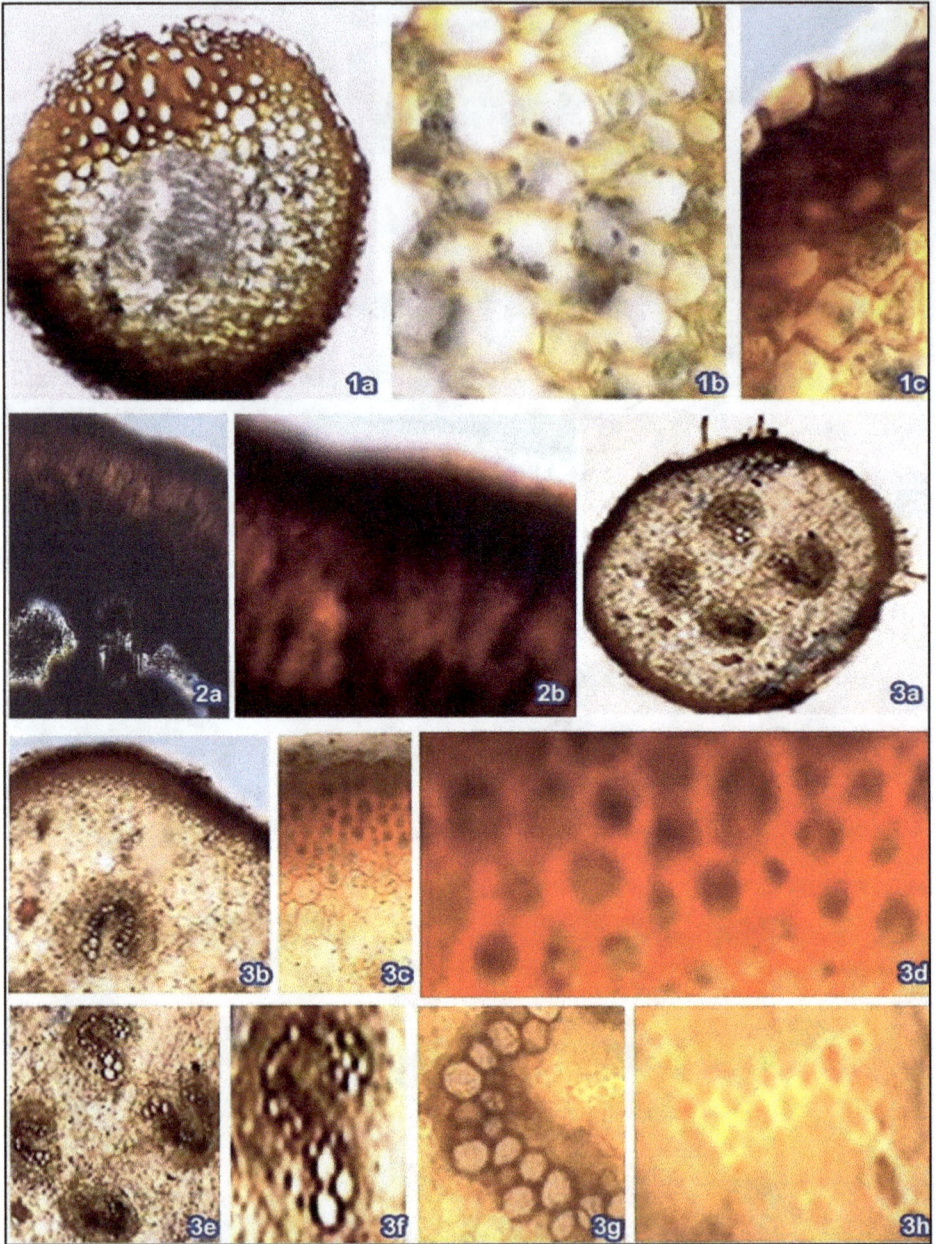

**Plate 4.18**: Results of histochemical analysis on Lignin.

1: *Trichomanes obscurum* Bl.—1a: T.S. of stipe- Entire view; 1b: Ground tissue enlarged; 1c: A portion enlarged showing epidermis and hypodermis; 2. *Dicranopteris linearis* (Burm.f.) Underwood var. *sebastiana*—2a, b: T.S. of stipe- Different portions enlarged; 3. *Cyathea nilgirensis* Holttum—3a: T.S. of stipe – Entire view; 3b, c: A portion enlarged showing epidermis, hypodermis and ground tissue; 3d: Hypodermis enlarged; 3e: Vascular bundles; 3f: Single vascular bundle enlarged; 3g. Xylem enlarged; 3h. Phloem enlarged.

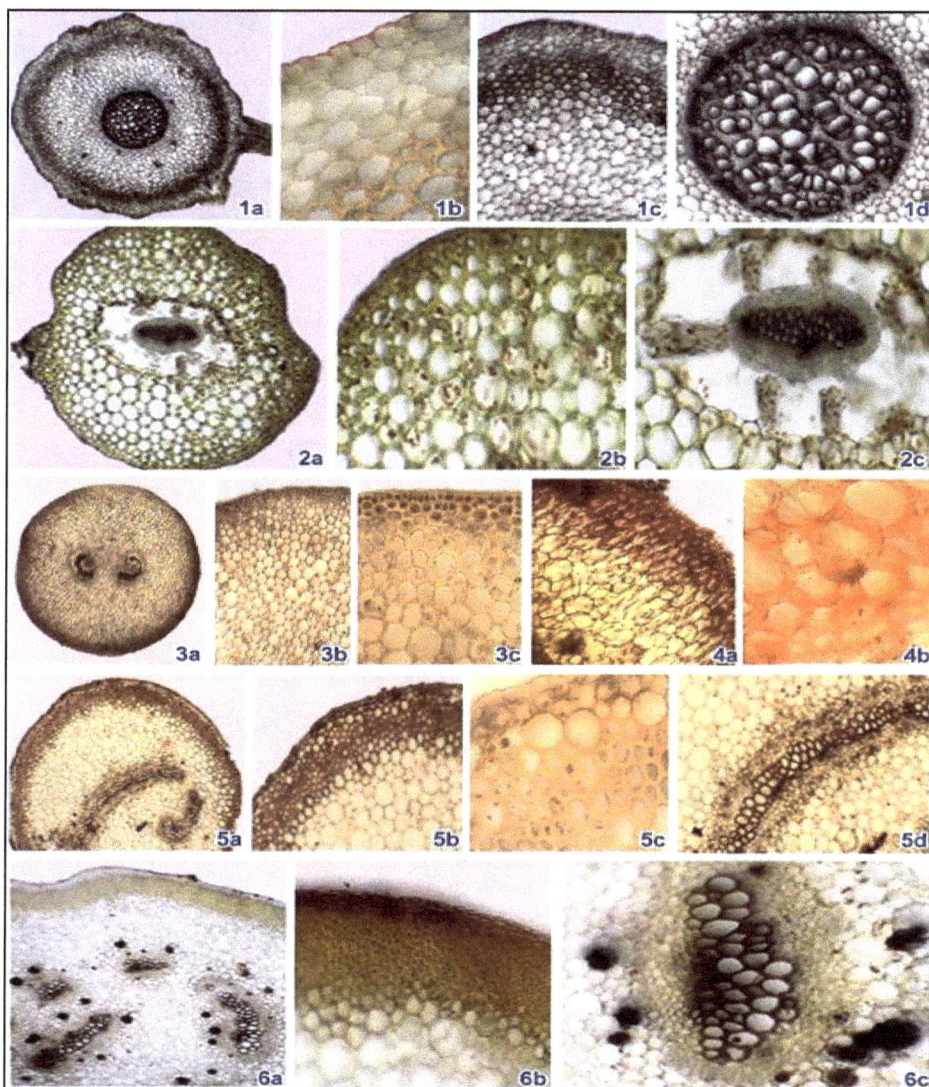

**Plate 4.19**: Results of histochemical analysis on Tannin.

1: *Lycopodiella cernua* (L.) Pic. Ser.—1a: T.S. of stem–Entire view; 1b,c: Portion enlarged showing epidermis and hypodermis and ground tissue; 1d: Vascular bundle; 2. *Selaginella inaequalifolia* (Hook. and Grev.) Spring—2a: T. S. of stem. Entire view; 2b: A portion enlarged showing epidermis, hypodermis and ground tissue; 2d: Vascular bundle with trabeculae; 3. *Selaginella tenera* (Hook.& Grev.) Spring—3a: T.S. of stem. Entire view; 3b,c: A portion enlarged at different magnifications; 4. *Selaginella* species.—4a: A portion of T.S. of stem showing epidermis, hypodermis and ground tissue; 4b: Hypodermis enlarged; 5. *Angiopteris evecta* (Forst.) Hoffm.—5a: T.S. of stipe – A portion; 5b,c: T.S. of stipe – A Portion enlarged at different magnifications showing epidermis, hypodermis and ground tissue; 5d: Portion of vascular bundle enlarged showing xylem and phloem; 6. *Marattia fraxinea* Sm.—6a: T.S. of stipe – A portion; 6b: Portion enlarged with epidermis, hypodermis and ground tissue, 6c: A vascular bundle enlarged showing xylem and phloem.

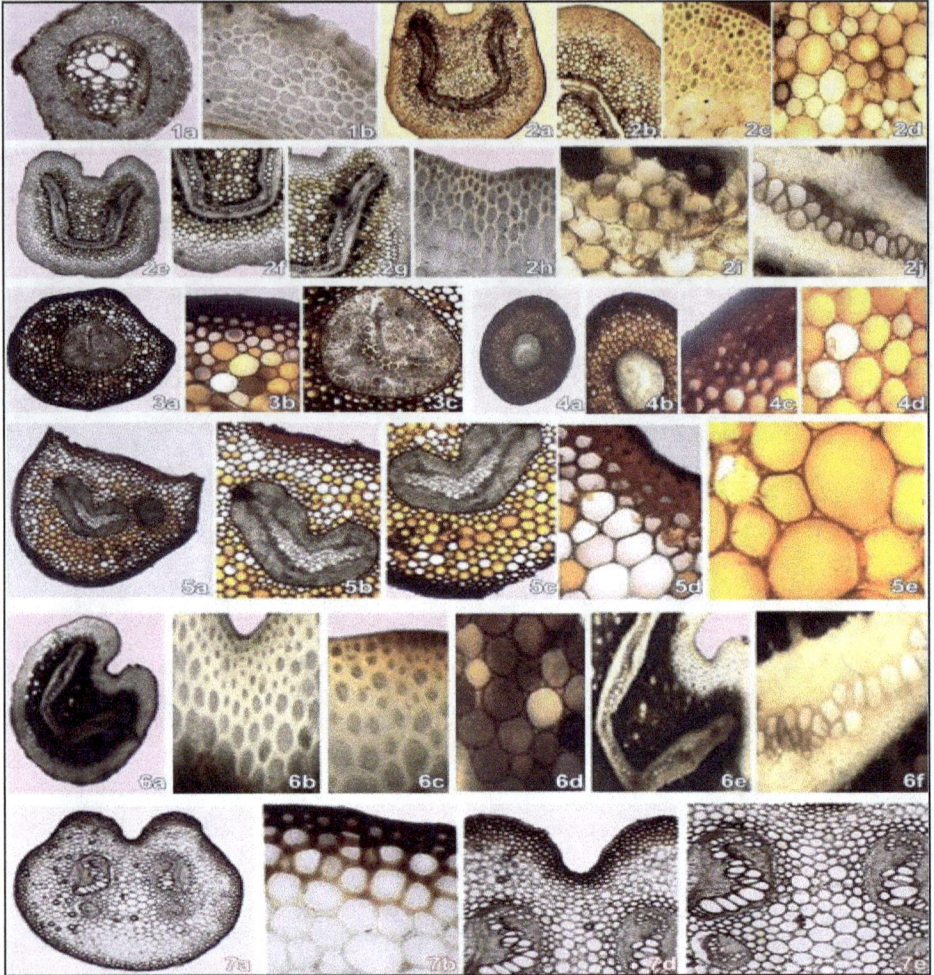

**Plate 4.20**: Results of histochemical analysis on Tannin.

1: *Lygodium micropyllum* (Cav.) R.Br; 1a: T. S. of stipe – Entire view; 1b: A portion enlarged showing epidermis and hypodermis; 2. *Pteris confusa* T.G. Walker—2a: T. S. of stipe – Entire view, 2b: A portion enlarged; 2c: A portion enlarged showing epidermis, hypodermis and ground tissue; 2d: Ground tissue enlarged; 3. *Doryopteris concolor* (Lagsd: *et* Fisch.) Kuhn—3a: T.S. of stipe–Entire view; 3b: A portion enlarged showing epidermis, hypodermis and ground tissue; 3c: Vascular bundle; 4. *Cheilanthes viridis* (Forssk.) Swartz—4a: T.S. of stipe- Entire view; 4b: A portion enlarged; 4c: A portion enlarged showing epidermis, hypodermis and ground tissue; 4c: Ground tissue enlarged; 5. *Pellaea boivini* Hook.— 5a: Transverse Section of stipe – Entire view; 5b,c: A portion enlarged with vascular bundle; 5d: A portion enlarged with epidermis, hypodermis and ground tissue; 5d: Ground tissue enlarged; 6. *Hemionitis arifolia* (Burm.) Moore—6a: T.S. of stipe–Entire view; 6b, c: Different portions enlarged with epidermis, hypodermis; 6d: Ground tissue enlarged; 6e: Vascular bundle; 6f: A portion of vascular bundle with Xylem and Phloem; 7. *Pityrogramma calomelanos* (L.) Link var. *calomelanos*—7a: T.S. of stipe- Entire view; 7b,d: A portion enlarged with epidermis, hypodermis and ground tissue; 7e: Vascular bundles.

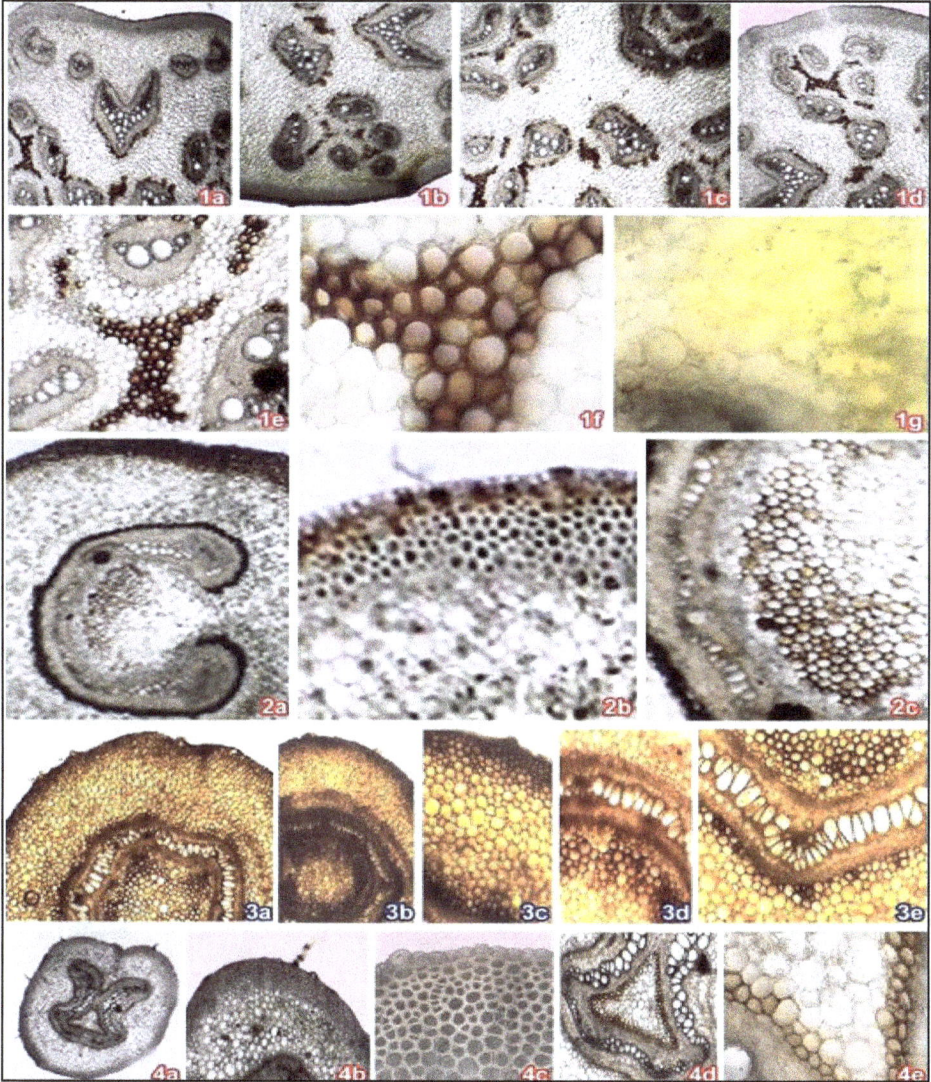

**Plate 4.21**: Results of histochemical analysis on Tannin.

1: *Pteridium aquilinum* (L.) Kuhn v.Deck—1a, b, c, d: T.S. of stipe- Different portions with meristeles; 1c: A portion enlarged showing epidermis, hypodermis and ground tissue; 1e: Vascular bundles with lignified tissue in the ground tissue; 1f: Lignified tissue (enlarged) in the ground tissue; 1g: Ground tissue enlarged; 2. *Microlepia speluncae* (L.) Moore— 2a: T.S. of stipe–A portion with C-shaped vascular bundle; 2b: A portion enlarged showing epidermis, hypodermis and ground tissue; 2c: Portion of vascular bundle showing xylem and phloem; 3. *Histiopteris incisa* (Thunb.) J. Sm; 3a, b: T.S. of stipe- Different portions; 3c: A portion enlarged showing epidermis, hypodermis and ground tissue; 3d, e: A portion of vascular bundle enlarged at different magnifications showing amphiphloic nature of the stele; 4. *Hypolepis glandulifera* Brownsey *et* Chinnock—4a: Transverse Section of stipe – Entire view; 4b: A portion enlarged; 4c: A portion enlarged with epidermis and hypodermis; 4d: Vascular bundle; 4e: Ground tissue with lignified cells towards vascular bundle.

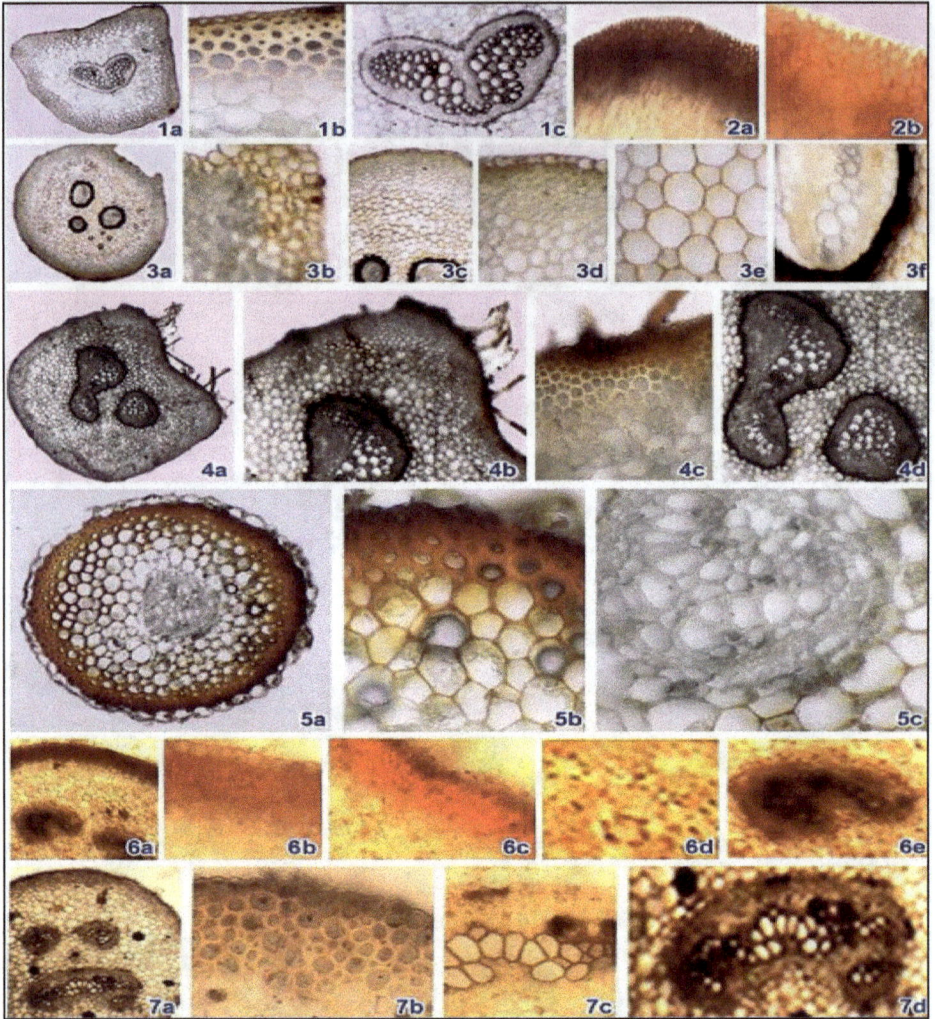

**Plate 4.22**: Results of histochemical analysis on Tannin.

1: *Lindsaea ensifolia* Sw.—1a: T.S. of stipe–Entire view; 1b: A portion enlarged showing epidermis and hypodermis; 1c: Vascular bundle enlarged; 2: *Odontosoria chinensis* (L.) J. Sm.—2a, b: T.S. of stipe-Different portions enlarged showing epidermis, hypodermis and ground tissue; 3: *Araiostegia hymenophylloides* (Bl.) Copel.—3a: T. S. of stipe – Entire view; 3b, c, d: Different portions enlarged showing epidermis, hypodermis and ground tissue; 3e: Ground tissue enlarged; 3f: A portion of vascular bundle enlarged showing endodermis, xylem and phloem enlarged; 4: *Nephrolepis multiflora* (Roxb.) Jarret—4a: T.S. of stipe – Entire view, 4b: A portion enlarged; 4c: A portion enlarged showing epidermis, hypodermis and ground tissue enlarged; 4d: Vascular bundles enlarged; 5: *Trichomanes obscurum* Bl.—5a: T.S. of stipe- Entire view; 5b: A portion enlarged showing epidermis, hypodermis and ground tissue; 5c: Vascular bundle enlarged showing xylem and phloem; 6 and 7: *Cyathea nilgirensis* Holttum—6a, 7a: T.S. of stipe – A portion; 6b, c, 7b: A portion enlarged showing epidermis, hypodermis and ground tissue; 7c: Portion of vascular bundle enlarged showing xylem and phloem; 7d: Single vascular bundle.

rachis of this species are used for catching fish by making fence or weir across a small steam. It is also used for binding, basket-making and plaiting (Manickam and Irudayaraj, 1992). All the other ferns and fern allies are erect forms for which the mechanical support is provided by the presence of several layered sclerenchymatous hypodermis.

The hypodermis is made up of strictly thick walled sclerenchyma in which lignin is present abundantly. Lignin is an aromatic heteropolymer that is deposited most abundantly in the secondary cell walls of vascular plants. It provides structural rigidity to the plant body and it also serves a defensive role against herbivores and pathogens. The presence/absence of lignin in different tissues of the stipe including hypodermis has been analysed by histochemical tests and the results has been given separately.

## Ground Tissue

Next to the sclerenchymatous, parenchymatous ground tissue is present up to the endodermis. The number of parencymatous layers strictly depends upon the size of the stipe. In *A. evecta,* the stipe is of more than 1 cm in diameter and so the entire stipe is almost filled by parenchymatous ground tissue. The stipe may be green or pale or dark brown even black in colour. In *S. inaequalifolia,* the stem is green and so the parenchymatous ground tissue is filled with chloroplasts [Plate 4.19(2a,c)]. Usually the ground tissue is more or less uniform and in some species like *M. fraxinea* and *P. aquilinum*, it is intermingled with some lignified cells [Plate 4.19(6a,c); Plate 4.21(1a–f)]. The individual cells of the ground tissue are circular or polygonal in outline and they usually enclose small intercellular spaces. The parenchymatous ground tissue is the region of stored food materials.

## Vascular Bundles

The features of vascular bundle in the stipe are very important and the number, size and shape are highly variable from genus to genus. In most of the species there is a single vascular bundle at the centre of the stipe. But in *S. inaequalifolia, A. evecta, M. fraxinea, P. calomelanos* var. *calomelanos, A. hymenophylloides, P. aquilinum, M. speluncae, N. multiflora* and *C. nilgirensis,* two or more vascular bundles are scattered in the ground tissue.

It is to be noted that the number and shape of the vascular bundles may vary from position to position within a stipe. Thus in *M. speluncae,* the number of vascular bundle is single [Plate 4.8(4a,b)] or many [Plate 4.15(2a)]. The number and arrangement of vascular bundles in the stipe is an important character which is used to delimit major groups of ferns. Thus usually two vascular strands are present at the base of the stipe in Thelypteroid ferns in contrast to Cyatheoid, Tectarioid and Dryopteroid ferns where several vascular strands are present at the base of the stipe. In *Cyathea,* the vascular strands in stipe are numerous and not in a single arc, while in Tectarioid and Dryopteroid ferns the vascular strands are in a single arc (Manickam and Irudayaraj, 1992).

The shape of the vascular bundles is highly variable. It may be of U-shape, V-shape, C-shape, boat-shape, circular or ovate. In genera like *Lycopodiella, Selaginella,*

*Cyathea* and *Nephrolepis*, the vascular bundles are almost circular while in genera like *Adiantum, Cheilanthes* and *Doryopteris* it is boat-shaped. In *Pteris* and *Dicranopteris*, it is always C or U-shaped. Regarding the type of vascular bundles, they are all of typically single protostelic type. When there are many vascular bundles as in *Angiopteris, Marattia, Pteridium* and *Nephrolepis*, they are all together called dictyostele and the individual stele is called meristele which is of typical protostele. Each and every vascular bundle has individual endodermis followed by pericycle which encloses the phloem and xylem. As it is known generally the xylem is made up of xylem-tracheids.

In *A. evecta*, transverse section of petiole shows single layered epidermis which consists of thin walled cells. The bulk of petiole is composed of ground tissue. It is differentiated into three zones. The outer most zones consist of 3-4 layers of cells which are made up of thin walled parenchymatous cells. The middle zone consists of 3-4 layers of cells made up of thick walled sclerenchymatous cells, being comparatively smaller in size than the cells of outer and inner zone inner zone consists of large, thin walled polygonal cells. Eleven widely separated vascular strands are present at the base of the petiole embedded in the parenchymatous ground tissue. Each strand has a single layered endodermis. Endodermis is followed by pericycle containing thin walled cells, which are 1-3 layers in thickness. Xylem lies in the centre of the vascular strand. It is plate like with several protoxylem points in exarch condition. Xylem is surrounded by phloem. The vascular strands in the petiole during their upward course gradually fuse with each other and at last become five vascular strands at the tip. During the fusion, first the endodermis and at slightly higher level the pericycle and ultimately the phloem and xylem bundles of the two strands also fuse together. At this stage vascular strand becomes somewhat 'C' shaped (Srivastava, 2008).

Srivastava (2008) has studied the anatomy of the petiole of *C. dentata*. It is cylindrical. On the adaxial side of the petiole there is a prominent groove running from the base up to the tip where the first pinna is attached. In transverse section of the petiole, epidermal cells throughout their length appear small, thick walled, dark brown in colour and covered by smooth and delicate hairs. The epidermis is followed by three or more layers of thick walled cells followed by several layers of thin walled parenchymatous cells which constitute the ground tissue. The petiole receives two widely separated vascular strands from the rhizome. Each vascular strand is enclosed by a single layered endodermis. The pericycle is made up of thin walled cells, which are one to three layers in thickness. The xylem is mesarch and surrounded by phloem. The xylem is hippocampus-shaped and the two vascular strands appear slightly elongated. The adaxial arms are comparatively more turned inwards and the two vascular stands appear almost reniform in shape. Soon after entering the stipe base, the two separate vascular strands start coming closer to each other and get fused somewhere near the middle of the petiole to form a single strand for further upward course. During the merger of the two vascular strands first the endodermis, the pericycle and at still higher levels the two phloem and xylem strands also join each other at their abaxial side. The distal arms of xylem and phloem however remain as such. Thus, the single vascular strand that resulted due to fusion of the two is almost 'U' shaped and free arms of xylem strands turned inwards more in the former as compared to the latter. Xylem consists of tracheids with protoxylem having annular or spiral thickenings and metaxylem with scalariform thickenings. Phloem consists of sieve cells with occasional parenchyma (Srivastava, 2008).

Origin of pith at the centre core of xylem of protostele is an important event in stellar evolution resulting in the origin of siphonostele. In the present observation the stipe in all the species studied are with protostele with the exception of Dennstaedtioid member *Histiopteris incisa* in which amphiphloic siphonostele is present. It is remarkable to note that most of the dennstaedtioid ferns like *Pteridium, Microlepia* etc. are with primitive type of sori i.e marginal or superficial exindusiate (*Hypolepis*). In *Histiopteris* the number and shape of the meristeles vary along the stipe. Meristeles are of two shapes *viz.*, irregular and u-shaped. In *H. incisa*, the meristele number remains constant at five throughout the stipe but becoming less enhanced in shape towards the middle. However, in *H. stipulacea* the meristele number tends to increase towards the middle to a maximum of 17, becoming less enhanced in shapes and smaller in size (Faridah-Hanum *et al.*, 2008). In the present study two different sections on different positions taken randomly for different histochmical tests show uniformly a single wavy ring of amphiphloic stele enclosing a considerable size of parenchymatous pith.

*Lycopodium* and *Selaginella* have built their adult vascular structure upon the basis of primary xylem which is purely tracheidal. They have met the demand for maintenance of presentation-surface, as the size increases, by elaboration of the form of the woody tracts, or even of the stele itself. Some *Selaginella* spp. have even adopted mouldings of the stele parallel to those seen in advanced Ferns. In the fossils the elaborations in relation to increasing size are for the most part based upon medullation usually combined with cambial thickening. The results find their parallel in many other land plants of dendroid type; but among the living Lycopods the cambial thickening appears in a highly specialized and anomalous form in the stock of *Isoetes*. There is hardly any class of plants in which such variety of resource in meeting the demands of increasing size is to be found (Bower, 1935).

In the construction of any ordinary vascular plant there are three limiting tissue-surfaces of prime importance: i. the outer contour of the plant or part, complicated though it usually is in sub-aerial plants by the added cell surfaces lining the ventilating channels. ii. The endodermal sheath, which delimits the conducting tracts from the tissues that envelop them, and iii. The collective surface by which the dead tracheal system faces the living tissues that surrounds or permeates it. Each of the three is a surface of physiological transit, and each independently of the others will be a suitable subject of observation from the point of view of the proportion of surface to bulk as the size increases. Of these the third is the most important in the comparative study of form in the conducting tracts of the Pteridophyta: for the woody tissues are those best preserved in the fossils, and being resistant they frequently retain their natural condition, and make it possible to contrast their form and dimensions with the corresponding tissues in their living correlatives. We may then expect that in the smallest, and particularly in their sporeling stage, the form of the conducting tracts will be simple, such as the cylinder. Moreover, this is frequently continued to the adult state in the most primitive pteridophytes. But with greater size its form tends to become more complicated, as it is seen to do even in the progressive stages of the individual life. The primary conducting tracts of the Pteridophyta offer the best examples of moulding in accordance with the Principle as above stated (Bower, 1935). Increasing morphological complexity involving any increased complication of form would tend to meet the physiological need consequent on increasing size, by leveling up the proportion of surface of bulk.

# Histochemical Studies

The occurrence and distribution of various chemicals (lipids, lipoprotein, polyphenols, lignins, tannins, starch) in different tissues of the stipe have been given in the Tables 4.1–4.3 and Plates 4.1–4.22.

## Lipids

Lipids are present in plants both as food materials and structural materials. In the present study, presence of lipids has been observed in almost in all the parts of the stipe in all the species studied. Thus it is the universal compound present in all the parts mainly as structural material like phospholipid in plasma membrane of the cell. The same result of the present test also indicates the presence of lipoprotein. When different tissues are concerned lipids are abundantly present in the epidermal cells. Out of 26 species studied lipids are present in high degree of concentration in the epidermal cells of 14 species. Usually the outer side of the epidermal cells is with thick cuticle, which are made up of several kinds of epicuticular waxes and this is the reason for the occurrence of lipids in higher degree and higher frequency in the epidermal cells. Next to epidermis, the ground tissues show the higher frequency of occurrence of lipids.

## Polyphenols

Polyphenols are mildly present in the xylem and phloem of twelve species and in ground tissue and endodermis of 18 species. High concentration of polyphenols has been observed in the epidermis of eleven species and hypodermis of twelve species. Next to lipids, polyphenols occur commonly in different tissues of the stipe in considerable amount.

## Lignin

Lignin is present at high concentration in the epidermis of 15 species and hypodermis of 16 species. It is averagely present in the ground tissue of 11 species. Since all the ferns and fern allies are herbaceous and non-woody species, lignin is present mainly in the epidermis and hypodermis which are the parts for mechanical support of majority of the ferns and fern allies.

## Tannin

Tannin is present in generally in lesser amount in lesser number of species. It is present in high concentration only in the epidermis of 4 species and hypodermis of 5 species. It is present mildly in the epidermis of 7 species and ground tissue of 7 species. It is absent in the endodermis, xylem and phloem of all the 26 species.

## Comparative Analysis Based on Species

The histochemical data present in the Table 4.1 may be divided symmetrically in various manners with various numbers of boxes. For example it may be divided into two vertical or horizontal halves or two halves on each side with four boxes or 2 x 3 or 3 x 2 giving six equal sized boxes. When the number of boxes is more, the resulting number of species and number of data in each box will be less and it may be easy to identify the species with uniform data. Table 4.1 is divided into 3 x 3 boxes, out of nine boxes the bottom left corner box has less uniform data for about 1/3 of the species studied. Species and data of this region have been indicated in bold letters. Specifically, it includes nine species namely, *M. speluncae, L. ensifolia, O. chinensis, A. hymenophylloides, N. multiflora, T. obscurum, D. linearis* var. *sebastiana* and *C. nilgirensis*. All the 26 species has been listed in a taxonomic hierarchy and the above nine species are the advanced species when compared with other species. Almost the first half the number of species are primitive group belonging to the orders Schizaeales and Pteridales and the next half the number of species are in somewhat advanced orders such as Dicksoniales, Davalliales, Hymenophyllales, Gleicheniales and Cyatheales. The present histochemical studies on lipids, polyphenols, lignins and tannins show the evolutionary progress of such chemicals from the primitive ferns towards the advance ferns. To compare the occurrence of different chemicals in different tissues of the stipe, the maximum score of each species has been counted and it has been given in the Table 4.3. If the comparison is made between different species, *C. viridis* of Pteridales shows the maximum occurrence of chemicals in high concentration in maximum regions of the stipe. It is with the maximum score of 13 followed by *H. glandulifera* for which the high concentration score is 11. The quantitative data for total phenol present in nearly 30 South Indian fern has been compiled by Karpagavinayagam (2005). The quantity ranges from 0.6 mg/g.d.wt in *C. parasitica* to 17.5 mg/g.d.wt in *H. glandulifera*. In the presence study also, the presence of maximum amount of different chemicals has been observed in *H. glandulifera* next to *C. viridis*. Based on this data, the 26 species can be divided into three groups with the number of high concentration score as follows: 1-5 with 16 species, 6-10 with 8 species and above 10 with 2 species. The occurrence of polyphenols in general, lignin and tannin in particular along with epicuticular waxes in high concentration as in the cases of *C. viridis* and *H. glandulifera* may indicate the high degree of defense mechanism.

Biochemical factors, both nutrient contents and toxic chemicals are the important factors to determine the herbivory (Karlson and Bode, 1969; Mukhopadhyay and Thapa, 2004). Herbivory is dependent on the presence of high concentration of nutrients like starch, proteins and lipids and low concentration of phenolic substances. Generally all the members of Athyriaceae from South India are without the phenolic compound tannin. So they are more susceptible to herbivores. Thus, out of six Athyrioid ferns observed from Upper Kothayar, five species have been infested by various herbivores (Karpagavinayagam, 2005). Karpagavinayagam (2005) has also observed severe herbivory by a hairy caterpillar in the tree fern *C. nilgirensis*. The reason for the severe herbivory in this species even with the presence of high concentration of total phenol (20.5 mg/g d wt) is attributed to the herbaceous nature of the leaves. It is also important to note that the total sugar in this species is very high when compared to other species of *Cyathea* in South India (Gopalakrishnan *et al.*, 1993). In the present study also high concentration of lignin in epidermis, hypodermis and Xylem and high concentration of polyphenols in hypodermis have been observed.

The origin of biosynthetic pathways of polyphenols, particularly lignin and tannin is an important event in the evolution of land plants to survive in the new environment. Lycophytes arose in the early Silurian (ca. 400 Million Years ago) and represent a major lineage of vascular plants that has evolved in parallel with the ferns, gymnosperms, and angiosperms. A hallmark of vascular plants is the presence of the phenolic lignin heteropolymer in xylem and other sclerified cell types. Although syringyl lignin is often considered to be restricted in angiosperms, it has been detected in lycophytes as well. Here we report the characterization of a cytochrome P450-dependent monooxygenase from the lycophyte *Selaginella moellendorffii*. Gene expression data, cross-species complementation experiments, and *in vitro* enzyme assays indicate that this P450 is a ferulic acid/coniferaldehyde/coniferyl alcohol 5-hydroxylase (F5H), and is capable of diverting guaiacyl-substituted intermediates into syringyl lignin biosynthesis. Phylogenetic analysis indicates that the *Selaginella* F5H represents a new family of plant P450s and suggests that it has evolved independently of angiosperm F5Hs.

Lignin is an aromatic heteropolymer that is deposited most abundantly in the secondary cell walls of vascular plants. It provides structural rigidity to the plant body while enabling individual tracheary elements to withstand the tension generated during water transport; it also serves a defensive role against herbivores and pathogens. Lignins are derived mainly from the phenylpropanoid monomers *p*-coumaryl, coniferyl, and sinapyl alcohol, which give rise to *p*-hydroxyphenyl, guaiacyl, and syringyl subunits when incorporated into the lignin polymer. In angiosperms, three cytochrome P450-dependent monooxygenases (P450s) are involved in the biosynthesis of lignin monomers, cinnamate 4-hydroxylase (C4H), *p*-coumaroyl shikimate/quinate 32-hydroxylase (C32H) and ferulic acid/coniferaldehyde/coniferyl alcohol 5-hydroxylase (F5H). C4H and C32H are responsible for phenylpropanoid 4 and 3-hydroxylation, respectively, whereas F5H catalyzes the 5-hydroxylation of coniferaldehyde and coniferyl alcohol, leading to the formation of syringyl lignin. Lignin monomer composition has been found to vary among major phyla of vascular plants. Generally, ferns and gymnosperms deposit lignins that are derived primarily from guaiacyl monomers together with a small proportion of *p*-hydroxyphenyl units, whereas angiosperm lignins are guaiacyl/syringyl copolymers that also can contain some *p*-hydroxyphenyl monomers. This distribution suggests that F5H may be a relatively recent addition to plants' biochemical repertoire. Nevertheless, there are older reports in the literature in which syringyl monomers have been detected in lignins from lycophytes, including species of *Selaginella*, by using histochemical reagents and by today's standards relatively crude chemical methods. These results have been verified recently by using more modern techniques. How species that diverged from angiosperms >400 Mya acquired the ability to synthesize syringyl lignin is unknown.

Jing Ke *et al.* (2008) found that, although guaiacyl lignin derivatives can be detected from all of the species examined, syringyl lignin derivative is only present in the three angiosperm species examined and *S. moellendorffii*. The lignin of *S. moellendorffii* has a high content of syringyl subunits, with a mole percentage of >70 per cent. Notably, a *Lycopodium* species, which represents another lycophyte lineage, does not deposit syringyl lignin. To be able to synthesize syringyl lignin, we hypothesized that the *Selaginella* genome encodes a phenylpropanoid 5-hydroxylase capable of diverting guaiacyl lignin precursors to syringyl lignin biosynthesis.

Lycophytes today comprise H"1,200 species in the three extant orders Lycopodiales, Selaginellales, and Isoetales, accounting for only a small and inconspicuous group of living vascular plants. In contrast, the ancestors of these plants once dominated the Earth's flora during the Carboniferous period and can be traced back to H"420 Mya, 280 million years earlier than the emergence of angiosperms. The distribution of syringyl lignin in the plant kingdom suggested two possible models for the evolution of F5H. First, the enzyme could have arisen early in plant evolution, was lost in ferns and gymnosperms, but was not lost in angiosperms or *Selaginella*. Alternatively, F5H could have evolved independently in lycophyte and angiosperm lineages after they had diverged. Our results suggest that the second model is correct and that F5H from *Selaginella* is functionally equivalent to, but phylogenetically independent from, angiosperm F5Hs. This conclusion is further supported by the observation that syringyl lignin derivatives are not detected in extant members of the Lycopodiaceae and have not been found in fossils of the extinct lycophyte *Sigillaria ovata* (order Lepidodendrales). Taken together, these data suggest that the Selaginellales may be the only lycophyte order that acquired the ability to synthesize syringyl lignin, although, if confirmed, early reports of syringyl lignin in *Isoetes* and *Huperzia* may indicate that the enzymatic activities required for syringyl lignin biosynthesis are more widespread within the lycophytes. Although independent occurrence of identical enzyme function in distinct lineages is not commonly observed, similar cases have been presented in the literature. For example, limonene synthase, a plant terpenoid synthase, has been characterized from angiosperm species and one gymnosperm species, *Abies grandis*. Despite their functional resemblance, phylogenetic analysis suggests that the genes that encode limonene synthase in angiosperms and gymnosperms evolved independently. In gibberellin biosynthesis, *ent*-kaurene oxidase and *ent*-kaurenoic acid oxidase from higher plants are encoded by P450s from CYP701 and CYP88 families, respectively, whereas the analogous enzymes in fungus *Gibberella fujikuroi* are encoded by very distinct P450s from CYP503 and CYP68 families. This phenomenon also has been attributed to be the result of convergent evolution (Jing Ke *et al.*, 2008).

It is interesting to consider what evolutionary advantages may have led to the independent invention of syringyl lignin in two lineages of vascular plants. For example, in angiosperms, syringyl lignin is often associated with fiber cells that have an important role in mechanical support. This correlation has led to the hypothesis that syringyl lignin may be superior to guaiacyl lignin in its ability to strengthen cell walls into which it is incorporated. Our study shows that, in *Selaginella*, syringyl lignin accumulates primarily in the sclerified cortical cells, suggesting that these cells may play an important role in support of the plant body. Alternatively, a recent study of resistance responses of wheat to pathogen attack revealed that syringyl lignin was hyperaccumulated in the plant cell wall in response to fungal penetration, suggesting that syringyl lignin also may provide a selective advantage in defense against pathogens (Jing Ke *et al.*, 2008).

In conclusion, they identified and characterized a unique cytochrome P450 from *Selaginella* that is capable of diverting guaiacyl-substituted intermediates into syringyl lignin biosynthesis. Our phylogenetic analysis suggested that the occurrences of syringyl lignin in lycophytes and angiosperms might be independent. The gene identified in this article also adds a potential tool for engineering lignin biosynthesis in gymnosperms where syringyl lignin is absent (Jing Ke *et al.*, 2008).

Polyphenols like tannin and lignin are not only act as anti-herbivory chemicals but also they act as antimicrobial chemicals. The rust fungi are obligate parasites, occurring on a wide range of angiosperms, gymnosperms and ferns. No rusts occur on fern allies. Most of these fungi are extremely host specific and it is argued that the evolutionary radiation of many rusts mirrors that of their hosts (Savile, 1967). Present histochemical studies on four fern allies (3- *Selaginella* and 1-*Lycopodiella*) show the presence of different chemicals at high concentration in maximum scores of eight or nine in three species (Table 4.3).

Two types of pheromones have been isolated so far in the Schizaeaceae, one from *Lygodium* and one from *Schizaea* (Yamane *et al.*, 1979). There are several questions remaining unanswered concerning the evolution in two closely related genera of two signals which, despite a cross-reaction between both pheromones, are chemically different.

In *A. evecta,* thin walled polygonal cells of ground tissue are filled with starch grains starch grains are usually large and spherical or oval in shape. The concentrations of these grains are more towards the base of petiole and gradually decrease towards the apex. At the top of the petiole and in the rachis cells are usually devoid of starch grains. Some of the cells of middle and inner zone contain tannin (Srivastava, 2008). Microchemical tests on the stipe of *C. dentata* reveal that the entire ground tissue is made up of cellulosic cells. Thin walled inner layers of cells usually have starch grains (Srivastava, 2008).

In the construction of any ordinary vascular plant there are three limiting tissue-surfaces of prime importance: i. The outer contour of the plant or part, complicated though it usually is in sub-aerial plants by the added cell surfaces lining the ventilating channels. ii. The endodermal sheath, which delimits the conducting tracts from the tissues that envelope them, and iii. The collective surface by which the dead tracheal system faces the living tissues that surrounds or permeates it. Each of the three is a surface of physiological transit, and each independently of the others will be a suitable subject of observation from the point of view of the proportion of surface to bulk as the size increases. The primary conducting tracts of the Pteridophyta offer the best examples of moulding in accordance with the principle as above stated (Bower,

1935). Increasing morphological complexity involving any increased complication of form would tend to meet the physiological need consequent on increasing size, by leveling up the proportion of surface of bulk. The present histochemical studies on lipids, polyphenols, lignins and tannins show the evolutionary progress of such chemicals from the primitive ferns towards the advance ferns. The occurrence of polyphenols in general, lignin and tannin in particular along with epicuticular waxes in high concentration as in the cases of *C. viridis* and *H. glandulifera* may indicate the high degree of defense mechanism.

# References

Bower, F.O. (1935). *Primitive Land Plants*. Macmillan and Co. Ltd. St. Martin's Street, London.

Chamberlain, C.J. (1924). Methods in plant histology. Univ. Chicago Press, Chicago.

Faridah-Hanum, I., Mustapa, M.Z., and Jaman Razali. (2008). Spore micromorphology and anatomy of the fern genus *Histiopteris* J.Sm. (Dennstaedtiaceae) in Peninsular Malaysia. *Internatl. J. Bot.,* **4**(2): 236-240.

Gahan, P.B. (1927). Plant Histochemistry and Cytochemistry: An introduction. Academic Press, Florida.

Gopalakrishnan, S., Rama, V., Angelin, S. and Manickam, V.S. (1993). Phytochemical studies on members of Dennstaedtiaceae and Cyatheaceae. *India fern Journal,* **10**: 146-151.

Haridass, E.T., and Suresh Kumar, N. (1985). Some techniques in the study of insect-host plant interactions. *In*: Pollen physiology and Fertilization. Ed. By Linsken, H.F., North-Holland Publ. Co., Amsterdam.

Hort, Sir, A. (1916). *Theophrastus. Enquiry into plants.* London. Heinamenn inducing factor of the fern *Anemia phyllitidis. J. Am. Chem. Soc.,* **93**: 5579-5581.

Jing-ke Weng, Xu Li, Jake Stout, and Clint Chapple. (2008). Independent origins of syringyl lignin in vascular plants. *Proc. Natl. Acad. Sci. USA.,* **105**(22): 7887-7892

Karlson, P. and Bode, P. (1969). Die inaktivierung des Ecdysones bei der Schmeissfliege *Calliphora erythrocephala* Meigen. *J. Insect Physiol.,* **15**: 111-118.

Karpagavinayagam, C. (2005). Preliminary survey on herbivory in South Indian ferns. M.Sc. Thesis submitted to Department of Plant Biology and Plant Biotechnology, St Xavier's College (Autonomous), Palayamkottai, Tamil Nadu, India.

Manickam, V.S., and Irudayaraj, V. (1992). Pteridophytic flora of the Western Ghats, South India, New Delhi.

Mukhopadhyay, A. and Thapa, D. (2004). Nutritional status of ferns and their relation to insect infestation from Darjeeling foothills and plains. *J. Bombay Nat. Hist. Soc.,* **101**(2): 224-226.

Ruthmann, A.C. (1970). Methods in cell Research. Cornell University Press, Ithaca, New York.

Savile, D.B.O. (1967). Evolution and relationships of the North American *Peducularis rusts* and their hosts. *Can. J. Bot.*, **45**: 1093-1103.

Srivastava, K. (2008). The petiolar structure of *Christella dentata* (Forssk.) Brownsey and Jermy (Thelypteridaceae, Pteridophyta). *Ethnobotanical Leaflets*, **12**: 96-102.

Yamane, H., Takahashi, H.N., Takeno, K. and Furuya, M. (1979). Identification of gibberellin $A_4$ methyl ester as a natural substance regulating formation of reproductive organs in *Lygodium japonicum*. *Planta*, **147**: 251-256.

Utilisation and Management of Medicinal Plants Vol. 2 (2014)    *Pages* **153–187**
*Editor-in-Chief:* **V.K. Gupta**
*Published by:* **DAYA PUBLISHING HOUSE, NEW DELHI**

# 5

# Histochemical Studies on Selected Medicinally Important Advanced Ferns of Western Ghats, South India

M. Johnson[1]*, V. Irudayaraj[1], D. Linson[1], S. Kalaiselvi[1],
N. Janakiraman[1] and A. Babu[1]

## ABSTRACT

*The present study was made to fill the lacuna with a detailed study of the stipe anatomy and histochemistry of the selected twenty one medicinally important ferns from Western Ghats, South India. The matured stipes of twenty one species was collected and used for anatomical and histochemical studies by taking free hand sections. For anatomical studies, the fresh free hand sections were stained using safranin. Histochemical tests were made on the fresh sections of the stipes treated with the following reagents to identify the presence or absence of chemicals like lipids, polyphenols, lignin and tannins. The anatomical characters such as shape, presence or absence of hairs, scales, groove and wing on the stipes; number of hypodermal layers, nature of ground tissue; number and shapes of vascular bundle have been illustrated. Lipids and polyphenols are abundantly present in the epidermal cells. Polyphenols and lignins are abundantly present in the hypodermis and they are also commonly present in ground tissue. Polyphenols are also present abundantly in the endodermis. Xylem and phloem shows the mild occurrence of polyphenols but in considerable frequency. In general, tannins are poorly represented in epidermis, hypodermis and ground tissue and they are almost absent in endodermis, xylem and phloem. The present histochemical studies on lipids, polyphenols, lignins and tannins show the evolutionary progress of such chemicals from the primitive ferns towards the advance ferns.*

*Keywords*: Histochemistry, Pteridophytes, Anatomy, Stipes.

---

1   Centre for Plant Biotechnology, Department of Plant Biology and Plant Biotechnology, St. Xavier's College (Autonomous), Palayamkottai – 627 002, Tamil Nadu, India.
*   *Corresponding author*: E-mail: ptcjohnson@gmail.com

# Introduction

Pteridophytes are the successful group of primitive land plants with the evolution of several morphological, anatomical and biochemical adaptations to survive in the new terrestrial environment. Out of 1250 species of pteridophytes occurring in India, 173 species are ethnomedicinally important (Singh, 2003). As far as south Indian ferns are concerned, taxonomic and cytological studies have been carried out almost completely. A good number of ferns have been subjected to preliminary phytochemical screening (Irudayaraj *et al.*, 2010; Haripriya *et al.*, 2010; Irene Pearl *et al.*, 2011; Paulraj *et al.*, 2011; Mithraja *et al.*, 2012; Pauline Vincent *et al.*, 2012). But all such studies are mainly on the aerial parts or entire plants. There is no effort to know the distribution of particular chemicals in particular part or tissues. Histochemistry is the branch of histology dealing with the identification of chemical components of cells and tissues (Krishnamurthy, 1998). Histochemical techniques have been employed to characterize structure and development, and to study time course of deposition and distribution of major storage compounds such as protein, lipid, starch, phytin and minerals such as calcium, potassium and iron in rice grains (Krishnan *et al.*, 2001; Krishnan and Dayanandan, 2003). The use of histochemical attributes of plants in solving critical taxonomic problems is now gaining wider popularity just as the use of other characters. A perusal through the botanical literature shows that the use of histochemistry in taxonomic conclusions is now a common practice (Heintzelman and Howard, 1948; Al-Rais *et al.*, 1971; Mathew and Shah, 1984; Edeoga and Okoli, 1995).

It is well known that floral characters are most widely used in the classification of angiosperms. In lower plants, flowers are lacking and taxonomists have always had to rely on other characters. In Pteridophytes, vegetative characters particularly rhizome, scales, cutting of fronds, venation pattern and reproductive characters *viz.*, soral characters, sori arrangement patterns, sporangium, number of chromosomes are commonly used in the classification (Manickam and Irudayaraj, 1992). The realization that morphology and anatomy are inseparable aspects, which led to the use of alternative pairs of terms such as macromorphology and micromorphology. In addition to the morphology and anatomy, histochemical studies can be used to study the chemical composition of the cells resulting from growth and development. These data can be used to make inferences about the importance of secondary metabolites in growth and development of plants. Several species of pteridophytes have already been proved for having higher degree of antimicrobial activity. Identification and localization of active compounds will enhance the high potential compound production in the pharmaceutical industries. With this knowledge, the present investigation was made to fill the lacuna with a detailed study of the stipe anatomy and histochemistry of the selected 21 ferns from Western Ghats, South India. The present anatomical and histochemical studies on ferns will not only give basic knowledge about the adaptive potential of this successful group of land plants, but also it will enumerate the qualitative features of important medicinal ferns.

# Materials and Methods

Stipes of different species *viz.*, *Pseudophegopteris pyrrhorhachis* (Kunze) Ching, *Macrothelypteris torresiana* (Gaudich.) Ching, *Trigonospora ciliata* (Wall. Ex Benth.)

Holttum, *Cyclosorus interruptus* (Willd.) H. Ito, *Amphineuron terminans* (Hook.) Holttum, *Sphaerostephanos arbuscula* (Willd.) Holttum, *Christella parasitica* (L.) H.Lev. and *Christella dentata* (Forssk.) Brownsey and Jermy belonging to the family Thelypteridaceae, *Asplenium cheilosorum* Kunze ex Mett. (Aspleniaceae), *Diplazium muricatum* (Mett.) Alderw., *Diplazium travancoricum* Beddome and *Diplazium brachylobum* (Sledge) Manickam and Irudayaraj belonging to the family Athyriaceae, *Tectaria paradoxa* (Fee) Sledge, *Arachniodes tripinnata* (Goldm.) Sledge, *Arachniodes amabilis* (Bl.) Kuntze, *Arachniodes aristata* (Forst. f) Tindale and *Dryopteris sparsa* (Buch. Hm. Ex D. Don) C. Chr. belonging to the family Dryopteridaceae, *Bolbitis appendiculata* (Willd.) K. Iwatz. (Lomariopsidaceae), *Blechnum orientale* L. (Blechnaceae), *Leptochilus decurrens* Bl. and *Pyrrosia porosa* var. *porosa* Hovenkamp belonging to the family Polypodiaceae were collected from Upper Kothayar, Tirunelveli hills, Western Ghats, South India. The selected specimens were identified based on "Pteridophyte Flora of the Western Ghats, South India" and confirmed by Dr. Irudayaraj, Department of Botany, St. Xavier's College, Palayamkottai (Manickam and Irudayaraj, 1992).

The fresh materials were used for anatomical and histochemical studies by taking free hand sections. For anatomical studies, the fresh free hand sections were stained using safranin. Histochemical tests were made on the fresh sections of the stipes treated with the following reagents to identify the presence or absence of chemicals like lipids, polyphenols, lignin and tannins. Sudan III is used to detect lipids (Ruthmann, 1970), Fast Blue BB salts for polyphenols (Gahan, 1927), Lugol's Iodine is used to detect lignin and tannin (Chamberlain, 1924; Haridass and Suresh Kumar, 1985). The stained sections were observed on Motic trinocular microscope (Japan). They were photographed at different magnifications and at different views. Based on the photographs taken, anatomical description and localization of tested chemicals were done.

# Results and Discussion

In the present study, anatomical and histochemical studies have been carried out in the stipes of selected 21 ferns. The stipe is somewhat easier when compared to other major parts such as leaves and rhizomes. The anatomical characters such as shape, presence or absence of hairs, scales, groove and wing on the stipes; number of hypodermal layers, nature of ground tissue; number and shapes of vascular bundle etc. have been illustrated in Table 5.1.

## Shape

In transverse section the outline of the stipe is circular in almost all the species except in *A. cheilosorum* and *P. porosa* var. *porosa* in which the outline is semicircular [Plates 5.3(1a) and 5.5(4a)]. The smooth adaxial side of the stipe was observed in *P. pyrrhorhachis, M. torresiana, A. terminans, S. arbuscula, T. paradoxa, B. appendiculata, L. decurrens* and *C. interruptus*; grooved in *T. ciliata, C. parastica, A. cheilosorum, D. brachylobum, D. travancoricum, D. muricatum, A. tripinnata, A. amabilis, A. aristata, D. sparsa* and *B. orientale*; flattened in *C. dentata* and *P. porosa* var. *porosa* [Plates 5.2(2a) and 5.5(4a)] and winged in *A. cheilosorum* and *P. porosa* var. *porosa* [Plates 5.3(1a) and 5.5(4a)]. The number and depth of groove may vary from position to position within a stipe.

**Table 5.1:** Comparative anatomical features of the selected ferns of the Western Ghats.

| Anatomical Characters | 1 | 2 | 3 | 4 | 5 | 6 | 7 | 8 | 9 | 10 | 11 | 12 | 13 | 14 | 15 | 16 | 17 | 18 | 19 | 20 | 21 |
|---|---|---|---|---|---|---|---|---|---|---|---|---|---|---|---|---|---|---|---|---|---|
| Circular | + | + | + | + | + | + | + |  | + | + | + | + | + | + | + | + | + | + |  | + | + |
| Acircular |  |  |  |  |  |  |  | + |  |  |  |  |  |  |  |  |  |  | + |  |  |
| Smooth | + | + |  | + | + |  | + |  |  |  |  |  |  |  |  | + | + | + |  | + | + |
| Grooved |  |  | + | + | + |  |  |  |  |  |  |  |  |  | + |  |  | + | + |  |  |
| Flattened |  |  |  |  |  | + |  |  |  |  |  |  |  |  |  |  |  |  |  |  |  |
| Winged |  |  |  |  |  |  |  |  |  |  |  |  |  |  |  |  |  |  |  |  |  |
| Smooth Epidermis | + |  |  |  | + |  |  | + |  | + | + | + | + | + |  |  | + | + |  | + | + |
| Unicellular Hairs |  | + |  | + | + | + | + |  | + |  |  |  |  |  | + |  |  |  |  |  |  |
| Multicellular Hair |  |  | + |  |  |  |  |  |  |  |  |  |  |  | + |  |  |  | + |  |  |
| Scaly |  |  |  |  |  |  |  |  |  |  |  |  |  |  |  |  |  |  | + |  |  |
| Few Hypodermis |  | + |  |  |  |  |  |  |  |  |  |  |  |  |  |  | + |  |  |  |  |
| Many Hypodermis | + | + | + | + | + | + | + | + | + | + | + | + | + | + | + | + |  | + | + | + | + |
| Continuous Hypodermis | + | + | + | + | + | + | + | + | + | + | + | + | + | + | + | + | + | + | + | + | + |
| Discontinuous |  |  |  |  |  |  |  |  |  |  |  |  |  |  |  |  |  |  |  |  |  |
| Paranchymatous Ground tissue | + | + | + | + | + | + | + | + | + | + | + | + | + | + | + | + | + | + | + | + | + |
| Chlorenchyma Ground Tissue |  |  |  |  |  |  |  |  |  |  |  | + |  |  |  |  |  |  |  | + |  |
| Single Vascular Bundle | + | + | + |  | + |  | + |  |  |  |  | + |  |  |  |  |  |  |  |  |  |
| Two Vascular Bundle |  |  |  | + |  | + |  |  | + | + |  |  |  |  |  |  |  |  |  |  |  |
| Three Vascular Bundle |  |  |  |  |  |  |  |  |  |  | + |  |  |  |  |  |  |  |  |  |  |
| More than Three |  |  |  |  |  |  |  | + |  |  |  |  | + | + | + | + | + | + | + | + | + |
| Circular VB |  |  |  |  |  |  |  |  |  | + | + |  |  | + | + | + | + | + | + | + | + |
| Boat VB |  |  |  |  |  |  |  |  |  |  |  |  |  |  |  |  |  |  |  |  |  |
| C shaped VB | + | + | + |  | + |  |  |  | + |  |  | + |  |  |  |  |  |  |  |  |  |
| Dumb bell VB |  |  |  | + |  | + |  |  |  |  |  |  |  |  |  |  |  |  |  |  |  |
| Ring |  |  |  |  |  |  |  |  |  |  |  |  |  |  |  |  |  |  |  |  |  |
| Single Xylem | + | + | + | + | + |  |  |  | + | + | + | + |  | + | + | + | + | + | + | + | + |
| Splitted Xylem |  |  |  |  |  |  | + | + |  |  |  |  |  |  |  |  |  |  |  |  |  |
| **Pith Absent** |  |  |  |  | + |  |  |  |  |  |  |  |  |  |  |  |  |  |  |  |  |

## Epidermis

In all the species studied, the epidermis is single layered with thick or thin cuticle on the outer side. Usually it is smooth without bearing any epidermal appendages. Rarely the epidermis bears unicellular or multicellular hairs or rarely scales. It also depends upon the position of the stipe. Usually the appendages are commonly seen towards the base of the stipe since the rhizome in all the species bears scales or hairs. Unicellular hairs were observed in *M. torresiana*, *A. terminans*, *C. parastica*, *C. dentata*, *S. arbuscula*, *D. brachylobum* and *T. paradoxa*; Multicellular hairs in *T. ciliata*, *A. amabilis* [Plate 5.1(3a), Plate 5.2(1a, 2a, 3a, 4b ) and Plate 5.4(4a)] and *P. porosa* var. *porosa* [Plate 5.5(4a)]. Other species were showed the smooth epidermis. The epidermal cells are generally rectangular or barrel-shaped and colourless. The epidermis is without any stomata. Generally in land plants, the epidermis, particularly the cuticle plays an important role in the protection of the plants against pathogens. Usually the cuticle is made up of several waxy compounds which play an important role in defense mechanisms.

## Hypodermis

Sclerenchymatous hypodermis is present in all the 21 studied and the number of layers varies from species to species. Usually it ranges from 3-10. In all the species, the sclerenchymatous hypodermis is present as a continuous ring next to the epidermis. The stipes of *P. pyrrhorhachis*, *M. torresiana* and *B. appendiculata* showed only few layers of sclerenchymatous layers. Other selected species were showed the multilayer of hypodermis [Plates 5.6(1e, 2c) and 5.8(2b)]. The sclerenchymatous hypodermis gives mechanical support to the stipe. Depends upon the number of hypodermal layers the rigidity of the stipe varies. Thus the stipe of *P. pyrrhorhachis* and *M. torresiana* is soft with the presence of three sclerenchymatous layers.

## Ground Tissue

Next to the sclerenchymatous, parenchymatous ground tissue is present up to the endodermis. The number of parencymatous layers strictly depends upon the size of the stipe. The individual cells of the ground tissue are circular or polygonal in outline and they usually enclose small intercellular spaces. The parenchymatous ground tissue is the region of stored food materials.

## Vascular Bundles

The features of vascular bundles in the stipe are very important and the number, size and shape are highly variable from species to species. In most of the species, there is a single vascular bundle at the centre of the stipe. But in the following species *M. torresiana* and *C. dentata*, two vascular bundles are scaterred. It is to be noted that the number and shape of the vascular bundles may vary from position to position within a stipe. The number and arrangement of vascular bundles in the stipe is an important character which is used to delimit major groups of ferns. The shape of the vascular bundles is highly variable. It may be of U-shape, V-shape, C-shape, boat-shape, circular or ovate. The vascular bundles are almost circular in *T. ciliata*, *C. dentata*, *A. cheilosorum*, *T. paradoxa*, *A. tripinnata*, *A. amabilis*, *D. sparsa*, *B. appendiculata*, *L. decurrens* and *P. porosa* var. *porosa*. In *P. pyrrhorhachis* [Plate 5.1(1a)], *C. interruptus*

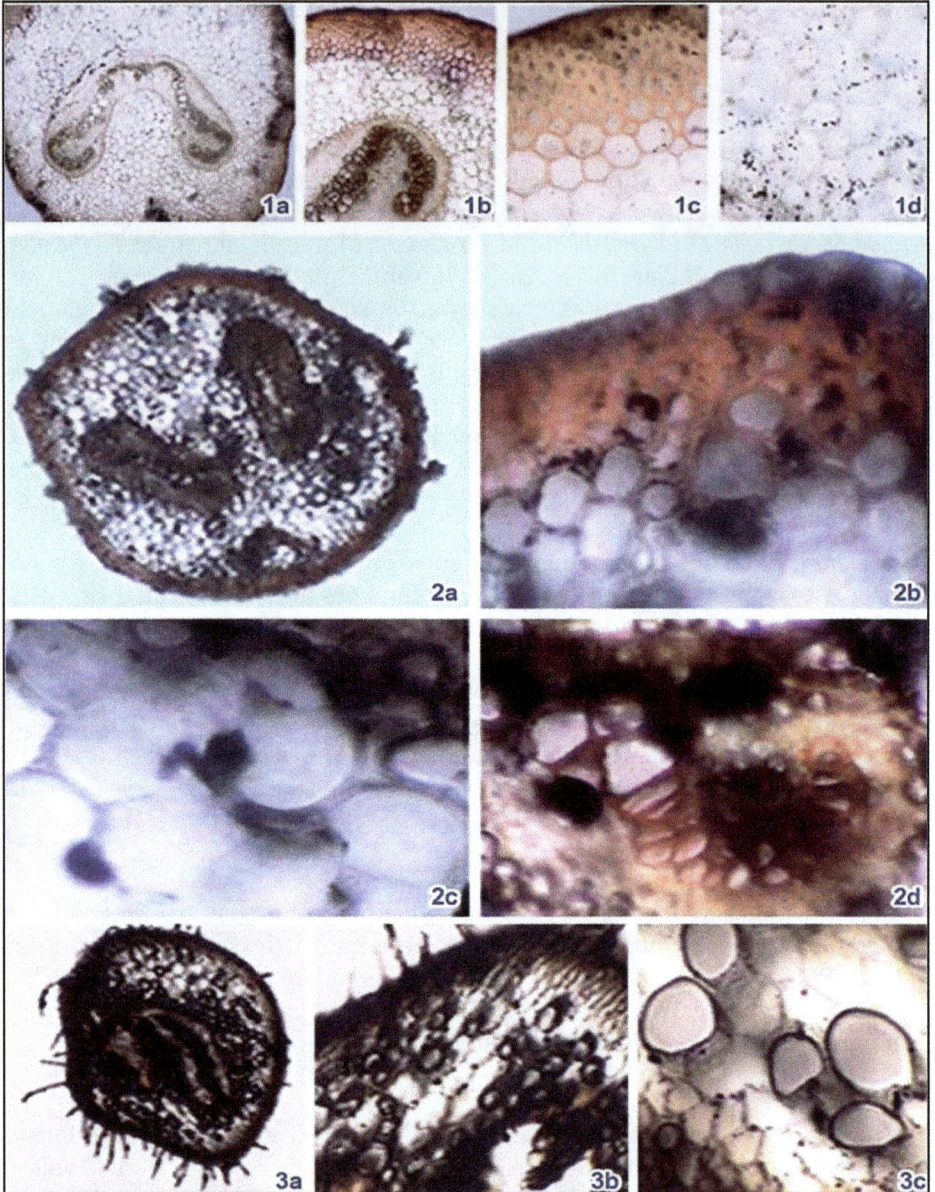

**Plate 5.1**: Results of histochemical analysis on lipids.

1: *Pseudophegopteris pyrrhorhachis* (Kunze) Ching—1a: T.S. of stipe–A portion; 1b: T.S. of stipe – A portion enlarged; 1c: A portion enlarged showing epidermis and hypodermis; 1d: Ground tissue enlarged; 2: *Macrothelypteris torresiana* (Gaudich.) Ching.—2a: T.S. of stipe–Entire view; 2b: A portion enlarged showing epidermis, hypodermis and ground tissue; 2c: Ground tissue enlarged; 2d: Portion of vascular bundle enlarged showing xylem and phloem; 3: *Trigonospora ciliata* (Wall. ex Benth.) Holttum—3a: T.S. of stipe – Entire view; 3b: A portion enlarged showing epidermis, hypodermis and ground tissue; 3c: Portion of vascular bundle enlarged showing xylem and phloem.

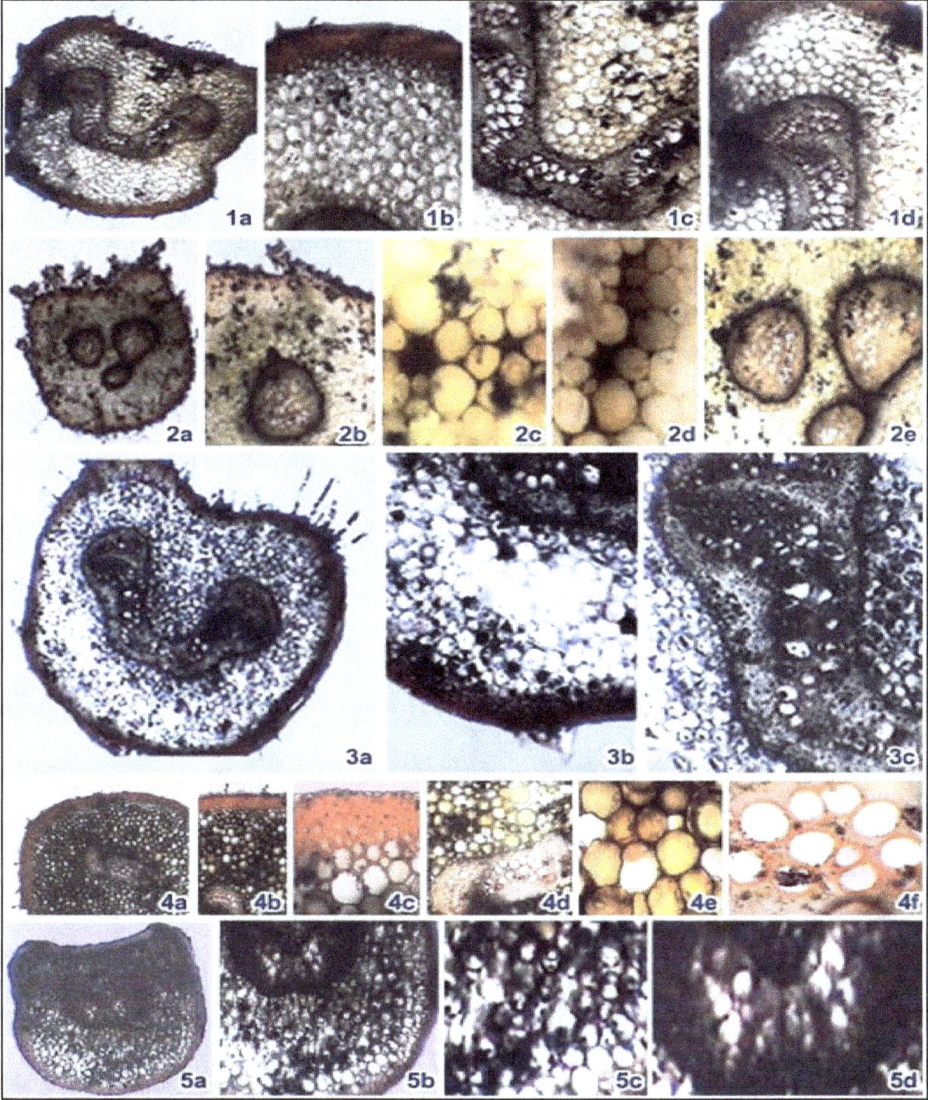

**Plate 5.2**: Results of histochemical analysis on lipids.

1: *Christella parasitica* (L.) H. Lev.—1a: T.S. of stipe–Entire view; 1b: T.S. of stipe–Portion enlarged; 1c, d: A portion of vascular bundle enlarged; 2: *Christella dentata* (Forssk.) Brownsey and Jermy—2a: T.S. of stipe–Entire view, 2b: T.S. of stipe–Portion enlarged; 2c: Ground tissue enlarged, 2d: Endodermis enlarged; 2e: Vascular bundles enlarged; 3: *Sphaerostephanos arbuscula* (Willd.) Holttum—3a: T.S. of stipe–Entire view; 3b: T.S. of stipe–Portion enlarged; 3c: Vascular bundle enlarged; 4: *Amphineuron terminans* (Hook.) Holttum—4a: T.S. of stipe–A portion, 4b: T.S. of stipe–Portion enlarged; 4c: A portion enlarged showing epidermis, hypodermis and ground tissue; 4d: Portion of vascular bundle enlarged, 4e: Ground tissue enlarged; 4f: Xylem enlarged; 5: *Cyclosorus interruptus* (Willd.) H. Ito—5a: T.S. of stipe–Entire view; 5b: T.S. of stipe–Portion enlarged, 5c: Ground tissue enlarged; 5d: Portion of vascular bundle enlarged.

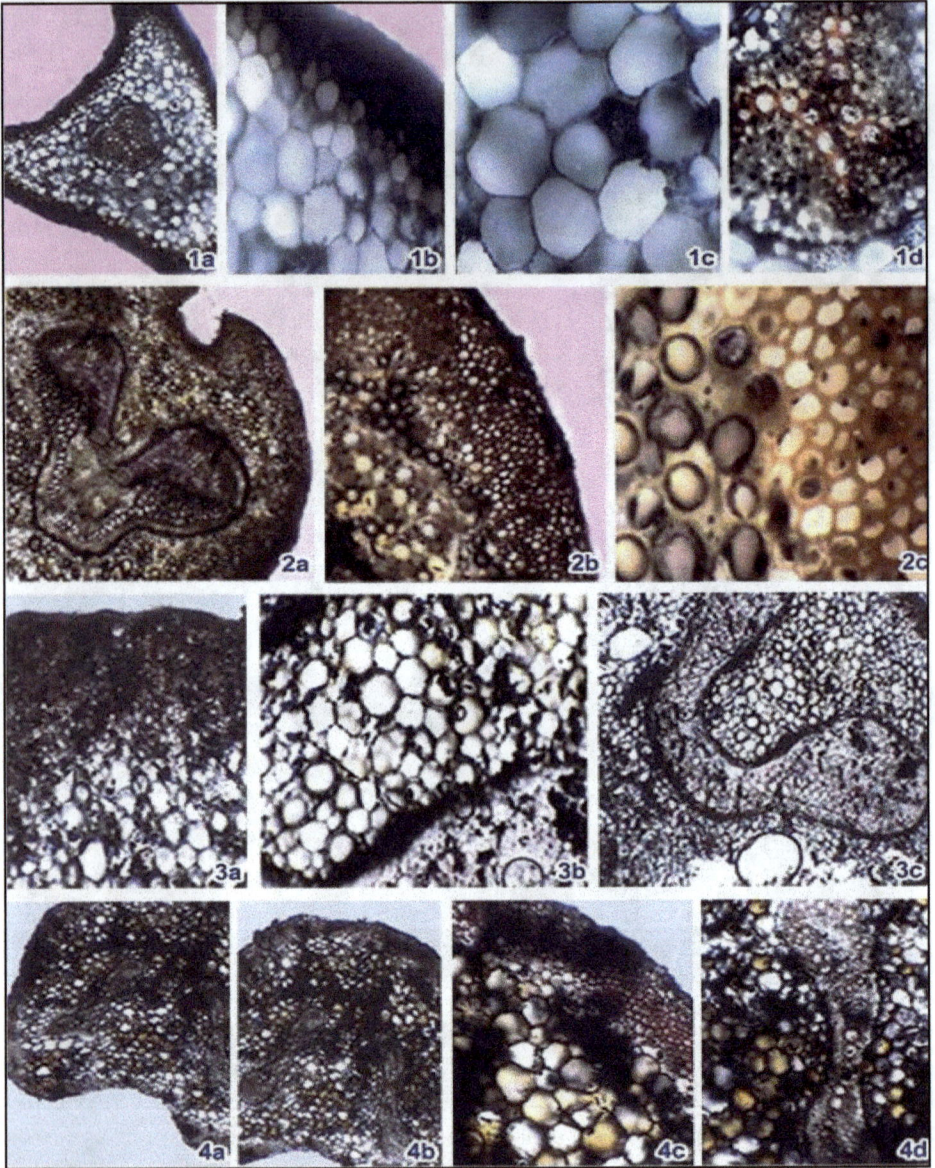

**Plate 5.3**: Results of histochemical analysis on lipids.

1: *Asplenium cheilosorum* Kunze ex Mett—1a: T.S. of stipe–A portion; 1b: A portion enlarged showing epidermis, hypodermis and ground tissue; 1c: Ground tissue enlarged; 1d: Portion of vascular bundle enlarged; 2: *Diplazium brachylobum* (Sledge) Manickam and Irudayaraj—2a: T.S. of stipe–A portion; 2b: A portion enlarged showing epidermis, hypodermis and ground tissue; 2c: A portion enlarged showing hypodermis and ground tissue; 3: *Diplazium muricatum* (Mett.) Alderw—3a: T.S. of stipe–A portion; 3b: A portion enlarged showing ground tissue; 3c: A portion of vascular bundle; 4: *Diplazium travancoricum* Beddome—4a, b: T.S. of stipe–Two different portions; 4c: A portion enlarged showing epidermis, hypodermis and ground tissue; 4d: Portion of vascular bundle enlarged.

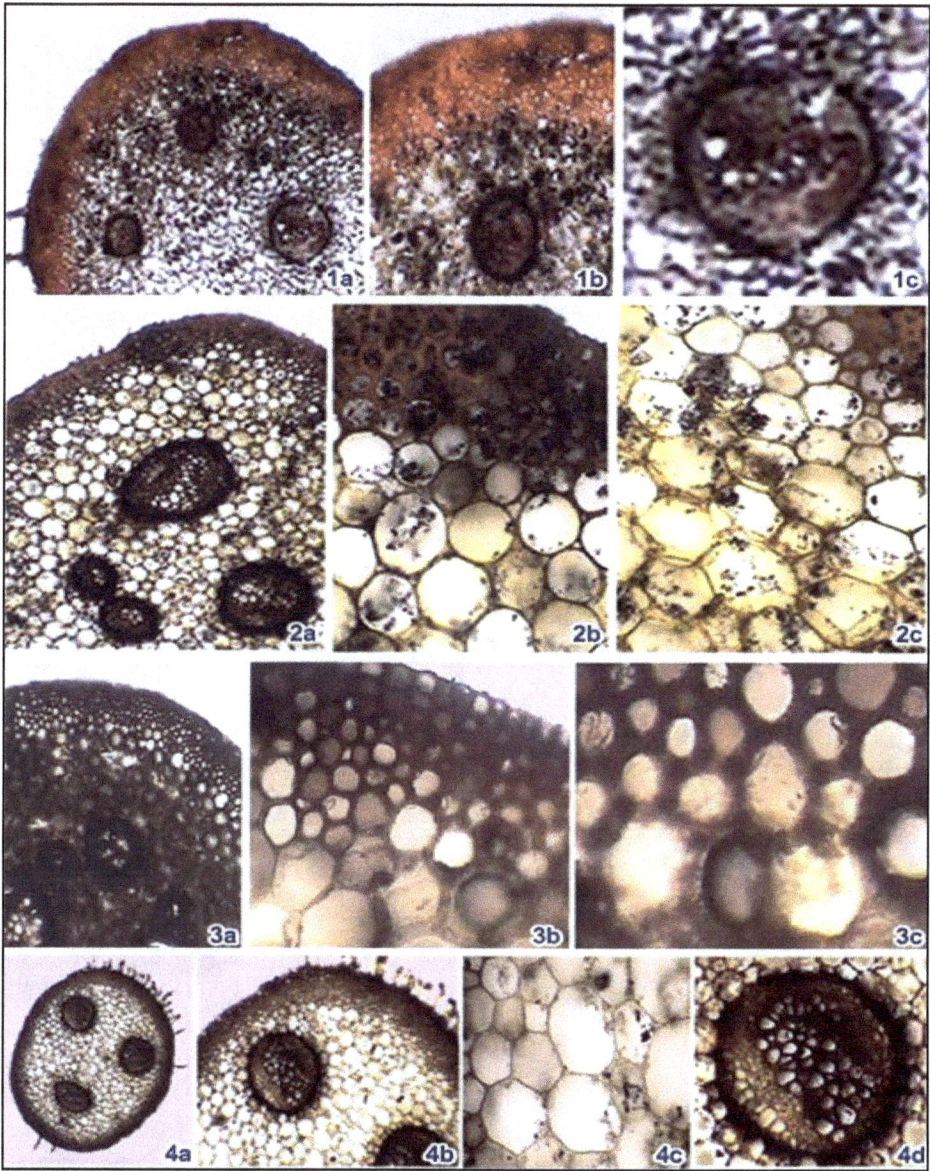

**Plate 5.4**: Results of histochemical analysis on lipids.

1: *Arachniodes amabilis* (Bl.) Kuntze—1a: T.S. of stipe–A portion; 1b: A portion enlarged showing epidermis, hypodermis and ground tissue; 1c: A single vascular bundle; 2: *Arachniodes tripinnata* (Goldm.) Sledge—2a: T.S. of stipe–A portion; 2b: A portion enlarged showing epidermis, hypodermis and ground tissue; 2c: Ground tissue enlarged; 3: *Dryopteris sparsa* (Buch. Hm. Ex D: Don) C.Chr.—3a: T.S. of stipe–A portion; 3b: A portion enlarged showing epidermis, hypodermis and ground tissue; 3c: Sclerenchymatous hypodermis enlarged; 4: *Tectaria paradoxa* (Fee) Sledge—4a: T.S. of stipe–Entire view; 4b: A portion enlarged showing epidermis, hypodermis and ground tissue; 4c: Ground tissue enlarged; 4d: A single vascular bundle enlarged.

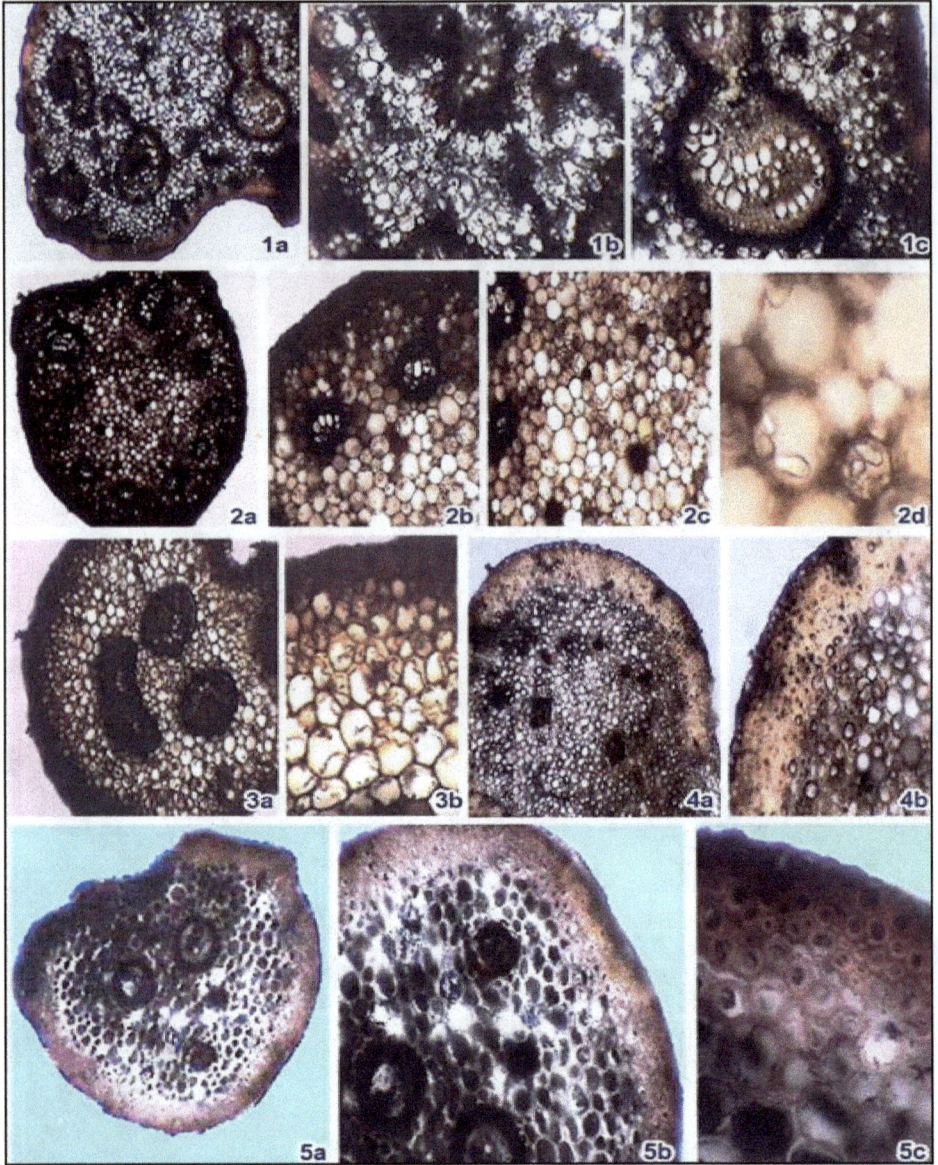

**Plate 5.5**: Results of histochemical analysis on lipids.

1 and 2: *Blechnum orientale* L.—1a: T.S. of stipe–A portion; 1b: A portion enlarged showing ground tissue and vascular bundles; 1c: A single vascular bundle enlarged; 2a: T.S. of stipe–Entire view; 2b: A portion enlarged showing ground tissue and vascular bundles; 2c, d: Ground tissue at different magnifications; 3: *Bolbitis appendiculata* (Willd.) K. Iwatz.—3a: T.S. of stipe–A portion; 3b: A portion enlarged showing epidermis, hypodermis and ground tissue; 4: *Pyrrosia porosa* var. *porosa* Hovenkamp—4a: T.S. of stipe–A portion with scattered vascular bundles; 4b: A portion enlarged showing epidermis, hypodermis and ground tissue; 5: *Leptochilus decurrens* Bl.—5a: T.S. of stipe–Entir view; 5b, c: A portion enlarged at different magnifications showing epidermis, hypodermis and ground tissue.

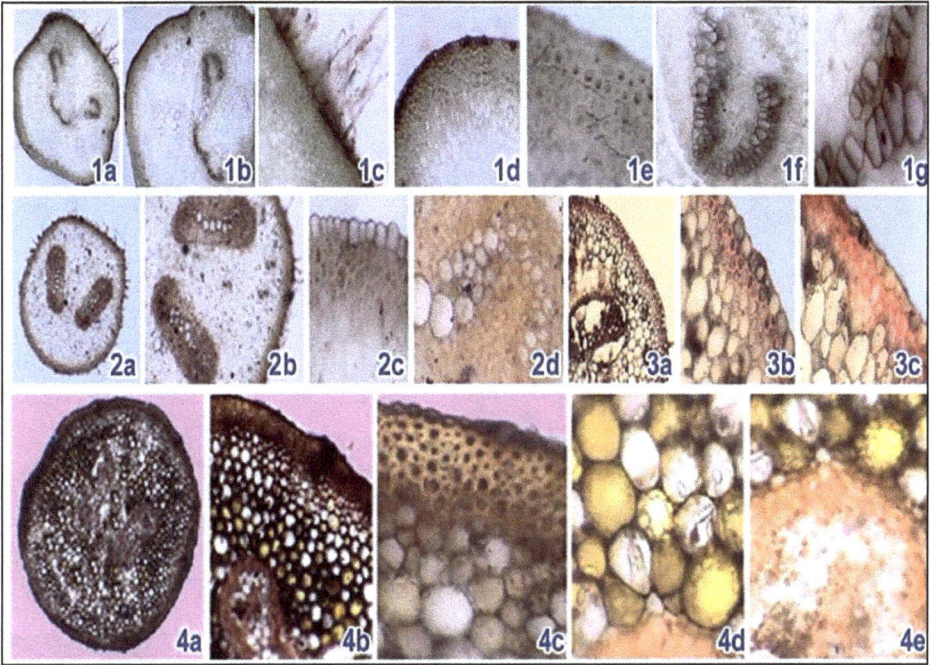

**Plate 5.6**: Results of histochemical analysis on polyphenols.

1: *Pseudophegopteris pyrrhorhachis* (Kunze) Ching—1a: T.S. of stipe–Entire view; 1b: A portion enlarged showing epidermis, hypodermis, ground tissue and portion of vascular bundle; 1c: Epidermis with unicellular-acicular hairs; 1d, e: A portion enlarged at two different magnifications showing epidermis (d, e), hypodermis (d, e) and ground tissue (d); 1f: A portion of vascular bundle enlarged, 1g. Xylem enlarged; 2: *Macrothelypteris torresiana* (Gaudich.) Ching—2a: T.S. of stipe – Entire view; 2b: A portion enlarged showing epidermis, hypodermis and ground tissue with vascular bundles; 2c: A portion enlarged showing epidermis, hypodermis and ground tissue; 2d: Xylem and phloem enlarged; 3: *Trigonospora ciliata* (Wall. ex Benth.) Holttum—3a: T.S. of stipe–A portion; 3b, c: A portion enlarged showing epidermis, hypodermis and ground tissue; 4: *Amphineuron terminans* (Hook.) Holttum—4a: T.S. of stipe–Entire view, 4b: T.S. of stipe–Portion enlarged; 4c: A portion enlarged showing epidermis, hypodermis and ground tissue; 4d: Ground tissue enlarged; 4e: Portion of vascular bundle enlarged showing xylem and phloem.

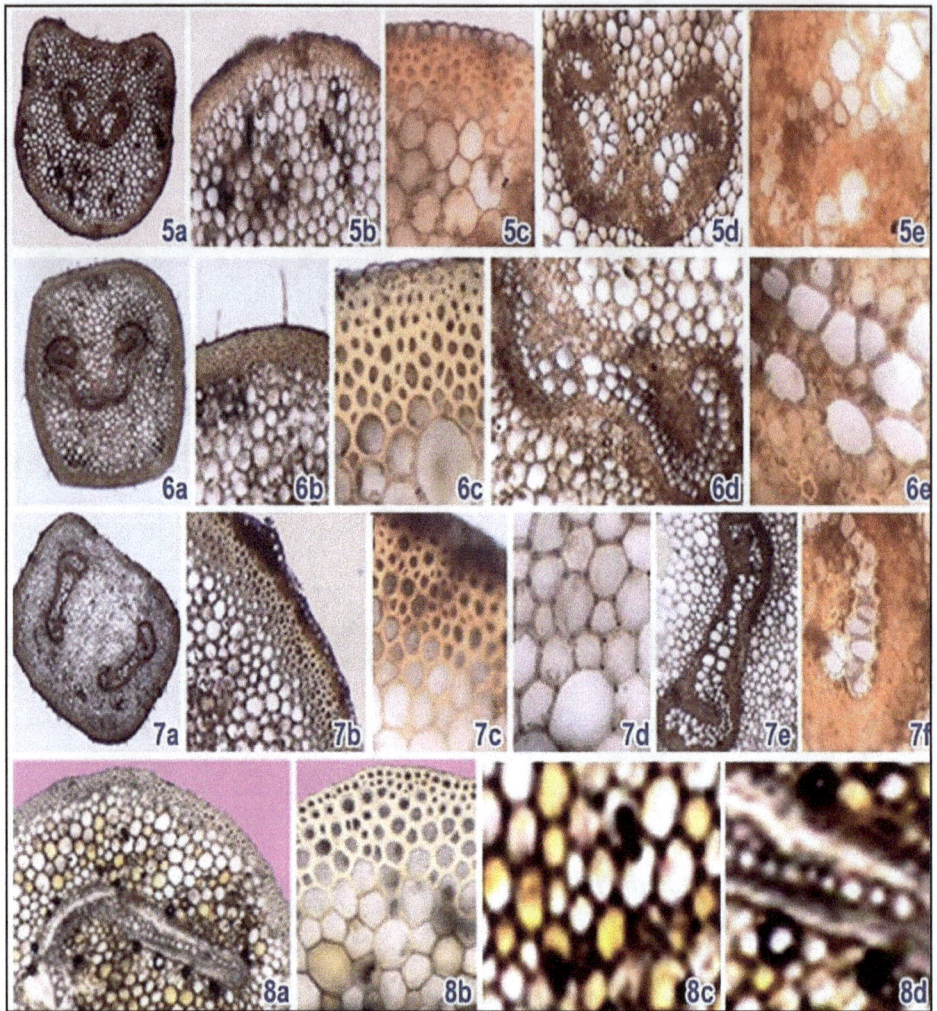

**Plate 5.6**: Results of histochemical analysis on polyphenols.

5: *Cyclosorus interruptus* (Willd.) H.Ito—5a: T.S. of stipe–Entire view; 5b, c: T.S. of stipe–A portion enlarged showing epidermis, hypodermis and ground tissue; 5d: Vascular bundle enlarged, 5d: Portion of vascular bundle showing xylem and phloem; 6: *Sphaerostephanos arbuscula* (Willd.) Holttum—6a: T.S. of stipe–Entire view, 6b: T.S. of stipe–Portion enlarged; 6c: A portion enlarged showing epidermis, hypodermis and ground tissue; 6d: Vascular bundle enlarged, 6e: Xylem and phloem enlarged; 7: *Christella dentata* (Forssk.) Brownsey and Jermy—7a: T.S. of stipe–Entire view, 2b: T.S. of stipe–Portion enlarged; 7c: A portion enlarged showing epidermis, hypodermis and ground tissue; 7d: Vascular bundles enlarged; 7f: Xylem and phloem enlarged; 8: *Christella parastica* (L.) H.Lev.—8a: T.S. of stipe–A portion; 8b: A portion enlarged showing epidermis, hypodermis and ground tissue; 8c: Ground tissue enlarged, 8d: A portion of vascular bundle enlarged.

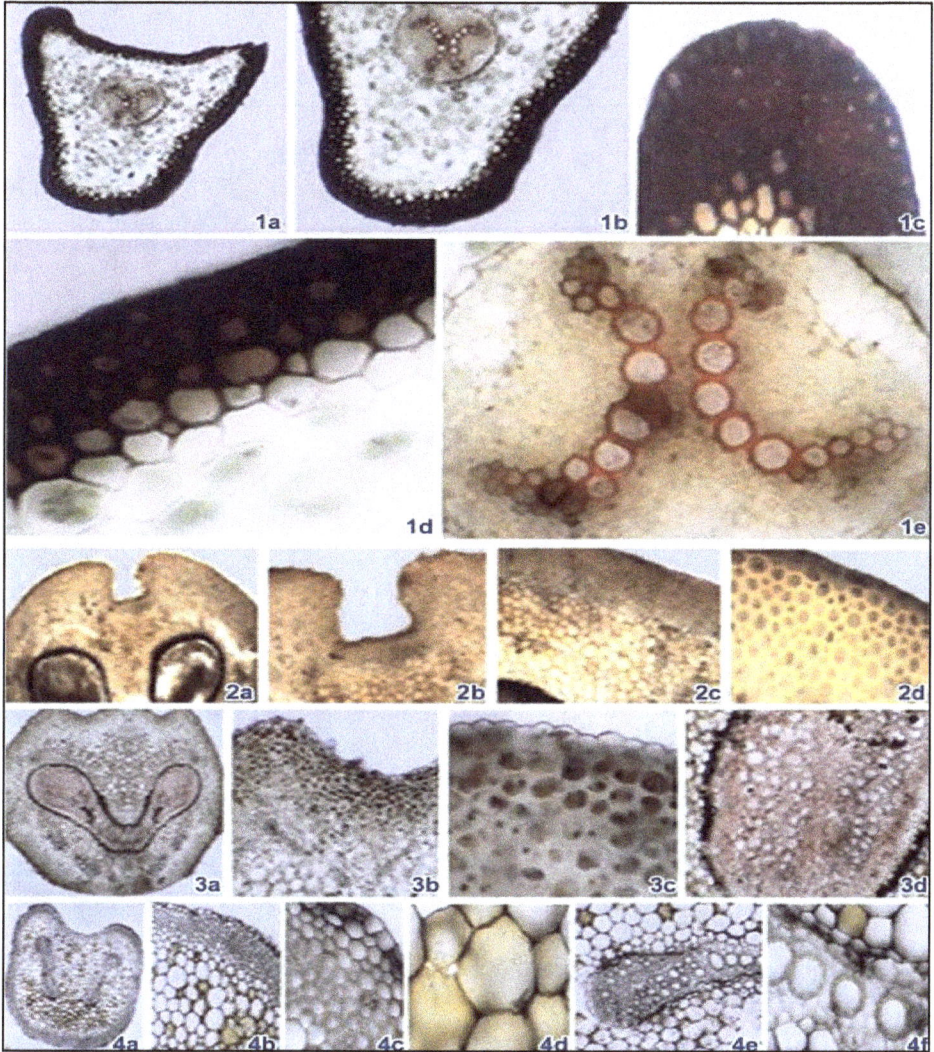

**Plate 5.7**: Results of histochemical analysis on polyphenols.

1: *Asplenium cheilosorum* Kunze ex Mett; 1a: T.S. of stipe – Entire view; 1b: A portion enlarged; 1c, d: A portion enlarged at different magnifications showing epidermis, hypodermis and ground tissue; 1e: Portion of vascular bundle enlarged showing xylem and phloem; 2: *Diplazium brachylobum* (Sledge) Manickam and Irudayaraj—2a, b: T.S. of stipe – Adaxial portion at different magnification showing deep groove; 2b, c: A portion enlarged showing epidermis, hypodermis and ground tissue; 3: *Diplazium muricatum* (Mett.) Alderw; 3a: T.S. of stipe–Entire view; 3b: A portion enlarged showing epidermis, hypdermis and ground tissue; 3c: A portion enlarged showing epidermis and hypodermis; 3d: A portion of vascular bundle enlarged showing endodermis, pericycle, xylem and phloem; 4: *Diplazium travancoricum* Beddome—4a: T.S. of stipe–Entire view; 4b: A portion enlarged at different magnification showing epidermis, hypodermis and ground tissue; 4c: Ground tissue enlarged; 4d, e: Portion of vascular bundle enlarged showing endodermis, pericycle, xylem and phloem.

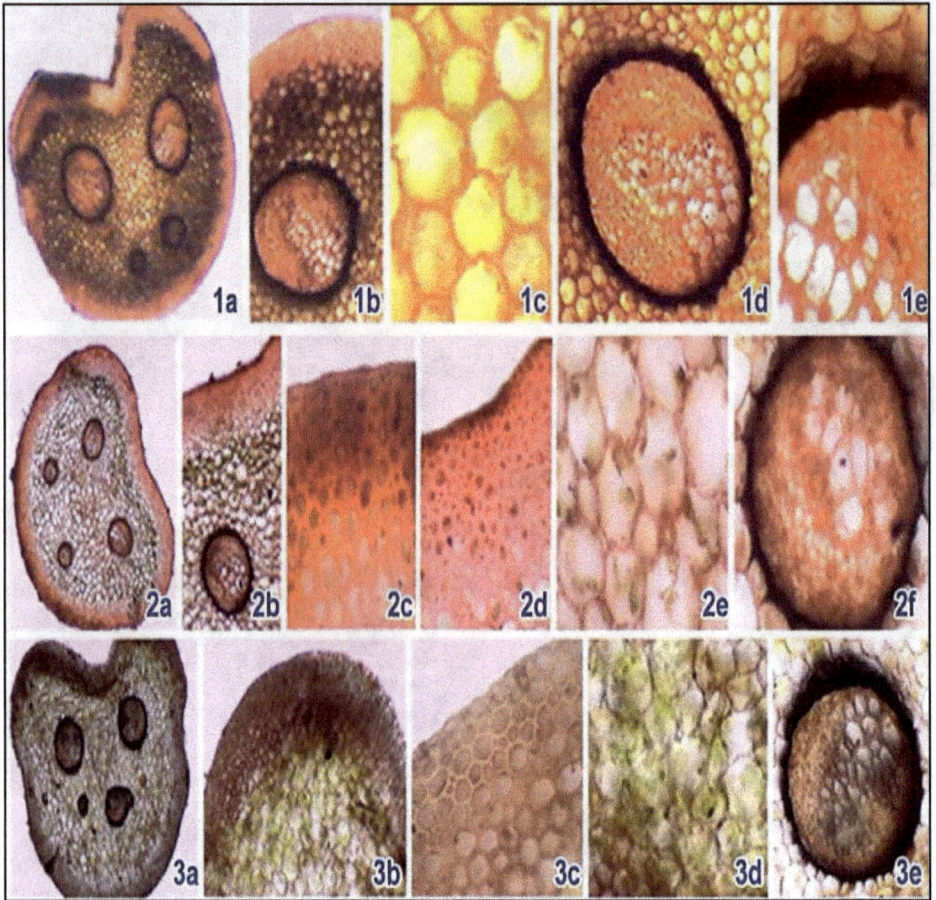

**Plate 5.8**: Results of histochemical analysis on polyphenols.

1: *Arachniodes tripinnata* (Goldm.) Sledge—1a: T.S. of stipe–Entire view; 1b: A portion enlarged, 1c: Ground tissue enlarged; 1d, e: Entire (d) and portion (e) of vascular bundle enlarged showing endodermis, pericycle, xylem and phloem; 2: *Arachniodes amabilis* (Bl.) Kuntze—2a: T.S. of stipe–Entire view; 2b: A portion enlarged showing epidermis, hypodermis and ground tissue; 2c, d: A portion enlarged showing epidermis and hypodermis; 2d: Ground tissue enlarged; 2e: A single vascular bundle enlarged; 3: *Dryopteris sparsa* (Buch. Ham. ex D: Don) C.Chr.—3a: T.S. of stipe–A portion; 3b: A portion enlarged showing epidermis, hypodermis and ground tissue; 3c: Sclerenchymatous hypodermis enlarged; 3d: Ground tissue enlarged; 3e: A single vascular bundle enlarged.

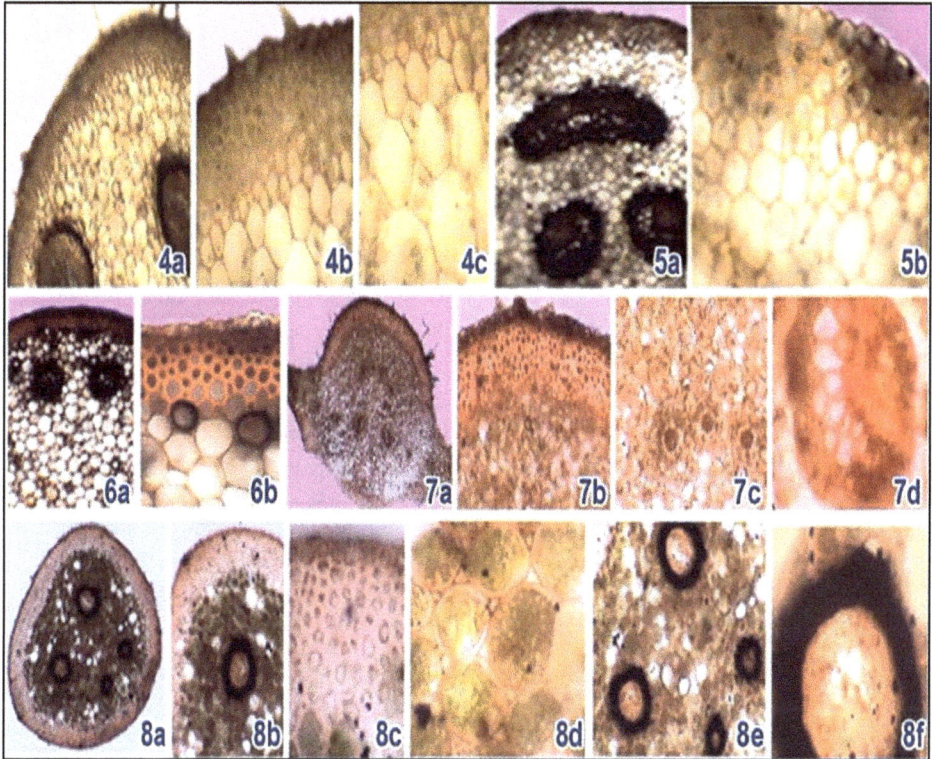

**Plate 5.8**: Results of histochemical analysis on polyphenols.

4: *Tectaria paradoxa* (Fee) Sledge—4a, b: T.S. of stipe–A portion at two different magnifications; 4c: Ground tissue enlarged; 5: *Bolbitis appendiculata* (Willd.) K. Iwatz.— 5a: T.S. of stipe–A portion; 5b: A portion enlarged showing epidermis, hypodermis and ground tissue; 6: *Blechnum orientale* L; 6a: T.S. of stipe–A portion; 6b: A portion enlarged showing epidermis, hypodermis and ground tissue; 7: *Pyrrosia porosa* var. *porosa* Hovenkamp—7a: T.S. of stipe–Entire view showing lateral wings; 7b: A portion enlarged showing epidermis, hypodermis and ground tissue; 7c: Ground tissue enlarged; 7d: Portion of vascular bundle with xylem and phloem; 8: *Leptochilus decurrens* Bl.—8a: T.S. of stipe– Entire view; 8b,c: A portion enlarged at different magnifications showing epidermis, hypodermis and ground tissue; 8d: Ground tissue enlarged, 8e: Vascular bundles; 8f: Single vascular bundle enlarged.

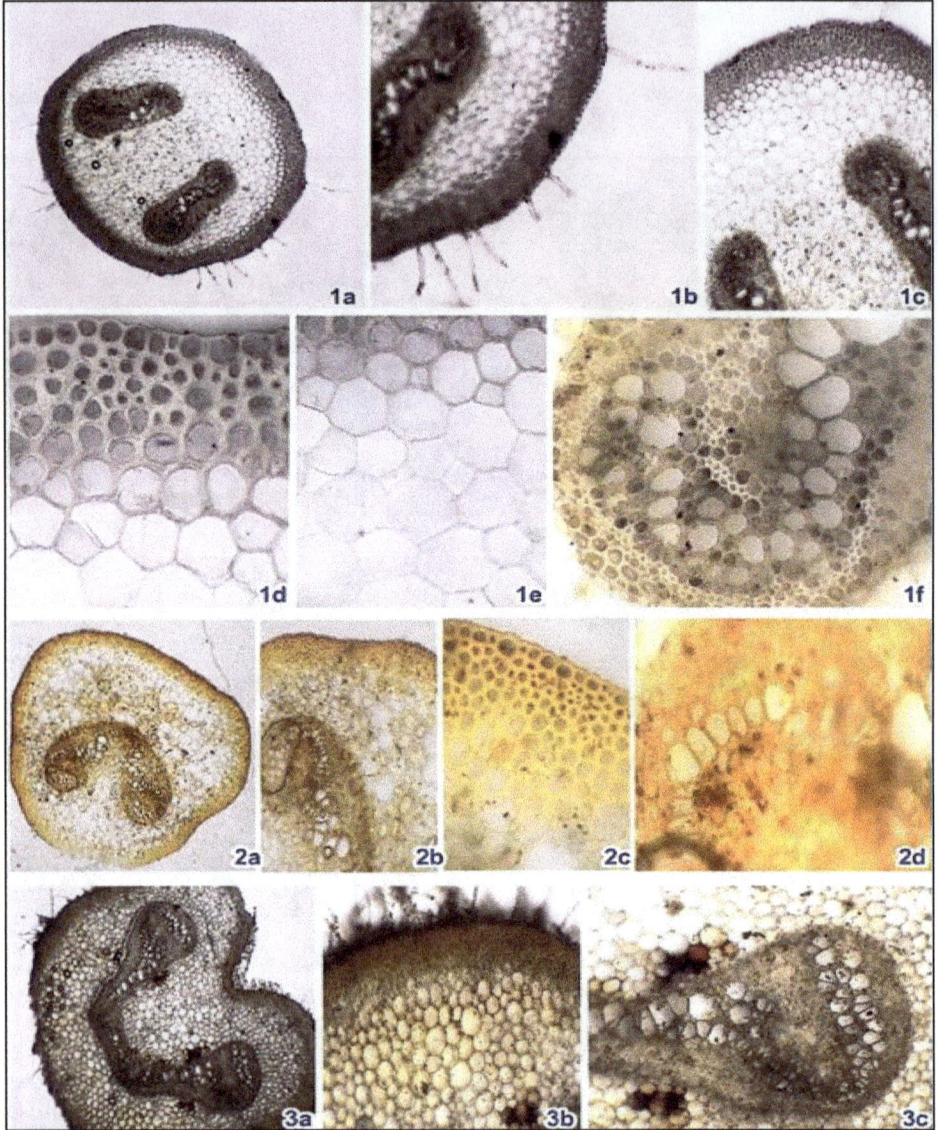

**Plate 5.9**: Results of histochemical analysis on lignin.

1: *Macrothelypteris torresiana* (Gaudich.) Ching—1a: T.S. of stipe – Entire view; 1b: A portion enlarged showing epidermis, hypodermis and ground tissue with vascular bundle and epidermal hairs; 1c, d: A portion enlarged showing epidermis, hypodermis and ground tissue; 1e: Ground tissue enlarged; 1f: Portion of vascular bundle enlarged showing xylem and phloem; 2: *Pseudophegopteris pyrrhorhachis* (Kunze) Ching—2a: T.S. of stipe–Entire view; 2b: A portion enlarged showing epidermis, hypodermis, ground tissue and portion of vascular bundle; 2c: A portion enlarged showing epidermis, hypodermis and ground tissue; 2d: A portion of vascular bundle enlarged showing xylem and phloem; 3: *Sphaerostephanos arbuscula* (Willd.) Holttum—3a: T.S. of stipe–Entire view; 3b: T.S. of stipe–Portion enlarged; 3c: Vascular bundle enlarged.

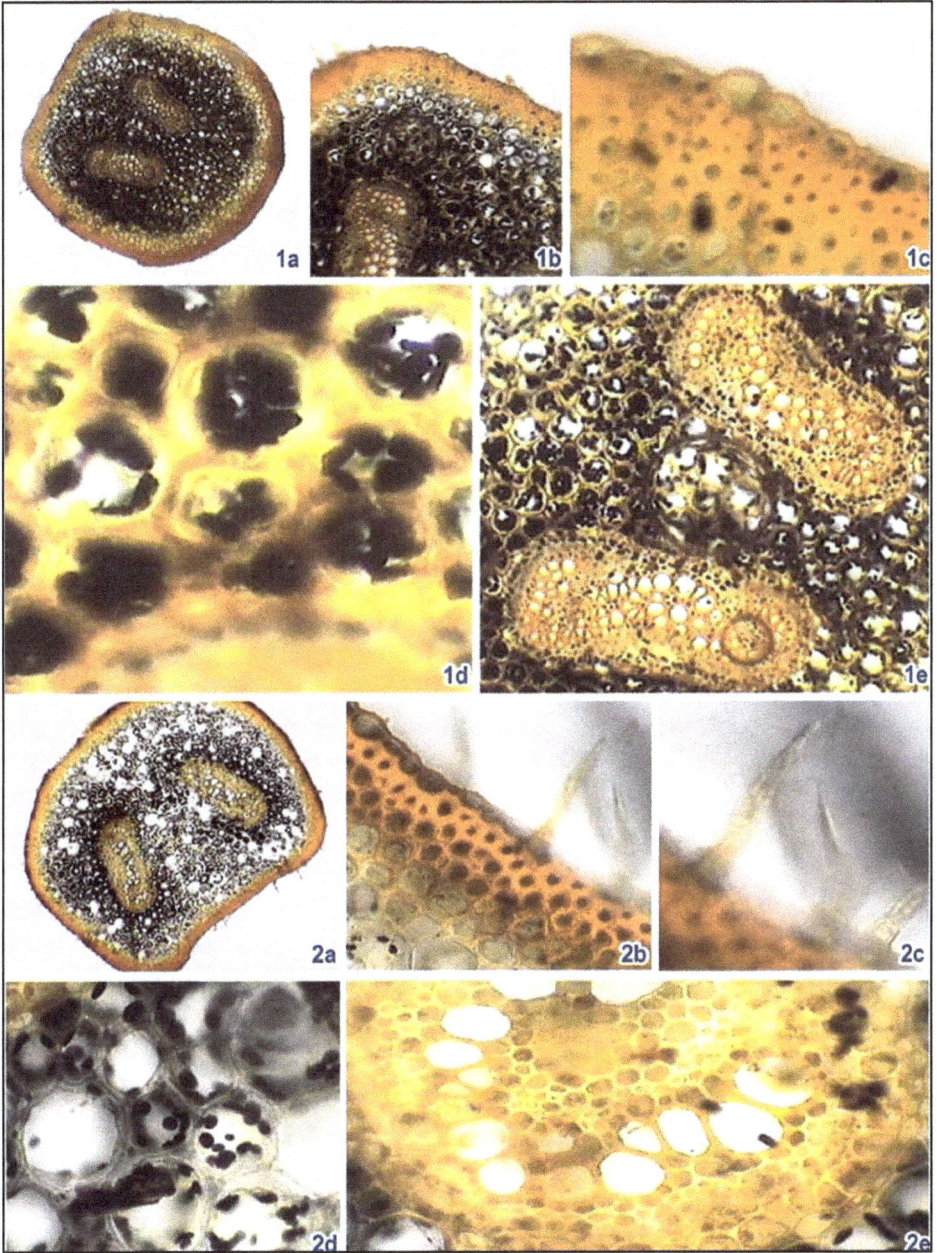

**Plate 5.10**: Results of histochemical analysis on lignin.

1: *Amphineuron terminans* (Hook.) Holttum—1a: T.S. of stipe–Entire view; 1b: T.S. of stipe–Portion enlarged; 1c: Ground tissue enlarged; 1d: A portion enlarged showing epidermis, hypodermis and ground tissue; 1e: Vascular bundles enlarged; 2: *Christella dentata* (Forssk.) Brownsey and Jermy—2a: T.S. of stipe–Entire view; 2b: T.S. of stipe–Portion enlarged showing epidermis, hypodermis and epidermal hairs; 2c: Ground tissue enlarged; 2d: Vascular bundles enlarged showing xylem and phloem.

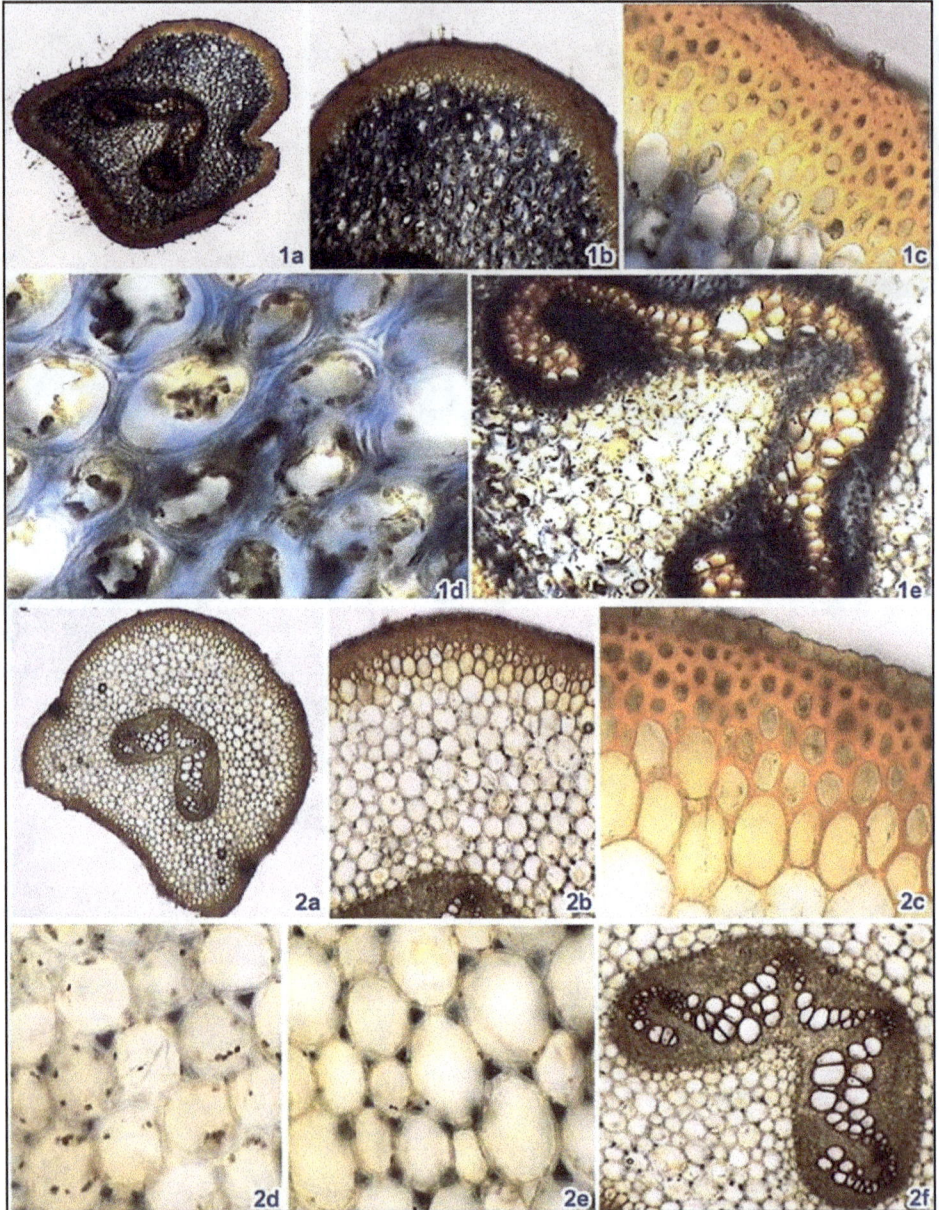

**Plate 5.11**: Results of histochemical analysis on lignin.

1: *Christella parasitica* (L.) H.Lev.—1a: T.S. of stipe–Entire view; 1b: T.S. of stipe–Portion enlarged showing epidermis with hairs, hypodermis and ground tissue; 1c: A portion enlarged showing epidermis, hypodermis and ground tissue; 1d: Ground tissue enlarged; 1e: Vascular bundle enlarged; 2: *Cyclosorus interruptus* (Willd.) H. Ito—2a: T.S. of stipe–Entire view; 2b, c: Portion enlarged at two different magnifications showing epidermis, hypodermis and ground tissue; 2d, e: Ground tissue enlarged; 2f: Vascular bundle enlarged showing zig-zag xylem.

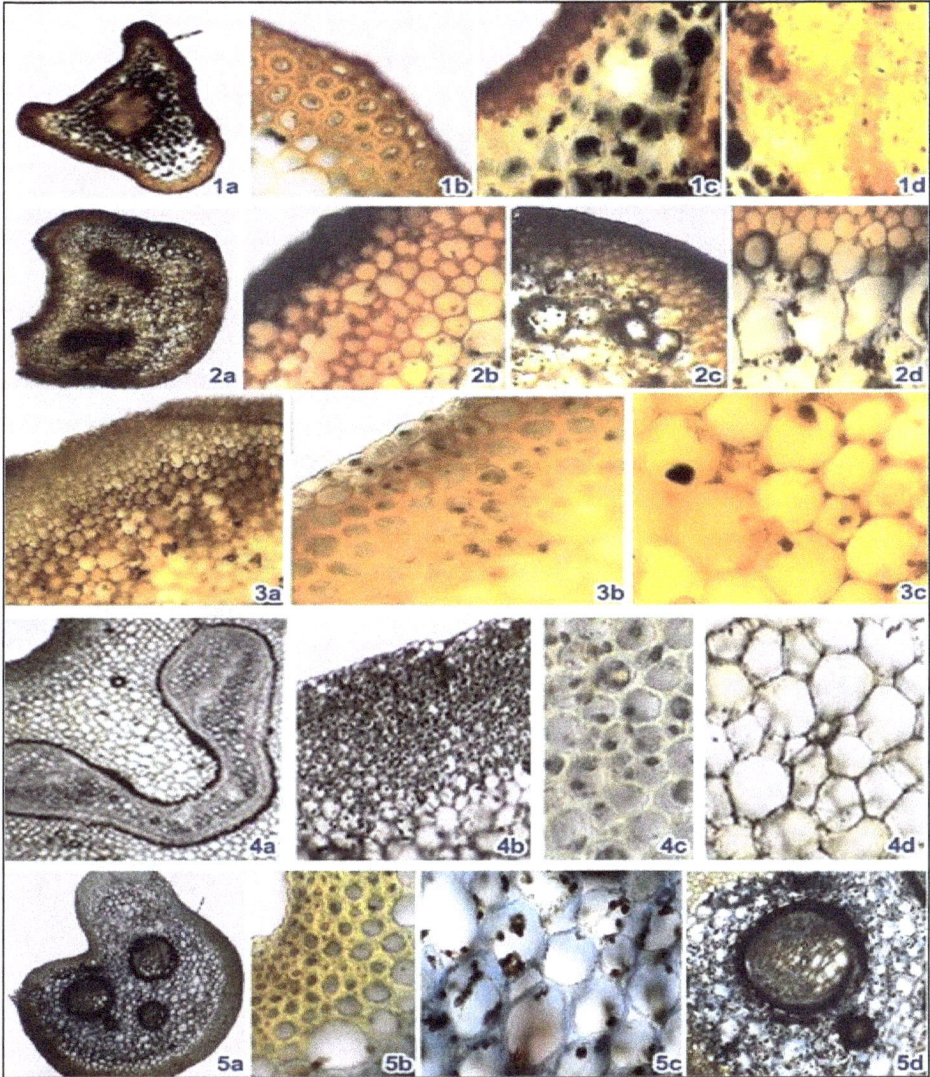

**Plate 5.12**: Results of histochemical analysis on lignin.

1: *Asplenium cheilosorum* Kunze ex Mett.—1a: T.S. of stipe – Entire view; 1b, c: A portion enlarged at different magnifications showing epidermis, hypodermis and ground tissue; 1d: Portion of vascular bundle enlarged showing xylem and phloem; 2: *Diplazium travancoricum* Beddome—2a: T.S. of stipe–Entire view; 2b, c: A portion enlarged showing epidermis, hypodermis and ground tissue; 2d: Ground tissue enlarged; 3: *Diplazium brachylobum* (Sledge) Manickam and Irudayaraj—3a, b: A portion enlarged showing epidermis, hypodermis and ground tissue; 3c: Ground tissue enlarged; 4: *Diplazium muricatum* (Mett.) Alderw—4a: T.S. of stipe – A portion; 4b: A portion enlarged showing epidermis, hypodermis and ground tissue; 4c.Hypodermis enlarged, 4d: Ground tissue enlarged; 5: *Dryopteris sparsa* (Buch. Ham. ex D: Don) C.Chr.—5a: T.S. of stipe–Entire view; 5b: A portion enlarged showing epidermis, hypodermis and ground tissue; 5c: Ground tissue enlarged; 5d: Vascular bundle enlarged.

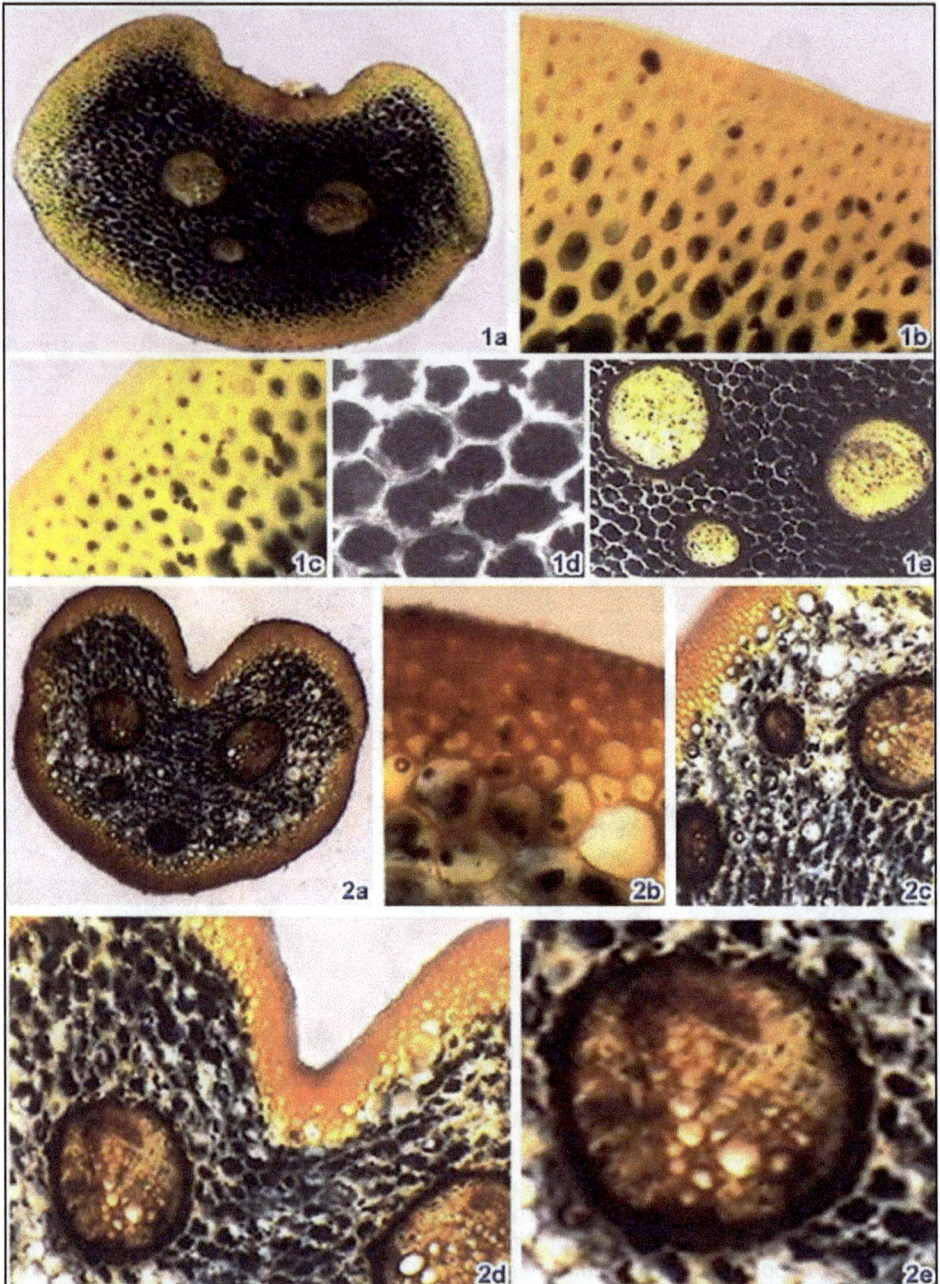

**Plate 5.13**: Results of histochemical analysis on lignin.

1: *Arachniodes tripinnata* (Goldm.) Sledge—1a: T.S. of stipe–Entire view; 1b, c: Hypodermis enlarged; 1d: Ground tissue enlarged; 1e: Vascular bundles; 2: *Arachniodes amabilis* (Bl.) Kuntze—2a: T.S. of stipe–Entire view; 2b: Hypodermis enlarged; 2c, d: A portion enlarged at two different magnification showing epidermis, hypodermis and ground tissue and vascular bundles; 2e: A single vascular bundle enlarged.

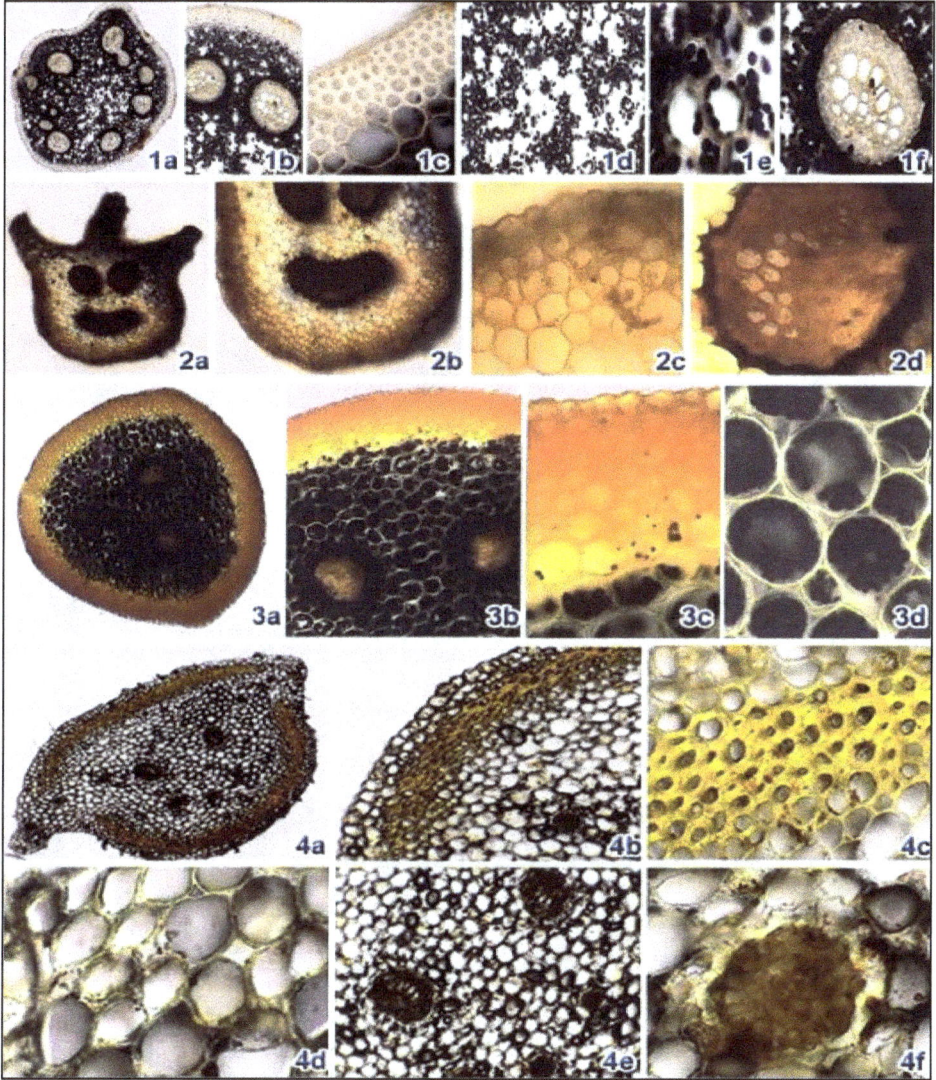

**Plate 5.14**: Results of histochemical analysis on lignin.

1: *Blechnum orientale* L.—1a: T.S. of stipe–Entire view; 1b: A portion enlarged showing ground tissue and vascular bundles; 1c: Hypodermis enlarged; 1d, e: Ground tissue enlarged at different magnifications; 1f: A single vascular bundle enlarged; 2: *Bolbitis appendiculata* (Willd.) K. Iwatz.—2a: T.S. of stipe–Entire view; 2b: A portion enlarged showing epidermis, hypodermis and ground tissue with vascular bundles; 2c: A portion enlarged showing epidermis, hypodermis and ground tissue; 2d: A portion of vascular bundle enlarged; 3: *Leptochilus decurrens* Bl.—3a: T.S. of stipe–Entire view; 3b, c: A portion enlarged at different magnifications showing epidermis, hypodermis and ground tissue; 3d: Ground tissue enlarged; 4: *Pyrrosia porosa* var. *porosa* Hovenkamp—4a: T.S. of stipe–Entire view showing lateral wings; 4b: A portion enlarged showing epidermis, hypodermis and ground tissue; 4c: Hypodermis enlarged, 4d: Ground tissue enlarged; 4e: Ground tissue with two vascular bundles; 4f: Single vascular bundle enlarged.

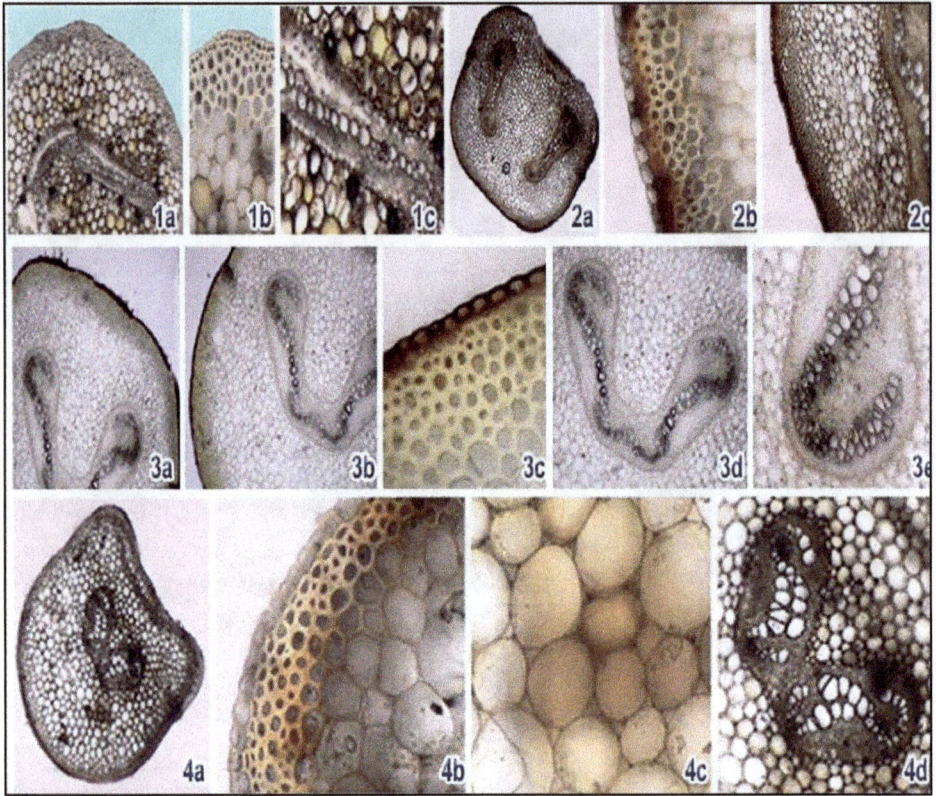

**Plate 5.15**: Results of histochemical analysis on tannin.

1: *Christella parasitica* (L.) H.Lev.—1a: T.S. of stipe – A portion; 1b: T.S. of stipe–Portion enlarged showing epidermis, hypodermis and ground tissue; 1c: Vascular bundle enlarged; 2: *Christella dentata* (Forssk.) Brownsey and Jermy—2a: T.S. of stipe–Entire view; 2b, c: T.S. of stipe–Portion enlarged showing epidermis, hypodermis and ground tissue; 3: *Pseudophegopteris pyrrhorhachis* (Kunze) Ching—3a, b: T.S. of stipe – Two different portions; 3c: A portion enlarged showing epidermis, hypodermis and ground tissue; 3d, e: Vascular bundle enlarged showing incurved arms; 4: *Cyclosorus interruptus* (Willd.) H. Ito—4a: T.S. of stipe–Entire view; 4b: Portion enlarged showing epidermis, hypodermis and ground tissue; 4c: Ground tissue enlarged; 4d: Vascular bundle enlarged.

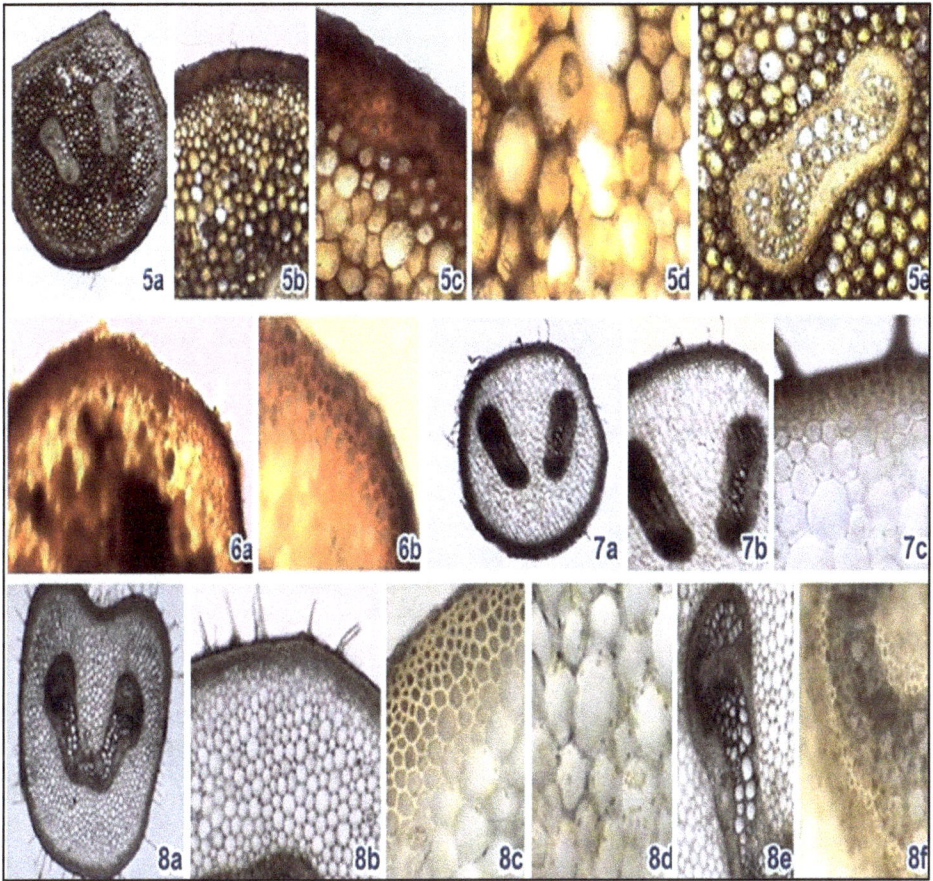

**Plate 5.15**: Results of histochemical analysis on tannin.

5: *Amphineuron terminans* (Hook.) Holttum—5a: T.S. of stipe–Entire view; 5b, c: A portion enlarged at different magnifications dhowing epidermis, hypodermis and ground tissue; 5d: Ground tissue enlarged; 5e: Vascular bundle enlarged showing 'E' shaped xylem; 6: *Trigonospora ciliata* (Wall. Ex Benth.) Holttum—6a: T.S. of stipe – A portion; 6b: A portion enlarged showing epidermis, hypodermis and ground tissue; 7: *Macrothelypteris torresiana* (Gaudich.) Ching—7a: T.S. of stipe – Entire view; 7b, c: A portion enlarged at different magnifications showing epidermis, hypodermis and ground tissue with vascular bundles (b) and epidermal hairs; 8: *Sphaerostephanos arbuscula* (Willd.) Holttum—8a: T.S. of stipe–Entire view, 8b: T.S. of stipe–Portion enlarged; 8c: A portion enlarged showing epidermis, hypodermis and ground tissue; 8d: Ground tissue enlarged; 8e, f: Portion of vascular bundle enlarged enlarged at different magnifications.

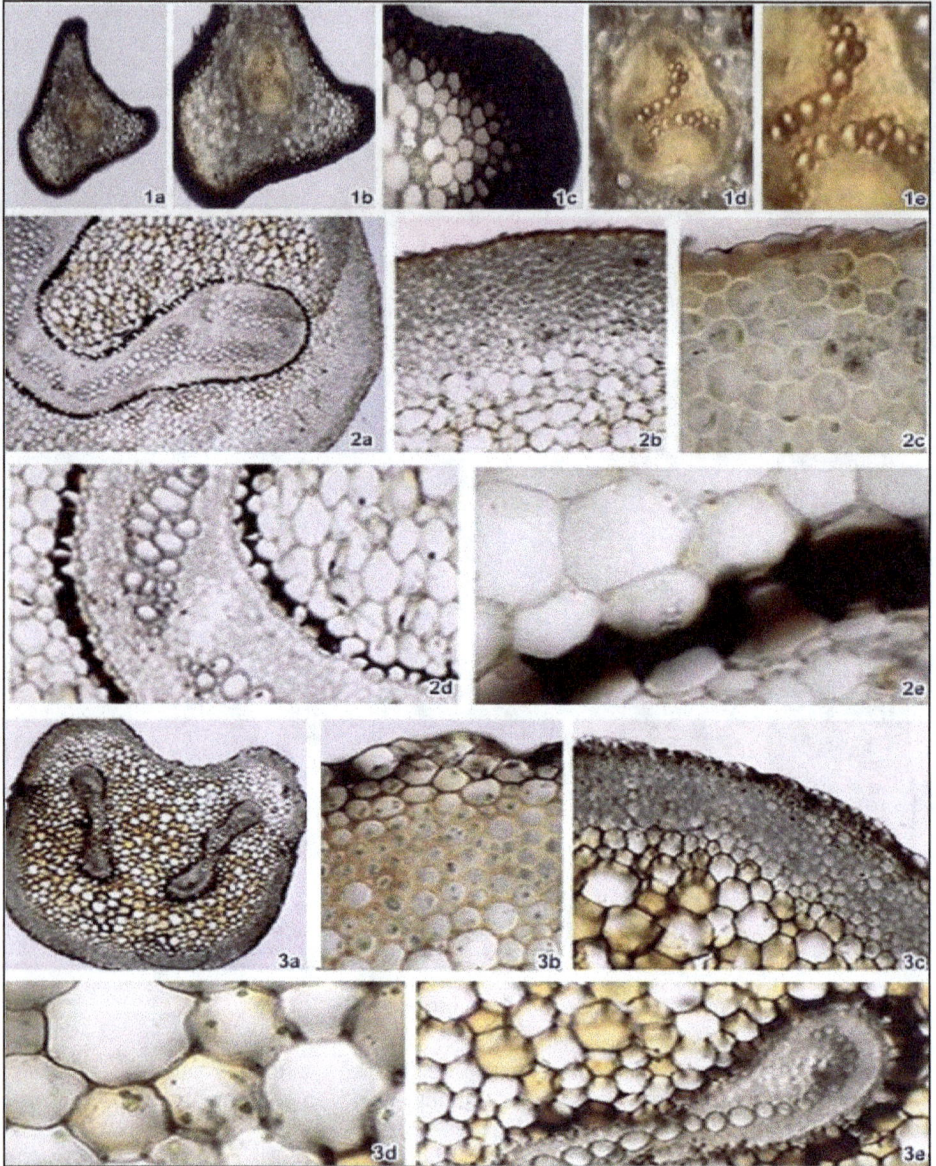

**Plate 5.16**: Results of histochemical analysis on tannin.

1: *Asplenium cheilosorum* Kunze ex Mett.—1a: T.S. of stipe – Entire view; 1b, c: A portion enlarged at different magnifications showing epidermis, hypodermis and ground tissue; 1d, e: Vascular bundle enlarged at different magnifications showing xylem and phloem; 2: *Diplazium muricatum* (Mett.) Alderw—2a: T.S. of stipe – A portion; 2b: A portion enlarged showing epidermis, hypodermis and ground tissue; 2c: Epidermis and hypodermis enlarged; 2d: Portion of vascular bundle enlarged; 2e: Endodermis enlarged; 3: *Diplazium travancoricum* Beddome—3a: T.S. of stipe–Entire view; 3b, c: A portion enlarged at different magnifications showing epidermis, hypodermis and ground tissue; 3d: Ground tissue enlarged; 3e: Portion of vascular bundle enlarged.

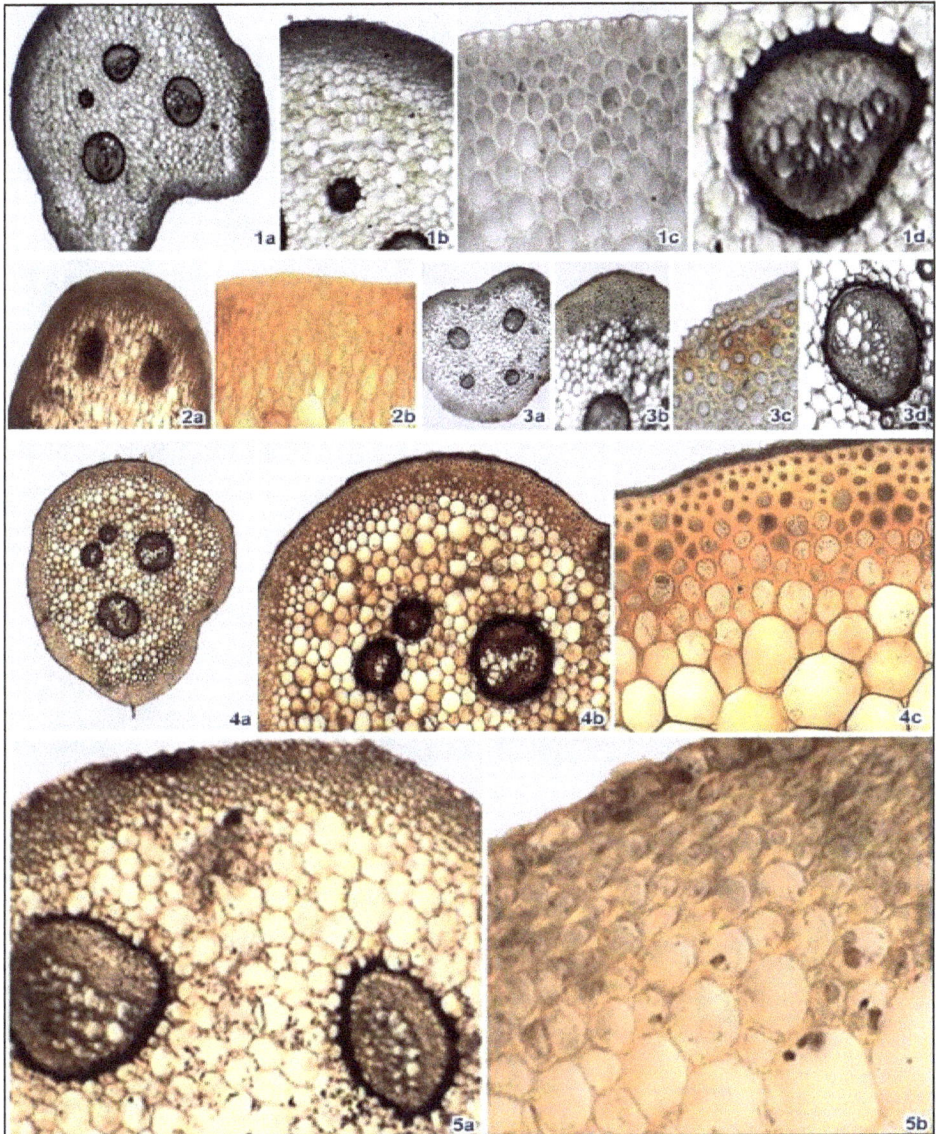

**Plate 5.17**: Results of histochemical analysis on tannin.

1: *Dryopteris sparsa* (Buch. Ham. ex D: Don) C.Chr.—1a: T.S. of stipe–Entire view; 1b, c: A portion enlarged at different magnifications showing epidermis, hypodermis and ground tissue; 1d: Vascular bundle enlarged; 2: *Arachniodes aristata* (Forst. f) Tindale— 2a: T.S. of stipe–A portion; 2b: A portion enlarged showing epidermis and hypodermis; 3: *Arachniodes amabilis* (Bl.) Kuntze—3a: T.S. of stipe–Entire view; 3b, c: A portion enlarged at two different magnifications showing epidermis, hypodermis and ground tissue; 3d: Vascular bundle enlarged; 4: *Arachniodes tripinnata* (Goldm.) Sledge—4a: T.S. of stipe–Entire view; 4b, c: A portion enlarged at two different magnifications showing epidermis, hypodermis and ground tissue; 5: *Tectaria paradoxa* (Fee) Sledge—5a: T.S. of stipe–A portion with two vascular bundles; 5b: A portion enlarged showing epidermis hypodermis and ground tissue enlarged.

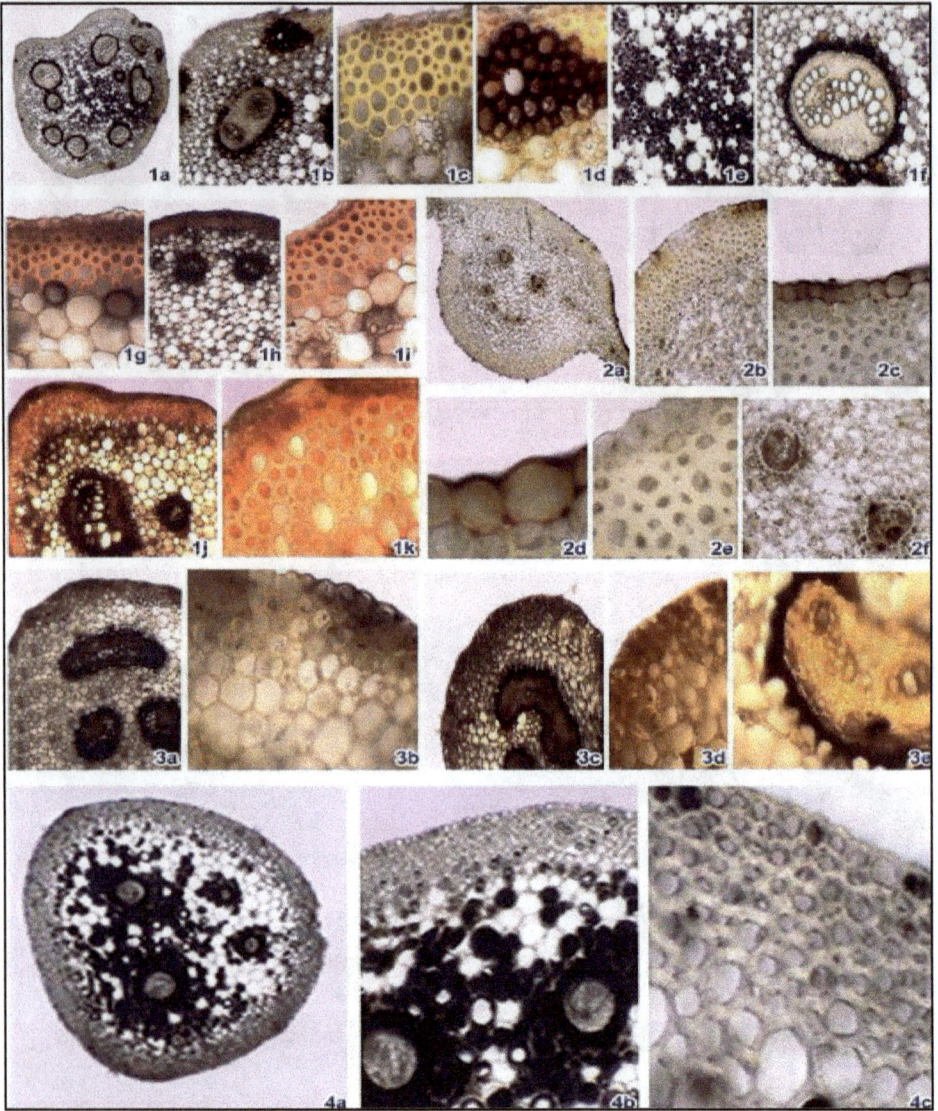

**Plate 5.18**: Results of histochemical analysis on tannin.

1: *Blechnum orientale* L.—1a: T.S. of stipe–Entire view; 1b, g, h, i. A portion enlarged showing epidermis, hypodermis, ground tissue; 1c: i. Hypodermis enlarged; 1d: Lignified cells intermingled with ground tissue; 1e: Ground tissue enlarged; 1f: A single vascular bundle enlarged; 2: *Pyrrosia porosa* var. *porosa* Hovenkamp—2a: T.S. of stipe–Entire view showing lateral wings; 2b: A portion enlarged showing epidermis, hypodermis and ground tissue; 2c, e: Epidermis and hypodermis enlarged; 2d: Epidermis enlarged; 2f: Vascular bundles; 3: *Bolbitis appendiculata* (Willd.) K. Iwatz.—3a, c: T.S. of stipe–Different portions; 3b, d: A portion enlarged showing epidermis, hypodermis and ground tissue; 3e: A portion of vascular bundle enlarged; 4: *Leptochilus decurrens* Bl.—4a: T.S. of stipe–Entire view; 4b: A portion enlarged showing epidermis, hypodermis and ground tissue; 4d: A portion enlarged showing epidermis and hypodermis.

[Plate 5.2(5b)], *D. muricatum* [Plate 5.3(3c)] and *D. brachylobum* [Plate 5.3(2a)], the vascular bundles are in V shaped. In *C. parasitica* [Plate 5.2(1a)] and *S. arbuscula* [Plate 5.2(3a)], the vascular bundles are in U shaped. In *M. torresiana* [Plate 5.6(2a)] and *D. travancoricum* [Plate 5.3(4d)], the vascular bundles are Dump bell shaped. In *B. orientale* [Plate 5.14(1a)], the vascular bundle is C shaped. Regarding the type of vascular bundles, they are all of typically single protostelic type. When there are many vascular bundles, as in *B. orientale*, they will look like dictyostele and the individual stele is called meristele which is of typical protostele. Each and every vascular bundle has individual endodermis followed by pericycle which encloses the phloem and xylem. As it is known, generally the xylem is made up of xylem-tracheids.

Srivastava (2008) has studied the anatomy of the petiole of *C. dentata*. It is cylindrical. On the adaxial side of the petiole there is a prominent groove running from the base up to the tip where the first pinna is attached. In a transverse section of the petiole, epidermal cells throughout their length appear small, thick walled, dark brown in colour and covered by smooth and delicate hairs. The epidermis is followed by three or more layers of thick walled cells followed by several layers of thin walled parenchymatous cells which constitute the ground tissue. The petiole receives two widely separated vascular strands from the rhizome. Each vascular strand is enclosed by a single layered endodermis. The pericycle is made up of thin walled cells, which are one to three layers in thickness. The xylem is mesarch and surrounded by phloem. The xylem is hippocampus-shaped and the two vascular strands appear slightly elongated. The adaxial arms are comparatively more turned inwards and the two vascular strands appear almost reniform in shape. Soon after entering the stipe base, the two separate vascular strands start coming closer to each other and get fused somewhere near the middle of the petiole to form a single strand for further upward course. During the merger of the two vascular strands first the endodermis, the pericycle, and at still higher levels the two phloem and xylem strands also join each other at their abaxial side. The distal arms of xylem and phloem however remain as such. Thus, the single vascular strand that resulted due to fusion of the two is almost 'U' shaped and free arms of xylem strands turned inwards more in the former as compared to the latter. Xylem consists of tracheids with protoxylem having annular or spiral thickenings and metaxylem with scalariform thickenings. Phloem consists of sieve cells with occasional parenchyma.

In the construction of any ordinary vascular plant there are three limiting tissue-surfaces of prime importance: i) the outer contour of the plant or part, complicated though it usually is in sub-aerial plants by the added cell surfaces lining the ventilating channels. ii) the endodermal sheath, which delimits the constructing tracts from the tissues that envelop them, and iii) the collective surface by which the dead tracheal system faces the living tissues that surrounds or permeates it. Each of the three is a surface of physiological transit, and each independently of the others will be a suitable subject of observation from the point of view of the proportion of surface to bulk as the size increases. Of these the third is the most important in the comparative study of form in the conducting tracts of the pteridophyta: for the woody tissues and those best preserved in the fossils, and being resistant they frequently retain their natural condition, and make it possible to contrast their form and dimensions with the

**Table 5.2:** Comparative histochemical analysis on the selected ferns of the Western Ghats.

| Name of the Species | Epidermis | | | | Hypodermis | | | | Ground Tissue | | | | Endodermis | | | | Xylem | | | | Phloem | | | | Remarks |
|---|---|---|---|---|---|---|---|---|---|---|---|---|---|---|---|---|---|---|---|---|---|---|---|---|---|
| | Li | P | Lig | T | Li | P | Lig | T | Li | P | Lig | T | Li | P | Lig | T | Li | P | Lig | T | Li | P | Lig | T | |
| P. pyrrhorhachis | - | 3+ | 2+ | + | - | + | 2+ | - | + | - | 2+ | - | - | - | 2+ | - | - | + | 2+ | - | - | - | 2+ | - | Hairs–Phenol |
| M. torressiana | 3+ | 3+ | * | - | - | 3+ | * | - | 3+ | + | * | - | - | * | * | - | + | + | * | - | + | + | + | - | Hairs Phenol |
| T. ciliata | 3+ | 3+ | * | 3+ | + | 3+ | 3+ | 3+ | + | 3+ | - | - | - | 3+ | - | - | + | 3+ | - | - | + | 3+ | - | - | |
| A. terminans | + | 3+ | 3+ | 2+ | - | 3+ | 3+ | 3+ | 3+ | 3+ | * | - | + | 3+ | 2+ | - | + | + | 2+ | - | - | 3+ | 2+ | - | |
| C. parasitica | 3+ | 3+ | 3+ | - | + | 3+ | 3+ | - | + | 3+ | - | - | 3+ | 3+ | 2+ | - | + | 3+ | 2+ | - | + | 2+ | 2+ | - | Hair–Phenol |
| C. dentata | 3+ | 3+ | 3+ | - | + | 3+ | + | - | + | 3+ | 2+ | - | 3+ | 3+ | 3+ | - | + | 3+ | 3+ | - | - | 3+ | 3+ | - | Hair–Lipid Phenol |
| S. arbuscula | 3+ | 3+ | + | - | + | 3+ | 3+ | - | 3+ | 3+ | - | - | 3+ | 3+ | + | - | 3+ | 3+ | + | - | 3+ | 3+ | + | - | |
| A. cheilosorum | 3+ | 3+ | 3+ | - | 3+ | 3+ | 3+ | - | 3+ | - | 3+ | - | 3+ | 3+ | 3+ | - | 3+ | 3+ | 3+ | - | 3+ | 2+ | 3+ | - | |
| D. brachylobum | 3+ | 3+ | 3+ | - | + | 3+ | 3+ | - | + | 2+ | 2+ | - | 3+ | 3+ | 3+ | - | + | 2+ | * | - | + | 3+ | * | - | |
| D. travancoricum | 3+ | 3+ | 2+ | - | 2+ | 3+ | * | - | 3+ | 2+ | - | - | 3+ | 3+ | 3+ | - | + | 3+ | 3+ | - | + | 2+ | 3+ | - | |
| D. muricatum | 3+ | 3+ | * | - | 3+ | 3+ | 2+ | - | 3+ | 2+ | - | - | 3+ | 3+ | 3+ | - | 3+ | 2+ | - | - | 3+ | + | - | - | Violet |
| D. sparsa | - | 3+ | 2+ | - | - | 3+ | - | - | - | 2+ | - | - | - | 3+ | 3+ | - | - | + | - | - | - | 2+ | - | - | Li-Absent, blue-ground |
| A. aristata | - | - | 2+ | 2+ | - | - | 3+ | 2+ | - | - | - | - | - | - | - | - | - | - | 2+ | - | - | - | - | - | Li- Absent |
| A. tripinnata | + | 3+ | 3+ | - | + | 3+ | 3+ | 2+ | + | 3+ | - | - | 3+ | 3+ | * | - | + | 3+ | - | - | + | 3+ | 2+ | - | Ground–black |
| A. amabilis | - | 3+ | 3+ | - | - | 3+ | * | + | + | 2+ | - | - | 3+ | 3+ | - | - | + | 3+ | 2+ | - | + | 3+ | 2+ | - | |
| T. paradoxa | - | 3+ | * | - | - | 3+ | 3+ | - | + | 2+ | - | - | - | 3+ | * | - | - | 3+ | 2+ | - | - | 3+ | - | - | Hairs–Phenol |
| B. appendiculata | + | 3+ | 3+ | - | - | 3+ | 2+ | - | - | 2+ | 2+ | - | - | 3+ | 3+ | - | - | 3+ | - | - | 3+ | 3+ | 3+ | - | |

Contd...

**Table 5.2**–Contd...

| Name of the Species | Epidermis | | | | Hypodermis | | | | Ground Tissue | | | | Endodermis | | | | Xylem | | | | Phloem | | | | Remarks |
|---|---|---|---|---|---|---|---|---|---|---|---|---|---|---|---|---|---|---|---|---|---|---|---|---|---|
| | Li | P | Lig | T | Li | P | Lig | T | Li | P | Lig | T | Li | P | Lig | T | Li | P | Lig | T | Li | P | Lig | T | |
| B. orientale | 2+ | 3+ | 2+ | 3+ | + | 3+ | 3+ | 3+ | 3+ | 3+ | – | + | 3+ | 3+ | * | – | – | 3+ | 3+ | – | – | 3+ | 2+ | – | |
| P. porosa | + | 3+ | 3+ | – | – | 3+ | 3+ | – | 2+ | 3+ | 3+ | – | * | 3+ | 2+ | – | * | 3+ | 2+ | – | * | 3+ | 2+ | – | Hairs and Scales–Phenol |
| L. decurrens | + | 3+ | 3+ | – | – | 3+ | 3+ | – | 3+ | 2+ | – | – | 3+ | 3+ | * | – | + | 2+ | 3+ | – | + | 2+ | + | – | Green Granules |
| Cyclosorus | 3+ | 3+ | 3+ | – | 3+ | 3+ | 3+ | 2+ | 3+ | 3+ | 3+ | – | 3+ | 3+ | 3+ | – | 3+ | 3+ | 3+ | – | 3+ | 3+ | 3+ | – | |

+: Mild; 2+: Average; 3+: High; *: Data not available.

**Table 5.3**: Frequency distribution of different chemicals in different parts of the stipe.

| Chemicals | Different Parts in the Stipe | | | | | | | | | | | |
|---|---|---|---|---|---|---|---|---|---|---|---|---|
| | Epidermis | | Hypodermis | | Ground Tissue | | Endodermis | | Xylem | | Phloem | |
| **Lipids** | M | 1 | M | 7 | M | 8 | M | 2 | M | 9 | M | 8 |
| | A | 1 | A | 1 | A | 1 | A | – | A | – | A | – |
| | H | 10 | H | 3 | H | 9 | H | 11 | H | 4 | H | 4 |
| **Polyphenol** | M | – | M | 1 | M | 1 | M | – | M | 4 | M | 2 |
| | A | – | A | – | A | 8 | A | – | A | 3 | A | 5 |
| | H | 20 | H | 19 | H | 9 | H | 18 | H | 13 | H | 12 |
| **Lignin** | M | 1 | M | 1 | M | – | M | 1 | M | 1 | M | 3 |
| | A | 4 | A | 2 | A | 4 | A | 4 | A | 7 | A | 7 |
| | H | 11 | H | 12 | H | 3 | H | 8 | H | 6 | H | 5 |
| **Tannin** | M | 1 | M | 1 | M | – | M | – | M | – | M | – |
| | A | 2 | A | 3 | A | – | A | – | A | – | A | – |
| | H | 2 | H | 3 | H | – | H | – | H | – | H | – |

M: Mild; A: Average; H: High.

**Table 5.4**: Number of scores with high concentration in each species.

| Name of the Species | Number of Scores with High Concentration |
|---|---|
| *Pseudophegopteris pyrrhorhachis* (Kunze) Ching | 1 |
| *Macrothelypteris torresiana* (Gaudich.) Ching | 4 |
| *Trigonospora ciliata* (Wall. Ex Benth.) Holttum | 10 |
| *Cyclosorus interruptus* (Willd.) H. Ito | 17 |
| *Amphineuron terminans* (Hook.) Holttum | 9 |
| *Sphaerostephanos arbuscula* (Willd.) Holttum | 12 |
| *Christella parastica* (L.) H.Lev. | 9 |
| *Christella dentata* (Forssk.) Brownsey and Jermy | 12 |
| *Asplenium cheilosorum* Kunze ex Mett. | 16 |
| *Diplazium muricatum* (Mett.) Alderw. | 10 |
| *Diplazium travancoricum* Beddome | 10 |
| *Diplazium brachylobum* (Sledge) Manickam and Irudayaraj | 9 |
| *Tectaria paradoxa* (Fee) Sledge | 6 |
| *Arachniodes tripinnata* (Goldm.) Sledge | 9 |
| *Arachniodes amabilis* (Bl.) Kuntze | 6 |
| *Arachniodes aristata* (Forst. f) Tindale | 1 |
| *Dryopteris sparsa* (Buch. Hm. Ex D. Don) C. Chr. | 4 |
| *Bolbitis appendiculata* (Willd.) K. Iwatz. | 8 |
| *Blechnum orientale* L. | 12 |
| *Leptochilus decurrens* Bl. | 8 |
| *Pyrrosia porosa* var. *porosa* Hovenkamp | 9 |

corresponding tissues in their living correlatives. We may then expect that in the smallest, and particularly in their sporeling stage, the form of the conducting tracts will be simple, such as the cylinder. Moreover, this is frequently continued to the adult state in the most primitive pteridophytes. But with greater size its form tends to become more complicated, as it is seen to do even in the progressive stages of the individual life. The primary conducting tracts of the pteridophyta offer the best examples of moulding in accordance with the principle as stated above (Bower, 1935). Increasing morphological complexity involving any increased complication of form would tend to meet the physiological need consequent on increasing size, by leveling up the proportion of surface or bulk.

In general, in all the members of Thelypteridaceae, the number of vascular bundle is single, with the presence of two vascular bundle in very few cases *M. torresiana* and *C. dentata*. In *Diplazium* also, two species are with single vascular bundle and one species with one vascular bundle. All the Dryopteriod species are with more than two vascular bundles. Remaining species *P. porosa* var. *porosa*, *B. appendiculata* and *L. decurrens* also show the presence of few vascular bundles but only the species *B. orientale* shows the presence of more than four vascular bundles in a Dictyostelic type.

The selected pteridophytes are anatomically more or less uniform with the presence of mostly single vascular bundle, occasionally with few vascular bundles, rarely with several vascular bundles. In all cases the vascular bundles are strictly of protostelic type. In general, all the species except very few like *P. pyrrhorhachis* and *M. torresiana* are growing on exposed or shaded road sites with rigid and hard stipes. But the stipes in *M. torresiana* and *P. pyrrhorhachis* are soft and the species are growing in wet places. Such ecological difference is also indicated in the anatomical characters.

## Histochemical Studies

The occurrence and distribution of various metabolites (lipids, polyphenols, lignin and tannins) in different tissues of the stipe have been demonstrated in Table 5.2 and Plates 5.1–5.18. Lipids, polyphenols and lignin are profusely present in the endodermis and the epidermal cells of the stipes. Polyphenols and lignin and abundantly present in the hypodermis. Polyphenols show large quantities in xylem and phloem, lignin shows the optimal occurrence and small quantity. In general, tannins are poorly represented in epidermis, hypodermis and ground tissue and they are almost absent in endodermis, xylem and phloem.

Among eight species of Thelypteridaceae ferns, there are grade variations in the presence of various metabolites in different quantities in different tissues of the stipes. It clearly shows very low occurrence and distribution of different metabolites in the primitive free veined species such as *M. torresiana* and *P. pyrrhorhachis* with the maximum score of 1 and 4 respectively. *C. interruptus* shows the maximum occurrence of metabolites in high quantity. It is to be noted that *C. interruptus* is a low to medium attitude fern in contrast to other seven species. So it may be interpreted the evolution of various chemical relatively in higher concentration in the species which have moved to the low altitude or plains to cope with new environment. Among Thelypteroid ferns, next to *C. interruptus* two species *S. arbuscula* and *C. dentata* show the maximum occurrence of different metabolites with the score of 12 each (Tables 5.3 and 5.4). As far as *C. dentata*, it is the most successful species by growing in different kinds of habitat throughout the world.

The single species of *A. cheilosorum* shows the maximum occurrence of different chemicals at high concentration next to *C. interruptus*. It usually grows in shaded rocks crevices at high altitude and it is not a common fern. In order to survive in such risky environment, varieties of chemicals are present at high quantity. In the Athyroid genus *Diplazium*, three species takes the position next to Thelypterioid and Asplenoid in the evolutionally linearity. All the three species investigated shows the presence of different chemicals at average level with the score of 9 or 10. Among the three species of *Diplazium*, tannin is present only in one species *D. muricatum* in mild quantity. Thus the present study is in accordance with Karpagavinayagam (2005) for the role of tannin in anti-herbivory among different species *Diplazium* present in the same locality.

Among the five Dryopteroid species studied, the result showed major difference between species to species and in the occurrence and distribution of various chemicals with the minimum score of one in *A. aristata* and with the maximum score of 9 in *A. tripinnata*. Other four species belonging to four genera showed the occurrence of various chemicals at average level with the score of 8 each in *B. appendiculata* and *L. decurrens*, 9 in *P. porosa* var. *porosa* and 12 in *B. orientale* (Tables 5.3 and 5.4). Among various species studied the species belongs to Thelypteroid, Asplenoid, Athyroid, Dryopteriod, Blechnoid and Polypodiod showed the polyphenols presence it shows the general evolutionary progress in chemical adaptation. The variation to the above general observation may be due to several other reasons.

Herbivory is dependent on the presence of high concentration of nutrients like starch, proteins and lipids and low concentration of phenolic substances. Karpagavinayagam (2005) reported the biochemical composition of *M. torresiana, T. ciliata, C. interruptus, A. terminans, S. arbuscula, C. parasitica, C. dentata, D. muricatum, D. travancoricum, T. paradoxa, A. tripinnata, A. aristata, B. appendiculata, B. orientale, L. decurrens* and *P. porosa* var. *porosa*. They showed that the starch, lipids, total phenols and tannin presence with varied percentage. Thus the present study is in accordance with Karpagavinayagam (2005) for the role of tannin in Anti-herbivory among different species *Diplazium* present in the same locality.

In the present study also, high concentration of lignin in epidermis, hypodermis and xylem and high concentration of polyphenols in hypodermis have been observed. Lignin is an aromatic heteropolymer that is deposited most abundantly in the secondary cell walls of vascular plants. It provides structural rigidity to the plant body while enabling individual tracheory elements to withstand the tension generated during water transport; it also serves a defensive role against herbivores and pathogens. Lignin monomer composition has been found to vary among major phyla of vascular plants. Generally, ferns and gymnosperms deposit lignins that are derived primarily from guaiacyl monomers together with a small proportion of *p*-hydroxyphenyl units, whereas angiosperm lignins are guaiacyl/syringyl copolymers that also can contain some *p*-hydroxyphenyl monomers. This distribution suggests that $F_5H$ may be a relatively recent addition to plants' biochemical repertoire. Nevertheless, there are older reports in the literature in which syringyl monomers have been detected in lignins from lycophytes, including species of *Selaginella*, by using histochemical reagents and by relatively standard crude chemical methods. These results have been verified recently by using more modern techniques (Jing Ke *et al.*, 2008).

Polyphenols like tannin and lignin are not only act as anti-herbivory chemicals but also they act as antimicrobial chemicals. The rust fungi are obligate parasites, occurring on a wide range of angiosperms, gymnosperms and ferns. No rusts occur on fern allies. Most of these fungi are extremely host specific and it is argued that the evolutionary radiation of many rusts mirrors that of their hosts (Saviele, 1967). Microchemical tests of *C. dentata* reveal that the entire ground tissue is made up of cellulosic cells. Thin walled inner layers of cells usually have starch grains (Srivastava, 2008).

The occurrence and distribution of various chemicals (lipids, lipoprotein, polyphenols, lignins, tannins, starch, chitin, suberin and protein) in different tissues of the stipe have been studied histochemically. Lipids and polyphenols are abundantly present in the epidermal cells. Polyphenols and lignins are abundantly present in the hypodermis and they are also commonly present the ground tissue. Polyphenols are also present abundantly in the endodermis. Xylem and phloem shows the mild occurrence of polyphenols but in considerable frequency. In general tannins are poorly represented in epidermis, hypodermis and ground tissue and they are almost absent in endodermis, xylem and phloem. The present histochemical studies on lipids, polyphenols, lignins and tannins show the evolutionary progress of such chemicals from the primitive ferns towards the advance ferns. Thus, *Cyclosorus interruptus* (Willd.) H. Ito shows the maximum occurrence of chemical in high quantity. It is to be noted that *Cyclosorus interruptus* (Willd.) H. Ito is a low to medium altitude fern in contrast to other seven species. As far as *Christella dentata* (Forssk.) Brownsey and Jermy it is the most successful species by growing in different kinds of habitats throughout the world so the presence of various defense chemicals relatively at higher concentration may be of reasonable.

# References

Al-Rais, A.H., Meyer, A., and Watson, L. (1971). The Isolation and properties of oxalate crystals from plants. *Ann. Bot.*, **35**: 1213-1218.

Chamberlain, C.J. (1924). Methods in plant histology. Univ. Chicago Press, Chicago.

Edeoga, H.O., and Okoli, B.E. (1995). Histochemical studies in the leaves of some *Dioscorea* L. (Dioscoreaceae) and the taxonomic importance. *Feddes Rep.*, **106**: 113-120.

Gahan, P.B. (1927). Plant Histochemistry and Cytochemistry: An introduction. Academic Press, Florida.

Haridass, E.T., and Suresh Kumar, N. (1985). Some techniques in the study of insect-host plant interactions. In: Pollen physiology and Fertilization. Ed. By Linsken, H.F., North-Holland Publ. Co., Amsterdam.

Haripriya, D., Selvan, N., Jeyakumar, N., Periasamy, R., Johnson, M., and Irudayaraj, V. (2010). The effect of extracts of *Selaginella involvens* and *Selaginella inaequalifolia* leaves on poultry pathogens. *Asian Pacific Journal of Tropical Medicine*, **3(9)**: 678-681.

Heintzelman, C.E., and Howard, R.A. (1948). The comparative morphology of the Icacinaceae V. The pubescence crystals. *Am. J. Bot.*, **35**: 42-52.

Irene Pearl, J., Syed Ismail, T., Irudayaraj, V., and Johnson, M. (2011). Pharmacognostical studies on anti-cancer spike moss *Selaginella involvens* (Sw.) Spring. *International Journal of Drug Formulation and Research*, **2(6)**: 195-211.

Irudayaraj, V., Janaky, M., Johnson, M., and Selvan, N. (2010). Preliminary phytochemical and antimicrobial studies on a spike-moss *Selaginella inaequalifolia* (Hook. and Grev.) Spring. *Asian Pacific Journal of Tropical Medicine*, 957-960.

Jing-ke Weng, Xu Li, Jake Stout, and Clint Chapple. (2008). Independent origins of syringyl lignin in vascular plants. *Proc. Natl. Acad. Sci. USA.*, **105(22)**: 7887-7892.

Karpagavinayagam, C. (2005). Preliminary survey on herbivory in South Indian ferns. M.Sc. Thesis submitted to Department of Plant Biology and Plant Biotechnology, St Xavier's College (Autonomous), Palayamkottai, Tamil Nadu, India.

Krishnamurthy, K.V. (1988). *Methods in plant histochemistry*. Ed. By Viswanathan, S. (Printers and Publishers) Madras, India.

Krishnan, S., and Dayanandan, P. (2003). Structural and histochemical studies on grain lling in the caryopsis of rice. *J. Biosci.*, **28**: 455-469.

Krishnan, S., Ebenezer, G.A.I., and Dayanandan, P. (2001). Histochemical localization of storage components in caryopsis of rice (*Oryza sativa* L.). *Curr. Sci.*, **80**: 567-571.

Manickam, V.S., and Irudayaraj, V. (1992). Pteridophytic flora of the Western Ghats, South India, New Delhi.

Matthew, L., and Shah, G.L. (1984). Crystals and their taxonomic significance in some Verbanaceae. *Bot. J. Linn. Soc.*, **83**: 279-289.

Mithraja, M.J., Johnson, M., Mony, M., Miller Paul, Z., and Jeeva, S. (2012). Inter-specific variation studies on the phyto-constituents of *Christella* and *Adiantum* using phytochemical methods. *Asian Pacific Journal of Tropical Biomedicine*, S40-S45.

Paul Raj, K., Irudayaraj, V., Johnson, M., Patric Raja, D. (2011). Phytochemical and anti-bacterial activity of epidermal glands extract of *Christella parasitica* (L.) H. Lev. *Asian Pacific Journal of Tropical Biomedicine*, **1(1)**: 8-11.

Pauline Vincent, C., Irudayaraj, V., and Johnson, M. (2012). Anti-bacterial efficacy of macroscopic, microscopic parts of sporophyte and *in vitro* cultured gametophyte of a fern *Cyclosorus interruptus* (Willd.) H. Ito (Thelypteridaceae–Pteridophyta). *Journal of Chemical and Pharmaceutical Research*, **4(2)**: 1167-1172.

Ruthmann, A.C. (1970). Methods in cell Research. Cornell University Press, Ithaca, New York.

Savile, D.B.O. (1967). Evolution and relationships of the North American *Peducularis rusts* and their hosts. *Can. J. Bot.*, **45:** 1093-1103.

Singh, H. (2003). Economically viable Pteridophytes of India. In: Pteridology in the New Millennium, Ed. By Chandra, S., and Srivastava, M, Kluwer Academic Publishers, London, Chapter 29, pp. 421-436.

Srivastava, K. (2008). The Petiolar Structure of *Christella dentata* (Forssk.) Brownsey and Jermy (Thelypteridaceae, Pteridophyta). *Ethnobotanical Leaflets,* **12**: 96-102.

Utilisation and Management of Medicinal Plants Vol. 2 (2014)    *Pages* **189–204**
*Editor-in-Chief:* **V.K. Gupta**
*Published by:* **DAYA PUBLISHING HOUSE, NEW DELHI**

# 6

# *Phyllanthus niruri* Linn.: A Versatile Herb

Satish Patel[1], Nagendra S. Chauhan[1,3]*, Durgesh Nandini[2],
Mayank Thakur[1], Vikas Sharma[1] and V.K. Dixit[1]

## ABSTRACT

*The paper review the literature regarding Phyllanthus niruri a commonly used herb in Ayurvedic medicine. Specifically, the literature was reviewed for articles pertaining to chemical constituents and therapeutic benefits. Studies indicate bhui amla possesses antidiabetic, anti-inflammatory, antitumor, antistress, antioxidant, immunomodulatory, and hepatoprotective properties. The Preliminary studies have found various constituents of bhui amla exhibit a variety of therapeutic effects with little or no associated toxicity. These results are very encouraging and indicate this herb should be studied more extensively to confirm these results and reveal other potential therapeutic effects.*

***Keywords***: *Phyllanthus niruri*, Ayurvedic medicine, Chemical constituents, Therapeutic benefits.

## Introduction

*Phyllanthus niruri* Linn. is a common weed, which grows well in moist, shady and sunny places (Cabieses *et al.*, 1993; Nanden *et al.*, 1998). The plants are

---

1    Department of Pharmaceutical Sciences, Doctor Hari Singh Gour Vishwavidyalaya, Sagar – 470 003, M.P., India

2    Sagar Institute of Pharmaceutical Sciences, Sagar – 470 003, M.P., India

3    Drugs Testing Laboratory and Research Centre, Government Ayurvedic College Campus, GE Road, Raipur – 492 010, C.G., India

*    *Corresponding author*: E-mail: vkdixit2011@rediffmail.com

monoecious or homogamous; leaves are simple, alternate or opposite, some are leathery; flowers are very small and diclinous, they cluster in cup-shaped structures, greenish, often with glands. The fruit is a three-lobed capsule extending from the cup and commonly the long stalk pendant (Lewis *et al.*, 1977). The name 'Phyllanthus' means "leaf and flower" because the flower, as well as the fruit, seems to become one with the leaf (Cabieses *et al.*, 1993). *Phyllanthus niruri* L is usually misidentified with the closely related *Phyllanthus amarus* in appearance, phytochemical structure and history of use. *Phyllanthus niruri* Linn reaches a length of 60 cm, the fruits are larger, and the seeds are dark brown and warty (Morton *et al.*, 1981).

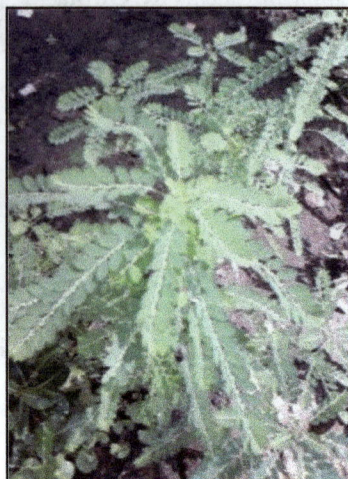

**Figure 6.1**: Whole plant of *P. niruri*.

## Taxonomical Hierarchy

| | |
|---|---|
| *Kingdom*: | Plantae |
| *Division*: | Magnoliphyta |
| *Class*: | Mangoliopsida |
| *Order*: | Malpighiales |
| *Family*: | Euphorbiaceae |
| *Genus*: | Phyllanthus |
| *Species*: | niruri |
| *Binomial name*: | *Phyllanthus niruri* Linn. |

## Biological Source

It consists of Aerial parts of *Phyllanthus niruri* Linn.

## Synonyms

*Phyllanthus carolinianus, P. sellowianus, P. fraternus, P. kirganella, P. lathyroides, P. lonphali, Nymphanthus niruri*

## Common Names

Chanca piedra, stone-breaker, arranca-pedras, punarnava, amli, bhonya, bhoomi amalaki, bhui-amla, bhui amla, bhuianvalah, bhuimy-amali, bhuin-amla, bhumyamalaki, cane peas senna, carry-me-seed, creole senna, daun marisan, derriere-dos, gale-wind grass, hurricane weed, jar-amla, quinine weed, sacha foster, cane senna.

## Habit and Habitat

It is indigenous to the rainforests of the Amazon and other tropical areas throughout the world, including the Bahamas, southern India, and China. *P. niruri* is quite prevalent in the Amazon and other wet rainforests, growing and spreading freely (much like a weed). *Phyllanthus amarus* and *Phyllanthus sellowianus* are closely related to *Phyllanthus niruri* in appearance, phytochemical structure, and history of use, but typically are found in the drier tropical climates of India, Brazil, and even Florida and Texas (Santos *et al.*, 1990).

## Morphological Details

*Phyllanthus niruri* is a monoecious annual herb that can grow to a height of 60 cm with phyllanthoid branching of 4–12 cm long with about 15–30 leaves. The leaves are subsessile and elliptical-oblong (5 -11 mm × 3–6 mm), obtuse to round at the base, obtuse or rounded and often apiculate at apex. The stipules are ovate-lanceolate to lanceolate. The proximal deciduous branchlets always covered with cymules of one to two male flowers and succeeding axils consists of one male and one female flower. The flowers are pale green with 5 to 6 calyx lobes with scarious margins.

The male flower consists of five disc-segmented stamens that make up of two to three filaments with free anthers that dehiscing obliquely to horizontally. The female flower was pale green with 5 petals with scarious margins. The flowers were shortly pedicellate with 5-lobed styles free 8 lokul, appressed to the ovary and bifid in the middle.

The fruits, an oblate capsule with 1.0–2.5 mm diameter, are obtusely trigonous smooth and the seeds are longitudinally ribbed on the back.

*Phyllanthus niruri* occurs as a weed in open ground, waste land, grossy scrub and dry deciduous forest and usually grows well on humid, sandy soils, up to 1000 m altitude. One of the characteristics of this genus is the phyllanthoid branches, which superficially appear to be compound leaves, but the presence of flowers and fruit of the base of each leaf indicates true branches (Unander and Blumberg, 1991).

## Chemical Constituents

The secondary metabolites present in *Phyllanthus niruri* Linn. are alkaloids, flavonoids, hydrolysable tannins, major lignans and polyphenols (Figures 6.2a and b).

### Alkaloids

Main alkaloid present in the *P. niruri* are securinine, norsecurinine, epibubbialine and isobubbialine (Houghton *et al.*, 1996), nirurine and phyllochrysine (Cuellar and Estevez, 1980; Mulchandani *et al.*, 1984).

### Flavonoids

Many flavonoids are also present in *P. niruri* like catechin, gallocatechin, quercetin, quercitoside, rutin (Morton *et al.*, 1981) and quercitrin (Bagalkotkar *et al.*, 2006), astragalin (Kale *et al.*, 2001), nirurin (Gupta *et al.*, 1984), epi-catechin and catechin-3-o-gallate (Ishimaru *et al.*, 1992)

### Lignans

Major lignans present are phyllanthin and hypophyllanthin (Morton, 1981; Chevallier, 2000). Leaf of *Phyllanthus niruri* Linn also have lignan lariciresinol, iso-

**Figure 6.2(a):** Structure of secondary metabolites present in *Phyllanthus niruri* Linn.

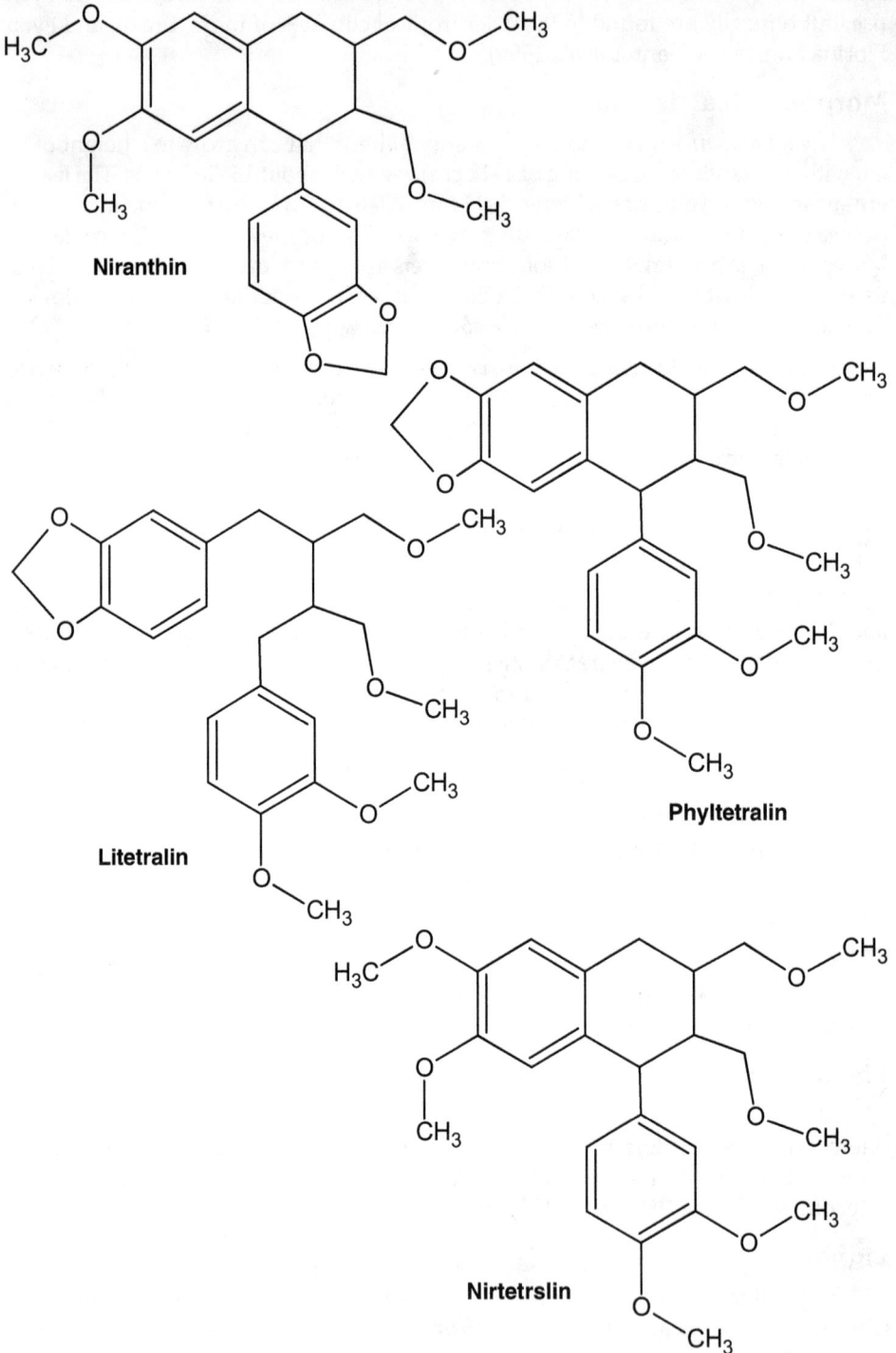

Niranthin

Litetralin

Phyltetralin

Nirtetrslin

*Contd...*

**Figure 6.2(a)**–*Contd...*

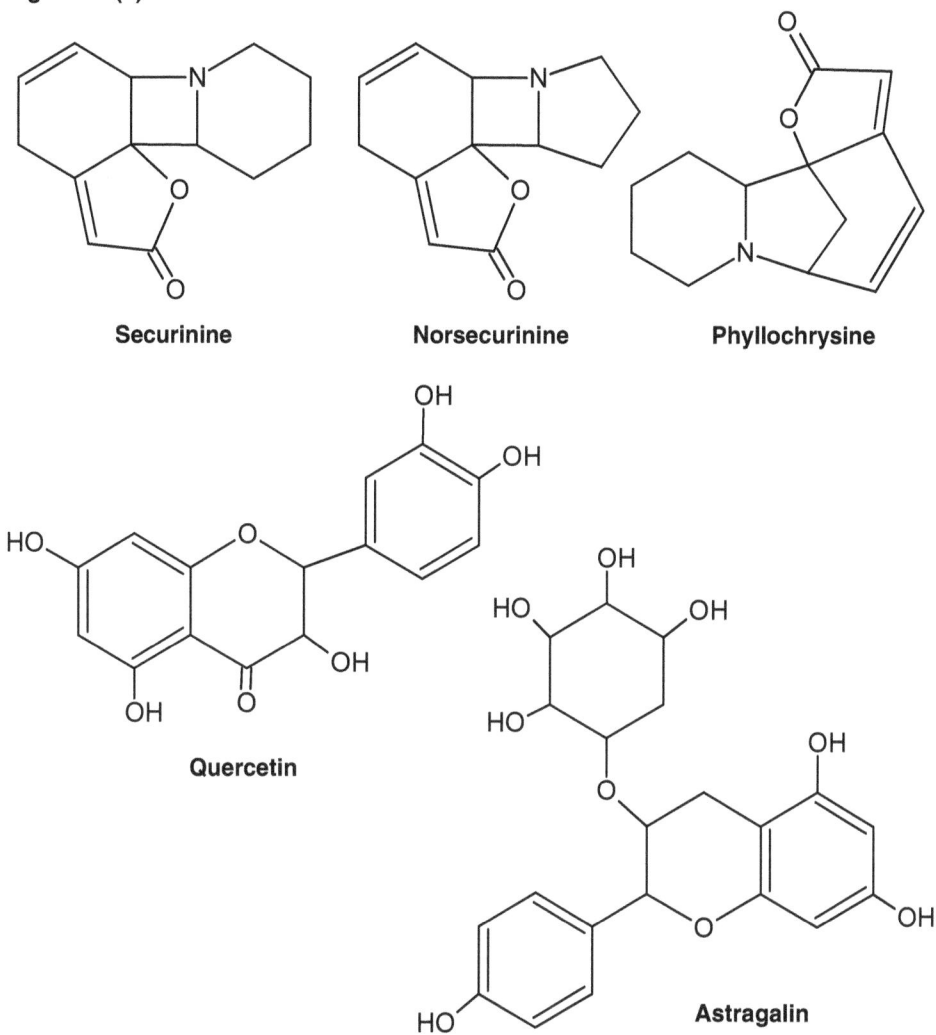

Securinine          Norsecurinine          Phyllochrysine

Quercetin

Astragalin

Seco- trimethyl ether butyrolactone, dibenzyl-hypophyllanthin, linnanthin, lintetralin, lintetralin, 4-hydroxy Seco-intetralin, iso-lintetralin, iso-2-3-demethoxyseco-kiacetate-niranthin, demethylenedioxy-niranthin, phyllanthin and phyltetralin (Satyanarayana *et al.*, 1991).

## Phenolics

Gallic acid and polyphenols: ellagic acid, phenazine and phenazine derivatives are also present ((Bagalkotkar *et al.*, 2006).

## Steroid

Estradiol also present in the entire plant of *Phyllanthus niruri* (Mannon *et al.*, 1976).

**Figure 6.2(b)**: Structure of secondary metabolites present in *Phyllanthus niruri* Linn.

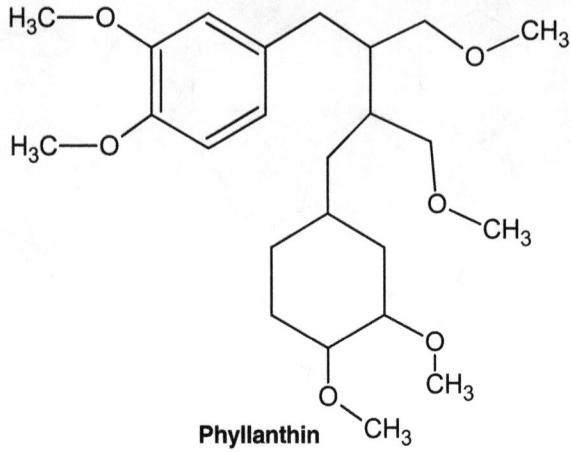

**Phyllanthin**

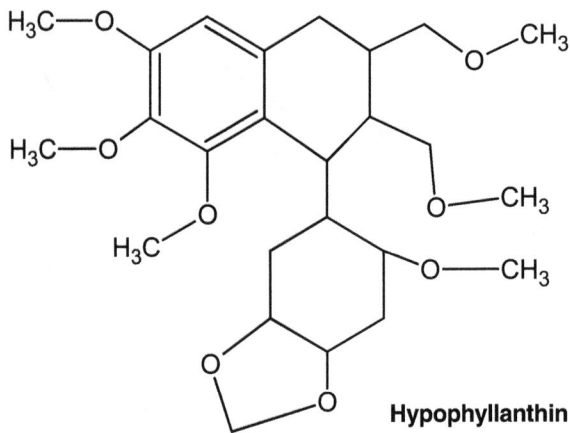

**Hypophyllanthin**

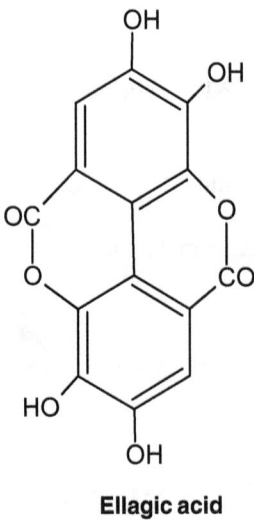

**Ellagic acid**

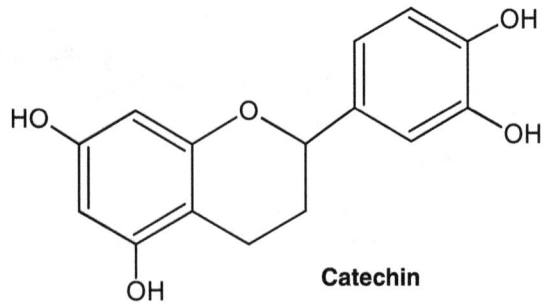

**Catechin**

*Contd...*

**Figure 6.2(b)**–*Contd...*

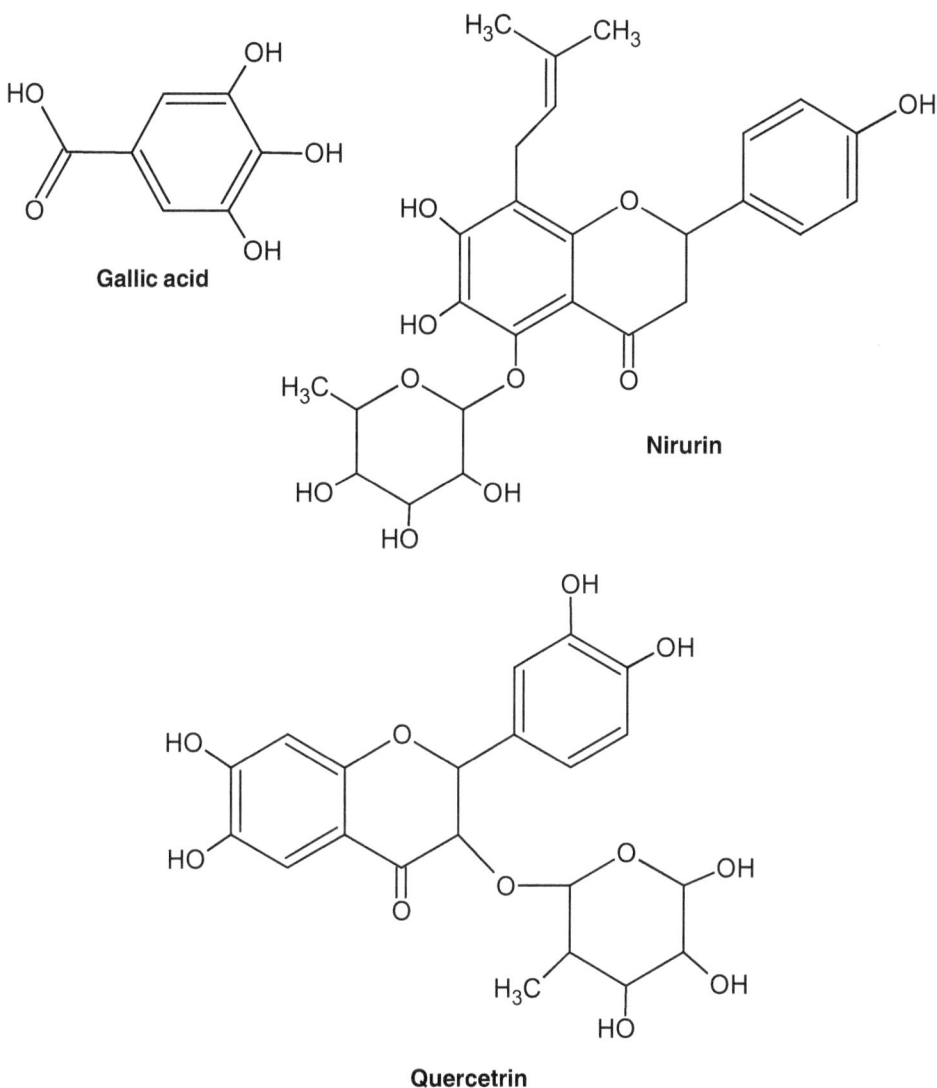

Gallic acid

Nirurin

Quercetrin

## Lipid

Linolenic Acid, Linolenic Acid and Ricinoleic acid are present in the seed oil of *Phyllanthus niruri* Linn (Ahamad *et al.*, 1981). Heptacosanoic acid derivative, alkanol c-5 pentacosanol ester, diterpene phyllanterpenyl ester are isolated from the root of plant *Phyllanthus niruri* Linn (Gupta *et al.*, 1999).

## Triterpenes

Phyllanthenol, phyllanthenone, and phyllantheol are isolated from aerial part of *Phyllanthus niruri* Linn (Singh *et al.*, 1989).

## Others

Alkanal C-5 Triacontan-1-al (Syamsundar *et al.*, 1985)

# Pharmacological Activity

Extracts of this herb have shown promise in treating a wide range of human diseases. Some of the medicinal properties suggested by numerous preclinical trials are anti-hepatotoxic, anti-lithic, anti hypertensive, anti-HIV, anti-hepatitis B, anti viral, immunostimulant, antihyperuricemic, stone breaker, antioxidant, hyperlipidemic and other activity.

## Antihepatotoxic Activity

Syamsundar *et al.* (1985) reported antihepatotoxic activity of Hypophyllanthin, phyllanthin from aerial part of *Phyllanthus niruri* Linn. Bhattacharjee *et al.* (2007) isolated a protein and suggested that the protein isolate of *P. niruri* protects hepatocytes against $CCl_4$-induced oxidative damage and may be used as an effective cytoprotector against $CCl_4$-induced hepatotoxicity. Boeira *et al.* (2011) showed that hydroalcohlic extract of *P. niruri* and compounds quercetin, rutin, and gallic acid significantly reduced CYP-induced liver lipid peroxidation. Pradhan, (2001) concluded that the ingredients in catliv, effectively helped in regeneration of hepatic cells and is an effective liver tonic for calves animals with carbon tetrachloride induced hepatotoxicity. Catliv contains extracts of *Swertia chirata, Eclipta alba, Fumaria vaillanti, Picorrhiza kurroa, Andrographis paniculata* and *Phyllanthus niruri*. Latha *et al.* (1999) showed the hepatoprotective effect of herbal preparation HPN – 12, an ayurvedic medicine. HPN-12 containing *Glycirrhiza glabra, Pichorhiza kurroa, Berberis aristata, Piper longum, Phyllanthus niruri, Solanum dulcamara, Zingiber officinale, Curculigo orchioides, Elettaria cardamomum, Tinospora cordifolia, Desmodium trifolium* and *Sacchrum officinarum*. It was orally administered to male albino rats at 1ml/100g body weight and were found to be effective against hepatic damage.

## Anti viral Activity

Alternative herbal medicine using extracts of *Phyllanthus niruri* and *Phyllanthus urinaria* have been reported to be effective against Hepatitis B and other viral infections. A study reports quantitative determination of the anti viral effect of these herbs in well-defined *in vitro* systems (Meixa *et al.*, 1995). *Phyllanthus niruri* has been reported to exhibit marked antihepatitis B virus surface antigen activity in *in vivo* and *in vitro* studies (Thyagarajan *et al.*, 1988). Aqueous extract of *Phyllanthus niruri* was reported to have inhibitory effect on human immunodeficiency virus. The alkaloidal extract of *Phyllanthus niruri* showed suppressing activity on strains of HIV-1 cells cultured on MT-4 cell lines. The alkaloidal extract of *Phyllanthus niruri* was thus found to exhibit sensitive inhibitory response on cytopathic effects induced by both the strains of human immunodeficiency virus on human MT-4 cells in the tested concentrations (Naik *et al.*, 2003).

## Immunostimulant

Nworu *et al.* (2010) studied the effect of the aqueous extract of *P. niruri* on the activation of murine lymphocytes and macrophages. They showed that the extract of

*P. niruri* is a potent murine lymphocytes mitogen, inducing significant (p < 0.01) increases in the expression of surface activation maker (CD69) and proliferation of B and T lymphocytes. These activities suggested that stimulation of the immune system by the extracts of *P. niruri* could be partly responsible for the ethnomedicinal applications in the management of infectious diseases. Nworu *et al.* (2010) studied the effect of lyophilized aqueous extract of *P. niruri* on structural and functional maturation of murine bone marrow-derived DCs (BM-DCs). They concluded that it enhances the structural and functional maturation of BM-DCs and their antigen-presenting function. These effects were relevant in immunodeficient conditions, tumor control, and in infectious diseases.

## Antihyperuricemic Activity

Murugaivah *et al.* (2006) showed that the methanol extract from the leaves of *Phyllanthus niruri* L. have oral antihyperuricemic activity in potassium oxonate- and uric acid-induced hyperuricemic rats. Further they also evaluated antihyperuricemic-guided purification of the fraction afforded three lignans, phyllanthin, hypophyllanthin and phyltetralin, of which phyllanthin significantly reversed the plasma uric acid level of hyperuricemic animals to its normal level in a dose-dependent manner, comparable to that of allopurinol, benzbromarone and probenecid which are used clinically for the treatment of hyperuricemia and gout. Murugaivah *et al.* (2009) showed that the antihyperuricemic effect of *Phyllanthus niruri* methanol extract may be mainly due to its uricosuric action and partly through xanthine oxidase inhibition, whereas the antihyperuricemic effect of the lignans was attributed to their uricosuric action.

## Kidney Stone Breaker

The plant has long been used in Brazil and Peru as an herbal remedy for Kidney stones. Research among sufferers of Kidney stones has shown that, while intake of *Phyllanthus niruri* Linn didn't lead to a significant difference in either stone voiding or pain levels, it may reduce urinary calcium, a contributing factor to stone growth (Nishiura *et al.*, 2004). In addition, one study conducted on rats showed that an aqueous solution of *Phyllanthus niruri* Linn. may inhibit kidney stone growth and formation in animals that already have stones (Freitas *et al.*, 2002). Barros *et al.* (2006) suggested that *P. niruri* extract interfered with the arrangement of the precipitating crystals, probably by modifying the crystal-crystal and/or crystal-matrix interactions. It concluded that *P. niruri* extract may have a therapeutic potential, since it was able to modify the shape and texture of calculi to a smoother and probably more fragile form, which could contribute to elimination and/or dissolution of calculi.

## Anti-tumor Activity

Sharma *et al.* (2009) studied to evaluate the anti-tumor activity of a hydro-alcoholic extract of the whole plant in male Swiss albino mice, on the two stage process of skin carcinogenesis induced by a single topical application of 7, 12-dimethylbenz (a) anthracene and two weeks later promoted by repeated application of croton oil. The oral administration of *P. niruri* caused significant reduction in tumor incidence, tumor yield, tumor burden and cumulative number of papillomas as compared to carcinogen-treated controls. Lee *et al.* (2011) investigated the antimetastatic activity of *Phyllanthus* on cancer cells. Prior to that, an effective dose which is non-toxic to the cells had to be determined. Hence, they evaluated the toxicity of both aqueous and methanolic extracts of four different species of *Phyllanthus* plants, namely *P. niruri*, *P. urinaria*, *P. watsonii*, and *P. amarus*, on two human cancer cell lines (A549 and MCF-7) and two normal human cell lines (184B5 and NL20). They showed that *Phyllanthus* exhibited selective cytotoxicity against MCF-7 and A549 human cancer cells, with IC50 values ranging from 50 mg/ml to 180 mg/ml and 65 mg/ml to 470 mg/ml, respectively for both methanolic and aqueous extracts while having minimal toxicity to the normal cell lines at the same concentrations. Kashiwada *et al.* (1992) investigated antitumor activity of tannins of *Phyllanthus niruri* Linn.

## Antioxidant Activity

Sarkar *et al.* (2010) carried out to investigated the mechanism of the protective action of a novel antioxidant protein molecule, isolated from the herb, *Phyllanthus niruri* against tertiary butyl hydroperoxide induced cytotoxicity and cell death. Sharma *et al.* (2011) evaluated that *P. niruri* extract has potentiality to reduce skin papillomas by enhancing antioxidant defence system. Thakur *et al.* (2011) studied the effects of aqueous and alcoholic extract of *P. niruri* on *in vivo* gamma radiation induced chromosome aberration and *in vitro* antioxidant activity. Radioprotective potential of alcoholic extract was founded to be more effective than the aqueous extract. Qualitative phytochemical investigation of both extract revealed the presence of sugars, flavonoids, alkaloid, lignans, polyphenols, tannins, coumarins and saponins. Higher radioprotective effect of the alcoholic extract may be attributed to rich presence of antioxidant polyphenolic compounds.

## C.V.S. Activity

Cheng *et al.* (1994) investigated Antihypertensive activity of Gerannin from *Phyllanthus niruri* Linn. Thippeswami *et al.* (2011) investigated the effect of the aqueous extract of *Phyllanthus niruri* against doxorubicin-induced myocardial toxicity in rats.

## Hyperlipidemic Activity

Lipid lowering activity of *Phyllanthus niruri* alcoholic extracts in triton induced hyperlipidemia was examine in rats (Chandra, 2000). Khanna *et al.* (2002) show the Lipid lowering activity of *Phyllanthus niruri* in hyperlipemic rats. Latha *et al.* (2010) tested the protective role of *Phyllanthus niruri* aqueous leaf extract on alcohol and heated sunflower oil-induced hyperlipidemia. And they concluded that the *P.niruri* leaf extract effectively protects the system against alcohol and δPUFA-induced hyperlipidemia and has a definite anti-hyperlipidemic potential.

## Other Activity

An alcoholic extract of *Phyllanthus niruri* was found to reduce significantly the blood sugar in normal rats and in alloxan diabetes rats (Raphael *et al.*, 2000). *Phyllanthus niruri* also have anti malarial activity (Neraliya *et al.*, 2004).

Santos *et al.* (1995) investigated the antinociceptive properties of hydroalcohlic extracts of new species of plants of the genus *Phyllanthus* (Euphorbiaceae). Odetola *et al.* (2000) investigated in mice the anti-diarrhoeal and gastro-intestinal protective potentials of aqueous extract of leaves of *Phyllanthus*.

Frietas *et al.* (2002) evaluated the effect of an aqueous extract of *Phyllanthus niruri*, a plant used in folk medicine to treat lithiasis, on the urinary excretion of endogenous inhibitors of lithogenesis, citrate, magnesium and glycosaminoglycans (GAGs).

Iizuka *et al.* (2007) isolated a platelet-aggregatory inhibitor from the 50 per cent MeOH extract of *Phyllanthus niruri* L. leaf. Its structure was determined to be methyl brevifolin carboxylate on the basis of the 1H-, 13C-NMR, and high-resolution mass spectral data. They confirmed that it is a potent inhibitor of platelet aggregation comparable to adenosine in spite of differences in the inhibitory mechanisms.

Shakil *et al.* (2008) isolated two prenylated flavanones from the hexane extract of *Phyllanthus niruri* plant. The structure of these flavanones were established as 8-(3-Methyl-but-2-enyl)-2-phenyl chroman-4-one and 2-(4-hydroxyphenyl)-8-(3-methyl-but-2-enyl)-chroman-4-one on the basis of spectral analysis. These were evaluated for nematicidal activity against root-knot, *Meloidogyne incognita*, and reniform, *Rotylenchulus reniformis*, nematodes.

## Traditional Uses

This Plant has a long history in herbal medicine systems in every tropical country where it grows. For the most part, it is employed for similar conditions worldwide. The natural remedy is usually just a standard infusion or weak decoction of the whole plant or its aerial parts. Its main uses are for many types of billiary and urinary conditions including kidney and gallbladder stones; for hepatitis, cold, flu, tuberculosis, and other viral infections; liver diseases and disorders including anaemia, jaundice and liver cancer; and for bacterial infections such as cystitis, prostatitis, venereal diseases and urinary tract infections. It is also widely employed for diabetes and hypertension as well as for its diuretic, analgesic, stomachic, antispasmodic, febrifugal, and cell protective properties in many other conditions. It is little wonder that chanca piedra is used for so many purposes in herbal medicine systems.

In many countries around the world plants in the genus *Phyllanthus* are used in folk remedies; therefore this genus is of great importance in traditional medicine (Foo *et al.*, 1993). The genus *Phyllanthus* has a long history of use in the treatment of liver, kidney and bladder problems, diabetes and intestinal parasites. Some related species in this region with medicinal significance are *P. epiphyllanthus, P. niruri P. urinaria, P. acuminatus and P. emblica* (Tirimana, 1987). *P. amarus, P. nururi* and *P. urinaria* are used in the treatment for kidney/gallstones, other kidney related problems, appendix inflammation, and prostate problems (Heyde *et al.*, 1990).

Hot water extract of the entire plant is administered orally, to reduce fevers, as a laxative, as a cholagogue, as a spasmolytic, for diabetes, as a diuretic, as an anti-inflammatory agent. Hot water extract of roots together with hot water extract of *Citrus aurantifolia* roots is taken orally to increase appetite (Mokkhasmit *et al.*, 1971). Decoction of dried leaves and roots is taken orally for fever, and for good health. Decoction of dried entire plant is administered orally to treat venereal diseases.

Decoction of dried leaf when taken orally is a treatment for diarrhoea (Weninger, 1986). Dried entire plant, grounded in buttermilk is administered orally for jaundice. Fresh leaf juice is used externally for cuts and bruises. For eye diseases, the juice is mixed with castor oil and applied to the eye. Infusion of dried leaves is administered orally for dysentery and diarrhea. Infusion of green root is taken orally to treat heavy menstrual periods. Fresh plant juice is taken orally for genitourinary disorders (Singh, 1986). The fruit is used externally for tubercular ulcers, scabies and ringworm. Fresh leaf juice or fresh root juice are taken orally for venereal diseases. *Phyllanthus niruri* is used as a diuretic in dropsically affection, gonorrhea and other troubles of genitourinary tract. Herb is bitter, astringent, diuretic and febrifuge, antiseptic. Fresh root is a remedy for jaundice. Infusion of young shoots given in dysentery. Leaves are popular remedy against fever. It can be used to increase the appetite and locally to relieve inflammations. It can also be used in case of anorexia (Sircar, 1984; Velazco, 1980; Loustalot *et al.*, 1949; Khan *et al.*, 1978).

## Medicinal Uses

*Phyllanthus niruri* Linn use as Analgesic, antibacterial, antihepatotoxic, anti-inflammatory, antilithic, antimalarial, antimutagenic, antinociceptive, antispasmodic, antiviral, aperitif, carminative, choleretic, deobstruent, digestive, diuretic, febrifuge, hepatotonic, hepatoprotective, hypoglycaemic, hypotensive, laxative, stomachic, tonic, vermifuge.

## References

Bagalkotkar, G., Sagineedu, S.R., Saad, M.S., and Stanslas, J. (2006). Phytochemicals from *Phyllanthus niruri* Linn. and their pharmacological properties: A review. *The Journal of Pharmacy and Pharmacology*, **58**: 1559-1570.

Barros, M.E., Lima, R., Mercuri, L.P., Matos, J.R., Schor, N., and Boim, M.A. (2006). Effect of extract of *Phyllanthus niruri* on crystal deposition in experimental urolithiasis. *Urological Research*, **34**: 351-357.

Bhattacharjee, R., and Sil, P.C. (2007). Protein Isolate from the Herb *Phyllanthus niruri* modulates carbon tetrachloride-induced cytotoxicity in hepatocytes. *Toxicology Mechanism and Methods*, **17**: 41-47.

Boeira, V.T., Leite, C.E., Santos, A.A., Edelweiss, M.I., Calixto, J.B., Campos, M.M., and Morrone, F.B. (2011). Effects of the hydroalcoholic extract of *Phyllanthus niruri* and its isolated compounds on cyclophosphamide-induced hemorrhagic cystitis in mouse. *Naunyn-Schmiedeberg's Archives of Pharmacology*, **384**: 265–275.

Cabieses, F. (1993). Apuntes de medicina traditional. La racionalizacion de lo irracional. "Notes of traditional medicine." Consejo Nacional de Ciencia y Technologia CONCYTEC Lima-Peru, 414.

Chevallier, A. (2000). Encyclopedia of Herbal Medicine: Natural Health, Second edition, Dorling Kindersley Book. USA, pp. 336.

Cuellar, A., and Estevez, P.V. (1980). A preliminary phytochemical study of Cuban plants. V. *P. niruri. Revista Cubana de Farmacia*, **14**: 63–68.

Foo, L.Y. (1993). Amariin, a di-dehydrohexahydroxydiphenoyl hydrolysable tannin from *Phyllanthus amarus. Phytochemistry*, **33**: 487-491.

Freitas, A.M., Schor, N., and Boim, M. A. (2002). The effect of *Phyllanthus niruri* on urinary inhibitors of calcium oxalate crystallization and other factors associated with renal stone formation. *British Journal of Urology International*, **89**: 829–834.

Gupta, J., and Ali, M. (1999). Four new seco-sterols of *Phyllanthus fraternus* roots. *Indian Journal of Pharmaceutical Science*, **61**: 90-96.

Gupta, D.R., Ahmed, B., and Shoyakugaku, Z. (1984). A new flavones glycoside from *Phyllanthus niruri. Journal of Natural Products*, **383**: 213–215.

Heyde, H. (1990). Medicijn planten in Suriname. (Den dresi wiwiri foe Sranan). "Medicinal Plants in Suriname." Uitg. Stichting Gezondheidsplanten Informaite (SGI) Paramaribo, p. 157.

Houghton, P.J., Woldemariam, T.Z., O'Shea S., and Thyagarajan, S.P. (1996). Two securinega type alkaloids from *Phyllanthus amarus. Phytochemistry*, **43**: 715-717.

Iizuka, T., Nagai, M., Taniguchi, A., Moriyama, H., and Hoshi, K. (2007). Inhibitory effects of methyl brevifolincarboxylate isolated from *Phyllanthus niruri* L. on platelet aggregation. *Biological and Pharmaceutical Bulletin*, **30**: 382-384.

Ishimaru, K., Yoshimatsu, K., Yamakawa, T., Kamada, H., and Shimomura, K. (1992). Phenolic constituents in tissue cultures of *Phyllanthus niruri. Phytochemistry*, **31**: 2015–2018.

Kale, K.U., Parag, D., and Vivek, C. (2001). Isolation and estimation of an antihepatotoxic compound from *Phyllanthus niruri. Indian Drugs*, **38**: 303–306.

Kashiwada, Y., Nonaka, G., Nishioka, I., Chang, J.J., and Lee, K.H. (1992). Antitumor agents 129. tannins and related compounds as selective cytotoxic agents. *Journal of Natural Products*, **55**: 1033-43.

Khan, M.R., Ndaalio, G., Nkunya, M.H.H., and Wevers, H. (1978). Studies on the rationale of African traditional medicine Part II. Preliminary screening of medicinal plants for antigonococci activity. *Pakistan Journal of Science and Industrial Research*, **27**: 189-192.

Khanna, A.K., Rizvi, F., and Chander, R. (2002). Lipid lowering activity of *Phyllanthus niruri* in hyperlipemic rats. *Journal of Ethnopharmacology*, **82**: 19-22.

Latha, P., Chaitanya, D., and Rukkumani, R. (2010). Protective effect of *Phyllanthus niruri* on alcohol and heated sunflower oil induced hyperlipidemia in wistar rats. *Toxicology Mechanisms and Methods*, **20**: 498-503.

Latha, U., and Rajesh, M.G. (1999). Hepatoprotective effect of an Ayurvedic medicine. *Indian Drugs*, **36**: 470-473.

Lee, S.H., Jaganath, I.B., Wang, S.M., and Sekaran, S.D. (2011). Antimetastatic effects of *Phyllanthus* on human lung (A549) and breast (MCF-7) cancer cell Lines. *PLoS ONE*, **6**: e20994.

Lewis, W.H. and Elvin-Lewis, P.F. (1977). Medical Botany Plants Affecting Man's Health. A Wiley-Inter science Publication. John Wiley and Sons, New York-London-Sydney-Toronto, p. 515.

Loustalot, A.J., and Pagan, C. (1949). Local "Fever" Plants tested for the presence of Alkaloids. EL Crisol, 3: 3.

Mannan, A., and Ahmad, K. (1976). A short note on the occurrence of sex hormones in Bangladesh plants. *Bangladesh Journal of Biological Sciences*, 5: 45.

Meixa, W., Haowei, C., Yanjin, L. *et al.* (1995). Herbs of the genus *Phyllanthus* in the treatment of chronic hepatitis B observation with three preparation from different geographic sites. *Journal of Laboratory and Clinical Medicine*, 126: 350.

Mokkhasmit, M.K., Swasdimongkol, W., Ngarmwathana, and Permphipat, U. (1971). Study of toxicity of Thai medicinal plants. *Journal of the Medical Association of Thailand*, 54: 490-504.

Morton, J.F. (1981). Atlas of Medicinal Plants of Middle America. Library of Congress.

Mulchandani, N.B., and Hasarajani, S. A. (1984). 4-methoxy-norsecurinine, a new alkaloid from *Phyllanthus niruri*. *Planta Medica*, 50: 104-105.

Murugaiyah, V., and Chan, K.L. (2006). Antihyperuricemic lignans from the leaves of *Phyllanthus niruri*. *Planta Medica*, 72: 1262-1267.

Murugaiyah, V., and Chan, K.L. (2009). Mechanisms of antihyperuricemic effect of *Phyllanthus niruri* and its lignan constituents. *Journal of Ethnopharmacology*, 124: 233-239.

Naik, A.D., and Juvekar, A.R. (2003). Effects of alkaloidal extract of *Phyllanthus niruri* on HIV replication. *Indian Journal of Medical Sciences*, 57: 387-393.

Nanden-Amattaram, T. (1998). Medicinale Planten: tips en simpele recepten voor een goede gezondheid. "Medicinal plants and simple recipes for a good health." Paramaribo-Suriname, p. 18.

Neraliya, S., and Gaur, R. (2004). Juvenoid activity in plant extracts against filarial mosquito *Culex quinquefasciatus*. *Journal of Medicinal and Aromatic Plant Sciences*, 26: 34-38.

Nworu, C.S., Akah, P.A., Okoye, F.B., and Esimone, C.O. (2010). Aqueous extract of *Phyllanthus niruri* (Euphorbiaceae) enhances the phenotypic and functional maturation of bone marrow-derived dendritic cells and their antigen presentation function. *Immunopharmacology and Immunotoxicology*, 32: 393-401.

Nworu, C.S., Akah, P.A., Okoye, F.B., Proksch, P., and Esimone, C.O. (2010). The effects of *Phyllanthus niruri* aqueous extract on the activation of murine lymphocytes and bone marrow-derived macrophages. *Immunological Investigation*, 39: 245-67.

Odetola, A.A., and Akojenu, S.M. (2000). Anti-diarrhoeal and gastrointestinal potentials of the aqueous extract of *Phyllanthus amarus* (Euphorbiaceae). *African Journal of Medical Sciences*, 29: 119-122.

Pradhan, N.R. (2001). Therapeutic effect of catliv on induced hepatopalthy in calves. *Indian Veterinary Journal*, **79**: 1104-1106.

Raphael, K.R., Sabu, M.C., and Kuttan, R. (2000). Antidiabetic activity of *Phyllanthus niruri*. *Amala research bulletin*, **20**: 19-25.

Santos, D.R. (1990). Chade "quebra-pedra" (*Phyllanthus niruri*) na litiase urinaria em humanos e ratos. Thesis, Escola Paulista de Medicina (S o Paulo, Brazil).

Santos, A.R., Filho, V.C., Yunes, R.A., and Calixto, J.B. (1995). Analysis of the mechanisms underlying the antinococeptive effects of the extracts of plants from the genus *Phyllanthus*. *General Pharmacology*, **26**: 1499–1506.

Sarkar, M.K., and Sil, P.C. (2010). Prevention of tertiary butyl hydroperoxide induced oxidative impairment and cell death by a novel antioxidant protein molecule isolated from the herb, *Phyllanthus niruri*. *Toxicology In Vitro*, **24**: 1711-1719.

Satyanarayana, P., and Venkateswarlu, S. (1991). Isolation, structure and synthesis of new diarylbutane lignans from *Phyllanthus niruri*: Synthesis of 52-desmethoxy niranthin and an antitumour extractive. *Tetrahedron*, **47**: 8931–8940.

Shakil, N.A., Kumar, P.J., Pandey, R.K., Saxena, D.B. (2008). Nematicidal prenylated flavanones from *Phyllanthus niruri*. *Phytochemistry*, **69**: 759.

Sharma, P., Parmar, J., Verma, P., Sharma, P., and Goyal, P.K. (2009). Anti-tumor activity of *Phyllanthus niruri* (a Medicinal Plant) on chemical induced skin carcinogenesis in mice. *Asian Pacific Journal of Cancer Prevention*, **10**: 1089-1095.

Sharma, P., Parmar, J., Verma, P., and Goyal, P.K. (2011). Modulatory influence of *Phyllanthus niruri* on oxidative stress, antioxidant defense and chemically induced skin tumors. *The Journal of Environmental Pathology, Toxicology and Oncology*, **30:** 43-53.

Singh, B., Agarwal, P.K., and Thakur, R.S. (1989a). Triterpenoids from *Phyllanthus niruri*. *Indian Journal of Chemistry*, **28**: 319–321.

Singh, B., Agrawal, P.K., and Thakur, R.S. (1989b) A new lignan and a new neo lignan from *Phyllanthus niruri*. *Journal of Natural Products*, **52**: 48–51.

Singh, Y.N. (1986). Traditional medicine in Fiji. Some herbal folk cures used by Fiji Indians. *Journal of Ethnopharmacol*, **15**: 57-88.

Sircar, N.N. (1984). Pharmacotherapeutics of Dasemani drugs. *Ancient Science of Life*, **3**: 132-135.

Symasundar, K.V., Singh, B., Thakur, R.S., Husain, A., Kiso, Y., and Hikino, H. (1985). Antihepatotoxic principles of *Phyllanthus niruri* herbs. *Journal of Ethnopharmacology*, **14**: 41–44.

Thakur, I., Uma Devi, P., and Bigoniya, P. (2011). Protection against radiation clastogenecity in mouse bone marrow by *Phyllanthus niruri*. *Indian Journal of Experimental Biology*, **49**: 704-710.

Thippeswamy, A.H.M., Shirodkar, A., Koti, B.C., Jaffar Sadiq, A., Praveen, D.M., Viswanatha Swamy, A.H.M., and Patil, M. (2011). Protective role of *Phyllantus niruri* extract in doxorubicin-induced myocardial toxicity in rats. *Indian Journal of Pharmacology*, **43**: 31–35.

Thyagarajan, S.P., Subramanian, S., Thirunalasundar, T. *et al.* (1988). Effect of *Phyllanthus niruri* on chronic carriers of hepatitis B virus. *The Lancet*, 2: 764-6.

Tirimana, A.S.L. (1987). Medicinal plants of Suriname. Uses and Chemical Constituents. Chemical Laboratory, Ministry of Agriculture, Animal Husbandry and Fisheries. Suriname, p. 92.

Unander, D.W. and Blumberg, B.S. (1991). *In vitro* activity of *Phyllanthus* (Euphorbiaceae) species against the DNA polymerase of hepatitis viruses: Effect of growing environment and inter- and intra-specific differences. *Economy Botany*, **45:** 225-242.

Velazco, E.A. (1980). Herbal and traditional practices related to material and child health care. *Rural Reconstruction Review*, pp. 35-39.

Weninger, B., Rouzier, R.M., Henrys, D.D., Henrys, J.H., and Anthon, R. (1986). Popular medicine of Plateau of Haiti. *Journal of Ethnopharmacology*, **17**: 13-30.

Utilisation and Management of Medicinal Plants Vol. 2 (2014)      *Pages* **205–231**
*Editor-in-Chief*: **V.K. Gupta**
*Published by*: **DAYA PUBLISHING HOUSE, NEW DELHI**

# 7

# Process Management in Herbal Medicine Research and Development in Accordance with International Industrial Standardization

Sunday J. Ameh[1]*, Timothy N. Abner[2], Garba Magaji[3]
and Karniyus S. Gamaniel[4]

## ABSTRACT

*Process management (also called "Product realization and service provision") according to the International Organization for Standardization (ISO) involves: planning of product realization; customer-related processes; design and development; purchasing; production and service provision; and control of measuring and monitoring equipment, as specified in the seventh (7th) clause of ISO 9001. The chapter aimed to provide a conceptual framework that can be used to facilitate the introduction of quality into herbal medicinal products as well as facilitate the provision of quality consultancy services to interested parties as per ISO/IEC 17025. The chapter reviewed; (i) the quality management system (QMS) processes involved in product realization (as per ISO 9001) and service provision (as per ISO/IEC 17025); and (ii) the critical stages of herbal drug research and development (R&D) from traditional herbal*

1   Department of Medicinal Chemistry and Quality Control (MCQC), National Institute for Pharmaceutical Research and Development (NIPRD), Idu Industrial Area, Abuja, Nigeria.

2   Directorate of Technical Services, Standards Organization of Nigeria (SON), Abuja, Nigeria.

3   Department of Pharmaceutical and Medicinal Chemistry, Faculty of Pharmaceutical Sciences, Ahmadu Bello University, Zaria, Nigeria.

4   Office of the Director General/CEO, NIPRD, Idu Industrial Area, Abuja, Nigeria.

*   *Corresponding author*: E-mail: sjitodo@yahoo.com

*medicine (THM) as planned in the Nigerian National Institute for Pharmaceutical Research and Development (NIPRD). The resulting framework was discussed in terms of its relevance and applicability to operations at NIPRD. The article concluded that ISO 9001's and ISO/ IEC 17025's provisions for process management can indeed be applied in introducing quality into herbal drug products and to facilitate the provision of standardized services.*

***Keywords***: Traditional herbal medicine (THM), International Organization for Standardization (ISO), National Institute for Pharmaceutical Research and Development (NIPRD), Process management, Quality management system (QMS), Research and development (R&D).

# Introduction

Herbal "process management" or "product realization" (as per the 7th clause of ISO 9001) embraces all phases of herbal product R&D including investigation and appreciation of traditional knowledge and technology (TKT), application of biomedical knowledge, concepts and formulation technology, pilot production, clinical trials and industrial production. Quality control consists of three key activities: sampling, analytical control, and inspection control. Analytical aspects of herbal quality control either as per ISO 9001 or as per ISO/IEC 17025 involves identification and qualification of raw materials for use in production; determining the production processes and the stages of in-process quality control; developing analytical methods for all materials including intermediates; and developing methods and criteria for accepting the finished herbal product. For a manufacturer, ISO 9001 certification alone adequately provides for in-house quality control, but ISO/IEC 17025 accreditation is required if the manufacturer is to provide quality control services to a second or third party. A drug regulatory agency (DRA) like the Nigerian National Agency for Food and Drug Administration and Control (NAFDAC) requires at least ISO 9001 or ISO/IEC 17025, but ideally both. That is: NAFDAC as organization needs ISO 9001, while its laboratories need ISO/IEC 17025. The same should apply to a drug R&D institution like NIPRD, since the Institute not only develops and test produces, but is expected to provide services to a second or third party. Among services that an R&D institution like NIPRD should provide a second or third party are: analytical control services; advising on good manufacturing practice (GMP)/inspection control; and advising on documentations including the preparation of product dossier required for regulatory purposes. ISO standards generally ensure that products and services are safe, reliable, of good quality and environment friendly. For profitability in business, the standards are strategic tools that reduce costs by minimizing waste and errors and by increasing productivity. ISO standards assist companies to access new markets, level the playing field for developing countries and facilitate free and fair global trade. In short, the standards provide practical tools for tackling most global economic challenges, and ought to be the best friend of countries desiring to enter the world market. This would particularly apply to countries like Nigeria and others that have a rich herbal tradition deserving modernization given the increasing demand for herbal products worldwide (Ameh *et al.*, 2009; 2010a; 2010b; 2010c;

2011a; 2012). Copies of ISO 9001:2008 and ISO/IEC 17025:2005, obtainable from ISO Headquarters, Basle, Switzerland, are required for this study.

# Methodology: Review of Relevant Concepts

## Process Management or Product Realization

Product realization (or Process management) is the term used to describe the work or processes that an organization undertakes or goes through to research, develop, manufacture, and deliver finished goods or services (Becker *et al.*, 2011). An effective QMS for product realization includes a comprehensive approach to getting from the product concept to the finished product. This approach, often called "quality plan", includes the following: planning of product requirements (or quality characteristics) and quality objectives; creation of the processes from concept to the finished product; documentation of all activities related to the product; mobilization of all the resources needed for realizing the product; and acquisition of the means for verification, monitoring, inspection, testing and determining the records to be kept for purposes of continual improvement (ISO 9001:2008; ISO/IEC 17025:2005). In a less technical parlance, product realization or process management may be defined as: the application of knowledge, skills, tools, techniques and systems to define, visualize, measure, control, report and improve processes with the goal of meeting customer requirements (Becker *et al.*, 2011). Both ISO 9001and ISO/IEC 17025 promote the process approach to managing an organization, including testing and calibration laboratories. Both standards promote the adoption of a process approach when developing, implementing and improving the effectiveness of a QMS for producing and delivering goods and services so as to enhance customer satisfaction by meeting customer requirements and expectations.

**Table 7.1**: The top 10 countries in ISO certification in 2009.

| Country | Ranking | No. Certificates | Pertinent Remark |
| --- | --- | --- | --- |
| China | 1 | 257,076 | Relies mostly on ISO standard |
| Italy | 2 | 130,066 | Relies mostly on ISO standard |
| Japan | 3 | 68,484 | Relies only partly on ISO standard |
| Spain | 4 | 59,576 | Relies substantially on ISO standard |
| Russia | 5 | 53,152 | Relies substantially on ISO standard |
| Germany | 6 | 47,156 | Relies only partly on ISO standard |
| UK | 7 | 41,193 | Relies only partly on ISO standard |
| India | 8 | 37,493 | Relies substantially on ISO standard |
| South Korea | 9 | 28,935 | Relies substantially on ISO standard |
| US | 10 | 23,400 | Relies only partly on ISO standard |

Source: ISO Survey (2009). Most countries have their own national standards in addition to ISO standards. For example the UK is well known for its industrial standards developed by the oldest standards institution in the world – the British Standards Institution (BSI); and for its accreditation services by UK Accreditation Services (UKAS).

## Worldwide Impact of ISO

Every year ISO performs a survey of certifications to its standards. The latest edition (2012) shows an increase of 6.23 per cent over the previous year to a worldwide total of 1,457,912 certificates and users of one or more of the standards in 178 countries (*i.e.*, ~1.5 million certificates/users in 178 countries). In 2009 ISO 9001 alone accounted for over 1 million. These figures clearly demonstrate the global relevance of ISO standards. The data for the top ten nation users of ISO 9001 in 2009 are given in Table 7.1.

## Certification/Accreditation

Certification is a confirmation that an organization meets certain specified characteristics (or abilities). Such confirmation is usually provided through an external review, assessment, or audit. Accreditation is a specific method or process by which an organization provides certification. The two terms are similar but not necessarily synonymous in practice. In common usage an organization is said to be certified (or registered) to ISO 9001; or accredited to ISO 17025. As earlier stated ISO 9001 certification allows only for in-house quality control, but not for providing quality control service to a customer. By contrast accreditation to ISO 17025 allows the accredited organization to provide quality control service to a customer. The term "product certification" refers to activities (or processes) intended to determine if a given product meets minimum standards. Its usage is similar to the usage of the "quality assurance". It is evident from the foregoing that different certification systems (*i.e.*, methods of certification) exist for different purposes.

## Reasons for Seeking Certification (or Accreditation) to an ISO Standard

Organizations often want to get certified to ISO standards (Example: ISO 9001) or accredited to ISO standards (Example: ISO 17025) although certification or accreditation is not a legal requirement to do business. The best reason for wanting to implement an ISO standard is to improve the efficiency and effectiveness of the organization. Thus, an organization may decide to seek certification or accreditation for one or more reasons, since certification or accreditation may be: (*i*) a contractual or regulatory requirement; (*ii*) necessary to meet customer preferences; (*iii*) fall within the context of a risk management programme, and (*iv*) help motivate staff by setting a clear goal for the development of its QMS.

## UKAS and SON: Examples of National Agencies that "Accredit Certification Bodies"

It is already seen that accreditation is a process in which authority or a certificate of competency is granted a "conformity assessment body" (CAB) - also called "accredited certification body" (ACB), which may be a public, but usually, a private body. CABs or ACBs that issue certificates against International Standards (like ISO 9001 and ISO 17025) must themselves be formally accredited by a higher "accrediting body" (AB) such as a national agency like the UK Accreditation Service (UKAS) or the Standards Organization of Nigeria (SON). A national AB like UKAS or SON performs the following functions:

1. Assesses CABs or ACBs for competence against Internationally Standards.
2. Accredits organisations to perform conformity assessment tasks.
3. Issues accreditation certificates and schedules showing the limits of the accreditation.
4. Permits the use of its logo on accredited certification/Accreditation Number.
5. Provides a website to enable the validity of an accreditation to be checked.

Examples of types of accreditation provided by ABs and by CABs or ACBs includes:

1. Accreditation of testing and/or calibration laboratories to ISO/IEC 17025.
2. Certification of organization to ISO 9001.
3. Certification of experts or specialist organizations that are permitted to issue official certificates of compliance with established technical standards.

Both ABs and CABs or ACBs usually operate according to ISO/IEC 17011, published by the Committee on Conformity Assessment (CASCO) Secretariat of ISO.

## The Role of ISO in Relation to Certification

ISO does not perform certification, but develops International Standards, including QMS standards such as ISO 9001 and ISO/IEC 17025. Certification is performed by external certification bodies (CABs or ACBs), which are largely private.

**Table 7.2**: Standards and projects under the direct responsibility of CASCO secretariat.

| Standard and Project | Purpose |
|---|---|
| ISO/EIC 17011:2004 | Conformity assessment – provides general requirements for peer review of conformity assessment bodies and accreditation bodies |
| ISO/IEC 17043:2010 | Conformity assessment – provides general requirements for proficiency testing |
| ISO/IEC 17021:2011 | Conformity assessment – provides requirements for bodies providing audit and certification of management systems |
| ISO/IEC 17024:2012 | Conformity assessment – provides general requirements for bodies operating certification of persons |
| ISO/IEC 17020:2012 | Conformity assessment – provides requirements for the operation of various types of bodies performing inspection |
| ISO/IEC 17001:2005 | Conformity assessment – provides the principles and requirements assessment of impartiality |
| ISO/IEC 17005:2008 | Conformity assessment – provides the principles and requirements for assessing the use of management systems by organizations |
| ISO/IEC 17025:2005 | Provides general requirements for the competence of testing and calibration laboratories |

The Table reaffirms that accreditation is the formal recognition by an independent body ("accreditation body") that another independent body ("certification body") is capable of carrying out certification. Accreditation is not legally obligatory but it adds another level of confidence, as 'accredited' means that the certification body itself has been independently checked to make sure that it operates according to CASCO standards.

When a company or organization is certified to an ISO standard they receive a certificate from the certification body (CAB or ACB). Even though the logo of ISO and the name of the ISO Standard appear on such a certificate, it is not issued by ISO. Although ISO does not perform certification, its Committee on Conformity Assessment (CASCO) has produced a number of standards (publications) that regulate certification processes. These publications contain voluntary criteria agreed internationally to ensure best practices in certification. The organizations (*i.e.*, public or usually private CABs or ACBs) that offer certification or accreditation services operate their activities in accordance with these CASCO publications. An abridged list of some CASCO standards and projects is shown in Table 7.2.

## Synopsis on ISO/IEC 17011

ISO/IEC 17011:2004 specifies the general requirements for accreditation bodies (ABs) that assess and accredit conformity assessment bodies (CABs or ACBs). It is also appropriate for the peer evaluation processes for mutual recognition arrangements between accreditation bodies (ABs). ABs operating in accordance with ISO/IEC 17011:2004 do not have to offer accreditation to all types of CABs. For the purposes of ISO/IEC 17011:2004, CABs or ACBs are organizations that provide the following conformity assessment services:

1. Testing – like quality control as per ISO 17025 or other standards like the British Pharmacopeia (BP).
2. Sampling – of items to be tested according to agreed procedures of sampling.
3. Inspection - of items or processes according to agreed procedures of inspection.
4. QMS certifications – like ISO 9001 certification and ISO/IEC 170250 accreditation -
5. Personnel certification – namely a certificate of competence to under a specific task
6. Product certification – like a certificate of analysis of a product
7. Calibration – of measuring and monitoring equipment

## Quality Assurance

Quality assurance (QA) refers to the planned and systematic activities implemented in a quality system so that quality requirements for a product or service will be fulfilled. It is the systematic measurement, comparison with a standard, monitoring of processes and an associated feedback loops that confer error prevention. This can be contrasted with quality control, which is focused on process outputs. QA has two implicit principles: "Fit for purpose" - the product should be suitable for the intended purpose; and "Right first time"– mistakes should be eliminated (Juran and Godfrey, 1999). QA includes management of the quality of raw materials, assemblies, products and components, services related to production, and inspection processes. Strictly speaking, "suitable quality" is determined by product users, clients or customers, not by the general public. It is not related to cost. Adjectives or descriptors like "high" and "low" are not applicable. For example, a low priced product may be viewed as having high quality because it is disposable (environmentally friendly); whereas, another may be viewed as having low quality because it is not disposable (environmentally unfriendly).

## Quality Control

Quality control (**QC**) refers to the process by which entities (eg: humans or robots) review the quality (characteristics) of all factors involved in production (Juran and Godfrey, 1999). This approach places an emphasis on three multifaceted aspects, namely: i) Elements such as inspection controls, job management, managed processes, performance and integrity criteria, and identification of records; ii) Competences, like knowledge, skills, experience, and qualifications; and iii) Soft elements, such as personnel integrity, confidence, organizational culture, motivation, team spirit, and quality relationships. As an example: aspects of inspection controls include product inspection, whereby every product is examined for fine detail before the product is sold or shipped. Inspectors will be provided with checklists and descriptions of unacceptable product defects like cracks or surface blemishes. The quality of the outputs is at risk if any of these three aspects is deficient in any way. Quality control emphasizes testing of products to uncover defects and reporting to management, which make the decision to allow or deny product release. By contrast, quality assurance attempts to improve and stabilize production (and associated processes) to avoid, or at least minimize, issues which led to the defect(s) in the first place. For contract work, particularly work awarded by government agencies or transnational agencies, quality control issues are among the top reasons for not renewing a contract.

## Analytical Quality Control

The term analytical quality control (AQC) is applicable to laboratories or similar settings. It refers to all those processes or procedures designed to ensure that the results of laboratory tests or analyses are consistent, comparable, accurate, and within specified limits of precision. In well managed laboratories, AQC processes are built into the routine operations of the laboratory often by the random introduction of known standards in to the sample stream or by the use of spiked samples (UN/ECE Task Force LQMA, 2002).

## Concluding Remarks on Quality Control/Quality Assurance Concepts

Organizations that engage in quality control typically have a team of workers who focus on testing a certain number of products or observing services being given. The products or services examined are usually chosen at random. The goal of the quality control team is to identify products or services that do not meet agreed specifications. If a problem is identified, the job of the QC team might involve halting production or service until the problem has been corrected. Depending upon the type of product or service (or the type of problem identified), production or services may not cease. Usually, it is not the job of the QC team to correct quality issues. Typically, other personnel are involved in the process of discovering the cause of quality issues and fixing them. Quality control might also involve evaluating people. If an organization has employees that do not have adequate skills or training, or have trouble understanding directions, or are misinformed, the quality of products or services is at risk. For service-oriented organizations, the personnel that interact with customers are, in a sense, the product that they provide to customers. The following conclusions are drawn:

1. Quality control is concerned with examining the product or service — the end result.

2. Quality assurance is concerned with examining the processes that lead to the end result.

3. An organization uses quality assurance to ensure that a product is manufactured in the right way, thereby reducing potential problems with the quality of the final product.

4. Inspection means checking the characteristics of a product to ensure that conformity to a set of specifications is met. Sometimes it means checking 100 per cent of a batch; sometimes it means checking only some samples (in that latter case, it is exactly the same as "statistical quality control".

5. Analytical quality control usually means laboratory tests designed to check the conformity of raw materials, intermediates and finished products. It comprises inspection and other tests such as laboratory analyses.

## Synopsis on ISO 9001:2008

### ISO 9001 as an Industrial Standard

ISO 9001 as an industrial standard or QMS is a document of about 30 pages with 8 clauses, published by ISO and obtainable from its headquarters in Basle, Switzerland, or from any of its national affiliates. The standard is designed to be met by any organization that: i) needs to demonstrate its ability to consistently provide product or service that meets both customer and applicable legal requirements; ii) aims to enhance customer satisfaction by effectively and continually improving its QMS; and iii) plans to provide continual assurance of conformity to customer and applicable legal requirements. These aims or approaches (often called "QMS requirements" or "quality procedures") are generic and are intended to be applicable to every organization irrespective of type, size and product it provides. Wherever any requirement cannot be applied due to the nature of an organization and its product, such can be considered for exclusion. But wherever exclusions are made, claims of conformity to the standard are not acceptable unless such exclusions are limited to requirements within the 7th clause of the standard, and such exclusions do not affect the organization's ability, or responsibility, to provide product that meets customer and applicable legal requirements. ISO 9001 defines the minimum requirements for a well managed organization. In other words, noncompliance to an ISO 9001 requirement puts at risk an organization's ability to consistently and efficiently satisfy the expectations of its customers/stakeholders.

### The Six QMS Requirements or "The Six Quality Procedures"

These procedures or requirements actually refer to sub-clause 4.1 (General requirements) under clause 4 (Quality Management System) of ISO 9001. The sub-clause prescribes that organizations shall establish, document, implement, and maintain a QMS, and continually improve its effectiveness. To do so means that the organization shall operate its QMS with a view to carrying out (or meeting) the following six procedures (or requirements):

1. Determine the processes needed for the QMS, and their application throughout the organization
2. Determine the sequence of the processes and their interactions
3. Determine the criteria and methods for operating and controlling the processes.
4. Determine and ensure the availability needed resources and supporting information.
5. Check, measure and analyze the processes, where applicable.
6. Implement actions to achieve planned results and continual improvement of the processes.

The processes needed for the QMS invariably include the processes for management activities (clause 5), provision of resources (clause 6), product realization (clause 7), and measurement, analysis, and improvement (clause 8). Philosophically, ISO 9001 is based on management by objectives (MBOs) and draws upon eight quality management principles. Ideally therefore, quality assurance (QA) or total quality management (TQM) covers activities in research, development, production and documentation. It embraces the rule: "do it right the first time". It involves regulating the quality of raw materials, the state of production line and works-in-progress, the product and related management processes. One of the most widely used paradigms for TQM or quality assurance management (QAM) is the "Shewhart cycle", also called "PDCA approach", meaning, "Plan-Do-Check-Act". The foregoing is illustrated in Figure 7.1 using NIPRD QMS processes as an example.

### The Eight Quality Management Principles that Underlie ISO 9001

All ISO standards including ISO 9004 - *Managing for Sustained Success* and ISO 9001are formulated on the bases of 8 quality management principles that are aligned with the philosophy and objectives of most quality award programmes in the world's most industrialized nations. The 8 principles are associated with the following themes:

1. Customer focus.
2. Leadership.
3. Involvement of people.
4. Process approach to management.
5. System approach to management.
6. Continual improvement.
7. Factual approach to decision making.
8. Mutually beneficial supplier relationships.

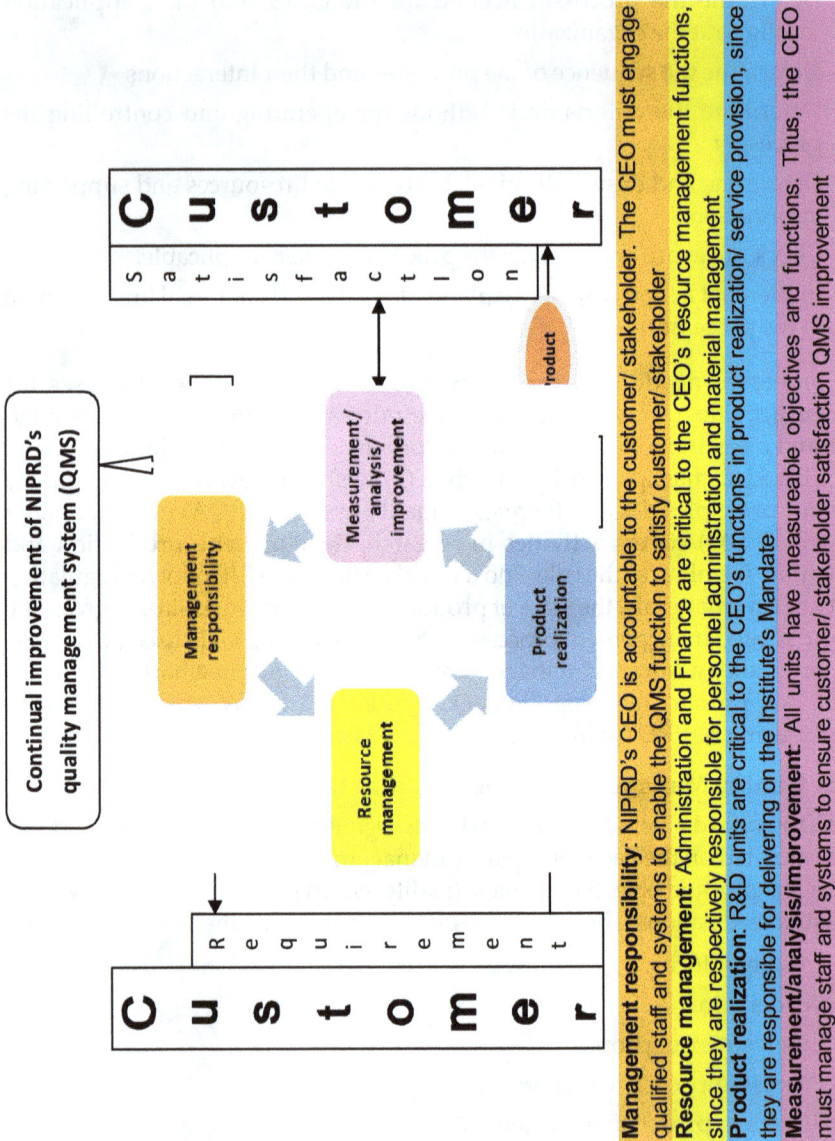

**Management responsibility:** NIPRD's CEO is accountable to the customer/ stakeholder. The CEO must engage qualified staff and systems to enable the QMS function to satisfy customer/ stakeholder

**Resource management:** Administration and Finance are critical to the CEO's resource management functions, since they are respectively responsible for personnel administration and material management

**Product realization:** R&D units are critical to the CEO's functions in product realization/ service provision, since they are responsible for delivering on the Institute's Mandate

**Measurement/analysis/improvement:** All units have measureable objectives and functions. Thus, the CEO must manage staff and systems to ensure customer/ stakeholder satisfaction QMS improvement

**Figure 7.1:** NIPRD's core business in the context of "Plan-Do-Check-Act" QMS process.

Management responsibility corresponds to clause 5 of ISO 9001; while Resource management, Product realization and Measurement/analysis/improvement correspond to clauses 6, 7 and 8 respectively.

## Synopsis on ISO/IEC 17025 : 2005

ISO/IEC 17025 (ISO 17025 hereafter) is a QMS standard jointly developed by ISO and the International Electrochemical Commission (IEC). The standard specifies the general requirements for the competence to carry out tests and/or calibrations. There are 15 management and 10 technical requirements. These requirements outline what a laboratory must do to become accredited. The standard applies to all organizations (like NIPRD and its five technical/R&D departments, Advanced Biology-Chemistry Laboratory (donated by NIH, USA) and NIPRD Research Clinic (NRC). The five R&D laboratories are: Medicinal Plants Research and Traditional Medicine (MPRTM); Pharmacology and Toxicology (P&T); Microbiology, Virology and Biotechnology (MVBT); Medicinal Chemistry and Quality Control (MCQC) and Pharmaceutical Technology and Raw Material Development (PTRMD) that perform tests and/or calibrations. ISO 17025 applies to all laboratories regardless of the number of personnel or the extent of the scope of testing and/or calibration activities covered by the Standard. Historically, the standard was first published in 1999 as a replacement to ISO Guide 25. ISO 17025 was revised and reissued in May 2005. The main reason for the replacement was that ISO Guide 25 did not have all the management requirements that were outlined in ISO 9001. However, ISO 17025:2005 includes the key requirements of ISO 9001 that stand it out as the most anticipated and most patronized standard in world history. As with ISO 9001, an organization or laboratory seeking ISO 17025 accreditation must document its QMS. The most basic requirement for accreditation includes: a quality manual; and the establishment and documentation of the procedures for how the QMS is to be implemented and maintained. These two documented requirement must accompany the laboratory's application for accreditation. The implementation period will require months or even years to establish the records that an accreditation body (or quality auditor) will review at an accreditation audit. Once the laboratory is ready, it finally undergoes the ISO17025 assessment by an accreditation body that is certified to perform laboratory accreditation.

## Key Terminologies of ISO 9001 and ISO/IEC 17025

☆ **Traceability:** Traceability is concerned with and refers to the fact that typically, recorded data are meant to show how and where raw materials and products were processed, in order to allow products and problems to be traced to their sources. As far as possible measurements are to be traceable to Standard International Units (SI units)

☆ **Product realization:** Product realization refers to the scenario in which, when developing a new product, an organization plans the stages of development, with appropriate testing at each stage. The organization tests and documents whether the product meets design requirements, legal requirements, and customer needs. Product realization is the subject of 7th clause of ISO 9001 and the main issue in this article as presented the Results and Discussion.

☆ **Quality plan:** Quality plan refers to a document specifying the QMS processes (including the product realization processes), and the resources to be applied to a specific product or project.

☆ **Monitoring and measurement**: Monitoring and measurement refer to the scenario in which an organization must regularly review its performance through meetings and internal audits, and determine whether the QMS is working and what improvements can be made. The organization must have a documented procedure for internal audits and a procedure for dealing with past problems and potential problems. It must keep records of these activities and the resulting decisions, and monitor their effectiveness. It must have documented procedures for dealing with actual and potential non-conformances (problems involving suppliers, customers, or internal problems).

☆ **Continual Improvement**: Continual Improvement refers to the scenario in which an organization 1) makes sure no customer uses a bad product, 2) determines what to do with a bad product, 3) deals with the root cause of problems, and 4) keeps records to use as a tool to improve the QMS.

☆ **Customer requirements:** Customer requirements refer to the attributes that the buyer of a product (or user of a service) wants. The core business of an organization is to determine customer requirements and to meet them – basis for "Customer focus".

## Relationships of ISO 17025 and ISO 9001: Implications for NIPRD

As already seen, ISO 9001 is the general standard which specifies the requirements for a quality management system. Laboratories which meet the requirements of ISO 17025 also operate in accordance with the requirements of ISO 9001 that are relevant to calibration and testing activities. What this means in practice is that an organisation which holds ISO 9001 certification may use a laboratory accredited against ISO 17025 as a supplier of test data without the need to carry out its own audit of the laboratory's quality system. The question often arises of whether laboratories should be accredited/certified to ISO 9001 or to ISO 17025. In general it is agreed that the appropriate accreditation for commercial testing and calibration laboratories is to ISO 17025. As a result of agreements with laboratory accreditation bodies, many ISO 9001 certification bodies will not allow their certification to be cited by commercial testing or calibration laboratories in support of their services to a third party. What this means in reality is that if an organization (*e.g.*, NIPRD) is ISO 9001 certified with an in-house laboratories, which form part of its QMS, such laboratories will be included in the ISO 9001 external audit. But, if NIPRD wants to sell the services of its laboratories to outsiders as a testing service it cannot advertise such services. NIPRD would need to obtain accreditation to ISO 17025 if such services are to be sold. This is the basis upon which an organization like NIPRD, NAFDAC or a Faculty of Pharmacy or a large corporation that intends to sell testing or calibration services needs to have its laboratories or departments individually ISO 17025 accredited. Such is done or recommended to enhance the organization's overall credibility and competence of its QMS. ISO 9001 external auditors will not usually do a detailed audit of such an internal laboratory if it holds a current ISO 17025 compliant accreditation. The QMS of such a laboratory is largely taken for granted for ISO 9001 purposes. Since laboratory accreditation procedures leading to accreditation to ISO

17025 are explicitly designed for laboratories, they are easier to interpret for the laboratory as opposed to the rather more diffuse requirements of ISO 9001, which are designed for a more general context. One other advantage of accrediting an internal quality control laboratory is that it will generally reduce the number of audits by customers and this is often a key reason for seeking ISO 17025 accreditation. Frequent audits by a range of customers can be disruptive to operations. There are a number of significant omissions from ISO 9001 as compared to ISO 17025, but there is a general ISO move to bring the two standards closer together. The additional requirements in ISO 17025 are:

1. Participation in proficiency testing
2. Adherence to documented and validated methodology.
3. Specification of technical competence, on the part of senior laboratory personnel.
4. The method of scrutiny of laboratories is more rigorous for ISO 17025 than ISO 9001.
5. ISO 17025 assessment bodies will always use technical assessors who are specialists and who carry out a peer review of the methods being used by the laboratory.
6. An ISO 9001 external audit to determine an organization's suitability for certification does not include the above peer review by technical specialists, only QMS auditor
7. From the viewpoint of laboratory's clients or subcontractors, a laboratory that meets the requirements of ISO 17025 fulfils all the relevant requirements of ISO 9001.

The practical import of the foregoing is that if an organisation that is certified to ISO 9001 is using an ISO 17025 accredited laboratory as a sub-contractor, it can treat it as an ISO 9001 certified sub-contractor for any work within the laboratory's scope of ISO 17025 accreditation. There will be no need to carry out quality audits of such a sub-contractor. The results of applying the foregoing principles and practices to herbal product realization and service provision, in accordance with the PDCA system (which applies to both ISO 9001 and ISO 17025 as it relate to the departments and special units of NIPRD) are discussed below.

# Results and Discussion

## Conceptual Framework for Herbal Product Realization as per ISO Standards

The framework generated from the foregoing is discussed in Tables 7.3–7.10 in the following the outline given below, bearing in mind the sequence of product realization and the roles of departments in NIPRD – an outfit mandated to develop drugs including herbal products:

1. Planning of product realization as per ISO 9001, supported by ISO 17025.
2. Customer-related processes as per ISO 9001, supported by ISO 17025.

**Table 7.3**: Roles of different departments in planning of product realization.

| Departmental Roles + ISO 9001 Requirements Under Sub-clause 7.1: Planning of Product Realization | Salient Points, Directing Principles and the Main Roles of Departments in Relation to the Application of Requirements for Planning of Product Realization as per ISO 9001, Supported by ISO 17025 |
|---|---|
| **Departments concerned**<br>Medicinal Plants Research and TM (MPRTM) ; Pharmacology and Toxicology (P&T); Microbiology, Virology and Biotechnology (MVBT); Medicinal Chemistry and Quality control (MCQC); and Pharmaceutical Technology and Raw Material Development (PTRMD) | Based on inputs from the departments, the CEO approves a material (*e.g.,*: aerial parts *Andrographis paniculata*) for development as anti-HIV (coded: AH1). Input may be an MPRTM report that the material has been in use as an immunostimulant since antiquity. The CEO may require further inputs (*e.g.,*: MVBT report that the material has antiviral effect against HIV1 or HIV2.). Once the CEO approves the material for AH1, a team led by a senior scientist (*e.g.*, a professor) is appointed, with a member or more from the departments. The Team Leader (TL) directs the research and reports to the CEO, with copies to all Heads of Department (HODs). Either the HOD or a representative on the team coordinates aspects of the study related to that department. The TL may for example direct as follows: |
| **Recap of ISO 9001 requirements**<br>1. Plan and develop the processes needed for product realization.<br>2. Keep the planning consistent with other requirements of the QMS and document it in a suitable form for organization.<br>3. Determine through the planning, as appropriate, the:<br>a) Quality objectives and product requirements.<br>b) Need for processes, documents, and resources.<br>c) Verification, validation, monitoring, measurement, inspection, and test activities.<br>d) Criteria for product acceptance.<br>e) Records needed as evidence that the processes and resulting product meet requirements. | 1. MPRTM: Confirm the name of the plant and determine how best to procure or cultivate/collect the need parts; determine if similar materials have the same or similar prospects; and suggest or determine a processing procedure based on knowledge gathered from ethnobotanical survey.<br>2. P&T: Determine the safety of use AH1 in laboratory animals and suggest suitable doses for further animal (or possibly human) studies.<br>3. MVBT: Determine the minimum inhibitory concentration of AH1 on HIV1 or HIV2 prepared as suggested by MPRTM or P&T; and suggest a line of action based on the results obtained.<br>4. MCQC: Determine the key physicochemical features of the AH1 raw material and establish parameters (*e.g.,*: loss on drying, extractive matter, chromatographic fingerprints and marker substance) essential for identification and for chemistry-manufacturing-control (C-M-C).<br>5. PTRMD: Determine and establish a suitable formulation based on confirmed findings and legal/customer requirements for the prospective product. |

A document specifying the processes of the QMS (including the product realization processes), and the resources to be applied to a specific product, project or contract, can be referred to as a quality plan. The requirements in sub-clause 7.3 (Design and Development) can also be applied to the development of product realization processes.

3. Design and development processes as per ISO, supported by ISO 17025.
4. Purchasing processes as per ISO 9001, supported by ISO 17025.
5. Production and service provision as per ISO 9001 and ISO 17025.
6. Control of measuring/monitoring equipment as per ISO 9001 and ISO 17025.

**Table 7.4**: Roles of different departments in determining customer-related processes.

| Departmental Roles + the 3 ISO 9001 Requirements Under Sub-clause 7.2: Customer-Related Processes | Salient Points, Directing Principles and the Main Roles of Departments in Relation to the Application of the QMS Requirements for Customer Related Processes |
|---|---|
| **Departments concerned**<br>Medicinal Plants Research and TM (MPRTM) ; Pharmacology and Toxicology (P&T); Microbiology, Virology and Biotechnology (MVBT); Medicinal Chemistry and Quality control (MCQC); and Pharmaceutical Technology and Raw Material Development (PTRMD)<br><br>**Recap of ISO 9001 requirements**<br>**1. Requirements related to the product**<br>Determine customer requirements:<br>1. Specified for the product (including delivery and post-delivery activities).<br>2. Not specified for the product (but needed for specified or intended use, where known).<br>3. Statutory and regulatory requirements applicable to the product.<br>4. Any additional requirements considered necessary by NIPRD.<br><br>**2. Review of the requirements related to product the**<br>Review the product requirements before committing to supply the product to the customer in order to:<br>1. Ensure product requirements are defined.<br>2. Resolve any requirements differing from those previously expressed.<br>3. Ensure its ability to meet the requirements.<br>4. Maintain the results of the review, and any subsequent follow-up actions.<br>5. When the requirements are not documented, they must be confirmed before acceptance.<br>6. If product requirements are changed, ensure relevant documents are amended and relevant personnel are made aware of the changed requirements.<br><br>**3. Customer Communication**<br>Determine and implement effective arrangements for communicating with customers on:<br>1. Product information.<br>2. Inquiries, contracts, or order handling (including amendments).<br>3. Customer feedback (including customer complaints). | If the TL's report to the CEO supports further action on AH1, the CEO directs TL to proceed with customer-related processes as per sub-clause 7.2. The TL may or may not reconstitute his team depending upon what is at stake. For example once it is decided that AH1 should be developed as a capsule or oral powder, MPRTM, MCQC and PTRMD will feature prominently in the tasks ahead. For example MPRTM, MCQC and PTRMD will need to concentrate on how best to provide AH1 in a suitable form efficiently and economically. The final design of the product rests on PTRMD in liaison with MCQC, which needs to develop procedures for qualifying the starting materials and the finished product. If anti-HIV assay of AH1 is a requirement for the finished product, the needed procedure must be developed by MVBT. Once PTRMD succeeds in producing trial sample of AH1, the CEO may direct that a clinical trial be conducted. The AH1 team may or may not be reconstituted, but the new direction of the research may call for a wider range of expertise from all departments/units or even from outside the Institute. |

Post-delivery activities include actions such as the need to institute a pharmacovigilance programme and the need to respond to reports of adverse effects. In situations where a formal review is not practical for each order, relevant product information such as catalogues or advertising material may be used as a basis for a review.

**Table 7.5**: Roles of different departments in determining customer-related processes.

| Departmental Roles + 3 of the 7 ISO 9001 Requirements Under Sub-clause 7.3: Design and Development Processes | Salient Points, Directing Principles and the Main Roles of Departments in Relation to the Application of the QMS Requirements for Design and Development |
|---|---|
| **Departments concerned**<br>Medicinal Plants Research and TM (MPRTM) ; Pharmacology and Toxicology (P&T); Microbiology, Virology and Biotechnology (MVBT); Medicinal Chemistry and Quality control (MCQC); and Pharmaceutical Technology and Raw Material Development (PTRMD) | Design and development can involve any department/unit depending on what is at stake. Example: once the decision is taken to continue with the development of AH1, the following scenarios may unfold: |
| **Recap of ISO 9001 requirements**<br>**1. Design and development planning**<br>Plan and control the product design and development such that the plan determines the:<br>1. Stages of design/development.<br>2. Appropriate review, verification, and validation activities for each stage.<br>3. Responsibility and authority for design/development.<br>4. Interfaces between the different groups involved must be managed to ensure effective communication/clear assignment of responsibility.<br>5. Update, as appropriate, the planning output during design and development. | 1. PTRMD strives to produce the most customer friendly and legally acceptable dosage form.<br>2. MCQC strives to provide the most efficient and economic procedures for qualifying the raw material and the finished product.<br>3. P&T strives to provide facilities for animal studies and discover the most suitable study model.<br>4. MVBT strives to provide efficient anti-HI assay and any other virology tests required.<br>5. The onus of writing up the AH1 dossier for purposes of registration with a regulatory agency rests on PTRMD, with assistance from departments/units like MPRTM, P&T, MCQC, MVBT and ABCL.<br>6. Study design for clinical trials rests with the Office of the CEO, who may choose to utilize expertise from within or outside NIPRD. |
| **2. Design and development inputs**<br>1. Determine product requirement inputs and maintain records.<br>2. The inputs must include: a) Functional and performance requirements. b) Applicable legal requirements. c) Applicable information derived from similar designs. d) Requirements essential for design and development.<br>3. Review these inputs for adequacy.<br>4. Resolve any incomplete, ambiguous, or conflicting requirements. | |
| **3. Design and development outputs**<br>1. Document the outputs of the design and development process in a form suitable for verification against the inputs to the process.<br>2. The outputs must: a) Meet design and development input requirements. b) Provide information for purchasing, production, and service. c) Contain or reference product acceptance criteria. d) Define essential characteristics for safe and proper use. e) Be approved before their release. | |

Design and development review, verification, and validation have distinct purposes. They can be conducted and recorded separately or in any combination. Information for production and service can include details for product preservation.

**Table 7.6**: Roles of different departments in design and development processes - II.

| Departmental Roles + 4 of the 7 ISO 9001 Requirements Under Sub-clause 7.3: Design and Development Processes | Salient Points, Directing Principles and the Main Roles of Departments in Relation to the Application of the QMS Requirements for Design and Development |
|---|---|
| **Departments concerned**<br>Medicinal Plants Research and TM (MPRTM); Pharmacology and Toxicology (P&T); Microbiology, Virology and Biotechnology (MVBT); Medicinal Chemistry and Quality control (MCQC); and Pharmaceutical Technology and Raw Material Development (PTRMD)<br><br>**Recap of ISO 9001 requirements**<br>**1. Design and development review**<br>1. Perform reviews of design and development at suitable stages in accordance with planned arrangements, so as to: a) Evaluate the ability of the results to meet requirements. b) Identify problems and propose actions.<br>2. Ensure the reviews include representatives of the functions concerned.<br>3. Maintain results of reviews and subsequent follow-up.<br><br>**2. Design and development verification**<br>1. Perform design and development verification in accordance with planned arrangements (Design and development planning) to ensure the output meets the design and development input requirements.<br>2. Maintain the results of the verification and subsequent follow-up actions.<br><br>**3. Design and development validation**<br>1. Perform validation in accordance with planned arrangements (Design and development planning) to confirm the resulting product is capable of meeting the requirements for its specified application or intended use, where known.<br>2. When practical, complete the validation before delivery or implementation of the product.<br>3. Maintain the results of the validation and subsequent follow-up actions<br><br>**4. Control of design and development changes**<br>1. Identify design and development changes and maintain records.<br>2. Review, verify, and validate (as appropriate) the changes and approve them before implementation.<br>3. Evaluate the changes in terms of their effect on constituent parts and products already delivered.<br>4. Maintain the results of the change review and subsequent follow-up actions. | Reviews of design and development are essential to discover the most economic/efficient procedure in the departments/units concerned with design and development. PTRMD, being the finishing department would particularly strive to produce the most customer friendly and legally acceptable dosage form. MCQC would strive to provide the most economic and efficient procedures for qualifying the raw material and the finished product. MVBT would similarly strive to provide the most economic and efficient anti-HIV assay and any other biological tests required in AH1 raw material and finished product. It is essential that every department/unit verifies the output of design and development against input in order to ensure that the fulfilment of the objective of the design. Designs need to be validated in order to confirm that product will perform as planned. When products or processes or service fail to perform as planned they must be re-designed, verified and validated. A design may be a new formulation or a new analytical methodology. |

Design Information for production and service can include details for product preservation.

**Table 7.7**: Roles of different departments in purchasing processes.

| Departmental Roles + the 3 ISO 9001 Requirements Under Sub-clause 7.4: Purchasing | Salient Points, Directing Principles and the Main Roles of Departments in Relation to the Application of the QMS Requirements for Purchasing |
|---|---|
| **Departments concerned**<br>Medicinal Plants Research and TM (MPRTM); Pharmacology and Toxicology (P&T); Microbiology, Virology and Biotechnology (MVBT); Medicinal Chemistry and Quality control (MCQC); and Pharmaceutical Technology and Raw Material Development (PTRMD)<br><br>**Recap of ISO 9001 requirements**<br>**1.  Purchasing process**<br>1.  Ensure that purchased product conforms to its specified purchase requirements, noting that the type and extent of control applied to the supplier and purchased product depends upon the effect of the product on the subsequent realization processes or the final product.<br>2.  Evaluate and select suppliers based on their ability to supply product in accordance with the requirements.<br>3.  Establish the criteria for selection, evaluation, and re-evaluation.<br>4.  Maintain the results of the evaluations and subsequent follow-up actions.<br><br>**2.  Purchasing information requirements**<br>1.  Ensure the purchasing information contains information describing the product to be purchased, including the requirements for: (a) Approval of product, procedures, processes, and equipment. (b) Qualification of personnel.<br>2.  Include QMS requirements in the purchasing information – *i.e.,* define and sequence the requirements.<br>3.  Ensure the adequacy of the specified requirements before communicating the information to the supplier.<br><br>**3.  Verification of purchased product**<br>1.  Establish and implement the inspection or other necessary activities for ensuring the purchased products meet the specified purchase requirements.<br>2.  If the organization or its customer proposes to verify the product at the supplier's location, state the intended verification arrangements and method of product release in the purchasing information. | Even though NIPRD's central purchasing unit is located in the department of Administration and Supplies, the criteria for purchases relevant to product realization are furnished by user R&D departments/units. For example in the development of AH1 the following procurement/purchase scenarios may apply:MPRTM would source or provide the criteria for the purchase of starting materials and other goods including reagents and equipment and accessories. For example, the seeds for propagating *Andrographis paniculata* in NIPRD were obtained from India by the Chief Botanist in MPRTM). P&T would source or provide criteria for all items (including animals and their feeds) required in toxicity, efficacy and other pharmacological studies. MVBT would source or provide criteria for all items (including microbial test organisms) and other goods like reagents and equipment. MCQC and PTRMD that must work hand in hand to develop the AH1 dosage form must source all the needed goods including analytical and manufacturing devices. The ABCL and NRC will similarly provide the criteria for all their requirements. Departments/units are responsible for verifying purchased items supplied to them. |

In view of the technical nature of some purchases it is necessary that the Purchasing Officer be familiar (or be specially assisted) with the technicalities involved and reasons behind a given purchase decision.

**Table 7.8**: Roles of different departments in production and service provision-I

| *Departmental Roles + 2 of the 5 ISO 9001 Requirements Under Sub-clause 7.5: Production and Service Provision. For departments to provides services other those needed for NIPRD's product development (ie: services for second and third parties) such departments need to acquire ISO/IEC 17025 accreditation* | *Salient Points, Directing Principles and Main Roles of the Departments in Relation to the Application of the QMS Requirements for Production and Service Provision* |
|---|---|
| **Departments concerned**<br>Medicinal Plants Research and TM (MPRTM) ; Pharmacology and Toxicology (P&T); Microbiology, Virology and Biotechnology (MVBT); Medicinal Chemistry and Quality control (MCQC); and Pharmaceutical Technology and Raw Material Development (PTRMD)<br><br>**Recap of ISO 9001 requirements in respect of production and service provision processes**<br>1. **Control of production and service provision**<br> The planning and implementation of production and service provision are conducted under controlled conditions to include, as applicable: availability of product characteristics information; availability of work instructions; use of suitable equipment; availability and use of monitoring and measuring equipment; implementation of monitoring and measurement activities; and implementation of product release, delivery, and post-delivery activities<br><br>2. **Validation of processes for production and service provision**<br> Wherever subsequent monitoring or measurement a product or service cannot verified, the processes involved should be validated before release of the product or provision of service. Such validation includes processes where deficiencies may become apparent only after product use or service delivery. The ability of processes to achieve the planned results should also be validated. Furthermore, the established validation arrangements should include, as applicable: criteria for process review and approval; approval of equipment; qualification of personnel; use of defined methods and procedures; requirements for records; and re-validation. | As far as the actual production of an herbal drug dosage form is concerned PTRMD is the last bus top. As for the provision of laboratory and related services, each departments can offer at least one such service. For example: MPRTM can provide herbalists with taxonomic data; P&T can provide herbalists with toxicity or efficacy data; MVBT can provide data on the comparative effect of an herb on difference cell species or the antiviral or antimicrobial potential of an herb; MCQC can furnish data essential for chemistry-manufacturing-control (C-M-C) and posology; and PTRMD can provide the recipe for producing the approved dosage form, and write up the dossier for registering the product with a regulatory agency. It must be stated that any function not directly captured by any of the 5 departments is assumed by the Office of the CEO, who may delegate such functions within the organization or contract them out. Examples of jobs that may be so handled include highly specialized services, including clinical trials |

Some pharmacopeial or compendial tests such disintegration and dissolution tests for tablets and capsules may be applied to herbal preparations as well.

**Table 7.9**: Roles of different departments in production and service-II.

| Departmental Roles + 3 of the 5 ISO 9001 Requirements Under Sub-clause 7.5: Identification and Traceability; Customer property; and Preservation of product. For departments to provides services other those needed for NIPRD's product development (i.e., services for second and third parties) such departments need to acquire ISO/IEC 17025 accreditation | Salient Points, Directing Principles and Main Roles of R&D Depts./Units in Relation to the Application of the QMS Requirements for Production and Service Provision. Measurement traceability requires that equipment used for tests or calibrations with significant effect on results of test, calibration or sampling shall be calibrated before use. ISO 17025 requires units to be traceable to International system of units (SI) |
|---|---|
| **Departments concerned**<br>Medicinal Plants Research and TM (MPRTM) ; Pharmacology and Toxicology (P&T); Microbiology, Virology and Biotechnology (MVBT); Medicinal Chemistry and Quality control (MCQC); and Pharmaceutical Technology and Raw Material Development (PTRMD)<br><br>**Recap of ISO 9001 requirements**<br>**1. Identification and Traceability**<br>1. Identify, where appropriate, the product by suitable means during product realization.<br>2. Identify the product status with respect to monitoring and measurement requirements throughout product realization.<br>3. Where traceability is a requirement, control the unique identification of the product and maintain records.<br><br>**2. Customer Property**<br>1. Exercise care with any customer property while it is under the control of, or being used by, NIPRD.<br>2. Identify, verify, protect, and safeguard customer property provided for use, or for incorporation into the product. Record and report any lost, damaged, or unsuitable property to the customer.<br><br>**3. Preservation of product**<br>Preserve the product during internal processing and delivery to the intended destination in order to maintain conformity to requirements. As applicable, preservation includes: 1) identification, 2) handling, 3) packaging, 4) storage, and 5) protection | A key objective of C-M-C is to establish route/ method for pilot scale production by PTRMD. As in the production of pharmaceuticals in-process quality control procedures are required. This means that MCQC and/or PTRMD must be able 1) identify the product by suitable means during product realization; and 2) identify the product status with respect to monitoring/measurement requirements throughout product realization. MCQC and/or PTRMD need to have the following: a) a defined reference active crude extract (RACE), b) a defined marker substance (DMS) and TLC, HPLC or GC-MS fingerprints of RACE and DMS. These strategies are essential for product realization and for regulatory purposes – they are the means by which problems can be traced to their sources. Department must exercise care with customer property under their control. They must record and promptly report any loss or damage. This approach is essential for fiscal accountability and for addressing specific regulatory concerns. |

Chromatographic fingerprints and the use of marker substances and the availability of reference crude extracts are essential as a means by which identification and traceability can be maintained in herbal drug production. Customer property can include the personal data and traditional knowledge revealed by an herbalist.

As noted earlier, the 7th clause of ISO 9001 is the bedrock of R&D, but many of the activities described therein are also covered by clauses 4 and of ISO 17025. ISO 17025's special provisions for proficiency testing, inter-laboratory comparisons and method development and validation are especially supportive of ISO 9001's 7th clause, hence outfits like NIPRD, drug regulatory agencies and pharmacy schools need both ISO 9001 and ISO 17025.

**Table 7.10**: Roles of departments in control of monitoring/measuring equipment.

| *Departmental Roles + 3 of the 5 ISO 9001 Requirements Under Sub-clause 7.6: The Equipment Most in Need of Calibration and re-calibration include: gravimetric instruments, volumetric wares, photometers, refractometers, and other electrochemical devices. ISO 17025's elaborate provisions for monitoring/ measuring equipment immensely those of ISO 9001* | *Salient Points, Directing Principles and Main Roles of Departments in Relation to the Application of the QMS Requirements for Control of Measuring and Monitoring Equipment* |
|---|---|
| **Departments concerned with** Medicinal Plants Research and TM (MPRTM) ; Pharmacology and Toxicology (P&T); Microbiology, Virology and Biotechnology (MVBT); Medicinal Chemistry and Quality control (MCQC); and Pharmaceutical Technology and Raw Material Development (PTRMD) | Standard practice requires all R&D departments/ units to calibrate their equipment as may be prescribed by operating procedures or other official compendia. In doing so, among other control measures, they need to: 1. Assess and record the validity of prior results if the equipment/method are found not to conform to requirements. 2. Maintain records of the results of calibration and verification. 3. Confirm or re-confirm the ability of any software or programme used for monitoring or measurement before its initial use. To ensure the validity of results, R&D departments/units would normally: |
| **Recap of ISO 9001 requirements** **Control of Measuring and Monitoring Equipment** 1. Determine the monitoring and measurements to be made, and the required equipment, to provide evidence of product conformity. 2. Use and control the monitoring and measuring devices to ensure that measurement capability is consistent with monitoring and measurement requirements. Where necessary to ensure valid results: a) Calibrate and/or verify the measuring equipment at specified intervals or prior to use. b) Calibrate the equipment to national or international standards (or record other basis). c) Adjust or re-adjust as necessary. d) Identify the measuring equipment in order to determine its calibration status. e) Safeguard them from improper adjustments. f) Protect them from damage and deterioration 3. Assess and record the validity of prior results if the device is found to not conform to requirements. 4. Maintain records of the calibration and verification results. 5. Confirm the ability of software used for monitoring and measuring for the intended application before its initial use (and reconfirmed as necessary). | 1. Calibrate and/or verify the measuring equipment at specified intervals or prior to use. 2. Calibrate the equipment to national or international standards (or record other appropriate basis). 3. Adjust or re-adjust as necessary. 4. Identify the measuring equipment in order to determine its calibration status 5. Safeguard equipment from improper adjustments. Protect equipment from damage and deterioration. |

Some calibrations are done daily, some whenever the equipment is to be used, some seasonally and some yearly. The frequency of calibration is normally stated in the relevant SOPs or compendia or equipment SOP or manual.

# The Rising Profile of Herbal Medicine and the Need for Standardization

We earlier identified and discussed the fact that Africa and Asia, especially China and India, have for long been the bastion of herbal medicine – the backbone of traditional medicine (Ameh *et al.*, 2010b; 2010c). We also identified 1978 as the turning point in the current global popularity of phytotherapy following WHO's declaration of support for TM. We further discussed the fact that the US Dietary Supplements Health Education Act (DSHEA, 1994) promoted herbalism in the United States, albeit indirectly, through the innovative provision it made for user information (Goldman, 2010; Chineseherbsdirect.com, 2010). Also described, was that a similar situation as in the US obtained in Europe, where the net effect of the European Directive on Traditional Herbal Medicinal Products (EDTHMP, 2004) had been to promote the production and use of herbal medicines (De Smet, 2005). In terms of economics, the following fact is notable: Although, Asia contributed only US$ 7.3 billion to herbal world trade in 1999 (Mapdb.com, 2003), by 2005, a mere 6 years later, China's contribution alone rose to US$ 14 billion (WHO, 2008). This stupendous growth was due to policies and programmes that favoured herbal medicine. Similar situations as in China held sway in Japan, South Korea and the Indian sub-continent, where government policies also favoured herbal medicine. The rising interest in phytotherapy is well exemplified by the data in Figure 7.2. It seems to us that this growing interest and trade in herbal drugs call for higher quality and better regulation, which in turn call for standardization of production as proposed by the 7th clause of ISO 9001, with the necessary support of clauses 4 and 5 of ISO/IEC 17025.

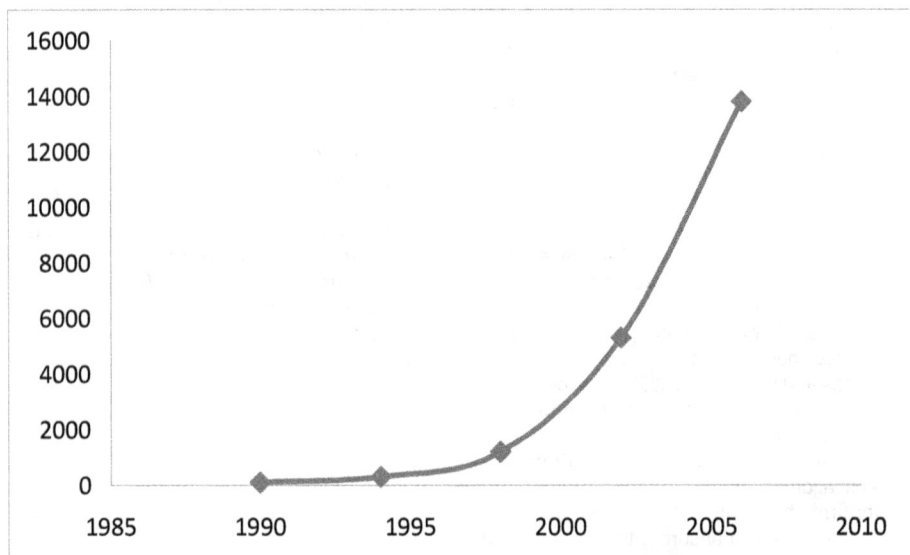

**Figure 7.2**: Number of articles listed on PubMed (1990-2006) containing "phytotherapy". Number of articles listed on PubMed (y-axis) from 1990-2006 (x-axis) containing the word "phytotherapy". Data derived from: Phytotherapy – Wikipedia (2009).

## Reasons for Deciding on ISO 9001 and ISO/IEC 17025 for Standardization

Herbal medicine is called other names in the literature such as botanical medicine, herbal drug, among others (European Pharmacopoeia, 2000; Gaedcke *et al.*, 2003; Bandaranayake, 2003). The term "herbal substance" had also been used by WHO (2005), and defined as "material derived from plant by extraction, mechanical manipulation, or some other process". The term specifically applies to whole preparations not isolated or purified components. In herbological terminology each "herbal substance" from a given plant is in its entirety regarded as the active substance, even though the preparation may contain several chemically defined entities (Bandaranayake, 2003). This is because it is conceived that the entities act cooperatively to achieve the pharmacological attribute of the plant. The practice of preparing herbal cocktails is termed "chemistry-manufacturing-control" (C-M-C) by WHO (2005), because it involves an understanding of physicochemical processes and how to control them. The C-M-C of a given "herbal product", defined by WHO (2005) as an "herbal material administered to clinical subjects", focuses on the fact that herbal substances are prone to contaminations by herbicides, pesticides, mycotoxins and others (Ameh *et al.*, 2009) and are subject to profound variations in physicochemical characteristics (WHO, 1998), such as moisture content, ash values, extractability and others (Ameh *et al.*, 2010a; 2010d; 2010e; 2010f). The cultivation, harvest, contaminations, process history and the physicochemical characteristic of an herbal substance are critical to its C-M-C evaluation, if good manufacturing practice (GMP) is to be applied in producing an herbal product (Ameh *et al.*, 2009; 2010d; 2010e; 2010f; 2011b). In NIPRD herbal drug research and development is guided by relevant WHO guidelines and quality control (WHO, 1993; 1998; 2000; 2005) on herbal drug research and development (R&D). ISO 9001 guidelines for "product realization" with necessary support from clauses 4 and 5 ISO/IEC 17025 have been selected for this study because: i) major purchasers require their suppliers to hold ISO 9001 certification (ISO Survey, 2009; Corbett *et al.*, 2005); ii) studies indicate significant financial benefits for organizations certified to ISO 9001 (ISO Survey, 2009; Heras *et al.*, 2002); and iii) similar superior operational performance of ISO certified firms has been severally confirmed (Naveh and Marcus, 2007; Sharma, 2005; Chow-chua *et al.*, 2002; Rajan and Tamimi, 2003). On the other hand ISO/IEC 17025 has been chosen because it immensely complements clauses 7 and 8 of ISO 9001, and because its provisions for proficiency testing and laboratory comparison are unique and highly supportive of the intents of clauses 7 and 8 of ISO 9001. As earlier stated, if an ISO 9001 certified organization has ISO 17025 accreditation it can legally supply services to a third party. It also avoids the need for disruptive external auditing of its operations by customers or sub-contractor requiring it to be ISO 17025 competent. On the whole the combined acquisition ISO 9001 and ISO 17025 certifications by an organization greatly enhances its ability to satisfy its customers

## Conclusion

ISO/IEC 17025 specifies the general requirements for the competence to carry out tests and/or calibrations, including sampling. It is for use by laboratories in developing their QMS, administrative setup and technical operations. Laboratory customers, regulatory authorities and accreditation bodies may also use it in confirming the competence of laboratories. It covers testing and calibration performed using standard methods, methods developed in-house or other defined methods. Therefore, ISO/IEC 17025 is applicable to organizations like NIPRD, drug regulatory agencies, pharmacy faculties and others that perform tests and/or calibrations, sampling, inspection and product certification. ISO/IEC 17025 is applicable to all laboratories regardless of the number of personnel or the extent of the scope of testing and/or calibration activities. Thus, taken side by side with ISO 9001, the general view to be taken of the foregoing is that ISO 9001 is the overall standard for quality management, whereas ISO 17025 provides specific guidance on the application of ISO 9001 principles to laboratories. This correspondence is becoming increasingly apparent with the development of both standards. An herbal drug outfit that is competent to both ISO 9001 and ISO 17025 certifications is more likely to satisfy its customer and boost its business than those competent to only one standard.

# References

Ameh, S. J., Obodozie, O. O., Afolabi, E. K., Oyedele, E. O., Ache, T. A., Onanuga, C. E., Ibe, M. C and Inyang, U. S. (2009). Some basic requirements for preparing an antisickling herbal medicine -NIPRISAN®. *African Journal of Pharmacy and Pharmacology*, **3 (5):** 259-264.

Ameh, S. J., Tarfa, F. D., Abdulkareem, T. M., Ibe, M. C., Onanuga, C. and Obodozie, O. O. (2010a). Physico-chemical analysis of the aqueous extracts of six Nigerian medicinal plants. *Tropical Journal of Pharmaceutical Research*, **9 (2):** 119-125.

Ameh, S. J., Obodozie, O., Inyang, U., Abubakar, M. and Garba, M. (2010b). Current phytotherapy: A perspective on the science and regulation of herbal medicine. *Journal of Medicinal Plants Research*, **4 (2):** 72-81.

Ameh, S. J., Obodozie, O., Inyang, U., Abubakar, M. and Garba, M. (2010c). Current phytotherapy: An inter-regional perspective on policy, research and development of herbal medicine. *Journal of Medicinal Plants Research*, **4 (15):** 1508-1516.

Ameh, S. J., Obodozie, O. O., Inyang, U. S., Abubakar, M. S. and Garba, M (2010d). A normative study of Nigerian grown "Maha-tita" (King of Bitters) - *Andrographis paniculata* Nees. *International Journal of Drug Development and Research*, **2(2):** 291-299.

Ameh*, S. J., Obodozie, O. O., Inyang, U. S., Abubakar, M. S. and Garba, M (2010e). Quality control tests on *Andrographis paniculata* Nees (Family: Acanthaceae) – an Indian 'Wonder' plant grown in Nigeria. *Tropical Journal of Pharmaceutical Research*, **9 (4):** 387-394.

Ameh, S. J., Obodozie, O. O., Gamaniel, S.K., Abubakar, M. S. and Garba, M (2010f). Physicochemical variables and real time stability of the herbal substance of Niprd-AM1®- an antimalarial developed from the root of *Nauclea latifolia* S.M. (Rubiaceae). *International Journal of Phytomedicine*, **2:** 332-340.

Ameh, S. J., Obodozie, O. O., Babalola, P. C. and Gamaniel, K. S. (2011a). Medical herbalism and herbal clinical research: A global perspective. *British Journal of Pharmaceutical Research*, **1(4)**: 99-123.

Ameh, S., Obodozie, O., Gamaniel, K., Abubakar M. and Garba, M (2011b). Herbal Drug Regulation Illustrated with Niprifan® Antifungal Phytomedicine. In: Eldin, A. B., (Ed) Modern Approaches to Quality Control, ISBN 978-953-307-329-3. INTECH Open Publisher, University Campus, STeP Ri Slavka Krautzeka 83/A, 51000 Rajeka, Croatia, pp. 367-382.

Ameh, S. J., Obodozie, O. O., Chindo, B. A., Babalola, P. C. and Gamaniel, K. S. (2012). Herbal clinical trials: Historical development and application in the 21st century. *Pharmacologia*, **3:** 121-131.

Bandaranayake, W. M. (2003). Quality Control, Screening, Toxicity, and Regulation of Herbal Drugs. In: *Modern Phytomedicine - Turning Medicinal Plants into Drugs*. Edited by Iqbal Ahmad, Farrukh Aqil, and Mohammad Owais. WILEY-VCH Verlag GmbH and Co. KGaA, Weinheim, Germany, 2003: 25-57.

Becker, J., Kugeler, M. and Rosemann, M. (2011). *Process Management*: A Guide for the Design of Business Processes. Springer, pp. 596. ISBN 3-540-43499-2

Chineseherbsdirect.com (2010). What are Chinese Herbs? [cited 2010 April 8]. Available from: http://www.chineseherbsdirect.com

Chow-chua, C., Goh, M. and Wan, T. B. (2002) Does ISO 9000 certification improve business performance? *The International Journal of Quality and Reliability Management*, **20 (8):** 936–953. Available: http://www.emeraldinsight.com/journals.htm?articleid=840633&show=abstract Accessed 2012 April 25.

Corbett, C. J., Montes-sancho, M. J. and Kirsch, D. A (2005). The financial impact of ISO 9000 certification in the United States: An empirical analysis. *Management Science*, **51 (7):** 1046–1059. Available: http://personal.anderson.ucla.edu/charles.corbett/paper/does_iso_pay.pdf Accessed 2012 April 25.

De Smet, P. (2005) Herbal medicine in Europe – relaxing regulatory standards. *N. Eng. J. Med.*, **352 (12):**1176-78.

DSHEA (1994). Dietary Supplements Health Education Act of 1994. [cited 2012 April 25]. Available from: http://fda/Food/DietarySupplements/ucm109764.htm.

EDTHMP (2004). European Directive on Traditional Herbal Medicinal Products. [cited 2010 April 8]. Available from: http://eur-lex.europa.eu/LexUriserv/LexUriserv.do?uri=CELEX:32004L0024:EN:NOT

European Pharmacopoeia (2000). European Pharmacopoeia (Supplement 2000). Technical Secretariat of the European Pharmacopoeia Commission.

Gaedcke, F. W., Steinhoff, S. K. and Blasius, H. R. (2003). *Herbal Medicinal Products: Scientific and Regulatory Basis for Development, Quality Assurance and Marketing Authorization*. Stuttgart: Medpharm Scientific Publishers, First Edition.

Goldman, P. (2010). Herbal medicines today and the roots of modern pharmacology. *Ann. Int. Med.*, **135 (8) Part 1):** 594-600.

Heras, I., Dick, G. P. and Casadesus, M. (2002). ISO 9000 registration's impact on sales and profitability - A longitudinal analysis of performance before and after accreditation. *International Journal of Quality and Reliability Management*, **19 (6):** 774–791. Available: http://eps.udg.es/oe/webmarti/p774.pdf Accessed 2012 April 25.

ISO 9001:2008. ISO Headquarter, Basle, Switzerland.

ISO Survey (2009). Available: http://www.iso.org/iso/survey2009.pdf Accessed 2012 April 25.

ISO/IEC 17025:2005. ISO Headquarter, Basle, Switzerland.

ISO/IEC 17011:2004. CASCO Secretariat, ISO Headquarter, Basle, Switzerland.

Juran, J. M. and Godfrey, A. B. (1999). *Juran's Quality Handbook: The Complete Guide to Performance Excellence*. 5th Edition, McGraw-Hill Professional, New York, USA.

Mapdb.com. Current status of medicinal plants 2003. [cited 2012 April 25]. Available from: http://www.mapbd.com/cstatus.htm

Naveh, E. and Marcus, A. (2007) "Financial performance, ISO 9000 standard and safe driving practices effects on accident rate in the U.S. motor carrier industry", *Accident Analysis and Prevention*, **39 (4):** 731–742.

Phytotherapy (2009). [cited 2009 June 2]. Available from: http://en.wikipedia.org/wiki/Phytotherapy

Pyzdek, T. (2003). Quality Engineering Handbook. 2nd Edition (Ed. Keller, P.). CRC Press, New York, USA.

Rajan, M. and Tamimi, N. (2003). Payoff to ISO 9000 registration. Available: http://www.iijournals.com/doi/abs/10.3905/joi.2003.319536 Accessed 2012 April 25.

Sharma, D. S. (2005). The association between ISO 9000 certification and financial performance. The international Journal of Accounting 40: 151–172. Available: http://masp.bus.ku.ac.th/files/ISO per cent 209000 per cent 20and per cent 20performamce.pdf Accessed 2012 April 25.

UN/ECE Task Force LQMA (2002). UN/ECE Task Force on Laboratory Quality Management and Accreditation. Technical Report: Guidance to operation of water quality laboratories. United Nations Publications, New York, USA.

WHO (1993). Research Guidelines for Evaluating the Safety and Efficacy of Herbal Medicines. World Health Organization, 1993.

WHO (1998). Quality control methods for medicinal plant materials. World Health Organization, 1998.

WHO (2000). General guidelines for methodologies on research and evaluation of traditional medicine (Document WHO/EDM/TRM/2000.1). World Health Organization.

WHO (2005). Information needed to support Clinical Trials of herbal products.TDR/GEN/Guidance/05.1Operational Guidance: Special Programme for Research and Training in Tropical Diseases.

WHO (2008). Traditional Medicine. WHO Fact sheet No. 134. Revised: December World Health Organization, Geneva.

Utilisation and Management of Medicinal Plants Vol. 2 (2014)     Pages 233–244
Editor-in-Chief: V.K. Gupta
Published by: DAYA PUBLISHING HOUSE, NEW DELHI

# 8

# Plants of the Holy Quran: Modulation of EMF-Induced Cells Hemostasis Deficiency by *Ocimum basilicum* Via its Antioxidant Action on Heart and Testis

Arash Khaki[1]*

## ABSTRACT

*With developing in electronic means and increase in using these, concern about the public health hazards of chronic exposure to EMF has gained more attention. Medicinal use of Ocimum basilicum dates back to ancient Iran, China and India. It has been used since ancient time as medicinal and food origin as an antioxidant's. Ocimum basilicum has a useful effect as a protective on EMF harmful effects. Wistar male rat (n=40) were allocated into four groups, control (n=10) and test groups (n=30), that subdivided into groups of 3, the extract group were received of Ocimum basilicum extract (1.5 g/kg body), second extract group were received of Ocimum basilicum extract (1.5 g/kg body) and EMF group that exposed to 50 Hz for 40 consequence day. Animals were kept in standard conditions. In the end of study, the testes, heart, tissues of rats in whole groups were removed and serum was prepared for biochemical analysis. Serum total testosterones, sperm concentration, percentage of sperm viability and motility, TAC and GPX significantly increased in experimental group that has received 1.5g/kg body Ocimum basilicum extract (p<0.05) in comparison to control and EMF groups Whereas, LH, FSH hormones, morphology and tissues weights in both experimental and control group were similar. Results revealed that administration of 1.5 g/kg body of Ocimum basilicum extract significantly increased cells antioxidants enzymes. This suggested that Ocimum basilicum extract may be promising in enhancing sperm healthy parameters. This research shows that references to Quran and its use in this age could be useful in improving community health.*

*Keywords*: Emf, GPX, *Ocimum basilicum*, Sperm, TAC.

---

1   Department of Veterinary Pathology, Islamic Azad University, Tabriz- Branch, Iran.

*   *Corresponding author*: E-mail: arashkhaki@yahoo.com

# Introduction

Basil is an aromatic herb that is adundant in Iran. Basil has also come up in Quran verses. Among them we can noted the Al-Rahman Sura, Verse 12 and Al-Waqiah Sura, Verse 89. *Ocimum basilicum* (Basil) is an annual herb of the Lamiacae family widely cultivated in Asia as a nourishing food and herbal medicine. It is widely used in folk medicine to treat wide range of diseases. Reyhan is derived from Arabic: rih or riha "odour, fragrance" and originally did not mean basil but myrtle. This is still so in North African Arabic (and Maltese); moreover, the word has been transferred to medieval Spanish as Arrayan. Rehan is India Tulsi Herb name (myrtle). For example the aerial part of *O. basilicum* is traditionally used as antispasmodic, aromatic, digestive, carminative, stomachic and tonic agents. They are also used externally for treatment acne, insect stings, snake bites and skin infections (Supawan *et al.*, 2007). Antioxidant capacity of phenolic compounds, flavonoids and food containing them has been repeatedly shown in various *in vitro* and *in vivo* systems (Alexandopoulou *et al.*, 2006). An electromagnetic field (also EMF or EM field) is a physical field produced by electrically charged objects. It affects the behavior of charged objects in the vicinity of the field. The electromagnetic field extends indefinitely throughout space and describes the electromagnetic interaction (Schüz *et al.*, 2009; Emre *et al.*, 2011). It is one of the four fundamental of nature (the others are gravitation, the weak interaction, and the strong interaction).The field can be viewed as the combination of an electric field and a magnetic field. With the increased use of power lines and modern electrical devices, concern about the public health hazards of chronic exposure to EMF has gained more attention. It has been shown that exposure to EMF adversely affects spermatogenic, sertoli and leydig cells (Martínez-Sámano *et al.*, 2010). Magnetic fields of 50 Hz may induce cytotoxic and cytostatic changes in the differentiating spermatogonia of mice (Khaki *et al.*, 2008). Little is known about the effect of EMF on the cytoarchitecture of the boundary tissue of the seminiferous tubules that perform a number of crucial functions, including the mechanical support and transport of nutrients for the spermatozoa (Roychoudhury *et al.*, 2009) and sperm discharge by maintaining pressure on the tubules (Iorio *et al.*, 2007). Other studies have been made of the transitory effects of EMF on the testes and no study has revealed, to date, the possibility of recovery from the potentially harmful effects of EMF exposure after an exposure-free time. Furthermore, the gonadal effects of EMF at the ultrastructural level have only infrequently been studied. The present study was aimed to investigate possible beneficial effects of *Ocimum basilicum* anamed in Quaran as a holy and as a source of natural antioxidant on testis and heart tissues on rats which exposed to the 50Hz EMF (non-ionising radiation).

# Material and Methods

## Preparation of Extract

Aerial parts of *O. basilicum* were purchased from local store. The explant was authenticated and fresh aerial parts of the plant were extracted by maceration with EtOH-Water (80: 20) to produce a total extract (Hydroalcoholic extract, HAE), which were included total phenols and flavonoids of the plant.

# Experimental Animals

## Animals and Maintenance

Total of 40 male Wistar rats were used for the study. Rats of the same sex were housed together (10 per cage). Rats were fed on compact food in the form of granules and water. This food consisted of all the essential ingredients, including vitamins and minerals. The environmental conditions (temperature and humidity) in all the animal holding areas were continuously monitored. Temperature was maintained in the range of 23°C and humidity was monitored at 35–60 per cent. Light was provided on a 12 h light/dark cycle and kept turned on from 7 A.M. till 7 P.M.

## EMF-Producing System

The equipment was based on the Helmholtz coil, which works following Fleming's right hand rule. This produced an alternate current of 50 Hz, creating an EMF of 80 G. The intensity of the EMF could be controlled by a transformer. The equipment had two main parts. In the first there were two copper coils placed one above the other and separated by a distance of 50 cm. Between the coils (the exposure area) there was a cylindrical wooden vessel, the interior of which had a chamber for holding the cages of the experimental animals. The second part was the transformer, which checked the input and output voltage with a voltmeter and the current with an ampere meter. To prevent increases in temperature inside the chamber a fan was utilized as necessary. Five cages at a time were placed within the chamber with seven or eight rats per cage.

## Surgical Procedure

In day 40, the pentobarbital sodium (40 mg/kg) was administered intra peritoneal for anesthesia, and the peritoneal cavity was opened through a lower transverse abdominal incision. Thereafter testis in control and experimental groups were immediately removed. The weights of testis in each group were registered. The animals were decapitated between 10: 00 AM and 12: 00 AM, and blood samples were obtained. Blood samples were centrifuged at 4°C for 10 min at 250Xg and the serum obtained was stored at -20°C until assayed.

## Epididymis Sperm Count, Viability and Motility

Sperms from the cauda epididymis were released by cutting into 2 ml of medium (Hams F10) containing 0.5 per cent bovine serum albumin. After 5 min incubation at 37°C (with 5 per cent $CO_2$), the cauda epididymis sperm reserves were determined using the standard hemocytometric method and sperm motility was analyzed with microscope (Olympus IX70) at 10 field and reported as mean of motile sperm according to WHO method.

## Serum FSH, LH Totals Testosterone Hormone Measurements

Serum concentration of FSH and LH were determined in duplicated samples using radioimmunoassay (RIA). Rat FSH/LH kits obtained from Biocode Company-Belgium, according to the protocol provided with each kit. The sensitivities of hormone detected per assay tube were 0.2 ng/ml and 0.14 ng/ml for FSH and LH respectively.

Serum concentration of total testosterone was measured by using a double antibody RIA kit from Immunotech Beckman Coulter Company-USA. The sensitivities of hormone detected per assay tube were 0.025 ng/ml.

## Total Antioxidant Capacity (TAC) and Malondialdehyde (MDA) Concentration Measurement in Serum

A TAC detecting kit was obtained from Nanjing Jiancheng Bioengineering Institute, China. According to this method, the antioxidant defense system, which consists of enzymatic and non-enzymatic antioxidants, is able to reduce $Fe^{3+}$ to $Fe^{2+}$. TAC was measured by the reaction of phenanthroline and $Fe^{2+}$ using a spectrophotometer at 520 nm. At 37°C, a TAC unit is defined as the amount of antioxidants required to make absorbance increase 0.01 in 1 mL of serum. Free radical damage was determined by specifically measuring malondialdehyde (MDA). MDA was formed as an end product of lipid peroxidation which was treated with thiobarbituric acid to generate a colored product that was measured at 532 nm (MDA detecting kit from Nanjing Jiancheng Bioengineering Institute-China).

## Glutathione Peroxidase (GPX) Activity Measurement in Serum

GPx activity was quantified by following the decrease in absorbance at 365 nm induced by 0.25 mM $H_2O_2$ in the presence of reduced glutathione (10 mM), NADPH, (4 mM), and 1 U enzymatic activity of GR (Yoshikawa *et al.*, 1993).

## TUNEL Analysis of Apoptosis

The *in-situ* DNA fragmentation was visualized by TUNEL method (Huang HFS *et al.*, 1995). Briefly, dewaxed heart tissue sections were predigested with 20 mg/ml proteinase K for 20 min and incubated in phosphate buffered saline solution (PBS) containing 3 per cent $H_2O_2$ for 10 min to block the endogenous peroxidase activity. The sections were incubated with the TUNEL reaction mixture, fluorescein-d UTP (*in-situ* Cell Death Detection, POD kit, Roche, Germany), for 60 min at 37°C. The slides were then rinsed three times with PBS and incubated with secondary antifluorescein-POD-conjugate for 30 min. After washing three times in PBS, diaminobenzidine-$H_2O_2$ (DAB, Roche, Germany) chromogenic reaction was added on sections and counterstained with hematoxylin. As a control for method specificity, the step using the TUNEL reaction mixture was omitted in negative control serial sections, and nucleotide mixture in reaction buffer was used instead. Apoptotic cells were quantified by counting the number of TUNEL stained nuclei per cross sections. Cross sections of 100 heart tissues per specimen were assessed and the mean number of TUNEL positive apoptotic cells per cross- section was calculated.

## Statistical Analysis

Statistical comparisons were made using the ANOVA test for comparison of data in the control group and the experimental groups. The results were expressed as mean ± S.E.M (standard error of means). Significant difference is written in parentheses.

# Results

## Weight of Individual Male Testis and Heart

The obtained results in this study are illustrated in Table 8.1. There was significant difference in testes weights between the groups (p<0.05).

## Results of Sperm Motility, Viability and Count

Administration of 1.5g/kg body *Ocimum basilicum* extract for forty consecutive days significantly increased sperm motility, viability and count significtly increased only in extract group (These changes were significant as p value less than 0.001),(P<0.001) and significantly decreased in EMF group when compared to control and other experimental groups (These changes were significant as p value less than 0. 05), (P<0.05) (Table 8.1).

In the EMF group that receive 1.5g/kg body *Ocimum basilicum* extract, sperm motility, viability and count significtly increased (P<0.05) (These changes were significant as p value less than 0. 05; Table 8.1).

## Results of Serum Total Testosterone, LH and FSH Hormones Measurement

Administration of 1.5 g/kg body *Ocimum basilicum* extract for forty consecutive days didn't show significant effect on LH and FSH concentration in the serum between the control and other experimental groups (Table 8.1), but it can significantly increased the serum total testosterone level in extract group when compared to control and other experimental groups (These changes were significant as p value less than 0.001),(P<0.001) (Table 8.1).

## Results of Total Antioxidant Capacity (TAC) and Malondialdehyde (MDA) Concentration Measurement in Serum

Administration of 1.5 g/kg body *Ocimum basilicum* extract for forty consecutive days significantly decreased concentration of Malondialdehyde (MDA) level in extract and EMF group that receive extract groups in compared with the control and EMF groups, These changes were significant as p value less than 0.05 (P<0.05), (Table I). Total antioxidant capacity (TAC) was significantly increased in extract group when compared to other groups these changes were significant as p value less than 0.05 (P<0.05) (Table 8.1).

## Results of Glutathione Peroxidase (GPX) Activity in Serum

Glutathione peroxidase (GPX) level in in EMF group was (93.90 ± 0.05) and in Ob, received group was (138.4 ± 0.7) and in Ob+EMF was (11.5 ± 0.01) and in control group was (125 ± 0.7) respectively. These changes were significant as p value less than 0.05 (P<0.05).Statistic analysis Dennett (one side) shows significant differences between experimental groups in comparison to control group (P<0.05) (Table 8.2).

## Results of Mitochondria Blebs in Myocardial Cells

Mitochondria blebs level in EMF group was (0.75 ± 0.05) and in Ob, received group was (0.01 ± 0.05) and in Ob+EMF was (0.55 ± 0.05) and in control group was

**Table 8.1:** The effect of the 50mg/kg/rat and 100mg/kg/rat Ocimum basilicum extract on sperm parameters, serum FSH, LH, total Testosterone and testis weight of control and experimental groups in the rats.

| Groups | Control (n=10) | HAEx (1.5g/kg body) (n=10) | EMF+HAE (1.5g/kg body) (n=10) | EMF |
|---|---|---|---|---|
| Testis (gr) | 1.40 ± 0.821 | 1.47 ± 0.373 | 1.40 ± 0.371 | 1.30 ± 0.001 |
| Sperm concentration (total count) (No of sperm/rat ×106) | 51.90 ± 5.36 | 68.60 ± 2.34** | 44.60 ± 2.34* | 39.90 ± 0.06* |
| Motility (Per cent) | 33.75 ± 6.88 | 73 ± 4. 35** | 37 ± 1.38 | 24.25 ± 0.05* |
| Viability (Per cent) | 60.25 ± 1.23 | 94.10 ± 1.68** | 58.10 ± 80* | 40.10 ± 1.38* |
| Serum Testosterone levels (ngr/ml) | 1.70 ± 0.01 | 2.99 ± 0.210** | 1.70 ± 0.01 | 1 ± 0.01 |
| LH levels (ngr/ml) | 1.51 ± 0.138 | 1.73 ± 0.164 | 1.31 ± 0.128 | 1 ± 0.128 |
| FSH levels (ngr/ml) | 22.17 ± 1.544 | 22.29 ± 1.545 | 22.17 ± 1.178 | 18.13 ± 1.134 |
| Total Antioxidant capacity (TAC) | 0.53 ± 0.666 | 0.91 ± 0.012* | 0.44 ± 0.240 | 0.38 ± 0.140* |
| Malondialdehyde (MDA) | 4.30 ± 0.212 | 2.55 ± 0.171* | 5.21 ± 0.122* | 6.30 ± 0.214** |

Data are presented as mean ± SE.

*: Significant different at $p < 0.05$ level, (compared with the control group).

**: Significant different at $p < 0.001$ level, (compared with the control group).

(0.01 ± 0.05) respectively. These changes were significant as p value less than 0.05 (P<0.05). Statistic analysis Dennett (one side) shows significant differences between experimental groups in comparison to control group (P<0.05) (Table 8.2).

**Table 8.2**: Myocardial cells apoptosis, GPX, mitochondria blebs, muscle fiber degeneration and heart weights of rats witch exposed to EMF and *O. basilicum* Extract.

| O. basilicum + (EMF) | O. basilicum (1.5 g/kg body weight) | EMF (50Hz) | Control | Groups |
|---|---|---|---|---|
| 9.05 ± 0.05* | 6.05 ± 0.05 | 16.03 ± 0.05* | 4.01 ± 0.05 | Myocardial cells apoptotic cell (per cent) |
| 3.40 ± 0.03 | 4.57 ± 0.03 | 3.00 ± 0.01* | 4.50 ± 0.05 | Heart weight's (Gram) |
| 11.5 ± 0.01* | 138.4 ± 0.7 | 93.90 ± 0.05* | 125 ± 0.7 | GPX (u/mg Hb) |
| 0.55 ± 0.05* | 0.01 ± 0.05 | 0.75 ± 0.05* | 0.01 ± 0.05 | Mitochondria blebs (per cent) |
| 5.05 ± 0.05* | 0.00 ± 0.01* | 8.55 ± 0.05* | 0.01 ± 0.01 | Muscle fiber degeneration (per cent) |

Data are presented as mean ± SE.

* Significantly different at p< 0.05 level (compared with the control group).

## Results of Ventricle Muscle Fiber Degeneration

Ventricle Muscle fiber degeneration in EMF group was (8.55 ± 0.05) and in Ob, received group was (0.00 ± 0.01) and in Ob+EMF was (5.05 ± 0.05) and in control group was 0.01 ± 0.01) respectively. These changes were significant as p value less than 0.05 (P<0.05).Statistic analysis Dennett (one side) shows significant differences between experimental groups in comparison to control group (P<0.05) (Table 8.2).

## Pathological Results

Heart ventricular section from a control rat group; shows the normal muscle tissue (arrow) and histological structure of the myocytes spaces, in EMF group hyperemia muscle fiber degeneration, enhanced in myocytes spaces were seen, ultra structural study of the myocardial tissue and sarcomere of this group, are shown by lose of area in sarcomeres and irregular structural of myocardial cells, sarcomeres were ruptures (in EMF + *O. basilicum* group), heart ventricular section from a EMF + *O. basilicum* rat group; shows the regeneration of muscle fiber (arrow) and histological structure of the myocytes spaces coming to normal form, also ultra-structural study of myocardial cells and sarcomeres showed these parts were backed to normal and fibrosis are seen in parts of myocardial cells.

## Discussion

Muslims turn to the Quran for guidance in all areas of life, which may include health and medical matters. Muslims are therefore encouraged to explore and use traditional and modern forms of medicine, with faith that any cure is from Allah. Traditional medicine in Islam is often referred to as medicine. Muslims often explore the medicine as an alternative to modern therapies, or as a supplement to modern medical treatment. Here are some traditional remedies that are a part of Islamic tradition such as Basil. Oxidative stress through $H_2O_2$ and the NO donor, N-acetyl-Snitroso-DL-penicillinaminamide (SNAP), induced apoptosis in ventricular cardiomyocytes isolated from a rat heart. In isolated rat cardiomyocytes, a correlation between the apoptotic effect of SNAP or YC-1 (a direct activator of soluble guanylyl cyclase) and the increased activity of soluble guanylyl cyclase (that is, the intracellular cGMP content) was seen (Taimor *et al.*, 2000). In recent years considerable researches have been reviewed about the hazards of these advantages that electromagnetic waves emitted by mobile phones are in their heads. Any electrical device can be a source of electromagnetic field (EMF). Radiofrequency (RF) energy is a type of nonionizing radiation that is not strong enough to cause ionization of atoms and molecules (Erogul *et al.*, 2006). Previous study showed two hours of 60Hz EMF exposure might immediately alter the metabolism of free radicals, decreasing SOD activity in plasma and GSH content in heart and kidney, but does not induce immediate lipid peroxidation (Koyu *et al.*, 2009). Researchers study on extremely low frequency electromagnetic fields (ELF-EMFs) with 50 Hz effects on human spermatozoon motility showed that exposure did not produce any significant effect on sperm motility and cause to improve spermatozoa motility but this effect didn't significant (Iorio *et al.*, 2007) and this is agree to our study. In same research that done on 50 hz extra low frequency electromagnetic field on spermatozoa motility and fertilization rates in rabbits by Roychoudhury and colleagues in 2009, revealed this wave affected spermatozoa motility and reduce fertility chance (Roychoudhury *et al.*, 2009). In fact, it was found to cause a slowdown in the embryo cleavage. In 2010, Bernabo *et al.*, research on low frequency electromagnetic field effects on swine fertilization, they demonstrated how and at which intensities ELF-EMF negatively affect early fertility outcome in a highly predictive animal model (Bernabo *et al.*, 2010). Other study about radiofrequency electromagnetic fields (RF-EMF) effects was done by Falzone and his colleagues on acrosome reaction revealed, the radiation did not affect sperm propensity for the acrosome reaction. Significant reduction in sperm head area and acrosome percentage of the head area was reported among exposed sperm compared with unexposed controls and this concludes that although RF-EMF exposure did not adversely affect the acrosome reaction, it had a significant effect on sperm morphometry. In addition, a significant decrease in sperm binding to the hemizona was observed. These results could indicate a significant effect of RF-EMF on sperm fertilization potential (Falzone *et al.*, 2011).

Exposure of biological systems to electromagnetic radiation as an ionizing radiation results in the formation of reactive oxygen species (ROS) and reactive nitrogen species (RNS). These reactive species inflict damage to the various bio-macromolecules like DNA, lipids and proteins present in the cell (Rajesh Arora, Herbal Radio Modulators Application in Medicine, Homeland Defence and Space, CABI, UK, 2008, p.2). These damages lead to early signs (*e.g.*, cataract induction, haemologic deficiencies, damages to skin and fertility impairment) or late sickness (*e.g.*, cancer appearing several year after exposure) of radiation injury(Schüz *et al.*, 2009).The harmful effects of EMF ionizing radiations (*e.g.*, X-rays and gamma rays) have previously been demonstrated on gonadal tissues (Khaki *et al.,* 2006). Since plants and natural products are extensively used in several traditional systems of medicine, screening of radio protective compounds from them has several advantages because usually they are considered non-toxic and are widely accepted by humans. Many natural antioxidants, whether consumed before or after radiation exposure, are able to confer some level of radioprotection. In addition to achieving beneficial effects from established antioxidants such as vitamins C and E and their derivatives, vitamin A, beta carotene, curcumin, *Allium cepa,* quercetin, caffeine, chlorogenic acid, ellagic acid and bixin, protection is also conferred by several novel molecules, such as flavonoids, eppigallocatechin and other polyphenols. (Rajesh *et al.,* 2008; Khaki *et al.,* 2009; khaki *et al.,* 2010). Basil (*Ocimum basilicum* L. Family Lamiaceae) is used as a kitchen herb and as an ornamental plant in house garden (Gülçin *et al.,* 2007). our results confirmed that EMF cause to increasing free radical and reactive oxygen spices(ROS) and this make cell injury and increase ability of sperm and this findings are same with other reports (Garip *et al.,* 2010; Sharma *et al.,* 2009), Malondialdehyde (MDA) levels in EMF group were significantly increased and this is agree with other researchers results which confirmed EMF cause to enhanced MDA content (indicating lipid peroxidation), and increased $H_2O_2$ accumulation thereby inducing oxidative stress and cellular damage (Sharma *et al.,* 2009; Grigor'ev *et al.,* 2010). But hydroalcoholic extract of *Ocimum basilicum* increase antioxidant capacity and have beneficial effect to neutralize free radicals in those EMF groups which receiving it and cause to decrease level of MDA and this results confirmed previous chemical studies on herbal antioxidant effects (Niwano *et al.,* 2011) such as *Ocimum basilicum* have shown the presence of flavonoids, phenylpropanoids and rosmarinic acid in aerial parts of the plant (Bors *et al.,* 1977; Peluso, 2006).These reports have also indicated the antioxidant and radical scavenging activity of *O. basilicum* (Jayasinghe *et al.,* 2003; Dorman *et al.,* 2010). In conclusion, many herbal such as *O. basilicum* well-known flavonoid and a strong antioxidant have beneficial effects on serum antioxidant levels and sperm health parameters by reduce oxidative stress (Khaki *et al.,* 2009, 2010), so it seems that long-term using of them can increase testosterone hormone and sperm parameters and increase chance of infertility. Also in this study, malondialdehyde (MDA) level in the basil extract used groups significantly decreased and Total Antioxidant Capacity (TAC) with Glutathione Peroxidase (GPX) levels in serum was increased, and programed cell death percentage was significantly decreased in *Osmium basilicum* groups. So we reach the conclusion that basil extract have beneficial effects on cardiovascular disorders such as apoptosis caused by electromagnetic field exposure, is significant.

# Acknowledgment

We would like to thank, Islamic azad university, Tabriz Branch-Iran for giving grant and financial support of this research. I hope this chapter could be showed more vision of Muslim holy book (Quran) in now-a-days sciences especial in medical sciences.

# References

Alexandopoulou, I., Komaitis, M., and Kapsokefakou, M. (2006). Effects of iron, ascorbate, meat and casein on the antioxidant capacity of green tea under conditions of *in vitro* digestion. *Food Chem.*, **94(3):** 359-365

Bors, W., Michel, C., and Stettmaier, K. (1977). Antioxidant effects of flavonoids. *Biofactors*, **6:** 399–402.

Dorman, H.J., and Hiltunen, R. (2020). *Ocimum basilicum* L.: phenolic profile and antioxidant-related activity. *Nat Prod Commun.*, **5(1):** 65-72.

Emre, M., Cetiner, S., Zencir, S., Unlukurt, I., Kahraman, I., and Topcu, Z.(2011). Oxidative stress and apoptosis in relation to exposure to magnetic field. *Cell Biochem Biophys.*, **59(2):** 71-77.

Erogul, 0., Oztas, E., Yildirim, 1., Kir, T., Aydur, E., Komesli, G., Irkilata, H.C., limak, M.K., and Peker, A.F. (2006). Effects of electromagnetic radiation from cellular phone on human sperm motility: an *in vitro* study. *Arc Med Res.*, **37:** 840-843.

Garip, A.I., and Akan, Z.(2010). Effect of ELF-EMF on number of apoptotic cells; correlation with reactive oxygen species and HSP. *Acta Biol Hung.*, **61(2):** 158-167.

Grigor'ev, G., Mikha-lov, V.F., Ivanov, A.A., Mal'tsev, V.N., Ulanova, A.M., Stavrakova, N.M., Nikolaeva, I.A., and Grigor'ev, O.A.(2010). Autoimmune processes after long-term low-level exposure to electromagnetic fields (the results of an experiment). Part 4. Manifestation of oxidative intracellular stress-reaction after long-term non-thermal EMF exposure of rats. *Radiats Biol Radioecol.*, **50(1):** 22-27.

Gülçin, I., Elmasta°, M., and Aboul-Enein, H.Y.(2007). Determination of antioxidant and radical scavenging activity of Basil (*Ocimum basilicum* L. Family Lamiaceae) assayed by different methodologies. *Phytotherapy Res.*, **21(4):** 354-361.

Iorio, R., Scrimaglio, R., Rantucci, E., Delle Monache, S., Di Gaetano, A., Finetti, N., Francavilla, F., Santucci, R., Tettamanti, E., and Colonna, R.(2007). A preliminary study of oscillating electromagnetic field effects on human spermatozoon motility. *Bioelectromagnetics*, **28(1):** 72-75.

Jayasinghe, C., Gotoh, N., Aoki, T., and Wada, S.(2003). Phenolics composition and antioxidant activity of sweet basil (*Ocimum basilicum* L.). *J Agric Food Chem.*, **51(15):** 4442-4449.

Khaki, A., Fathiazad, F., Nouri, M., Khaki, A.A., Abassi maleki, N., Ahmadi, P., and Jabari-kh, H.(2010). Beneficial effects of Quercetin on sperm parameters in streptozotocin -induced diabetic male rats. *Phytotherapy Research journal*, **24(9):** 1285-1291.

Khaki, A., Fathiazad, F., Nouri, M., Khaki, A.A., Chelar c,ozanci., Ghafari-Novin, M., and Hamadeh, M.(2009). The Effects of ginger on spermatogenesis and sperm parameters of rat. *Iranian Journal of Reproductive Medicine*, **7(1):** 7-12.

Khaki, A., FathiAzad, F., Nouri, M., Khaki, A.A., Jabbari-kh, H., and Hammadeh, M.(2009). Evaluation of androgenic activity of *Allium cepa* on spermatogenesis in rat. *Folia Morphologica,* **68(1):** 45-51.

Khaki, A.A., Tubbs, R.S., Shoja, M.M., Rad, J.S., Khaki, A., Farahani, R.M., Zarrintan, S., and Nag, T.C.(2006).The effects of an electromagnetic field on the boundary tissue of the seminiferous tubules of the rat: A light and transmission electron microscope study. *Folia Morphol (Warsz).,* **65(3):** 188-194.

Khaki, A.A., Zarrintan,S., Khaki, A., and Zahedi, A.(2008). The effects of electromagnetic field on the microstructure of seminal vesicles in rat: a light and transmission electron microscope study. *Pak J Biol Sci.,* **11(5):** 692-701.

Koyu, A., Ozguner, F., Yilmaz, H., Uz, E., Cesur, G., and Ozcelik, N.(2009).The protective effect of caffeic acid phenethyl ester (CAPE) on oxidative stress in rat liver exposed to the 900 MHz electromagnetic field. *Toxicol Ind Health,* **25(6):** 429-434.

Martínez-Sámano, J., Torres-Durán, P.V., Juárez-Oropeza, M.A., Elías-Vi as, D., and Verdugo-Díaz, L.(2010). Effects of acute electromagnetic field exposure and movement restraint on antioxidant system in liver, heart, kidney and plasma of Wistar rats: a preliminary report. *Int J Radiat Biol.,* **86(12):** 1088-1094.

Niwano, Y., Saito, K., Yoshizaki, F., Kohno, M., and Ozawa, T.(2011). Extensive screening for herbal extracts with potent antioxidant properties. *J Clin Biochem Nutr.,* **48(1):** 78-84.

Peluso, M.R. (2006). Flavonoids attenuate cardiovascular disease, inhibit phosphodiesterase, and modulate lipid homeostasis in adipose tissue and liver. *Exp Biol Med (Maywood).,* **231(8):** 1287-1299.

Rajesh, Arora (2008). Herbal radiomodulators application in medicine, Homeland Defence and Space, CABI, UK, 26.

Roychoudhury, S., Jedlicka, J., Parkanyi, V., Rafay, J., Ondruska, L., Massanyi, P., and Bulla, J.(2009).Influence of a 50 hz extra low frequency electromagnetic field on spermatozoa motility and fertilization rates in rabbits. *J Environ Sci Health A Tox Hazard Subst Environ Eng.,* **44(10):** 1041-1047.

Schüz, J., Lagorio, S., and Bersani, F.(2009).Electromagnetic fields and epidemiology: an overview inspired by the fourth course at the International School of Bioelectromagnetics. *Bioelectromagnetics,* **30(7):** 511-524.

Sharma, V.P., Singh, H.P., Kohli, R.K., and Batish, D.R.(2009). Mobile phone radiation inhibits *Vigna radiata* (mung bean) root growth by inducing oxidative stress. *Sci Total Environ.,* **407(21):** 5543-5547.

Somosy, Z., Forgács, Z., Bognár, G., Horváth, K., and Horváth, G.(2004). Alteration of tight and adherens junctions on 50-Hz magnetic field exposure in Madin Darby canine kidney (MDCK) cells. *Scientific World Journal,* **2:** 75-82.

Supawan, B., Chanida, P., and Nijsiri, R.(2007).Chemical compositions and antioxidative activities of essential oils from four *Ocimum* species endemic to Thailand. *J Health Res.*, **21(3):** 201-206.

Taimor, G., Hofstaetter, B., Piper, H.M. (2000). Apoptosis induction by nitric oxide in adult cardiomyocytes via cGMP-signaling and its impairment after simulated ischemia. *Cardiovasc Res.*, **45**: 588–594.

Utilisation and Management of Medicinal Plants Vol. 2 (2014)    *Pages* **245–259**
*Editor-in-Chief:* **V.K. Gupta**
*Published by:* **DAYA PUBLISHING HOUSE, NEW DELHI**

# 9

# Aniseed (*Pimpinella anisum* L.): An Overview

R.I. Shobha[1]* and B. Andallu[1]

## ABSTRACT

*Aniseed (Pimpinella anisum L.), a plant with potential health benefits, belonging to the Umbelliferae family is cultivated widely in many parts of the world. The fruits are extensively used as a spice for flavoring bakery products. Traditionally, aniseed is used as a mouth freshener, to relieve toothache and also to relieve digestive problems. The main constituents of this plant include trans-anethole, anisaldehyde, estragole, coumarins, scopoletin, umbelliferone, estrols, terpene hydrocarbons, polyenes, and polyacetylenes found mainly in the essential oil of aniseed which is reported to possess antiseptic, anti-microbial, antispasmodic, carminative, diuretic, expectorant, stimulant and fumigant effects and is also good for bronchitis, colds, cramps, emotional balancing, muscular aches and pains, muscular spasm, rheumatism and stress. The seeds have been found to have anti-diabetic effects proving to be hypoglycemic and hypolipidemic and have been effective in decreasing the absorption of glucose. It has estrogenic effects and hence, increases milk secretion, and promotes menstruation. Also, the plant has been found to possess anticonvulsant, antioxidant, radical scavenging, reducing potential, anti-lipid peroxidative and anti-inflammatory effects and also has direct effects on the gastrointestinal system.*

*Keywords:* Aniseed, essential oil, anti-microbial, antispasmodic, anti-inflammatory, anti-diabetic.

1    Sri Sathya Sai Institute of Higher Learning, Anantapur Campus, Anantapur – 515 001, Andhra Pradesh, India

*    *Corresponding author*: E-mail: rshobhaiyer@gmail.com

# Introduction

Aniseed (*Pimpinella anisum* L), commonly known as 'saunf', an annual grassy herb with 30–50 cm high, white flowers, and small green to yellow seeds, belonging to the Umbelliferae family, is one of the oldest medicinal plants (Figure 9.1) (Surmaghi, 2010). Aniseed (*Pimpinella anisum* L.) is a plant with great potential to accrue health benefits. It is an important spice, a native of the Eastern Mediterranean region, cultivated widely in many parts of the world, mainly in Southern Europe, India, China, Japan, Argentina and Mexico. In India, it is grown as a culinary herb, and a large quantity of aniseed is exported from India. In traditional medicine, aniseed is used as a mouth freshener and also to relieve toothache. The seeds are the parts of the plant used; these contain anethole, anisaldehyde, methyl-chavicol, and are reported to be antiseptic, antispasmodic, carminative, diuretic, expectorant and have stomachic activities (Figure 9.2). In the food industry, aniseed is used as flavorant for fish products, ice creams, sweets, and gums (Ozcan and Chalchat, 2006). The chief active constituent of aniseed oil is trans-anethole, which is used as an ingredient in cough lozenges in combination with liquorice (Andallu and Rajeshwari, 2011). In traditional medicine, the drug is used internally for bronchial catarrh, pertussis, spasmodic cough, and flatulent colic, and externally for pediculosis and scabies. Furthermore, it is used as an estrogenic agent. It increases milk secretion, and promotes menstruation (Barnes *et al.*, 2002).

# Chemical Composition

Aniseeds contain about 9-13 g per cent moisture, about 18 per cent protein, 1.5–6.0 mass per cent of a volatile oil and 12-25 g per cent of crude fiber. *Pimpinella anisum*

**Figure 9.1**: Aniseed (*Pimpinella anisum* L.) plant and seeds.

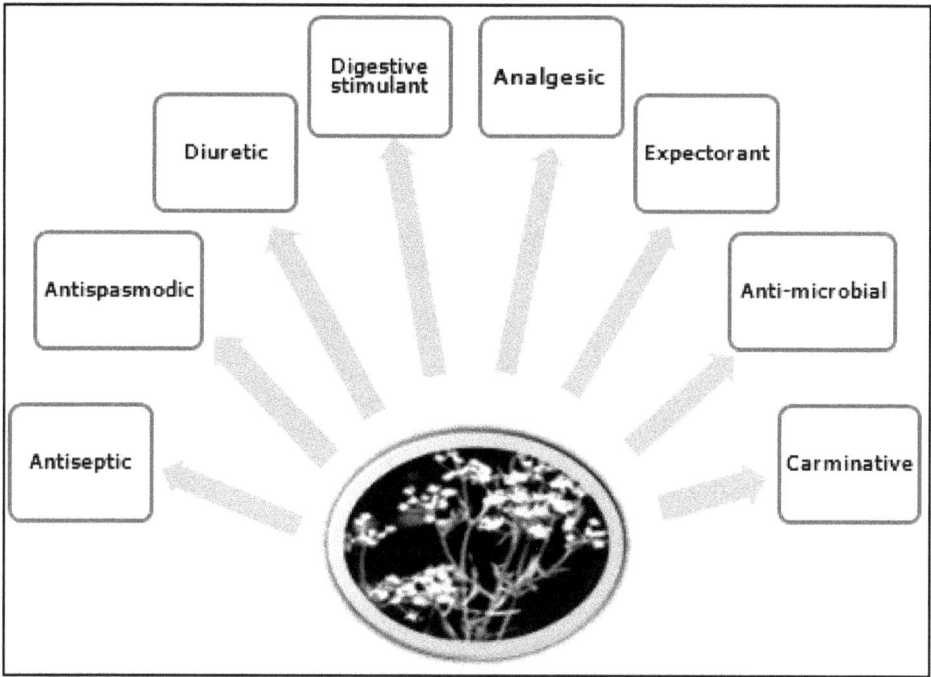

**Figure 9.2**: Beneficial effects of aniseed (*Pimpinella anisum* L.).

L. has as chemical representative, namely the aniseed oil (1-4 per cent). The major component of aniseeds oil, trans-anethole (75-90 per cent), is responsible for its characteristic taste and flavour, as well as for its medicinal properties (DerMarderosian and Beutler, 2002). Few studies have confirmed the presence of eugenol, *trans*-anethole, methylchavicol, anisaldehyde, estragole, coumarins, scopoletin, umbelliferone, estrols, terpene hydrocarbons, polyenes, and polyacetylenes as the major compounds of the essential oil of aniseeds (Gulcin *et al.*, 2003). Other constituents include coumarins (umbelliferone, umbelliprenine, bergapten, and scopoletin), lipids (fatty acids, beta-amyrin, stigmasterol and its salts), flavonoids (flavonol, flavone, glycosides, rutin, isoorientin, and isovitexin), protein and carbohydrate (Bisset, 1994).

*Pimpinella anisum* L. has been well studied, with the principal components being the phenol ethers: transanethole, t-anol, methyl chavicol (estragole), p-methoxyphenyl acetone and trans-pseudoisoeugenyl 2-methyl-butyrate, which is characteristic for aniseed. In minor concentrations, also included are aromatic aldehydes such as p-anisaldehyde, benzaldehyde and acetaldehyde and phenols such as p-cresol, eugenol, Isoeugenol, creosol, hydroquinine or caffeic acid, ferulic acid, coniferyl alcohol, estrols, coumarins, and terpene hydrocarbons (Jurado *et al.*, 2007; Orav *et al.*, 2008) (Figure 9.3).

In a study conducted to obtain chemical constituents by Supercritical extraction using $CO_2$ by GC-MS, the compounds isolated were anethole (~90 per cent), γ-himachalene (2-4 per cent), *p*-anisaldehyde (<1 per cent), methylchavicol (0.9–1.5

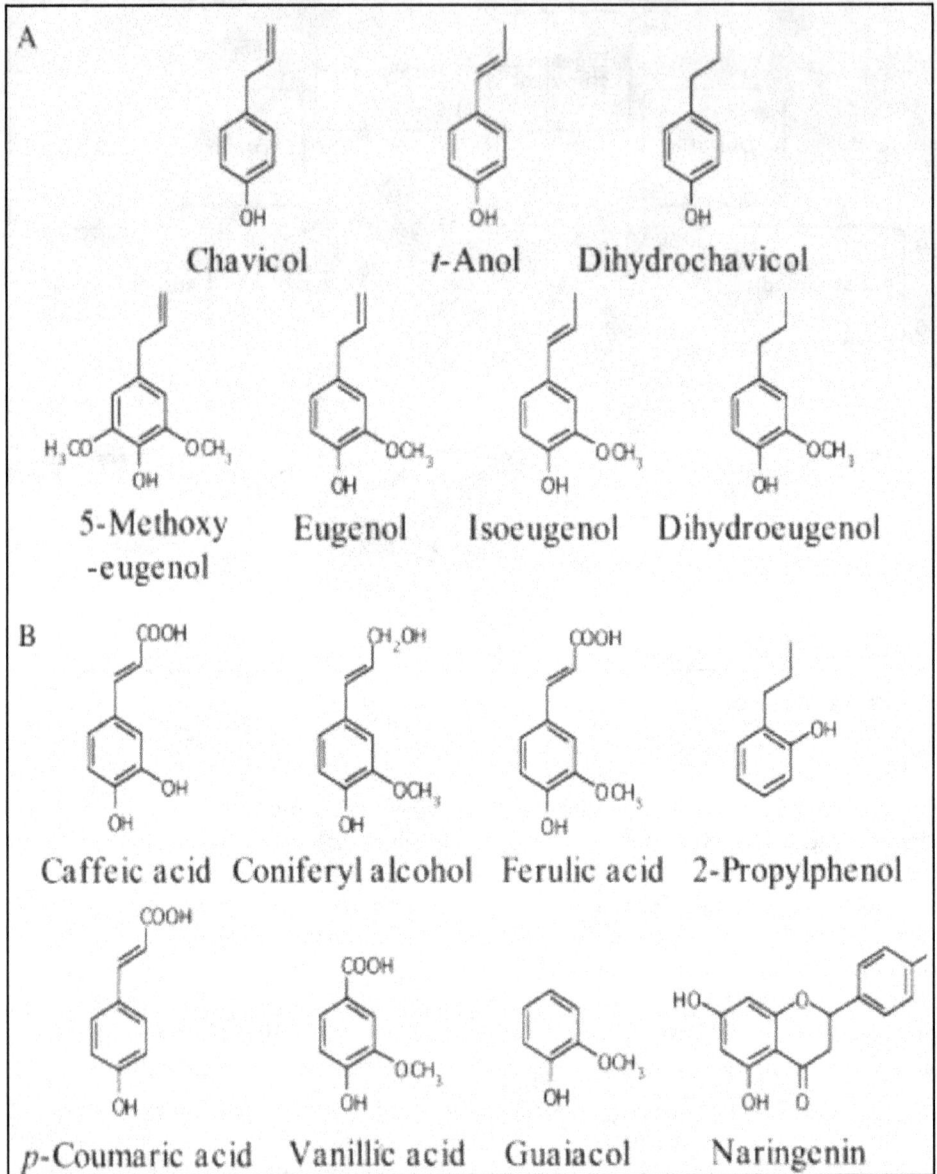

**Figure 9.3**: Chemical compounds in the essential oil of aniseed ( *Pimpinella anisum* L.).

per cent), *cis*-pseudoisoeugenyl 2-methylbutyrate (~3 per cent), and *trans*-pseudoisoeugenyl 2-methylbutyrate (~1.3 per cent) (Rodrigues *et al.*, 2003). Four aromatic compound glucosides, an alkyl glucoside, and a glucide, isolated as new compounds from the polar portion of methanolic extract of anise fruits were clarified as (E)-3-hydroxy-anethole β-D-glucopyranoside, (E)-10-(2-hydroxy-5-methoxyphenyl) propane β-D-glucopyranoside, 3-hydroxyestragole β-D-glucopyranoside, methyl syringate 4-*O*-β-D-glucopyranoside, hexane-1,5-diol 1- *O*-β-D-glucopyranoside, and 1-deoxy-l-erythritol 3-*O*-β-D-glucopyranoside (Fujimatu *et al.,* 2003; Shojaii and Fard, 2012).

A silver ion HPLC procedure used to determine the fatty acid composition of aniseed oil revealed the presence of the positionally isomeric 18: 1 fatty acids oleic acid (*cis* 9–18: 1), petroselinic acid (*cis* 6–18: 1), and *cis* vaccenic acid (*cis* 11–18: 1) in aniseed oil (Denev *et al.,* 2011).

Chemical compounds in aniseed and the therapeutic effects are shown in Figure 9.4. (www.ars_grin.gov/duke)

## Antibacterial and Antifungal Effects

The essential oil of aniseed is described as exerting fungicidal activity in a concentration dependent manner (Soliman and Badeaa, 2002). The aqueous decoction of aniseed exhibited maximum antibacterial activity against *Micrococcus roseus.* Singh *et al.* (2002) reported that the essential oil has strong antibacterial activity against eight human pathogenic bacteria. In addition, essential oil of aniseed possesses anticonvulsant activity in the mouse (Pourgholami *et al.,* 1999). Acetone extract of aniseed inhibited the growth of bacteria, including *Escherichia coli* and *Staphylococcus aureus*, and also exhibited antifungal activity against *Candida albicans* and other

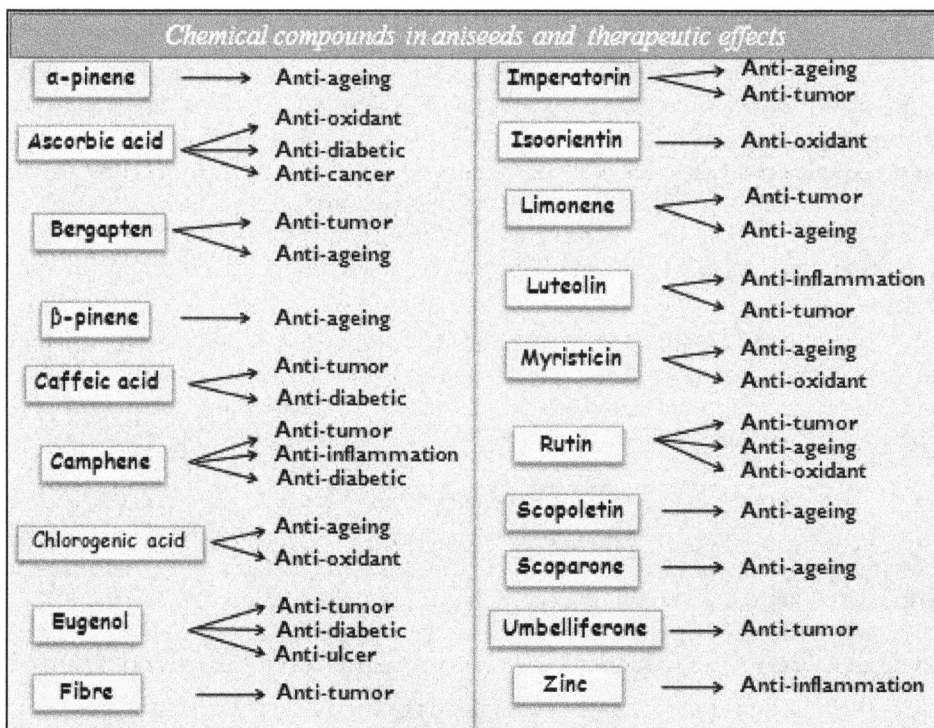

**Figure 9.4**: Chemical compounds in aniseed (*Pimpinella anisum* L.) and therapeutic effects. (www.ars_grin.gov/duke).

organisms (Maruzzella and Freundlich, 1959; Andallu and Rajeshwari, 2011). The aniseeds and oil were found to have a high antibacterial activity against *Staphylococcus aureus* (responsible for boils, sepses, and skin infections), *Streptococus haemoliticus* (throat and nasal infections), *Bacillus subtilis* (infection in immuno-compromised patients), *Pseudomonas aeruginosa* (hospital-acquired infection), and *Escherichia coli* (urogenital tract infections and diarrhoea) (Singh *et al.*, 2002).

Synergic antibacterial activity of methanol extract and the essential oil of *Thymus vulgaris* and *Pimpinella anisum* in 9 pathogenic bacteria exhibited maximum antibacterial effect against *Staphylococcus aureus*, *Bacillus cereus*, and *Proteus vulgaris* whereas combination of essential oil and methanolic extracts of these plants showed an additive effect against *Pseudomonas aeruginosa* (Al-Bayati, 2008).

Antimicrobial effects of water and ethanolic extracts of aniseeds against 10 bacterial species and also *Candida albicans* with disc diffusion method revealed ethanolic extract to have more inhibitory activity against all tested bacteria but not against *Candida albicans* (Gulcin *et al.*, 2003). Akhtar *et al.* (2008) studied antibacterial activities of 4 extracts of aniseeds against 4 pathogenic bacteria (*Staphylococcus aureus*, *Streptococcus pyogenes*, *Escherchia coli*, and *Klebsiella pneumoniae*) by disc diffusion method, the aqueous extract was found to be more effective than methanolic extract, whereas acetone and petroleum ether extracts could not inhibit the growth of the pathogenic test bacteria.

## Secretolytic and Expectorant Effects

The volume of respiratory secretion of anesthetized rabbits was increased dose-dependently from 19 to 82 per cent following administration of aniseed oil by inhalation (in steam) in doses of 0.7-5 g/kg b.w. via a vaporizer, but signs of tissue damage and a mortality rate of 20 per cent were observed at the highest dose level. Inhalation of anethole did not affect the volume, but produced a dose-dependent (1-9 mg/kg) decrease in the specific gravity of respiratory tract fluid in urethanized rabbits in doses from 1 to 243 mg/kg b.w. (Boyd and Sheppard, 1968; Shojaii and Fard, 2012).

The effect of a mixture of herbal extract containing aniseed was compared to that of the flavonoid-quercetin. The clinical results showed significant reduction in sleep discomfort, cough frequency, and cough intensity in the herbal tea-using subjects compared to the placebo tea-using subjects (Haggag *et al.*, 2003).

## Antioxidant Effects

A study conducted on the antioxidant activities of aniseed oil and its methanolic oleoresin, showed that both essential oil and the methanolic oleoresin had higher antioxidant activity in terms of inhibition of lipid peroxidation, scavenging of 2,2-diphenyl-1-picryl hydrazyl (DPPH) radical than synthetic antioxidants butylated hydroxy toluene (BHT) and butylated hydroxy anisole (BHA) (Singh *et al.*, 2008).

Al-Ismail and Aburjai (2004) investigated the antioxidant activities of water and alcohol extracts of chamomile flowers, dill seeds and aniseeds, where the water and alcohol extracts of aniseeds showed lower activity in both linoleic acid and liposome model systems than both the extracts of chamomile flowers and dill seeds.

*In vivo* study supported antioxidant role of aniseed as evidenced by elevated serum non-enzymatic antioxidants and raised activities of erythrocyte antioxidant enzymes and very effectively decreased lipid peroxidation and protein oxidation in diabetics treated with aniseeds (Rajeshwari *et al.*, 2011).

## Radical Scavenging Effect (*In vitro*)

*In vitro* study conducted on methanolic extract and various fractions (hexane, benzene, ethyl acetate, n-butanol and aqueous) of methanolic extract of aniseed effectively scavenged ABTS, DPPH, hydroxyl, nitric oxide and superoxide radicals. The radical scavenging effect could be attributed to the polyphenols, flavonoids and flavonols present in different fractions of aniseed, which indicate that aniseeds are rich source of compounds that can scavenge free radicals (Shobha and Andallu, 2011).

## Anti-Lipid Peroxidative Effect in the *In vitro* Model Systems

In an *in vitro* study conducted to evaluate the anti-lipid peroxidative effect of methanolic extract and various fractions of methanolic extract of aniseed (*Pimpinella anisum* L.) in terms of inhibition of lipid peroxidation, marker of oxidative stress using *in vitro* model systems *viz*. linoleate model, β-carotene linoleate model, liver homogenate and hemoglobin as models, methanolic extract and all the fractions dose dependently inhibited lipid peroxidation *in vitro*. Ethyl acetate fraction exhibited the highest inhibition of lipid peroxidation in all the four model systems due to the higher quantities of phenolic compounds, flavonoids, flavonols etc. extracted in ethyl acetate fraction which are reported to be antioxidants (www.ars_grin.gov/duke). Thus, aniseeds possess potential to prevent peroxidation of lipids there by combat oxidative stress, reported to be the root cause of various diseases (Shobha *et* al., 2012, Shobha *et al.*, 2013).

## Reducing Potential

Methanolic extract and various fractions of aniseeds also exhibited tremendous reducing potential in terms of Cupric reducing antioxidant capacity (CUPRAC), Ferric reducing antioxidant potential (FRAP), iron chelating and reducing power based on the reduction of $Cu^{++}$ and $Fe^{+++}$ to $Cu^+$ and $Fe^{++}$ respectively by the reductants (antioxidants) present in the sample *i.e.* aniseeds. Highest reducing potential exhibited by ethyl acetate fraction and benzene fractions could be due to the polyphenolic compounds and other compounds *viz*. aglycones, terpenoids, flavonols get extracted by ethyl acetate and benzene respectively (Shobha and Andallu, 2011).

## Antidiabetic Effects

A study conducted to assess *in vivo* anti-diabetic and hypolipidemic effects of aniseeds in type 2 diabetes patients by supplementation of aniseeds (5g/day) to the selected group for 60 days, helped to counter the effects of oxidative stress as substantiated by decreased lipid peroxidation, protein oxidation and decreased activity of erythrocyte catalase (CAT), increased serum β carotene, vitamin A, E and C in diabetics treated with aniseeds. Besides, the treatment increased the activity of erythrocyte antioxidant enzyme *i.e.* glutathione-S-tranferase (GST) and reduced glutathione content in aniseeds-treated diabetics. Also, there was a significant decrease in the serum cholesterol, triglycerides, VLDL and LDL cholesterol levels and an increase in the HDL cholesterol levels in aniseed-treated diabetics (Rajeshwari *et al.*, 2011).

A study conducted on the effect of aniseeds oil on the absorption of glucose from the jejunum and water from the colon and kidney tubules and also on its mechanism of action showed that aniseed oil increased significantly glucose absorption in the rat jejunum, because the oil enhanced the activity of $Na^+$-$K^+$ ATPase which increases sodium gradient that gears the mucosal glucose transport. Also addition of aniseed oil to drinking water, reduced the volume of urine produced in the rat and increased the activity of the renal $Na^+$-$K^+$ATPase even at extremely low concentrations (Kreydiyyeh *et al.*, 2003).

## α-amylase and α-glucosidase Inhibitory Effects (*In vitro*)

Dietary management of hyperglycemia linked to type 2 diabetes can be targeted through foods or botanical supplements that have high α-amylase and α-glucosidase inhibitory activities. Many natural resources have been investigated with respect to suppression of glucose production from carbohydrates in the gut or glucose absorption from the intestine (Matsui *et al.*, 2007). α- amylase catalyses the hydrolysis of α-1,4-glucosidic linkages of starch, glycogen and various oligosaccharides and α-glucosidase further breaks down the disaccharides into simpler sugars, readily available for the intestinal absorption. The inhibition of their activity, in the digestive tract of humans, is considered to be effective to control diabetes by diminishing the absorption of glucose formed from starch by these enzymes (Hara and Honda, 1990). *In vitro* α-amylase and α-glucosidase inhibitory activities of various fractions (hexane, benzene, ethyl acetate, n-butanol and water) of aniseeds (*Pimpinella anisum* L.) obtained by sequential fractionation of methanolic extract exhibited potential α-amylase and α-glucosidase inhibitory effects, among which, highest inhibitory effect shown by ethyl acetate fraction is attributed to the phenolic compounds in aniseeds. Hence, aniseed may be prescribed as adjunct to dietary therapy to combat oxidative stress in NIDDM patients (Shobha and Andallu, 2011).

## Anti-inflammatory Effects (*In vitro*)

Anti-inflammatory activity evaluated *in vitro* for methanolic extract and various solvent fractions of aniseeds in terms of inhibition of the activities of inflammatory enzymes *viz.* lipoxidase and xanthine oxidase revealed the highest lipoxidase and xanthine oxidase inhibitory activities for ethyl acetate fraction which was almost equal to the activity exhibited by quercetin, a positive control indicating the presence of quercetin in aniseeds (Shobha and Andallu, 2011, Shobha *et al.*, 2013).

## Anticonvulsant Effect

The aqueous extract of leaves and stem of few Arab medicinal plants such as *Pimpinella anisum, Matricaria chamomile,* and *Artemisia vulgaris* interrupt the inception of picrotoxin-induced seizures (Abdul-Ghani *et al.*, 1987). Anticonvulsant effect studied by Pourgholami *et al.* (1999) revealed that *Pimpinella anisum* possesses anticonvulsant effect against tonic seizures induced by maximal electroshock (MES) and also block tonic convulsions induced by the intra peritoneal injection of pentylenetetrazole (PTZ) in male mice. In another study, among the different doses of methanolic extract of aniseeds used, 200 mg/kg dosage of the extract significantly (p<0.05) increased the delay of the onset of picrotoxin-induced seizure in mice which was more satisfactory than the Phenobarbital (40 mg/kg) in delaying the death time (Heidari and Ayeli, 2005).

A study conducted to investigate the cellular mechanisms of anticonvulsant effect of aniseeds helped to determine whether the fruit essential oil of aniseeds affect the bioelectrical activity of snail neurons in control condition or after pentylenetetrazol-(PTZ-) induced epileptic activity. Also, the results signify that aniseed oil augmented $Ca^{2+}$ channel activity or inhibited voltage and/or $Ca^{2+}$ dependent $K^+$ channel activity underlying post-hyperpolarization potential (Janahamadi *et al.*, 2008).

## Analgesic Effects

In a study to assess analgesic effect conducted by Twaij *et al.* (1988) on some Iraqi medicinal plants, the extracts of *Tribulus terrestris* and *Pimpinella anisum* exhibited significant analgesic activity versus benzoquinone-induced writing and in thermal tests. The essential oil of aniseeds showed significant analgesic activity similar to morphine and aspirin (Tas, 2009). Also the fixed oil of aniseeds showed effective anti-inflammatory activity comparable to indomethacin and analgesic effects similar to morphine and aspirin in mice (Sahraei *et al.*, 2002; Shojaii and Fard, 2012).

## Antispasmodic and Relaxant Effect

The relaxant effect of aniseeds studied on the isolated tracheal chains of guinea pigs revealed bronchodilatory effect of aqueous and ethanol extracts of aniseeds and the essential oil. The results showed that aqueous and ethanol extracts displayed significant relaxant effects, when compared to controls, which was similar to that of theophylline, but the essential oil displayed slightly lower effect than theophylline (Boskabady and Ramazani-Azari, 2001).

In another study, antispasmodic effect on the rat anococcygeus smooth muscle was investigated using three hydroalcoholic extracts of the aerial parts of *Pimpinella anisum* (ethanol: water, 40:60, 60:40, 80:20) which showed that all the three extracts attenuated acetylcholine-induced contraction. Among the extracts, only 60 per cent ethanol extract showed concentration dependant attenuation of contraction when compared to the other extracts (Tirapelli *et al.*, 2007).

## Digestive Stimulant Effect

Antiulcer effect of aniseed studied against acute gastric ulceration in rats produced by various poisonous chemicals and indomethacin showed that aniseeds significantly inhibited gastric mucosal damage induced by necrotizing agents and indomethacin which was confirmed histologically (Al Mofleh *et al.*, 2007).

A study was conducted by Picon *et al.* (2010) to evaluate the laxative efficacy of a phytotherapic compound containing *Pimpinella anisum, Foeniculum vulgare, Sambucus nigra,* and *Cassia augustifolia* for the treatment of chronic constipation in randomized

20 patients. The primary endpoint- colonic transit time (CTT) was measured radiologically. Secondary endpoints included number of evacuations per day, perception of bowel function, adverse effects, and quality of life. Number of evacuations per day increased during the use of active tea; significant differences were observed as of the second day of treatment (p < 0.001). Patient perception of bowel function was improved (p < 0.01), but quality of life did not show significant differences among the study periods. Except for a small reduction in serum potassium levels during the active treatment, no significant differences were observed in terms of adverse effects throughout the study period. The findings of this randomized controlled trial showed that the phytotherapic compound assessed has laxative efficacy and is a safe alternative option for the treatment of constipation.

## Fumigant Effect

In a study, plant essential oils from 40 plant species which included horseradish (*Armoracia rusticana*), aniseed (*Pimpinella anisum*) and garlic (*Allium sativum*) oils were tested for their insecticidal activities against larvae of *Lycoriella ing'enue* (Dufour) using a fumigation bioassay. Horseradish, aniseeds and garlic oils showed the most potent insecticidal activities among the plant essential oils. At 1.25 µl/l, horseradish, aniseeds and garlic oils caused 100, 93.3 and 13.3 per cent mortality, but at 0.625 µl/l air this decreased to 3.3, 0 and 0 per cent respectively. Analysis by gas chromatography-mass spectrometry (GC-MS) led to the identification of one major compound from horseradish, and three each from aniseeds and garlic oils such as *m*-anisaldehyde and *o*-anisaldehyde, two positional isomers of *p*-anisaldehyde, which were tested individually for their insecticidal activities against larvae of *L. ingénue* and produced good insecticidal effects (Park *et al.*, 2006).

Aniseed essential oils exhibited highly effective larvicidal and ovicidal effect against three mosquito species and also showed repellency effect against mosquito *Culex pipiens* (Prajapati *et al.*, 2005; Erler *et al.*, 2006). The acaricidal activity of *p*-anisaldehyde derived from aniseed oil and commercially available components of aniseed oil studied against the house dust mites, *Dermatophagoides farina,* and *D. pteronyssinus* showed that the compounds most toxic to these dermatophagoides were *p*-anisaldehyde followed by benzyl benzoate and, therefore, may be useful as a lead compound for the selective control of house dust mites (Lee, 2004).

## Effect on Milk Production

The effects of diet supplementation with aniseeds and fenugreek seeds on the performance does and kits were studied. Findings revealed that the daily milk intake of kits in aniseed-fenugreek group was equivalent to that of control rabbits. Also, the 17 days body weight did not differ significantly between two groups. At 35 days of lactation, the differences between aniseed and fenugreek group and control groups were not significant in litter size, litter weight, kit weight and 1–35 day weight gain (Eiben *et al.*, 2004).

## Effect on Menopausal Hot Flashes

In a double blind clinical trial, the effect of aniseed extract on menopausal hot flashes in 72 postmenopausal women was examined. Consumption of 3 capsules of aniseed extract (each capsule contains 100 mg of extract) for 4 weeks lead to significant reduction in hot flash frequency and intensity in postmenopausal women (Nahidi *et al.*, 2008).

## Effect on Dysmenorrhea

A study by Khoda Karami *et al.* (2008) on the effectiveness of a herbal capsule containing dried extracts of celery, saffron, and aniseeds was compared with mephnamic acid capsule in 180 female students (with age 17–28yrs) with primary dysmenorrhea showed significant reduction in pain intensity in both herbal and mephnamic acid group compared to placebo group. Also the results revealed that the effectiveness of herbal capsule was better than mephnamic acid in pain relief and can be a suitable alternative in primary dysmenorrhea.

## Conclusions

*Pimpinella anisum* L., largely grown for culinary purpose, is an important spice with many potential health benefits. The main constituent of aniseeds is the essential oil which contains t-anethole, anisaldehyde, methyl chavicol, exhibits versatile pharmacological properties such as anti-microbial, carminative and stomachic activities. Also, the essential oil of aniseeds is found to have gastroprotective, antispasmodic, antioxidant and fumigant effects. Traditionally, aniseeds are used for pertussis, flatulent colic, spasmodic cough, bronchial catarrh and also as a remedy for insomnia and constipation. Aniseeds possess anticonvulsant and analgesic effects which are found to be similar to morphine and aspirin. Aniseeds have estrogenic effects and thereby show a great effect on the milk production and dysmenorrhea and are effective in the reduction of menopausal hot flashes. Aniseeds are hypoglycemic and hypolipidemic in diabetic patients, controlled lipid peroxidation, a marker of oxidative stress. Aniseeds contain a good amount of phenolic compounds, flavonoids, flavonols responsible for radical scavenging, reducing potential, anti-lipid peroxidative, and anti-inflammatory effects. Thus, aniseeds, possessing varied phytochemicals showing versatile therapeutic potential can be considered as a functional food.

## References

Abdul-Ghani, A.S., El-Lati, S.G., and Sacaan, A.I. (1987). Anticonvulsant effects of some Arab medicinal plants. *International Journal of Crude Drug Research*, **25(1):** 39–43.

Akhtar, A.A., Deshmukh, A.A. and Bhonsle, A.V. *et al.* (2008). *In vitro* Antibacterial activity of *Pimpinella anisum* fruit extracts against some pathogenic bacteria. *VeterinaryWorld*, **1(9):** 272–274.

Al Mofleh, A.A., Alhalder, J.S., Mossa, M.O., Al-Soohalbani, and Rafatullah, S. (2007). Aqueous suspension of anise *"Pimpinella anisum"* protects rats against chemically induced gastric ulcers. *World Journal of Gastroenterology*, **13(7):** 1112–1118.

Al-Bayati, F.A. (2008). Synergistic antibacterial activity between *Thymus vulgaris* and *Pimpinella anisum* essential oils and methanol extracts. *Journal of Ethnopharmacology*, **116(3):** 403–406.

Al-Ismail, K.M., and Aburjai, T. (2004). Antioxidant activity of water and alcohol extracts of chamomile flowers, anise seeds and dill seeds. *Journal of the Science of Food and Agriculture*, **84(2):** 173–178.

Andallu, B., and Rajeshwari, C.U. (2011). Aniseed: Aniseed (*Pimpinella anisum* L) in Health and Disease. *In*: Nuts and Seeds in Health and Disease Prevention'. Chapter 20, Ed. Preedy V.K., Elsevier Limited, Oxford, UK. pp. 175-181.

Barnes, J., Anderson, L.A., and Phillipson, J.D. (2002). Aniseed. *In:* Herbal Medicines: A guide for healthcare professionals. Pharmaceutical Press, London-Chicago, pp. 51-54.

Bisset, N.G. (1994). Herbal drugs and phytopharmaceuticals: a handbook for practice on a scientific basis, Stuttgart, Medpharm Scientific Publisher, pp. 351-352.

Boskabady, M.H., and Ramazani-Assari, M. (2001). Relaxant effect of *Pimpinella anisum* on isolated guinea pig tracheal chains and its possible mechanism (s). *Journal of Ethnopharmacology*, **74(1):** 83–88.

Boyd, E. M., and Sheppard, E. P. (1968). The effect of steam inhalation of volatile oils on the output and composition of respiratory tract fluid. *Journal of Pharmacology and Experimental Therapeutics*, **163(1):** 250-256.

Denev, R.V., Kuzmanova, I.S., Momchilova, S.M., and Nikolova-Damyanova, B.M. (2011). Resolution and quantification of isomeric fatty acids by silver ion HPLC: fatty acid composition of aniseed oil (*Pimpinella anisum*, Apiaceae). *Journal of AOAC International*, **94(1):** 4–8.

DerMarderosian, A., and Beutler, J. A. (Eds.), (2002). The Review of Natural Products, 3rd ed. Facts and Comparisons, St. Louis, MO, pp. 824–826.

Eiben, C.S., Rashwan, A.A., Kustos, K. Gódor-Surmann K.L., and Szendr , Z.S. (2004). Effect of *anise* and *fenugreek* suplementation on performance of rabbit does In: *Proceedings of the 8th World Rabbit Congress*, Puebla, Mexico.

Erler, F., Ulug, I., and Yalcinkaya, B. (2006). Repellent activity of five essential oils against *Culex pipiens*. *Fitoterapia*, **77(7- 8):** 491–494.

Fujimatu, E., Ishikawa, T., and Kitajima, J. (2003). Aromatic compound glucosides, alkyl glucoside and glucide from the fruit of anise. *Phytochemistry*, **63(5):** 609–616.

Gulcin, M., Oktay, E., Kirecci, S., and Kufrevioglu, O.I. (2003). Screening of antioxidant and antimicrobial activities of anise (*Pimpinella anisum* L.) seed extracts. *Food Chemistry*, **83(3):** 371–382.

Haggag, E. G., Abou-Moustafa, M. A., Boucher, W., and Theoharides, T. C. (2003). The effect of a herbal water-extract on histamine release from mast cells and on allergic asthma. *Journal of Herbal Pharmacotheraphy*, **3(4):** 41-54.

Hara, Y., and Honda, M. (1990). The inhibition of α-amylase by tea polyphenols. *Agricutural and Biological Chemistry*, **54**: 1939–1945.

Heidari, M.R., and Ayeli, M. (2005). Effects of methyl alcoholic extract of *Pimpinella anisum* on picrotoxin induced seizure in mice and its probable mechanism. *Scientific Journal of Kurdistan University of Medical Science*, **10(3)**: 1–8.

http://www.ars_grin.gov/duke//(accessed on June, 2012).

Jurado, J.M., Ballesteros, O., Alcázar, A., Pablos, F., Martín, M.J., Vílchez, J.L., and Navalón, A. (2007). Characterization of aniseed-flavoured spirit drinks by headspace solid-phase microextraction gas chromatography-mass spectrometry and chemometrics. *Talanta*, **72**: 506–511.

Janahmadi, M., Farajnia, S., Vatanparast, J., Abbasipour,V., and Kamalinejad, M. (2008). The fruit essential oil of *Pimpinella anisum* L. (Umblliferae) induces neuronal hyperexcitability in snail partly through attenuation of after-hyperpolarization. *Journal of Ethnopharmacology*, **120 (3)**: 360–365.

Khoda Karami, N., Moattar, F., and Ghahiri, A. (2008). Comparison of effectiveness of an herbal drug (celery, saffron, anise) and mephnamic acid capsule on primary dismenorrhea. *Ofoghe Danesh*, **14(1)**: 11–19.

Kreydiyyeh, S.I., Usta, J., Knio, K., Markossian, S., and Dagher S. (2003). Aniseed oil increases glucose absorption and reduces urine output in the rat. *Life Sciences*, **74(5)**: 663–673.

Lee, H.S. (2004). *p*-anisaldehyde: acaricidal component of *Pimpinella anisum* seed oil against the house dust mites *Dermatophagoides farinae* and *Dermatophagoides pteronyssinus*. *Planta Medica*, **70(3)**: 279–281.

Maruzzella, J. C., and Freundlich, M. (1959). Antimicrobial substances from seeds. *Journal of the American Pharmacists Association*, **48**: 356-358.

Matsui, T., Tanaka, T., Tamura, S., Toshima, A., Tamaya, K., Miyata,Y., Tanaka, K. and Matsumoto, K. (2007). Alpha glucosidase inhibitory profile of catechins and theaflavins. *Journal of Agricultural and Food Chemistry*, **55**: 99-105.

Nahidi, F., Taherpoor, M., Mojab, F., and Majd, H. (2008). Effect of anise extract on hot flush of menopause. *Pajoohandeh*, **13(3)**: 167–173.

Orav, A., Raal, A., and Arak, E. (2008). Essential oil composition of *Pimpinella anisum* L. fruits from various European countries. *Natural Product Research*, **22**: 227–232.

Ozcan, M.M., and Chalchat, J.C. (2006) Chemical composition and antifungal effect of anise (*Pimpinella anisum* L.) fruit oil at ripening stage. *Annals of Microbiology*, **56(4)**: 353–358.

Park, I.K., Choi, K.S., Kim, D.H., Choi, I.H., Kim, L.S., Bak, W.C., Choi, J.W., and Shin, S.C. (2006). Fumigant activity of plant essential oils and components from horseradish (*Armoracia rusticana*), anise (*Pimpinella anisum*) and garlic (*Allium sativum*) oils against *Lycoriella ingenua* (Diptera: Sciaridae). *Pest Management Science*, **62(8)**: 723–728.

Picon, P.D., Picon, R.V., Costa, A.F., Sander, G.B., Amaral, K.M., Aboy, A.L., and Henriques, A.T. (2010). Randomized clinical trial of a phytotherapic compound containing *Pimpinella anisum, Foeniculum vulgare, Sambucus nigra,* and *Cassia augustifolia* for chronic constipation. *BMC Complementary and Alternative Medicine,* **10(17):** 1-9.

Pourgholami, M.H., Majzoob, S., Javadi, M., Kamalinejad, M., Fanaee, G.H.R., and Sayyah, M. (1999). The fruit essential oil of *Pimpinella anisum* exerts anticonvulsant effects in mice. *Journal of Ethnopharmacology,* **66(2):** 211–215.

Prajapati, V., Tripathi, A.K., Aggarwal, K.K., and Khanuja, S.P.S. (2005). Insecticidal, repellent and oviposition-deterrent activity of selected essential oils against *Anopheles stephensi, Aedes aegypti* and *Culex quinquefasciatus. Bioresource Technology,* **96(16):** 1749–1757.

Rajeshwari, C.U., Abirami, M., and Andallu, B. (2011). *In vitro* and *in vivo* antioxidant potential of aniseed (*Pimpinella anisum*). *Asian Journal of Experimental Biological Sciences,* **2(1):** 80–89.

Rajeshwari, U., Shobha, I., and Andallu, B. (2011). Comparison of aniseed and coriander seeds for antidiabetic, hypolipidemic and antioxidant activities. *Spatula DD,* **1(1):** 9–16.

Rodrigues, V.M., Rosa, P.T.V., Marques, M.O.M., Petenate, A.J., and Meireles, M.A.A. (2003). Supercritical extraction of essential oil from aniseed (*Pimpinella anisum* L) using $CO_2$: solubility, kinetics, and composition data. *Journal of Agricultural and Food Chemistry,* **51(6):** 1518–1523.

Sahraei, H.H., Ghoshooni, S., Hossein Salimi *et al.* (2002). The effects of fruit essential oil of the *Pimpinella anisum* on acquisition and expression of morphine induced conditioned place preference in mice. *Journal of Ethnopharmacology,* **80(1):** 43–47.

Shobha, R.I., and Andallu, B. (2011). Therapeutic potential of aniseed (*Pimpinella anisum* L.): An *in vitro* assessment, M. Phil Thesis submitted to Sri Sathya Sai Institute of Higher Learning, Anantapur Campus.

Shobha, R.I., Rajeshwari, C.U., Mekha, M.S., and Andallu, B. (2012). Radical scavenging, anti-peroxidative and anti-inflammatory activities of aniseed: An *in vitro* assessment, *Proceedings of International Conference on Bioactive Natural Compounds from Plant Food in Nutrition and Health,* September 20-22, Kottayam, Kerala, pp. 239-245.

Shobha, R.I., Rajeshwari, C.U. and Andallu, B. (2013). Antihemolytic and anti-inflammatory activities of Aniseed (*Pimpinella Anisum* L.). *Journal of Advance Pharmaceutical Research and Bioscience,* **1(2):** 52-59.

Shojaii, A., and Fard, M.A. (2012). Review of pharmacological properties and chemical constituents of *Pimpinella anisum, ISRN Pharmaceutics,* 1-8.

Surmaghi, S.M.H. (2010). Medicinal Plants and Phytotherapy, Vol. 1, Donyay Taghziah Press, Tehran, Iran, pp. 81-82.

Singh, G., Kapoor, I. P.S., Pandey, S. K., Singh, U. K., and Singh, R. K. (2002). Studies on essential oil: Part 10; Antibacterial activity of volatile oils of some spices. *Phytotherapia Research*, **16:** 680-682.

Singh, G., Kapoor, I.P.S., Singh, P. de Heluani, C.S., and Catalan, C.A.N. (2008). Chemical composition and antioxidant potential of essential oil and oleoresins from anise seeds (*Pimpinella anisum* L.). *International Journal of Essential Oil Therapeutics*, **2(3):** 122–130.

Soliman, K. M., and Badeaa, R. I. (2002). Effect of oil extracted from some medicinal plants on different mycotoxigenic fungi. *Food and Chemical Toxicology*, **40(11):** 1669-1675.

Tas, A. (2009). Analgesic effect of *Pimpinella anisum* L. essential oil extract in mice. *Indian Veterinary Journal*, **86(2):** 145–147.

Tirapelli, C. R., de Andrade, C. R., Cassano, A. O., De Souza, F.A., Ambrosio, S.R., da Costa, F.B., and de Oliveira, A.M. (2007). Antispasmodic and relaxant effects of the hydroalcoholic extract of *Pimpinella anisum* (*Apiaceae*) on rat anococcygeus smooth muscle. *Journal of Ethnopharmacology*, **110(1):** 23–29.

Twaij, H.A.A., Elisha, E.E., Khalid, R.M., and Paul, N.J. (1988). Analgesic studies on some Iraqi medicinal plants. *International Journal of Crude Drug Research*, **25(4):** 251–254.

Utilisation and Management of Medicinal Plants Vol. 2 (2014)    Pages 261–272
Editor-in-Chief: V.K. Gupta
Published by: DAYA PUBLISHING HOUSE, NEW DELHI

# 10

# Wound Healing Activity of *Achyranthes aspera* L. Leaf Extract in Aged Animal

C.C. Barua[1]*, S.A. Begum[2], D.C. Pathak[2] and R.S. Bora[3]

## ABSTRACT

*Achyranthes aspera Linn. is one of the most common and versatile medicinal plants used in various disease conditions in the North East region of India which include a remedy for piles, renal dropsy, pneumonia, cough, kidney stone, skin eruptions, snake bite, gonorrhea, dysentery etc. The methanol extract of Achyranthes aspera Linn (MEAA) leaf was investigated for its wound healing effect using excision wound (in vivo) in aged Sprague Dawley rats. In excision wound model, contraction of wound was significantly (p< 0.01) higher in A. aspera (5 per cent w/w ointment) treated group in comparison to the control group. The collagen, elastin and hydroxyproline content of the granulation tissue in A. aspera treated animals were increased significantly (p< 0.01) than the control group. These findings were also confirmed by histopathological examination. The results suggested that MEAA possesses significant wound healing potential in aged animal wound model.*

*Keywords:* Achyranthes aspera, Excision wound, Incision wound, Wound healing.

## Introduction

Tissue repair and wound healing are complex processes that involve clotting, fibrin-fibronectin (FN) deposition, inflammation, fibroplasia, neovascularization, wound contraction, and reepithelialization (Sidhu *et al.*, 1998). The floral richness of

1   Department of Pharmacology and Toxicology,

2   Department of Veterinary Pathology,

3   Department of Livestock production and Management,

   College of Veterinary Science, Khanapara, AAU, Guwahati – 781 022, Assam, India

*   *Corresponding author*: E-mail: chanacin@gmail.com; chanacin@satyam.net.in

the North East region cannot be neglected in context to its medicinal importance. Considering the rich diversity of this region, it is expected that screening and scientific evaluation of plant extracts for their diverse activity may provide new drug molecule that can combat various side effects of the commercially available synthetic drugs, reducing the cost of medication as well.

*Achyranthes aspera* Linn. (Amaranthaceae) (Prickly Chaff flower), locally known as Apang, is an annual, biennial, lower portion perennial, erect under shrub or rather stiff herb growing up to 0.3 to 1.0 m in height. It grows in tropical and warmer regions. *Yunani* doctors and local *kabiraj* use the stem, leaves and fruits as a remedy for piles, renal dropsy, pneumonia, cough, kidney stone, skin eruptions, snake bite, gonorrhea, dysentery etc. The plant has anti-bacterial (Aziz *et al.*, 2005), anti-tumor (Chakraborty *et al.*, 2002), anti-inflammatory (Vetrichelvan *et al.*, 2003), abortifacient activity (Workineh *et al.*, 2006), increases pituitary and uterine wet weights in ovarectomised rats and produces reproductive toxicity in male rats (Sandhyakumari *et al.*, 2002). *A. aspera* extract elevates thyroid hormone level and reported to have anticoagulant, antiarthritic, antitumor and anti hepatocarcinogenic activity (Kartik *et al.*, 2010). We have reported antidepressant (Barua *et al.*, 2009, Barua *et al.*, 2010a) and analgesic (Barua *et al.*, 2010b) activity of this plant.

In our earlier study on the wound healing activity of *A. aspera* using excision and incision wound model (*in vivo*) in normal Sprague Dawley rats and in chorioallantoic membrane (CAM) model (*in vitro*) in 9 day old embryonated chicken eggs, promising wound healing activity was reported (Barua *et al.*, 2010c). The present study involves healing efficacy of *A. aspera* in excision wound model in aged rats. Any type of wound takes longer to heal in aged individual due to their poor health condition combined with other age related ailments than younger individual. Bed sore is a commonly encountered complicacy in geriatrics as they are bedridden due to various old age ailments like cancer, fracture to name a few. This has prompted us to study the healing pattern in aged animals using the same plant extract which showed promising wound healing activity in excision, incision wounds and impaired wounds like diabetic, burn wounds or in immunocompromised animals.

## Materials and Methods

### Plant Material

The leaves of the plants were collected from the medicinal garden of the Department of Pharmacology, College of Veterinary Science, Khanapara during the month of Feb–June, 2010. It was identified by Taxonomist Dr. S.C. Nath, Taxonomist, NEIST, Jorhat, and a voucher specimen (AAU/CVSC/PHT/02) was deposited.

### Preparation of Methanol Extract

Fresh leaves of the plant were cleaned from extraneous materials, washed, shade dried, powdered mechanically, weighed and stored in air tight container. About 250 g of powdered material was soaked in 1000 ml methanol for 72 hours in beaker and mixture was stirred every 18 hour using a sterile glass rod. Filtrate was obtained 3 times with the help of Whatman filter paper no 1 and the solvent was removed by rotary evaporator (Roteva, Equitron, Medica Instrument Mfg. Co.) under reduced pressure at <45°C temperature leaving a dark brown residue. It was stored in air tight container at 4°C until use. Recovery was 6.89 per cent (w/w).

## Phytochemical Screening

The methanol extract of *A. aspera* was subjected to phytochemical screening as per the method of Harborne (1991), for presence of different phytoconstituents as described below:

## Steroids

The extract (5 mg) was dissolved in 3 ml of chloroform and then shaken with about 3 ml concentrated sulfuric acid. Development of red colour indicates the presence of steroid.

## Alkaloids

The extract (5 mg) was dissolved in 5 ml of ammonia and then extracted with equal volume of chloroform. To this, 5 ml dilute HCl acid was added. The acid layer obtained was tested separately with Mayer's, Wagner's, Hager's and Dragendroff's reagent for alkaloids.

## Phenolic Compounds

About 5 mg of extract was dissolved in 1 ml of water and 5 drops of 10 per cent ferric chloride was added to it. Development of dark blue color indicates the presence of phenolic compounds.

## Tannins

About 0.5 g of extract was mixed with few drops of 1 per cent gelatin solution containing 10 per cent sodium chloride. White precipitation indicates the presence of tannins.

## Flavonoids

To 2 ml of alcoholic solution of the extract (0.5 g extract in 10 ml methanol), few drops of neutral ferric chloride solution was mixed. Development of green colour indicates the presence of flavonoids.

## Glycosides

To about 1 ml of the extract (0.5 g extract in 1 ml of water), 5 ml of Benedict's reagent was added and boiled for 2 min. Development of brown to red colour indicates the presence of glycoside.

## Diterpenes

About 5 mg of the extract was dissolved in 3 ml of copper acetate solution (5 per cent). Development of green colour indicates the presence of diterpenes.

## Triterpenes

The extract (5 mg) was dissolved in 3 ml of chloroform and then it was shaken with 3 ml of concentrated sulfuric acid. Development of yellow colour in lower layer indicates the presence of triterpenes.

## Saponins

The extract (5 mg) was shaken with 3 ml water. The formation of foam that persists for 10 min indicates the presence of saponins.

## Ointment Preparation of *A. aspera*

Ointment of three different concentrations were prepared by mixing 2.5, 5.0 and 7.5g of methanol extract of *A. aspera* with 97.5, 95 and 92.5g of white soft petroleum jelly (S. D. Fine Chemicals, India) to prepare 2.5, 5.0 and 7.5 per cent ointment (w/w), respectively.

## Animals

Nineteen month old healthy Sprague Dawley rats of either sex, weighing between 150-200g were used for the study. They were housed under controlled conditions of temperature ($25 \pm 2^\circ$C), humidity ($50 \pm 5$ per cent) and 12: 12 h light–dark cycles with food and water *ad libitum.* Animals were housed individually in polypropylene cages containing sterile paddy husk bedding. The experiments were performed as per guidelines of the Institutional Animal Ethical Committee (770/03/ac/CPCSEA/FVScAAU/IAEC/06/21) and conform to the national guidelines on the care and use of laboratory animals, India. Animals were periodically weighed before and after experiments. All the animals were closely observed for any infection and those which showed signs of infection were separated and excluded from the study.

## Determination of $LD_{50}$ and Acute Toxicity

$LD_{50}$ of *A. aspera* was estimated by following up-and-down stair-case method in mice using OECD TG- 425 guidelines. Doses were adjusted by a constant multiplication factor *viz.* 4 for this experiment. The dose for each successive animal was adjusted depending on the previous outcome. The acute toxicity and gross effect of crude methanolic extract of *A. aspera* was studied in albino mice by using $1/2\,LD_{50}$ dose. A total of six numbers of male albino mice were selected for each experiment. Animals were observed hourly for 6 h and again after 24 h. The parameters for motor activity and gross effect were determined after administration of *A. aspera* orally at a dose of 2.0 g kg$^{-1}$ body weight.

## *In vivo* Model

Excision wound model was used for studying wound healing activity of the plant in aged animals. The animals were randomly allocated into two groups of six animals each. Group I was assigned as control, Group II received topical application of 5 per cent, 2.5 per cent or 7.5 per cent ointment of *A. aspera* (w/w).

## Excision Wound Model in Aged Rat

Wounds were made by excising the full thickness circular skin (approximately 500mm$^2$) on the shaved back of the animal under ether anesthesia (Lee, 1950). Wound contraction was assessed by tracing the wound area on polythene paper first and then subsequently transferred to 1 mm$^2$ graph paper on day 7, 14 and 21. The percentage of wound contraction (taking the initial size of the wound, 500 mm², as 100 per cent) was calculated by using the formula:

[(Initial wound size – specific day wound size)/Initial wound size] × 100

## Experimental Design

A preliminary study was conducted for selection of the most effective concentration of methanol extract of *A. aspera* ointment by using 2.5, 5.0 and 7.5 per cent (w/w) ointment for topical application. As 5 per cent (w/w) ointment showed optimum wound healing activity, it was selected for further detail study.

The experimental animals (rats) were randomly allocated into groups of 6 animals each. Group I served as control and the rats received topical application of the vehicle, *i.e.* soft white petroleum jelly, and animals of group II received topical application of 5 per cent (w/w) ointment of *A. aspera* twice daily from day 1 till complete healing or 21ˢᵗ post-operative day, whichever is earlier.

## Estimation Biochemical Parameters

The collagen, elastin and hydroxyproline content in the granulation tissue were estimated as per standard method described below:

## Collagen

The collagen contents in tissues were estimated as per the method described by Robert *et al*. (1950). Five hundred mg (A) of finely mixed wet tissue with 10 ml of 0.1 N NaOH was kept overnight. It was then centrifuged and the supernatant was removed. To this, 10 ml distilled water was added and the pH was adjusted to 7.0 with 0.1 N HCl. It was again centrifuged and supernatant was removed. A mixture of 3 parts of 95 per cent alcohol and one part of the sample was allowed to stand for 10 minutes, centrifuged, supernatant removed, dried in oven at 100°C for overnight and allowed to cool. The tube was weighed with its contents (B), 5 ml of water was added and the tube was plugged with non absorbent cotton. It was autoclaved at 30 lb pressure for 4 hours, centrifuged and the supernatant was removed. The tube was dried at 100°C overnight, cooled and weighed (C). The quantity of the collagen was calculated as follows:

$$= \frac{(B - C) \text{ Weight of dry residue}}{A \text{ (Weight of the wet tissue)}} \times 100$$

## Elastin

The total elastin in tissues was estimated as per the method described by Robert *et al*. (1950). Five hundred mg (A) of finely mixed wet tissues were suspended in 0.1 N NaOH. The tube was placed in boiling water bath for 10 minutes. It was centrifuged and the supernatant was removed. The residue was washed with 10 ml distilled water, centrifuged, the supernatant was again removed and dried overnight at 100°C. It was cooled and weighed (B), the contents were taken out of the test tube and the empty tube was weighed (C). Quantity of the elastin was calculated as per the following formula:

$$= \frac{(B-C) \text{ Weight of dry residue}}{A \text{ (Weight of the wet tissue)}} \times 100$$

## Hydroxyproline

Hydroxyproline content in tissues was estimated using the method of Amma *et al.* (1969). Ten mg of dry tissue was placed in an ampoule. Two ml of 6 N HCl was added and it was incubated at 110°C for 18 hours. The ampoule was broken and a pinch of activated charcoal was added and allowed for 30 minutes. It was then filtered and neutralized with $NaHCO_3$ solution (pH 6.5 to 7.0). One ml of neutralized solution was taken in test tubes along with blank and 2 ml isopropyl alcohol was added to all the test tubes and mixed well. One ml of Chloramin T (7 per cent) and 2 ml of Ehrlich reagent (0.8g of para-dimethyl-aminobenzaldehyde + 30 ml of concentrated HCl + 30 ml of ethyl alcohol) was added. It was incubated at 60° C in hot water bath for 25 minutes and then allowed to cool. The optical density was measured at 560 nm in spectrophotometer (Chemito).

## Histopathological Study

For histological studies, granulation tissues collected on 21st day was fixed in 10 per cent neutral formalin solution and dehydrated with a sequence of ethanol-xylene series of solution. The materials were processed by conventional paraffin embedding method. Microtome sections were prepared at 6 µ thicknesses, mounted on glass slides, stained with hematoxylin and eosin (Lee and Luna, 1968) and Vangeison's stain, followed by observation for histopathological changes under light microscope.

## Statistical Analysis

Data were expressed as mean ± SE and statistical significance between experimental and control values were analyzed by one way ANOVA followed by Dunnett's test using Graph Pad Prism 2.01 (Graph Pad Software Inc., La Jolla, CA, USA). $P<0.05$ was considered statistically significant.

# Results

## Phytochemical Screening

Phytochemical screening of the methanolic extract of the plant revealed the presence of alkaloid by Wagner's and Dragendroff's test, steroid by Salkowski's and Lieberman Burchardt's test and triterpenes by Salkowski's and Lieberman Burchardt's test (Table 10.1).

## Determination of $LD_{50}$ and Acute Toxicity

In acute toxicity study, there was no change in motor activity and gross behaviour during 24 h of observation and the extract was found to be safe up to 2000 mg kg$^{-1}$ body weight, p.o. The low toxicity of the plant observed in this study suggests that the plant extract is safe and did not affect any of the parameters studied.

**Table 10.1**: Phytoconstituents present in the methanol extract of *A. aspera*.

| Active Principle | Tests Applied | A. aspera (MEAA) |
|---|---|---|
| Glycosides | NaOH test | − |
| | Benedict's test | − |
| Saponins | Foam test | − |
| Triterpenes | Salkowski test | + |
| | Liberman Burcharadt test | + |
| Alkaloids | Mayers test | − |
| | Waghner's test | + |
| | Hagers test | − |
| | Dragendorff's test | + |
| Phenolic compounds | | − |
| Tannins | a. $FeCl_2$ | − |
| | b. Gelatin test | − |
| Flavonoids | a. $FeCl_2$ test | − |
| | b. Lead acetate test | − |
| Diterpenes | | − |
| Steroids | Salkowski test | + |

## Excision Wound Model in Aged Rat

Significant contraction of the wound area was noticeable in the test group and healing was complete by 21st day. The percent wound contraction increased from $41.83 \pm 2.44$ and $77.24 \pm 5.95$ on day 7 to $87.83 \pm 3.35$ and $97.41 \pm 0.22$ on day 21, in control and *A. aspera* treated group respectively. Wound contraction was higher in *A. aspera* (5 per cent w/w) ointment treated group compared to the control group on different days of observations, indicating better wound healing activity of the test plant (Table 10.2).

**Table 10.2**: Effect of 5 per cent (w/w) ointment of *A. aspera* on percent wound contraction in excision wound model in aged rat (n =6, Mean ± SE).

| Treatment Groups | Per cent Wound Contraction at different Days of Observation | | |
|---|---|---|---|
| | 7 day | 14 day | 21 day |
| Control | A 41.83 ± 2.44 | A 76.33 ± 1.86 | A 87.83 ± 3.35 |
| A. aspera | A 77.24 ± 5.95 | B 94.78 ± 0.65 | B 97.41 ± 0.22 |

\* Mean in a column bearing same subscript and mean in a row bearing same superscript do not differ significantly (P<0.01), (n =6, Mean ± SE).

## Estimation of Biochemical Parameters

The percent collagen content of the granulation tissue in the treated rat increased from day of wounding till 21st day. The collagen content of the control animals as well as the animals treated with 5 per cent (w/w) ointment of *A. aspera*, increased from 7.83 ± 0.48 and 8.67 ± 0.49 per cent on day 7 to 13.17 ± 0.48 and 16.33 ± 0.49 per cent on day 21, respectively. Increase collagen content in the wound tissue in the *A. aspera* treated groups was evident in comparison to the control group (Table 10.3).

**Table 10.3**: Collagen, elastin and hydroxyproline content (per cent) in excision wound model in aged rats following topical application of 5 per cent (w/w) ointment of *A. aspera*.

| Treatment Groups | Days of Observations | Per cent Collagen | Per cent Elastin | Per cent Hydroxyproline |
|---|---|---|---|---|
| Control | 7 | $_A$7.83$^a$ ± 0.48 | $_A$2.67$^a$ ± 0.42 | $_A$29.67$^a$ ± 2.57 |
| | 14 | $_A$9.54$^a$ ± 0.43 | $_A$2.67$^a$ ± 0.42 | $_A$29.67$^a$ ± 2.57 |
| | 21 | $_A$13.17$^b$ ± 0.48 | $_A$2.67$^a$ ± 0.42 | $_A$29.67$^a$ ± 2.57 |
| A. aspera | 7 | $_A$8.67$^a$ ± 0.49 | $_B$3.33$^a$ ± 0.42 | $_A$34.83$^a$ ± 2.47 |
| | 14 | $_B$12.67$^b$ ± 0.42 | $_B$3.33$^a$ ± 0.42 | $_A$34.83$^a$ ± 2.47 |
| | 21 | $_B$16.33$^c$ ± 0.49 | $_B$3.33$^a$ ± 0.42 | $_A$34.83$^a$ ± 2.47 |

* Mean in a column bearing same subscript and mean in a row bearing same superscript do not differ significantly ($P<0.01$), (n =6, Mean ± SE).

Likewise, elastin content of the granulation tissue increased in the test group from day of wounding till 21$^{st}$ day. The elastin content of the control animals as well as the animals treated with 5 per cent (w/w) ointment of *A. aspera*, increased from 2.67 ± 0.42 and 3.33 ± 0.42 per cent on day 7 to 5.33 ± 0.49 and 6.33 ± 0.49 per cent on day 21. It was observed that the elastin content of the wound tissue was higher in the *A. aspera* treated groups compared to the control group (Table 10.3).

Another important parameter *i.e.* hydroxyproline content of the granulation tissue also increased from day of wounding till 21$^{st}$ day in *A.aspera* treated group. Hydroxyproline content of the control animals and *A. aspera* (5 per cent w/w ointment) treated group increased from 29.67 ± 2.57 and 34.83 ± 2.47 (mg/g) on day 7 to 36.17 ± 2.82 and 54.17 ± 3.16 (mg/g) on day 21, respectively (Table 10.3).

## Histopathological Study

Histopathological examination revealed necrosis and hemorrhage in the 7 day old granulation tissue in *A. aspera* treated group (Figure 10.1A), whereas 14 day old granulation tissue of the control group showed extensive necrosis on the wound surface (Figure 10.1B). On day 14, the granulation tissue of the treated group showed fibroblast and collagen fibers (Figure 10.1C). Collagen fibers and development of epidermis was observed on day 21 of the granulation tissue in *A. aspera* treated group (Figures 10.1D and E).

## Discussion

The present study clearly demonstrated that methanolic extract of *A. aspera* possessed a definite prohealing action in normal healing as observed by significant increase in the rate of wound contraction. Tissue repair and wound healing are

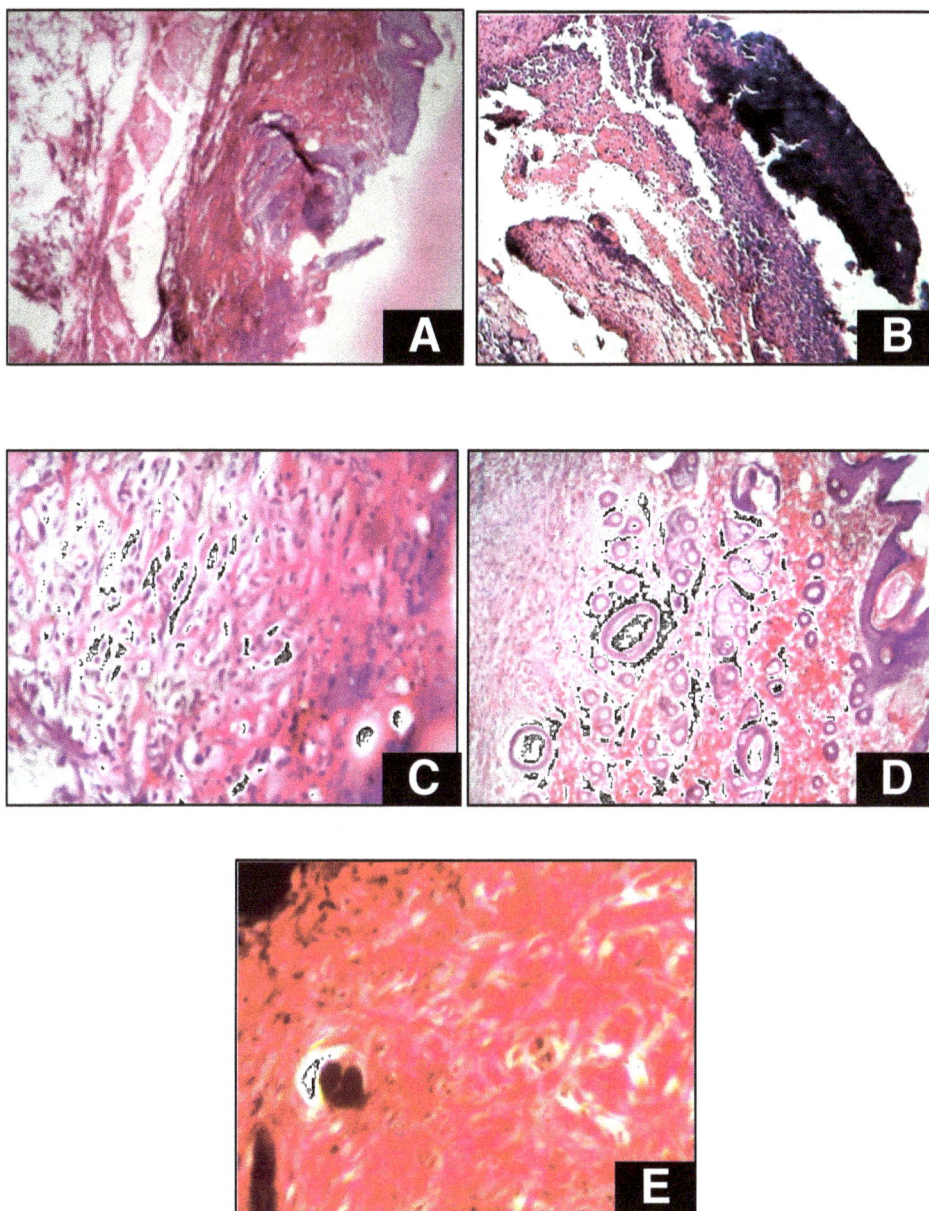

**Figure 10.1**: Photomicrograph showing histopathological changes of granulation tissue following topical application of 5 per cent (w/w) ointment of *Achyranthes aspera*.

(**A**) 7 day old wound showing necrosis and hemorrhage (H and E × 100); ( **B**) 7 day old wound of control group showing extensive necrosis on the wound surface (H and E × 100); (**C**) 14 day old wound showing fibroblast and collagen fibers (H and E × 400); ( **D**) 21 day old wound showing collagen fibers and epidermis (H and E × 100) and ( **E**) Photomicrograph of 21 day old wound showing collagen fibers (Vangeison's × 400).

complex processes that involve inflammation, fibroplasia, neovascularization, wound contraction, and resurfacing of the wound defect with epithelium. (Sidhu *et al.*, 1998). Wound contraction involves a complex and superbly orchestrated interaction of cells, extracellular matrix and cytokines. In the present study, extract treated wounds were found to contract much faster. Increased rate of wound contraction in *A. aspera* treated wounds might be due to increase in proliferation and transformation of fibroblast cells into myofibroblasts. Granulation, collagen maturation and scar formation are some of the many phases of wound healing, which run concurrently, but independent of each other (Udupa *et al.*, 2006). Collagen is a major extra cellular matrix protein which confers strength and integrity to the tissue matrix and plays an important role in homeostasis and in epithelialization at the later phase of healing (Upadhyay *et al.*, 2009). Collagen is also chemotactic for fibroblasts. (Sidhu *et al.*, 1998). Elastin-derived peptides significantly improve dermal regeneration. Increase in hydroxyproline content of the granulation tissue indicates rapid collagen turnover thus leading to rapid healing of wounds. In the present study, collagen, elastin and hydroxyproline content of the wound tissue was higher in the *A. aspera* treated groups compared to the control group on different days of observations indicating better wound healing activity of test plant.

Wound healing activity of methanol extract of *Achyranthes aspera* was studied by us in both normal as well as impaired wound models like diabetic or immunocompromised animals and in burn wound where healing is usually delayed. The plant showed significant and promising wound healing activity in all the models used which was evident by wound contraction, increase collagen, elastin and hydroxyproline content or elevated anti oxidant parameters and supported by histopathological observation better than the standard drug Himax, in the diabetic wound model (Barua *et al.*, 2011), burn wound model, (Barua *et al.*, 2011) and in immunocompromised model (accepted) as well. The test plant showed almost similar rate of percent wound contraction in diabetic, burn and immunocompromised wounds. Protein content of *A. aspera* treated granulation tissue was slightly higher in diabetic wound than in burn and immunocompromised wound. Antioxidant enzymes of *A. aspera* treated group *viz.* catalase, ascorbic acid, were comparatively more in immunocompromised wound model compared to burn and diabetic wound. But SOD level was more in burn wound model than in diabetic and immunocompromised wound model. Histopathological findings revealed that the test plant promoted wound healing in diabetic wound better than burn and immunocompromised wound. Thus our study on the test plant in various models of wound provide sufficient basis to develop this plant as a topical application for various types of impaired wounds as mentioned above.

In conclusion, from the present study, it can be interpreted that topical application of *Achyranthes aspera* has a positive influence on different phases of wound healing, including wound contraction, fibroblastic deposition and therefore, has a beneficial role even in healing of wound in aged rats as shown in the present study. Identification and elucidation of the active constituents of the plant might unveil many hitherto unknown molecules for drug development.

# Acknowledgements

The authors are grateful to Defence Research Development Organization (DRDO), Govt. of India, New Delhi for financial help and the Director of Research (Vety.), Khanapara for providing necessary facilities to carry out the research work.

# References

Amma, M.K., and Tandon, H.K. (1969). A micro method for hydroxyproline estimaton. *Indian Journal of Medical Research*, **57**: 1115–1121.

Aziz, A., Rahman, M., Mondal, A.K., Muslim, T., Rahman, A., and Quader, A. (2005) 3-Acetoxy-6-benzoyloxyapagamide from *Achyranthes aspera*. *Pharmaceutical Journal*, **4**: 1816-1820.

Barua, C.C., Begum, S.A., Talukdar, A., Barua, A.G., Borah, P., and Lahkar, M. (2010a). Effect of *Achyranthes aspera* on modified forced swimming in rats. *Pharmacologyonline*, **1**: 183-191.

Barua, C.C., Begum, S.A., Talukdar, A., Pathak, D.C., Sarma, D.K., and Bora, R.S. (2010c). Wound healing activity of methanolic extract of leaves of *Achyranthes aspera* Linn using *in vivo* and *in vitro* model–a preliminary study. *Indian Journal of Animal Science*, **80**: 969-972.

Barua, C.C., Begum, S.A., Talukdar, A., Pathak, D.C., Sarma, D.K., Bora, R.S., and Gupta, A. Influence of *Achyranthes aspera* Linn on healing of dermal wounds in diabetic rats. (2011). *Indian Journal of Veterinary Pathology*, **35**: 66.

Barua, C.C., Talukdar, A., Begum, S.A., Buragohain, B., Roy, J.D., Borah, R.S., and Lahkar, M. (2009). Antidepressant-like effects of the methanol extract of *Achyranthes aspera* Linn. in animal models of depression. *Pharmacologyonline*, **2**: 587-594.

Barua, C.C., Talukdar, A., Begum, S.A., Pathak, D.C., Sarma, D.K., Saikia, B. and Gupta A. (2011). *In vivo* wound healing efficacy and anti oxidant activity of *Achyranthes aspera* in experimental burns. *Pharmaceutical Biology*, DOI: 1031-109/12880209-2011-642885.

Barua, C.C., Talukdar, A., Begum, S.A., Sarma, D.K., Pathak, D.C., and Borah, P. (2010b). Antinociceptive activity of methanolic extract of leaves of *Achyranthes aspera* Linn. (Amaranthaceae) in animal models of nociception. *Indian Journal of Experimental Biology*, **48**: 817-821.

Chakraborty, A., Brantner, A., Mukuinaka, T., Nobukuni, Y., Kuchide, M., Konoshima, T., Tokuda, H., and Nishino, H. (2002). Cancer chemo preventive activity of *Achyranthes aspera* leaves on Epstein-Barr virus activation and two stage mouse skin carcinogenesis. *Cancer letters*, **177**: 1-5.

Harborne, J.B. (1991). Phytochemical methods-Guide to modern techniques of plant analysis. 2ⁿᵈ edn. Chapman and Hall, India.

Kartik, R., Rao, C.V., Trivedi, S.P., Pushpangadan, P., and Reddy, G.D. (2010). Amelioration effects against N-nitrosodiethylamine and $CCl_4$-induced hepatocarcinogenesis in Swiss albino rats by whole plant extract of *Achyranthes aspera*. *Indian Journal of Pharmacology*, **42:** 370-375.

Lee, G., and Luna, H.T. (1968). Manual of Histological Staining Methods of the Armed Forces Institute of Pathology. 3$^{rd}$ edn. American Registry of Pathology, The Blakiston Division, New York, Toronto, London, Sydney.

Robert, S., and Logan, A.M. (1950). The determination of collagen and Elastin in tissue. *Journal of Biological Chemistry*, **186:** 549–556.

Sandhyakumari, K., Boby, R.G., and Indira, M. (2002). Impact of feeding ethanolic extracts of *Achyranthes aspera* Linn. on reproductive functions in male rats. *Indian Journal of Experimental Biology*, **40:** 1307-1309.

Sidhu, G.S., Singh, A.K., Thaloor, D., Banaudha, K.K., Patnaik, G.K., Srimal, R.C. and Radha, K.M. (1998). Enhancement of wound healing by curcumin in animals. *Wound Repair Regeneration*, **6:** 167-177.

Udupa, S.L., Shetty, S., Udupa, A.L., and Somayaji, S.N. (2006). Effect of *Ocimum sanctum* Linn. on normal and dexamethasone suppressed wound healing. *Indian Journal of Experimental Biology*, **44:** 54.

Upadhyay, N.K., Kumar, R., Mandotra, S.K., Meena, R.N., Siddiqui, M.S., Sawhney, R.C., and Gupta, A. (2009). Safety and healing efficacy of Sea buckthorn (*Hippophae rhamnoides* L.) seed oil on dermal burn wounds. *Food and Chemical Toxicology*, **47:** 1146–1153.

Vetrichelvan, T., and Jegadeesan, M. (2003). Effect of alcohol extract of *Achyranthes aspera* Linn. on acute and sub-acute inflammation. *Phytotherapia Research*, **17:** 77-79.

Workineh, S., Eyasu, M., Legesse, Z., and Asfaw, D. (2006). Effect of *Achyranthes aspera* L. on fetal abortion, uterine and pituitary weights, serum lipids and hormones. *African Health Science*, **6:** 108-112.

Utilisation and Management of Medicinal Plants Vol. 2 (2014)     *Pages* **273–287**
*Editor-in-Chief:* **V.K. Gupta**
*Published by:* **DAYA PUBLISHING HOUSE, NEW DELHI**

# 11

# Chemoprofiling and Antioxidant Activity Studies of Shilajit: A Herbomineral Composition

Beena Joy[1]* and Zeena Pilla[2]

## ABSTRACT

*Studies on Shilajit had proved its potential as a source for bioactives. An attempt has been made to investigate in vitro antioxidant activities of different extracts of shilajit. Shilajit showed valid effects on the $\alpha$-glucosidase inhibition, but no marked effects on $\alpha$-amylase inhibition. This showed that the mechanism of action of shilajit on diabetics is not through $\alpha$-amylase inhibition mechanism. In vitro antioxidant studies showed that water extract of shilajit is a powerful free radical quencher. The chemoprofiling of the active extracts of shilajit was performed using HPTLC. The HPTLC chromatogram showed the presence of different highly polar organic compounds in the active extracts.*

***Keywords:*** Shilajit, Chemoprofiling, HPTLC, Free radical quencher, Antioxidant activity, Antidiabetic effects.

## Introduction

Shilajit (Figure 11.1) is the most potent rejuvenator and anti aging block buster ever known to mankind. Attributed with many magical properties, shilajit is found predominantly in Himalayan region bordering India, China, Tibet and parts of central

1   Agroprocessing and Natural Product Division, National Institute for Interdisciplinary Science and Technology, Thiruvananthapuram, Kerala, India.
2   Department of Chemistry, Amrita School of Arts and Sciences, Amrita Vishwa Vidyapeetham, Amritapuri, Kollam – 690 525, Kerala, India.
*   *Corresponding author*: E-mail: bjoy1571@gmail.com

**Figure 11.1**: Rock shilajit in its raw form.

Asia. The existence and use of shilajit was a closely guarded secret of Yogis of Himalayas for many centuries. The Indian Yogis considered shilajit as God's gift and nectar of longevity. Ancient Indian scriptures mention the wonderful powers of shilajit in tackling most of the ailments of body and mind.

## Origin of Shilajit (Suraj *et al.*, 2007)

Shilajit, is a concentrated historic plant life found commonly in the Himalayan region, encompassing India, China, Tibet and parts of Central Asia. The plants absorb various nutrients and minerals from the soil to form rich and green vegetation. The process continues over a period of many centuries. These remains of the plant life in the specific climatic condition and altitude of Himalayas formed the mineral pitch known as shilajit.

In Sanskrit the literal meaning of shilajit is "Rock Like"–the power to make our body like a rock enabling it to withstand the ravages of time. Shilajit has the unparalleled powers of arresting and reversing the aging process. Indian Yogis on seeing the powers of shilajit considered it to have divine powers capable of healing the body from ailments and above all preserve youthfulness of mankind. Shilajit commonly called, as shilajitu in ayurvedic terms is a kind of resin that oozes out from Himalayan Mountains due to heating effect of sun in summer. It is pale brown to blackish brown in colour. This resin is soft in texture, slimy to touch, pure and heavy. It is soluble in water. Other names by which shilajit is famous are asphaltum, mineral pitch, and girij.

Shilajit contains more than 85 minerals in ionic form and fulvic acids. The fulvic acid concentration in shilajit is between 60 to 75 per cent. Shilajit is an end product of plant matter that has decomposed centuries ago and got dumped in mountains and due to pressure, got preserved in mountains. Shilajit is found in Himalayan region especially in Nepal, Bhutan and Tibet and Kumaon area in India. Srivastava *et al.*

(1988) reported the presence of fulvic acid, humic acid, uronic acids, hippuric acid, benzopyrones, phenolic glycosides and amino acids in shilajit.

Mukhtar *et al*. (1999) reported that shilajit is widely used in preparation of ayurvedic medicines and is regarded as one of the most important ingredient in ayurvedic system of medicine. It is a part of the famous ayurvedic medicines like chandraprabha vati, arogya vardhani vati, and most important ingredient of all Chawanprashas. It works as a powerful antioxidant thereby delaying aging. Shilajit is an asphalt-like substance found embedded in rocky sediments in the Himalayas in western Nepal at altitudes between 2500-5000 m. It is popularly used in Nepal as a tonic. Chemical analysis of shilajit revealed that two-thirds by weight of this medicinal material was extractable by warm 50 per cent alcohol. Repeated crystallization of the hydroalcoholic extract has led to the isolation of crystals, which were subsequently identified as calcium benzoate. Mukhtar *et al*. (1999) reported the antiseptic properties of benzoates may account for the antiseptic effects of shilajit in places where hygiene remains at a low level.

A process was developed for the bioconversion of 3-OH-DBP to $3,8(OH)_2$-DBP using microorganisms. The biotransformation of 3-OH-DBP is achieved using *Aspergillus niger*, which was involved in the humification process on sedimentary rocks leading to "shilajit" formation. Aminul *et al*. (2008) narrated the isolation and identification *Aspergillus* species from native "shilajit".

Fulvic acids (FA) and 4'-methoxy-6-carbomethoxybiphenyl (MCB), two major organic compounds isolated from shilajit were screened for anti-ulcerogenic activity by Shibnath *et al*. (2006) in albino rats.

The effect of shilajit and its main active constituents fulvic acids, 42-methoxy-6-carbomethoxybiphenyl and 3,8-dihydroxy-dibenzo-alpha-pyrone were studied in relation to the degranulation and disruption of mast cell against noxious stimuli. Shilajit and its active constituents provided satisfactory significant protection to antigen-induced degranulation of sensitized mast cells markedly inhibited the antigen induced spasm of sensitized guinea-pig ileum and prevented mast cell disruption. Ghosal *et al*. (1989) reported the therapeutic use of shilajit in the treatment of allergic disorders. Tiwari *et al*. (2001) studied the development of tolerance to morphine induced analgesia in Swiss mice using shilajit. The effect of shilajit (a herbomineral preparation) on blood glucose and lipid profile in euglycemic and alloxan-induced diabetic rats and its effects on the above parameters in combination with conventional antidiabetic drugs were studied. Trivedi *et al*. (2004) reported the effectiveness of shilajit in controlling blood glucose level and the lipid profile.

Wang *et al*. (1996) reported that shiajith act as an immunomodulatory agent. White blood cells activity was studied and monitored prior to and intervals after receiving the shilajit extract. Recent investigations by Carlos *et al*. (2012) pointed to an interesting medical application of shilajit towards the control of cognitive disorders associated with aging, and cognitive stimulation. Anti-microbial, anti-oxidant and anti-ulcerogenic effects of shilajit on gastric ulcer in rats were studied by El-Sayed *et al*. (2012). Modern scientific research by Kishor *et al*. (2012) has proved that shilajit is truly a panacea in cancer therapy. It is used as a fertility agent. The effects of shilajit on spermatogenesis and ovogenesis were studied using male and female rats.

Jeong-Sook *et al.* (2006) reported that shilajit had both a spermiogenic and ovogenic effect in mature rats. Anuya *et al.* (2012) studied the antioxidant activity of aqueous extract of shilajit using three *in vitro* parameters, and anti-arthritic activity by proteinase inhibitory assay. El-Sayed *et al.* (2012) studied on shilajit samples collected from different locations to evaluate their possible role as antiulcerogenic and antiinflammatory agents.

Jaiswal and Bhattacharya (1992) studied the effect of shilajit for putative, nanotropic and anxiolytic activity in Charles Foster strain of albino rats. The biochemical studies carried out for the level of monoamines indicated that acute treatment with shilajit had an insignificant effect on rat brain monoamines and monoamine metabolite levels. The observed neuro chemical studies on shilajit indicate a decrease in rat brain 5-hydroxytryptamine turnover, associated with an increase in dopaminergic activity leading to an increase in memory and anxiolytic activity in albino rats.

The recent review on shilajit by Bhaumik *et al.* (1993) focused on the cancer chemopreventive and therapeutic properties of shilajit and humic compounds. Shilajit and HA possess anti-inflammatory, antioxidant, antimutagenic, antitoxic, antiviral, heavy metal chelating, antitumor, apoptotic and photo-protective properties. These properties make shilajit useful agents for cancer therapy and prevention.

# Materials and Methods

## Material Collection

Shilajit was bought with the help of agents from the mountain areas of Himalaya.

## Extraction Procedures

About 8g of the sample (shilajit) was subjected to cold extraction using various solvents such as water and methanol and stored below 4 °C in a refrigerator.

## Chemicals and Reagents

Chemicals and reagents used were Gallic acid, sodium carbonate, 2,2-diphenyl-2-picryl hydrazyl (DPPH), 2,2'-Azino-bis-ethyl benzothiazoline-6-sulfonic acid (ABTS), Trolox, hexane, ethyl acetate, methanol, Folin-ciocalteu reagent, concentrated sulphuric acid, Aluminium Chloride, Quercetin, sodium nitrate, sodium hydroxide, Dinitrosalicylic acid, Phenazine metho sulphate, Nitrotetrazolium Blue Chloride, Nicotinamide, Adenine Dinucleotide Disodium salt.

## Antioxidant Studies of Shilajit

### Total Phenolic Content (TPC)

The total phenol content was determined by Folin-Ciocalteu method. To the different concentration of the samples (aqeous solutions) 0.5 mL of Folin-Ciocalteu reagent and 1mL $Na_2CO_3$ was added. The mixture was vortexed for 2 minutes and absorbance was determined after 90 min at 760 nm against standard gallic acid. Determinations were done in triplicate. The linearity of the method was checked from 10 to 250 µg/mL of gallic acid (dissolved in methanol/water 1: 1). Folin-Ciocalteu's reagent does not contain phenol. Rather, the reagent will react with phenols and non-phenolic reducing substances to form chromogens that can be detected spectrophotometrically (Shimadzu UV –Vis-Spectrophotometer, Model 2450).The Total Phenolic Content was expressed as:

Per cent TPC = (Observed concentration/Actual concentration) × 100

## DPPH Radical Scavenging Capacity

The scavenging activity against DPPH radical was evaluated according to the method of Brand Williams. The assay mixture contained 2.5 mL of 130 µM DPPH (final concentration 83.3 µM) dissolved in absolute methanol. Samples at different concentrations were added to DPPH and the final volume was adjusted to 3 mL.The mixture was shaken vigorously on a Vortex mixer then incubated for 90 min at ambient temperature in dark, after which the absorbance of the remaining DPPH was determined at 517 nm against a blank. The scavenging activity was expressed as $IC_{50}$ (µg/mL).

The per cent of scavenging activity was concluded as:

Per cent scavenging activity= [(AC-AS)/AC] × 100

Where AC is the absorbance of control (without extract) and AS is the absorbance of the sample. Percentage of radical scavenging activity was plotted against corresponding concentration of the extract to obtain $IC_{50}$ value. $IC_{50}$ is defined as the amount of antioxidant material required to scavenge 50 per cent of the free radical in the assay system. The $IC_{50}$ values are inversely proportional to the antioxidant activity.

## ABTS Cation Decolorisation Capacity

A graph was plotted with concentration of ABTS radical cation generated by adding 7 mM ABTS and 2.4 mM potassium persulphate together in 10 mL ethanol and was allowed to stand for overnight in the dark at room temperature and the absorbance was maintained at 0.65-0.75 range. Several concentrations of the sample solution were prepared to 25 µL with adequate amount of methanol and 225 µL of ABTS was added. Control was prepared without sample. Trolox was used as the standard. The solutions were allowed to stand for 7 minutes. Absorbances were noted at 734 nm and a graph was plotted with concentration along x-axis and absorbance along y-axis. The percentage radical scavenging activity was determined using the formula,

Per cent RSA= [(A$_0$-A$_S$)/A$_0$] × 100

Where A$_0$ is the absorbance of control without tested samples and A$_S$ is the absorbance of tested samples. A graph was plotted with concentration along x-axis and percentage radical scavenging activity along y-axis and $IC_{50}$ values were calculated.

## Superoxide Radical Scavenging Activity

Super oxide anions were generated in a non-enzymatic phenazine methosulphate-nicotinamide adenine dinucleotide (PMS-NADH) system through the reaction of PMS, NADH and oxygen.100 mM Tris HCl buffer ($P^H$ 7.4) (1.58 g tris HCl buffer in 100 mL distilled water), 300 mM NBT (Nitroblue tetrazolium) (4.9 mg/

20 mL distilled water), 120μM, PMS (0.74 mg/20 mL tris HCl buffer), 936 mM NADH (13.28 mg/20 mL tris HCl buffer) etc were prepared. Various concentrations of the samples were taken and made up to 1 mL using methanol and added 250 μl NBT, 250μl NADH, and 250 μl PMS. Absorbance was taken at 560 nm using a spectrophotometer. The percentage radical scavenging capacity was determined with the formula,

$$\text{Per cent RSC} = [(A_0 - A_s)/A_0] \times 100$$

Where $A_0$ is the absorbance of control without tested samples and $A_s$ is the absorbance of tested samples. A graph was plotted with concentration along x-axis and percentage scavenging activity along y-axis and $IC_{50}$ values were calculated from the graph.

## Anti-diabetic Studies

### α-Glucosidase Inhibitory Assay

The assay is a modification of the procedure of Pistia-Brueggeman and Hollingsworth. α-glucosidase was premixed with 20μl of EtOAc extract and fractions of that extract at varying concentrations made up in 50mM phosphate buffer at pH 6.8 and incubated for 5 min at 37°C.1mM PNPG (20μl) in 50 mM of phosphate buffer was added to initiate the reaction and the mixture was further incubated at 37°C for 20 min. The reaction was terminated by the addition of 50 μl of 1M $Na_2CO_3$, and the final volume was made up to1500 μl. α-glucosidase activity was determined spectrophotometrically at 405 nm by measuring the quantity of p-nitrophenol released from PNPG. In this assay the enzyme used is α-glucosidase and the substrate is PNPG. α-glucosidase reacts on PNPG and release p-nitrophenol. If our sample possesses inhibitory activity on the enzyme then the amount of p-nitrophenol formed will be low and hence the absorbance is also low. The assay was performed in triplicate. The concentration of the extract required to inhibit 50 per cent of α-glucosidase activity under the assay conditions was defined as the $IC_{50}$ value.

### α-Amylase Inhibition Assay

0.5 mg ml$^{-1}$ of α-amylase was incubated with 250 μL samples. 250 μL of one percent starch was used as substrate. The samples without α-amylase were used as controls and the test reading were subtracted from the absorbance of these controls. The reducing sugar was estimated using DNS assay at 540 nm. The percentages of inhibition shown by the samples were calculated.

## HPTLC Profiling of Various Extracts of Shilajit

Readymade HPTLC plates (kieselgel 60 F 254, 20 cm x 20 cm, 0.2 mm thickness, Merck, Darmstadt Germany) were kept at 60°C for 5 minutes. The various extracts of shilajit were spotted in the form of bands of width 6 mm with Hamilton micro liter syringe using a Camang Linomat V (Switzerland). A constant application rate of 0.1 ml/s was employed and the space between two bands were maintained as 5 mm.The plates were developed in an ascending manner with solvent system, Ethyl acetate: Methanol: Water (7.5: 2: 0.5), Butanol: Acetic acid: Water (5: 4: 1) mixture in a developing chamber presaturated with the mobile phase. The developed phase were dried and scanned with in the wavelength range of 250-300 nm (TLC scanner 3,Camag,Switzerland). Data processing was performed using the software WinCATS planar chromatography manager.

# Results and Discussion

## Total Phenolic Content (TPC)

The total phenolic content of the extracts were determined by Foiln-Ciocalteau method. Gallic acid was used as the standard. The total phenolic content was expressed as gallic acid equivalents (GAE) per gram of dry material. Polyphenolic compounds are known to have antioxidant activity and it is likely that the activity of the extracts is due to these compounds (Djeridane *et al.*, 2006)

The two extracts of shilajit namely methanol and water were analysed for its total phenolic content. In the case of shilajit, water extract has higher total phenolic content than methanol extract and the values are 2.4121mg/g, 1.7298 mg/g respectively and are shown in Table 11.1 and in Graph 11.1.

**Table 11.1**: Total phenolic content of extracts of shilajit.

| Sl.No. | Name of the Sample | Name of the Extract | TPC (mg/g Equivalent of Gallic Acid) |
|--------|--------------------|--------------------|--------------------------------------|
| 1. | Shilajit | Methanol | 1.7298 |
| | | Water | 2.4121 |

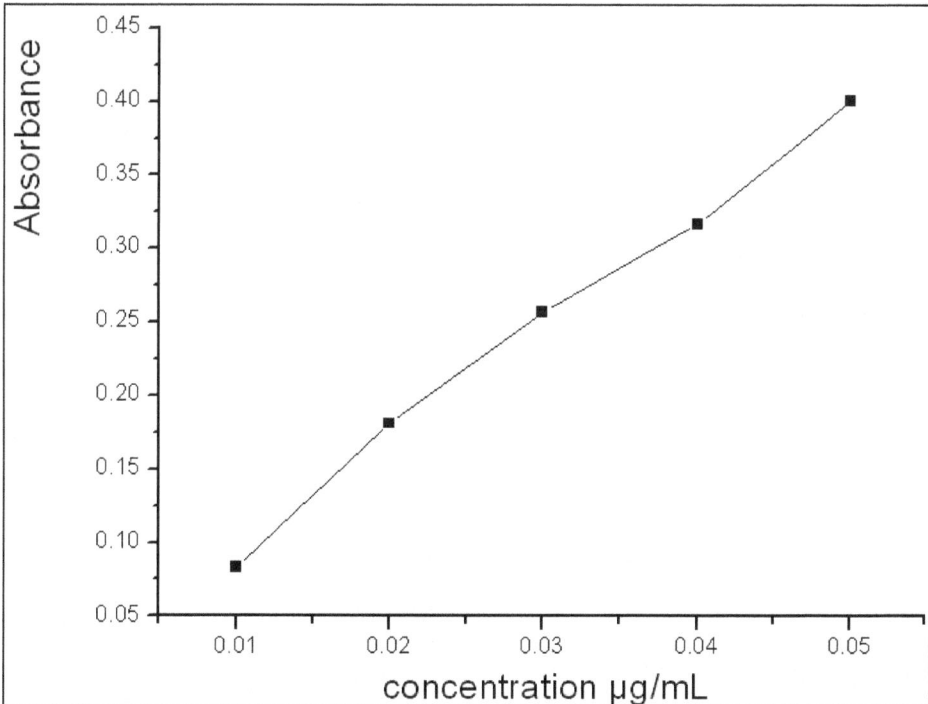

**Graph 11.1**: Total phenolic content for standard gallic acid.

## DPPH Scavenging Activity

The experiment revealed the concentration of shilajit required for the 50 per cent inhibition ($IC_{50}$) of DPPH radical. In the case of shilajit extracts $IC_{50}$ values are 1308 µg/mL and 323.66 µg/mL. Comparison of the scavenging activity with the standard Gallic acid was summarized in Table 11.2 and in Graphs 11.2 and 11.3.

**Table 11.2**: DPPH evaluation of different extracts of shilajit.

| Sl.No. | Name of the Sample | Name of the Extract | $IC_{50}$ Values (µg/mL) |
|---|---|---|---|
| 1. | Gallic acid | | 0.7674 |
| 2. | Shilajit | Methanol | 1308 |
| | | Water | 323.66 |

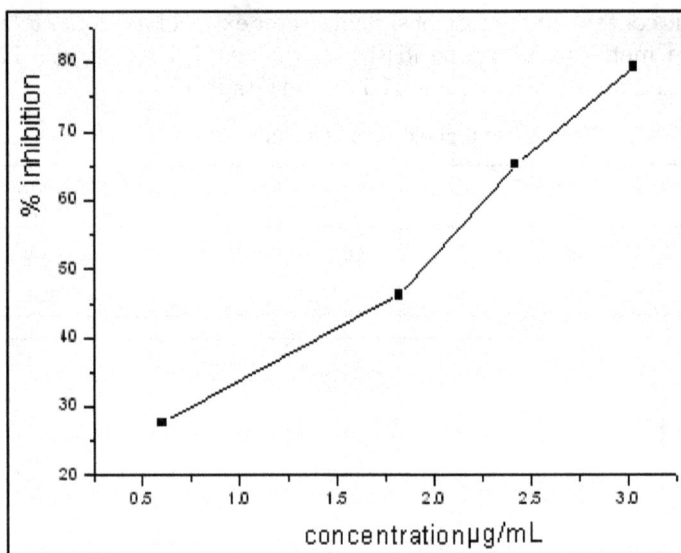

**Graph 11.2**: DPPH assay for the MeOH extract of shilajit.

## ABTS Cation Decolorisation Capacity

The experiment yielded the concentration of various extracts of shilajit required for the 50 per cent inhibition ($IC_{50}$) of ABTS. Shilajit methanol and water extracts showed $IC_{50}$ values 232.822 µg/mL and 86.115 µg/mL The details of the results are shown in Table 11.3 and Graphs 11.4 and 11.5.

**Table 11.3**: Evaluation of ABTS cation decolourisation capacity of extracts of shilajit.

| Sl.No. | Name of the Sample | Name of the Extract | $IC_{50}$ Values (µg/mL) |
|---|---|---|---|
| 1 | Shilajit | Methanol | 232.82 |
| | | Water | 286.115 |

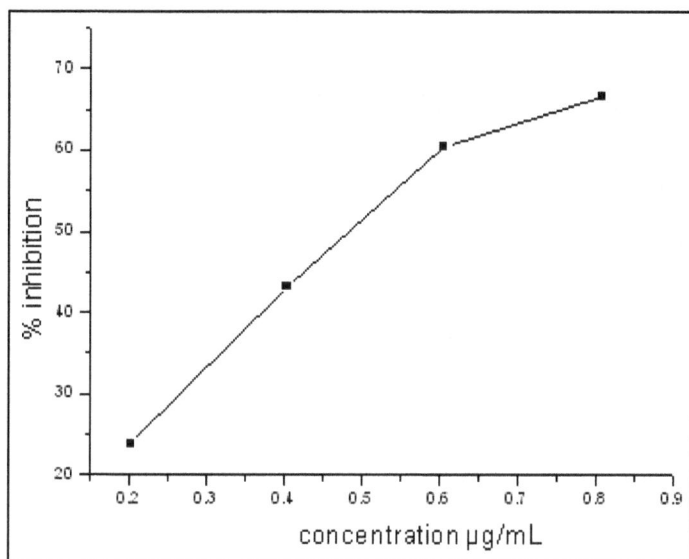

**Graph 11.3**: DPPH assay for the water extract of shilajit.

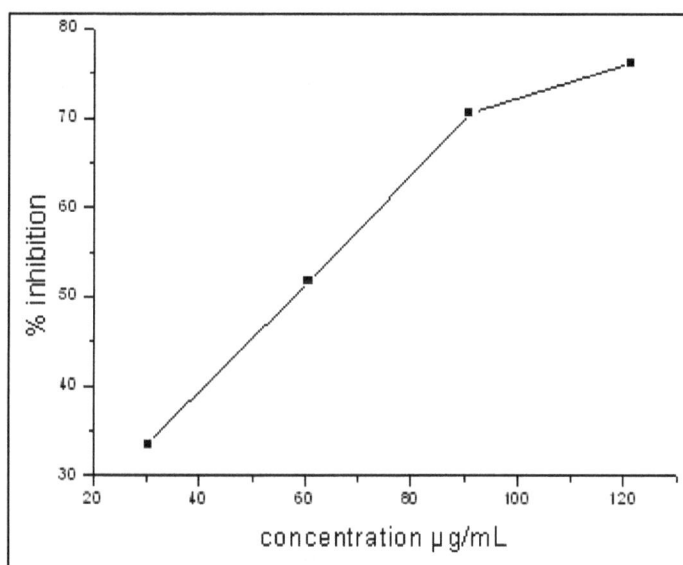

**Graph 11.4**: ABTS assay for the MeOH extract of shilajit.

## Superoxide Assay

Water extract of shilajit shows less $IC_{50}$ value (141.0882 µg/mL) than methanol extract (269.177 µg/mL). This implies that water extract shows high superoxide radical scavenging activity than methanol extract. In all the different antioxidant assays performed, water extract of shilajit is more active than methanol extract. And thus it is evident that a water soluble component of shilajit is responsible for its efficacy. The results are shown in Table 11.4 and Graphs 11.6 and 11.7.

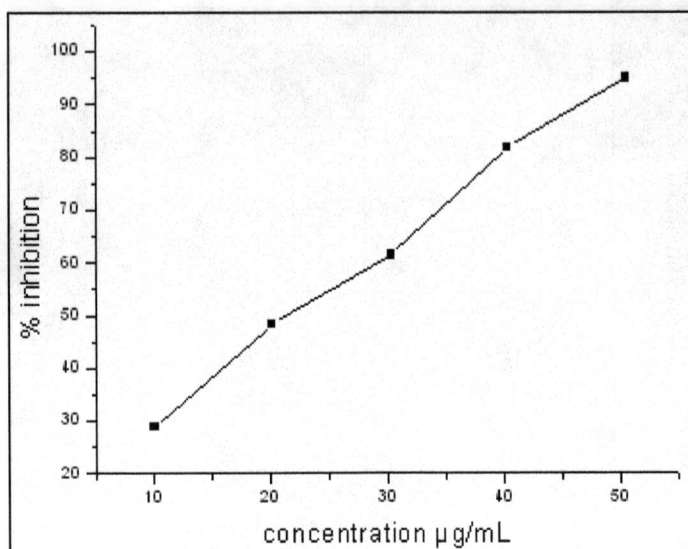

**Graph 11.5**: ABTS assay for the water extract of shilajit.

**Table 11.4**: Evaluation of superoxide radical scavenging activity of different extracts of shilajit.

| Sl.No. | Name of the Sample | Name of the Extract | $IC_{50}$ Values (µg/mL) |
|---|---|---|---|
| 1. | Shilajit | Methanol | 269.177 |
| | | Water | 141.0882 |

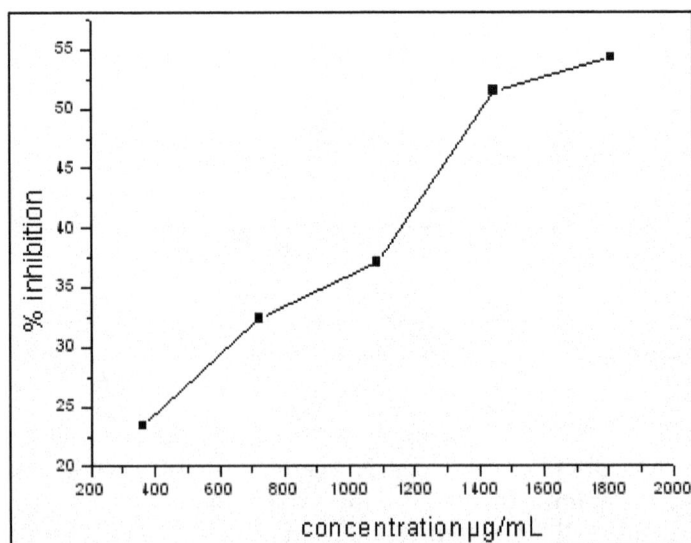

**Graph 11.6**: Superoxide assay for the MeOH extract of shilajit.

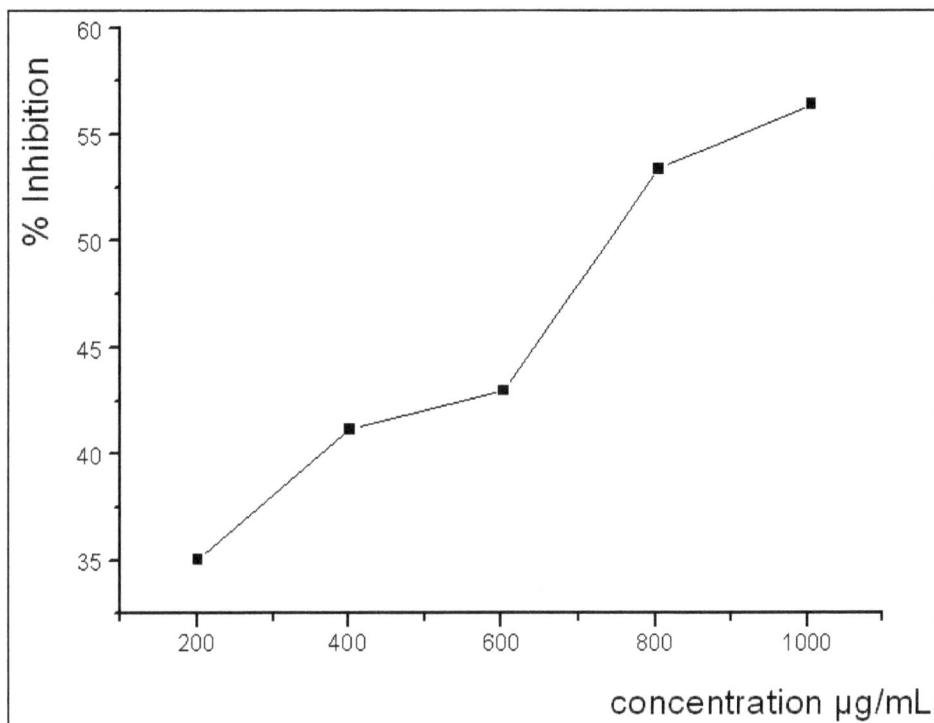

**Graph 11.7**: Superoxide assay for the water extract of shilajit.

## α-glucosidase Inhibitory Assay

Only water extract of shilajit is positive towards α-glucosidase inhibitory assay. The $IC_{50}$ is very low, 5.33 μg/mL. And hence shilajit is a potent antidiabetic natural herbomineral product. The results are summarized in Table 11.5 and in Graph 11.8.

**Table 11.5**: α-Glucosidase inhibition of extracts of shilajit.

| Sl.No. | Name of the Sample | Name of the Extract | $IC_{50}$ Values (μg/mL) |
|--------|--------------------|--------------------|--------------------------|
| 1 | Shilajit | Methanol | – |
| | | Water | 5.33 |

## α-Amylase Inhibition Assay

No significant result was obtained in the case of α-Amylase Inhibition of shilajit (Table 11.6).

## HPTLC Profiling of Shilajit

In order to have an idea of the chemical components present in the active extract HPTLC analysis was performed. The stadardisation of the HPTLC developing solvents for best separation of the constituents of shilajit extracts were carried out.

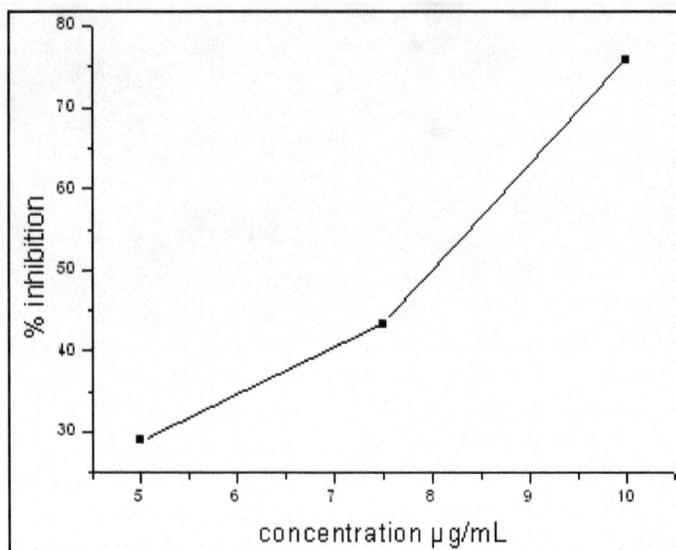

**Graph 11.8**: α-Glucosidase assay for the water extract of shilajit.

HPTLC profiling of both methanol and water extract of shilajith were optimized using different solvent systems. Ethyl acetate: Methanol: Water (7.5: 2: 0.5), Butanol: Acetic acid: Water (5: 4: 1) mixture were found to be the best mobile phases for best separation of the components of shilajit. The HPTLC absorption chromatogram for methanol extract of shilajit (Figure 11.2) and water extract (Figure 11.3) showed the presence of different highly polar compounds

**Table 11.6**: α-Amylase inhibition of extracts of shilajit.

| Sl.No. | Name of the Sample | Name of the Extract | $IC_{50}$ Values (µg/mL) |
|--------|--------------------|--------------------|--------------------------|
| 1. | Shilajit | Methanol | -- |
| | | Water | |

## Conclusions

Shilajit is an ancient medicine and wonder drug of the Ayurveda used by the Ayurveda physicians for the treatment of several diseases. Charaka Samhita recommends Shilajit as a cure for numerous human ailments and is highly recommended by the Ayurvedic practitioners. Shilajit has immense therapeutic properties and certainly can be proved to be a boon for mankind possibly in cancer prevention and therapy. Earlier studies on shilajit had proved its potential as a source for bioactives. An attempt has been made to investigate *in vitro* antioxidant activities of different extracts of shilajit. Shilajit gave valid effects on the α-glucosidase inhibitory properties, but no marked effects on α-amylase inhibitory assay. This showed that the mechanism of action of shilajit on diabetic is not through α -amylase

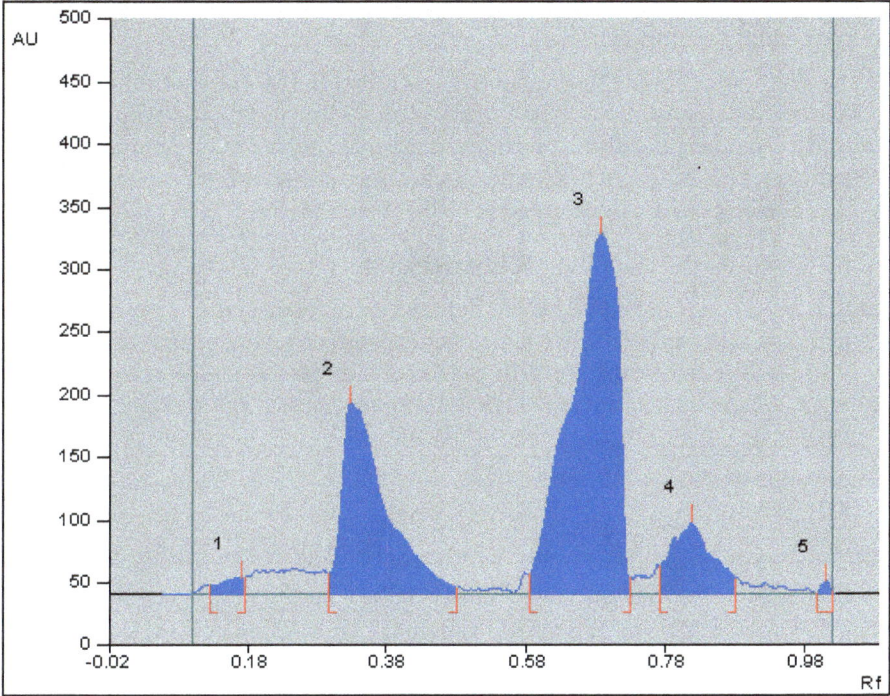

**Figure 11.2**: HPTLC profiling of the MeOH extract of shilajit.

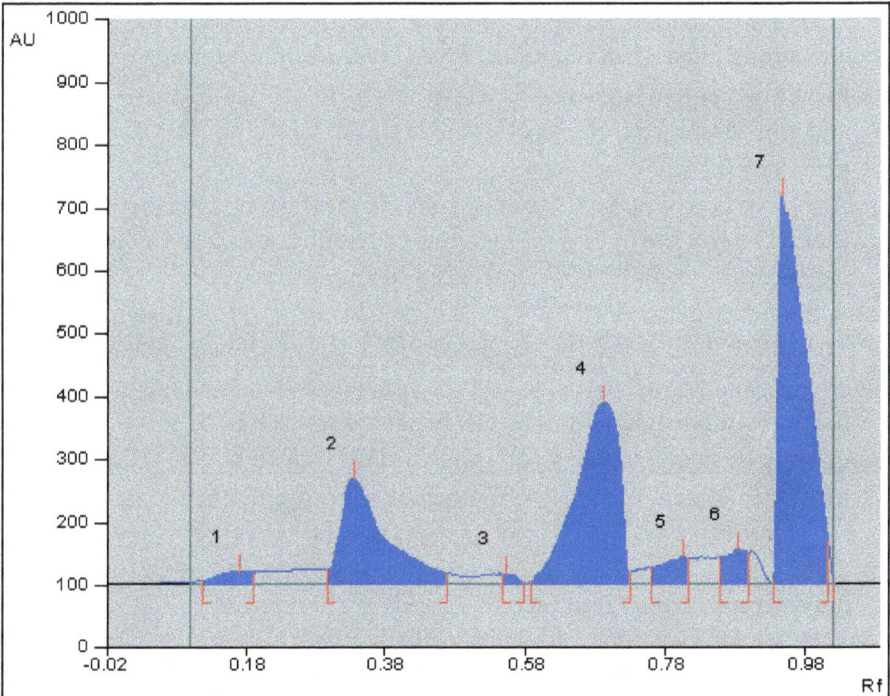

**Figure 11.3**: HPTLC profiling of the water extract of shilajit.

inhibition. In all the *in vitro* antioxidant studies performed, water extract of shilajit showed the highest efficacy.

The chemoprofiling of these active extracts by HPTLC showed the presence of different organic compounds. The isolation of these compounds was tried using column chromatography and it is experienced that high grade separating material like Sephadex LH- 20 is needed for the separation of these highly polar bioactives. The work currently is under progress at NIIST, Trivandrum.

# References

Aminul Islam., Runa Ghosh., Dipankar Banerjee., Piali Nath., Upal Kanti Mazumder., and Shibnath Ghosal. (2008). Biotransformation of 3-hydroxydibenzo-α-pyrone into 3,8 dihydroxydibenzo-α-pyrone and aminoacyl conjugates by *Aspergillus niger* isolated from native"shilajit". *Electronic Journal of Biotechnology*, **11**: 3.

Anuya Rege., Juvekar, P., and Juvekar, A. (2012). *In Vitro* antioxidant and anti arthritic activities of shilajit. *International Journal of Pharmacy and Pharmaceutical Sciences*, **4(2)**: 650.

Bhaumik, S., Chattapadhay, S., and Ghosal, S. (1993).Effects of shilajith on mouse peritoneal macrophages. *Phytother.*, **7**: 425.

Carlos Carrasco-Gallardo., Leonardo Guzmán., and Ricardo B. Maccion (2012). Shilajit: a natural phytocomplex with potential procognitive activity. *International Journal of Alzheimer's Disease*, Article ID 674142, 1-4.

Djeridane, A., Yousfi, M., Nadjemi, B., Boutassouna, D., Stocker, P., and Vidal, N. (2006). Antioxidant activity of some algerian medicinal plants extracts containing phenolic compounds. *Food Chemistry*, **97**: 654–660.

El-Sayed, M.I.K., Amin, H.K., and Al-Kaf, A.G. (2012). Anti-microbial, anti-oxidant and anti-ulcerogenic effects of shilajit on gastric ulcer in rats. *Am. J. Biochem. Biotechno.*, **8**: 26-39.

Ghosal, S., Lal, J., Singh, S.K., Dasgupta, G., Bhaduri, J., Mukopadhyay, M., and Bhattacharya, S.K. (1989). Mast cell protecting effects of shilajit and its constituents. *Phytotherapy Res.*, **3**: 249-252.

Ghosal, S., and Mukherjee, B. (1993). Shilajit Its origin and vital significance. In Traditional Medicine. Ed6. Oxford and IBH, New Delhi, pp. 308–319.

Jaiswal, A.K., and Bhattacharya, S. K.(1992). Effects of Shilajit on memory, anxiety and brain monoamines in rats. *Indian J Pharmacol*, **24**: 12–17.

Jeong-Sook, Park., Gee-Young, Kim., and Kun, Han. (2006). The spermatogenic and ovogenic effects of chronically administered shilajit to rats. *Journal of Ethnopharmacology*, **107**: 349-53.

Kishor, Pant., Bimala, Singh., and Nagendra, Thakur. (2012) Shilajit: A humic matter panacea for cancer. *International Journal of Toxicological and Pharmacological Research*, **4** : 17-25.

Mukhtar, H., and Ahmad, N. (1999). Cancer chemoprevention: future holds in multiple agents: contemporary issues in toxicology. *Toxicol Appl Pharmacol.*, **158**: 207-10.

Shibnath,Ghosal., Sushil, K., Singh,Yatenthra., Kumar, Radheyshyam., Srivastava, Raj. K.Goel., Radharaman, Dey., and Salil, K. Bhattacharya. (2006). Antiulcerogenic activity of fulvic acids and 4'-methoxy-6-carbomethoxy biphenyl isolated from shilajit. *Phytotherapy Research*, **2**: 187-191.

Srivastava, R.S., Kumar, Y., Singh, S.K., and Ghosal, S. (1988). Shilajit, its source and active principles. Proc 16 IUPAC (Chemistry of Natural Products), p. 524.

Suraj, P. Agarwal., Rajesh, Khanna., Ritesh, Karmarkar., Md. Khalid, Anwer., and Roop, K. Khar. (2007). Shilajit: a review. *Phytotherapy Research*, **21**: 401 – 405.

Tiwari, P., Ramarao, P., and Ghosal, S. (2001). Effects of Shilajit on the development of tolerance to morphine in mice. *Phytother Res.*, **15**: 177-179.

Trivedi, N.A., Mazumdar, B., Bhatt, J. D., and Hemavathi, K.G. (2004). Effect of shilajit on blood glucose and lipid profile in alloxan induced diabetic rats. *Indian J Pharmacol.*, **36**: 373–376.

Wang, C., Wang, Z., Peng, A., Hou, J., and Xin, W. (1996). Interaction between fulvic acids of different origins and active oxygen radicals. *Sci China C Life Sci.*, **39**: 267–275.

Utilisation and Management of Medicinal Plants Vol. 2 (2014)    *Pages* **289–296**
*Editor-in-Chief:* **V.K. Gupta**
*Published by:* **DAYA PUBLISHING HOUSE, NEW DELHI**

# 12

# Antioxidative Potential of Two Darjeeling Himalayan *Marchantia* sp.: *M. paleacea* and *M. papillata*

Souryadeep Mukherjee[1]*, Abhijit Dey[2], Arijit De[3] and Pinky Ghosh[3]

## ABSTRACT

*Bryophytes are considered as a natural storehouse of novel biomolecules with an array of diverse bioactivity. Presently, an investigation is performed to elucidate the antioxidative activity of two species of the liverwort genus Marchantia viz. M. paleacea and M. papillata (Marchantiaceae) collected from Darjeeling Himalaya, West Bengal, India. The crude methanolic extracts of the two hepatics were evaluated for potential antioxidative activity by using DPPH (2,2-diphenyl-1-picrylhydrazyl) radical scavenging activity and by estimating the total phenolic and flavonoid contents. M. paleacea and M. papillata showed highest DPPH activities of 73.28 ± 4.51 per cent and 71.35 ± 3.25 per cent respectively. The total phenolic content in 1mg of the extracts of M. paleacea and M. papillata were 47 ± 0.81 µg and 35 ± 0.74 µg of gallic acid equivalent respectively, and the flavonoid content in 1mg of the extracts were 150 ± 1.3 µg and 40 ± 0.67 µg of quercetin equivalent respectively. The present study indicates the antioxidative properties of two Himalayan Marchantiophyta members for the first time and the probable use of the same as a natural antioxidant with therapeutic and cosmetic values.*

*Keywords*: Antioxidant activity, DPPH, Flavonoids, *M. paleacea, M. papillata*, Methanolic extract, Phenolics.

1   Assistant Professor, Department of Zoology, Presidency University, 86/1, College Street, Kolkata – 700 073, West Bengal, India

2   Assistant Professor, Department of Botany, Presidency University, West Bengal, India

3   Researchers, Department of Zoology, Presidency University, West Bengal, India

*   *Corresponding author*: E-mail: souryadeep@rediffmail.com

# Introduction

Reactive oxygen species (ROS) are generated in the body due to a number of internal and external responses and the damage caused by the free radicals is prevented by the antioxidants. Natural antioxidants are reported for therapeutic ability in a number of investigations. Biomolecules of herbal origin have been used in medical as well as cosmetic purpose (Bhattarai *et al.*, 2008). Natural products such as phenolics (Hansakul *et al.*, 2011), flavonoids (Pietta, 2000), carotenoids (Carranco Jáuregui *et al.*, 2011) etc. are reported to possess antioxidative potential. Dietary source of antioxidants have been evaluated in a number of investigations (Sharmin *et al.*, 2011; Avignon *et al.*, 2012; Abe *et al.*, 2012) and the neutraceutical value of certain edibles has been established. Recent trends to evaluate the antioxidants against a number of ailments such as cancer (Miura *et al.*, 2012), diabetes (Wu *et al.*, 2012), Parkinson's disease (Sutachan *et al.*, 2012), atherosclerosis (Lönn *et al.*, 2012), kidney disease (Rojas-Rivera *et al.*, 2012) and many others indicate the therapeutic ability of these compounds.

Medicinal plants are used to treat a number of diseases (Dey and De, 2012a, b) and have shown to possess pharmacological efficacy (Dey *et al.*, 2011; Mukherjee *et al.*, 2012). Medical herbalism have gain popularity because of comparatively cheap price, less or no side effects and less occurrence of drug resistance which are otherwise present in case of using conventional synthetic drugs. Bryophytes, classified into liverworts, hornworts and mosses, are known to house a number of secondary metabolites with diverse bioactivity. This tiny and apparently insignificant group of non vascular plants are known to exhibit antimicrobial (Dey and De, 2011), antioxidative (Krzaczkowski *et al.*, 2009), cytotoxic (Dey and De, 2012d) and other activities. Bryophytes are known to store a number of secondary metabolites because of their constant combat against a number of biotic and abiotic stresses (Xie and Lou, 2009). In the present investigation, the authors report the antioxidant activity of two species of eastern Himalayan *Marchantia viz. M. paleacea* and *M. papillata* for the first time as a part of their present venture involving pharmacological exploration of the bryophytes present in the same area.

# Materials and Methods

## Plant Material

Two species of *Marchantia viz. M. paleacea* and *M. papillata*, were collected from Darjeeling district of the eastern Himalayas (Lava, West Bengal, India, altitude 7200 ft, coordinates 27°5'11"N and 88°39'47"E) at different seasons during the years 2010-2011. The thallus of the two species was washed with tap water immediately to remove the soil. The voucher specimen was identified from the key to the specimen (Singh and Singh, 2009) and deposited at the Department of Zoology and Molecular Biology, Presidency University, Kolkata, India.

## Extract Preparation

Before extraction, the plant materials were first washed with detergent Teepol® followed by Bavistin to remove microbial contamination. Final rinsing was done in sterilized distilled water. The plant material was air dried and powdered (40 mesh size) using an electrical mixer grinder. Extraction was done by soaking 1 gm of dried powder in 80 per cent methanol for 96 hours. Filtration was done using Whatmann's no.1 filter paper. The extract was concentrated using a rotary evaporator and was diluted in 80 per cent methanol to make a final solution of 10mg/ml and stored at +4°C.

## Free Radical Scavenging Activity using DPPH

The ability of the samples to scavenge the free radicals was estimated by *in vitro* method using a stable nitrogen centered radical *viz.* 2,2-diphenyl-1-picrylhydrazyl (DPPH). Determination of the free radical scavenging capacity or antioxidant potential of the extracts may be correlated to their effectiveness in the prevention, interception and repair mechanism against injury in a biological system. 0.5 ml of extract (0.0625 to 1 mg/ml) was added to 3 ml of 0.1 mM solution of DPPH. The mixture was incubated at room temperature for 20 minutes in dark. Optical density was measured at 517 nm with a Systronics PC Based Double Beam Spectrophotometer 2202. $IC_{50}$ value (the concentration required to scavenge 50 per cent DPPH free radicals) was also calculated. DPPH scavenging activity (per cent) was determined by using the formula:

$$\text{DPPH scavenging activity} = [(A_C - A_S)/A_C] \times 100$$

*where*, $A_C$ is the absorbance value of the control and $A_S$ is the absorbance value of the added test samples solution.

## Estimation of Total Phenolics

Total phenolics content of the 80 per cent methanolic extract of *M. paleacea* and *M. papillata* was determined by Folin-Ciocalteu method as described by Liu *et al.* (2008) with the following modifications. 0.5ml of extract (1mg/ml) was mixed with 5 ml of Folin-Ciocalteu reagent (1: 10 dilution) and 4 ml of 1(M) $Na_2CO_3$ and incubated at room temperature for 15 mins. Optical density was measured at 765nm (Systronics PC Based Double Beam Spectrophotometer 2202). The obtained result was plotted against the standard curve prepared using a concentration gradient of Gallic acid (10μg/ml – 100 μg/ml) and total phenolics contents of the extracts were expressed as mgGAE (gallic acid equivalent).

## Estimation of Total Flavonoid Content

Total flavonoid was estimated following the method described by Jia *et al.* (1999) with some modification. 250 μl of the plant extract (1mg/ml) was mixed with 1.25 ml double distilled $H_2O$ followed by the addition of 75 μl 5 per cent sodium nitrate solution. The mixture was left undisturbed for 5 min., 150 μl 10 per cent ammonium chloride solution was added and incubated at room temperature for another 6mins. 500μl of 1M NaOH was added and the resultant mixture was diluted with 275 μl of distilled water. Optical density of the solution was measured at 510 nm (Systronics PC Based Double Beam Spectrophotometer 2202). The same protocol was followed to obtain a quercetin standard curve using a concentration gradient of 100 to 500 μg/ml of quercetin in methanol. The obtained data was plotted against this standard curve and total flavonoid contents of the extracts were expressed as mgQE (quercetin equivalent).

## Statistical Analysis

All the experiments were carried in triplicates. The experimental results were expressed as mean ± standard error of mean (SEM) of the three replicates. Linear regression analysis was used to calculate the $IC_{50}$ values. Result processing was done by using the Microsoft Excel worksheet, 2007.

# Results

## Free Radical Scavenging Activity using DPPH

DPPH scavenging activity of the 80 per cent methanolic extracts of *M. paleacea* and *M. papillata* showed considerable amounts of radical scavenging activity. *M. paleacea* and *M. papillata* showed highest DPPH activities of 73.28 ± 4.51 per cent and 71.35 ± 3.25 per cent respectively at a concentration of 1 mg/ml and beyond (Figure 12.1). The $IC_{50}$ values of the two extracts were 0.25 ± 0.01 mg and 0.31 ± 0.02 mg respectively. *M. paleacea* showed slightly higher DPPH scavenging activity than that of *M. papillata*.

## Estimation of Total Phenolics and Total Flavonoid Content

The total phenolic content in 1 mg of the 80 per cent methanolic extracts of *M. paleacea* and *M. papillata* were 47 ± 0.81 µg and 35 ± 0.74 µg of gallic acid equivalent respectively, and the flavonoid content in 1mg of the extracts were 150 ± 1.3 µg and 40 ± 0.67 µg of quercetin equivalent respectively (Figure 12.2). The results indicate considerable amounts of phenolic and flavonoid contents in the both the tested extracts, which may be interpreted as the reason for the significant radical scavenging activity of the same. However, in both the experiments, *M. paleacea* showed a higher content of both phenolics and flavonoids than that of *M. papillata*.

# Discussion

The present investigation indicates the antioxidatve potential of two Darjeeling Himalayan *Marchantia* sp. *viz. M. paleacea* and *M. papillata*. Earlier, *M. paleacea* has been reported for allelopathic potential (Wang *et al.*, 2012) and vasorelaxant effects (Morita *et al.*, 2011). Furthermore, bioactive macrocyclic bis(bibenzyls) are reported from the species (So *et al.*, 2002; Morita *et al.*, 2011). Bis(bibenzyls) and bibenzyls are common in liverworts (Nagashima and Asakawa, 2011; Asakawa, 2012) and are known to show an array of biological activities (Labbé *et al.*, 2007; Qu *et al.*, 2007) due to which they are considered as one of the probable candidates to be used as potential medicinal source (Asakawa, 2008). On the other hand, *M. papillata* has got almost no attention regarding pharmacological and phytochemical investigations as reflected from the well known literature. *M. polymorpha*, considered as the most exploited species of the genus, is also known for its bioactivity (Kámory *et al.*, 1995; Kanamoto *et al.*, 2009).

Keeping in mind the tremendous phytochemical diversity and pharmacological efficacy of the genus, it is going to be exciting to evaluate the related liverworts species for potential bioactivity and bioactivity guided purification and isolation of novel compounds with therapeutic ability. Our experimentation gains momentum from

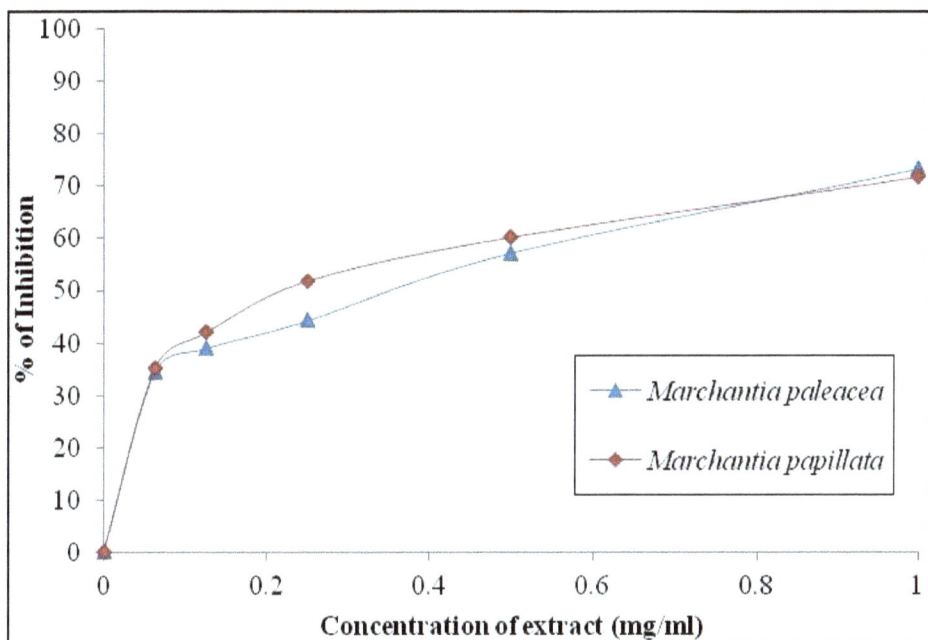

**Figure 12.1**: DPPH Radical scavenging activity of methanolic extracts of *M. paleacea* and *M. papillata*.

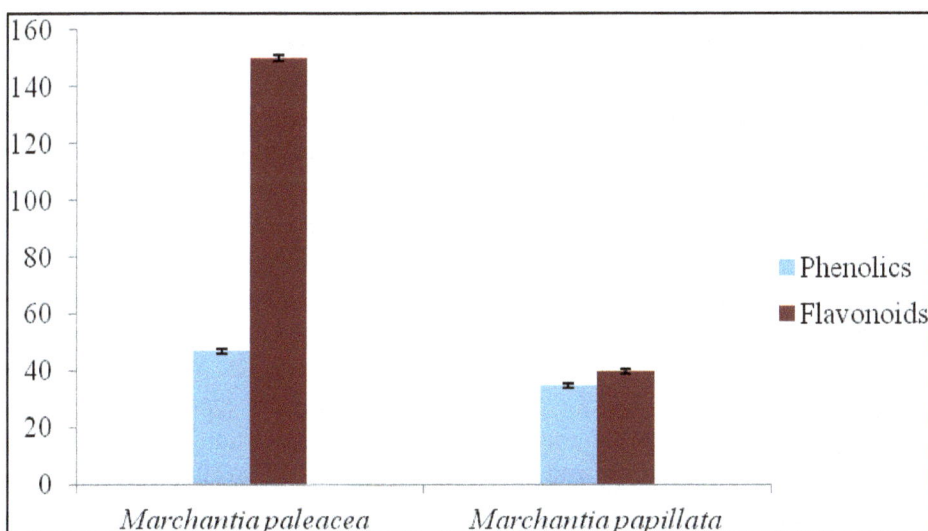

**Figure 12.2**: Total phenolics and total flavonoid contents of methanolic extracts of *M. paleacea* and *M. papillata*.

the fact that, although *M. polymorpha* is well known for its biological properties, so far so little has been found in the other species of the same genus. Moreover, the Indian Himalayas are known as the natural habitat for a number of bryophytes, most of which are unknown both phytochemically as well as pharmacologically. In the modern era of alternative and complementary treatment and medical herbalism, the plant based wealth may serve as a natural source for novel biomolecules as potential candidates to alleviate human morbidity and mortality.

# References

Abe, L.T., Lajolo, F.M., and Genovese, M.I. (2012). Potential dietary sources of ellagic acid and other antioxidants among fruits consumed in Brazil: jabuticaba (*Myrciaria jaboticaba* (Vell.) Berg). *J Sci Food Agric.*, **92**: 1679-1687.

Asakawa, Y. (2008). Liverworts-potential source of medicinal compounds. *Curr Pharm Des.*, **14**: 3067-3088.

Asakawa, Y., Ludwiczuk, A., and Nagashima, F. (2012). Phytochemical and biological studies of bryophytes. *Phytochemistry*, online published.

Avignon, A., Hokayem, M., Bisbal, C., and Lambert, K. (2012). Dietary antioxidants: Do they have a role to play in the ongoing fight against abnormal glucose metabolism? *Nutrition*, **28**: 715-721.

Carranco Jáuregui, M.E., Calvo Carrillo Mde, L., and Romo, F.P. (2011). Carotenoids and their antioxidant function: a review. *Arch Latinoam Nutr.*, **61**: 233-241.

Dey, A., and De, J.N. (2011). Antifungal bryophytes: a possible role against human pathogens and in crop protection. *Research Journal of Botany*, **6**: 129-140.

Dey, A., and De, J.N. (2012a). Anti snake venom botanicals used by the ethnic groups of Purulia District, West Bengal, India. *Journal of Herbs, Spices and Medicinal Plants*, **18**: 152-165.

Dey, A., and De, J.N. (2012b). Ethnomedicinal plants used by the tribals of Purulia district, West Bengal, India against gastrointestinal disorders. *Journal of Ethnopharmacology*, **143**: 68-80.

Dey, A., and De, J.N. (2012c). Antioxidative potential of Bryophytes: Stress tolerance and commercial perspectives: A Review. *Pharmacologia*, **3**: 151-159.

Dey, A., Das, T., and Mukherjee, S. (2011). *In vitro* antibacterial activity of n-Hexane fraction of methanolic extract of *Plumeria rubra* L. (Apocynaceae) stem bark. *Journal of Plant Sciences*, **6**: 135-142.

Hansakul, P., Srisawat, U., Itharat, A., and Lerdvuthisopon, N. (2011). Phenolic and flavonoid contents of Thai rice extracts and their correlation with antioxidant activities using chemical and cell assays. *J Med Assoc Thai.*, **94**: S122-S130.

Jia, Z., Tang, M., and Wu, J. (1999). The determination of flavonoid contents in mulberry and their scavenging effects on superoxide radicals. *Food Chem.*, **64**: 555-599.

Kámory, E., Keserü, G.M., and Papp, B. (1995). Isolation and antibacterial activity of marchantin A, a cyclic bis(bibenzyl) constituent of Hungarian *Marchantia polymorpha*. *Planta Med.*, **61**: 387-388.

Kanamoto, H., Takemura, M., and Ohyama, K. (2009). Detection of 5-lipoxygenase activity in the liverwort *Marchantia polymorpha* L. *Biosci Biotechnol Biochem.*, **73**: 2549-2551.

Krzaczkoswki, L., Wright, M., Reberioux, D., Massiot, G., Etievant C. and Gairin J.E. (2009). Pharmacological screening of bryophyte extracts that inhibit growth and induce abnormal phenotypes in human HeLa cancer cells. *Fundamental and Clinical Pharmacology,* **23**: 473-482.

Labbé, C., Faini, F., Villagrán, C., Coll, J., and Rycroft, D.S. (2007). Bioactive polychlorinated bibenzyls from the liverwort *Riccardia polyclada. J Nat Prod.,* **70**: 2019-2021.

Liu, X., Zhao, M., Wang, J., Yang, B., and Jiang, Y. (2008). Antioxidant activity of methanolic extract of emblica fruit (*Phyllanthus emblica* L.) from six regions in China. *J Food Compt Anal.,* **21**: 219-228.

Lönn, M.E., Dennis, J.M., and Stocker, R. (2012). Actions of "antioxidants" in the protection against atherosclerosis. *Free Radic Biol Med.,* online published.

Morita, H., Zaima, K., Koga, I., Saito, A., Tamamoto, H., Okazaki, H., Kaneda, T., Hashimoto, T., and Asakawa, Y. (2011). Vasorelaxant effects of macrocyclic bis(bibenzyls) from liverworts. *Bioorg Med Chem.,* **19**: 4051-4056.

Mukherjee, S., Dey, A., and Das, T. (2012). *In vitro* antibacterial activity of n-hexane fraction of methanolic extract of *Alstonia scholaris* L. R.Br. stem bark against some multidrug resistant human pathogenic bacteria. *European Journal of Medicinal Plants,* **2**: 1-10.

Nagashima, F., and Asakawa, Y. (2011). Terpenoids and bibenzyls from three Argentine liverworts. *Molecules,* **16**: 10471-8.

Pietta, P.G. (2000). Flavonoids as antioxidants. *J Nat Prod.,* **63**: 1035-1042.

Qu, J., Xie, C., Guo H., Yu, W., and Lou, H. (2007). Antifungal dibenzofuran bis(bibenzyl)s from the liverwort *Asterella angusta. Phytochemistry,* **68**: 1767-1774.

Rojas-Rivera, J., Ortiz, A., and Egido, J. (2012). Antioxidants in kidney diseases: the impact of bardoxolone methyl. *Int J Nephrol.,* **2012**: 321714.

Sharmin, H., Nazma, S., Mohiduzzaman, M., and Cadi, P.B. (2011). Antioxidant capacity and total phenol content of commonly consumed selected vegetables of Bangladesh. *Malays J Nutr.,* **17**: 377-383.

Singh, S.K., and Singh, D.K. (2009). Hepaticae and Anthocerotae of Great Himalayan National Park and its environs (HP), India. Botanical Survey of India.

So, M.L., Chan, W.H., Xia, P.F., and Cui, Y. (2002). Two new cyclic bis(bibenzyl)s, isoriccardinquinone A and B from the liverwort *Marchantia paleacea. Nat Prod Lett.,* **16**: 167-171.

Sutachan, J.J., Casas, Z., Albarracin, S.L., Stab, B.R. 2nd, Samudio, I., Gonzalez, J., Morales, L., and Barreto, G.E. (2012). Cellular and molecular mechanisms of antioxidants in Parkinson's disease. *Nutr Neurosci.,* **15**: 120-126.

Suzuki, K., Ohno, S., Suzuki, Y., Ohno, Y., Okuyama, R., Aruga, A., Yamamoto, M., Ishihara, K.O., Nozaki, T., Miura, S., Yoshioka, H., and Mori, Y. (2012). Effect of green tea extract on reactive oxygen species produced by neutrophils from cancer patients. *Anticancer Res.*, **32**: 2369-2375.

Wang, L., Wang, L.N., Zhao, Y., Lou, H.X., and Cheng, A.X. (2012). Secondary metabolites from *Marchantia paleacea* calluses and their allelopathic effects on Arabidopsis seed growth. *Nat Prod Res.*, online published.

Wu, H., Xu, G., Liao, Y., Ren, H., Fan, J., Sun, Z., and Zhang, M. (2012). Supplementation with antioxidants attenuates transient worsening of retinopathy in diabetes caused by acute intensive insulin therapy. *Graefes Arch Clin Exp Ophthalmol.*, online published.

Utilisation and Management of Medicinal Plants Vol. 2 (2014)    *Pages* **297–307**
*Editor-in-Chief:* **V.K. Gupta**
*Published by:* **DAYA PUBLISHING HOUSE, NEW DELHI**

# 13

# Economic Botany of the Amazonian Copaiba Oil (*Copaifera* spp.) from the Late Twentieth Century to the Beginning of the Twenty-First Century

F.E.B. Herculano[1,2], F.S. Folhadela[1,2], F.R. Silva[1,2,4],
F.A. Pieri[2,3], and V.F. Veiga Junior[2]*

## ABSTRACT

*Since the time of Brazilian colonization, copaiba oil (Copaifera spp) has been part of the production calendar of the native peoples, who used the plant for a wide range of purposes, particularly medicinal, and who still use it today, centuries later. The oil is extracted from the tree trunks, and in economic terms, provides a livelihood for families that still survive using resources from the Amazon rainforest. Based on a survey methodology and analysis of socioeconomic data involved in the economic base of botanical oil in Brazil, this study investigates the theme over a twenty-one year period, from 1990 to 2010. Based on the data, it was observed that oil production from the genus Copaifera did not occur throughout the country during this period, or even throughout the states of the Brazilian Amazon, the region where this industry was highest. For the twenty-one year period as a whole, Brazil produced a total of 7,172t of the oil, of which 6,986t (97.4 per cent) came from the North region, with the state of Amazonas accounting for 90.4 per cent of this amount, mainly trough the production*

1    Center for Research Analysis and Technological Innovation Foundation, FUCAPI, Manaus-AM, Brazil.

2    Institute of Exact Sciences, Federal University of Amazonas, Av. Rodrigo Octavio, 6200-Japiim, 69077-000, Manaus-AM, Brazil,

3    Leonidas and Maria Deane Institute, Oswaldo Cruz Foundation, Manaus-Am, Brazil

4    In Memoriam.

*    *Corresponding author*: E-mail: valdirveiga@ufam.edu.br

*of two of its municipalities (Apuí and Novo Aripuan ), which together contributed 84.5 per cent of Brazil's total production of the oil. From 1990 to 2010, domestic production of the oil leapt from 93 to 580 tons, an increase of 523.7 per cent, with the highest annual growth occurring in 1996 when there was a sudden 287.5 per cent increase over the previous year. In 2010, the total value of Brazilian extraction of the oil reached U.S. $ 4.9 million ($ 2.9 million), of which 87.3 per cent was produced by the state of Amazonas. This gives an idea of the economic scale of this industry in that state, particularly in the two municipalities where production is highest.*

*Keywords*: Brazil, Amazon, *Copaifera*, Oilresin, Botanic economy, Vegetable production.

## Introduction

Historically–the process of European occupation of the Brazilian Amazon is closely linked to the exploitation of the resources of the biodiversity of the region, particularly rubber, but also other products such as "spices, drugs from interior, turtle lard, medicinal plants, timber, fruits, animals, fish [.] "(Benchimol, 2009).

In the beginning, the indigenous peoples probably made use of the resources of the forests and rivers to meet their basic survival needs, then began to generate a surplus with economic value, as society became monetized, in order to facilitate the trade of goods and services with other communities.

Today, the range of economic assets of the Amazon biodiversity that are generating income for the region is vast, and includes many species of timber, fish, fruit, rubber, essential and fixed oils, fibers, dyes, spices and medicinal plants, among others.

In this context, copaiba oil has a role to play, therefore it is important to study the scale of this industry in the region, as one of the bases of support for the populations of the Brazilian Amazon.

More than two hundred and fifty studies have been published on the *Copaifera* genus, in several languages, with about two hundred of these being published in the last ten years, highlighting the importance of the genus, and the current interest in its properties among the scientific community. The genus comprises seventy-two species, which are distributed across the tropical regions of the world, mainly in Latin America and West Africa. Of these, sixteen (22.2 per cent) species have only been identified in Brazil (Index Kewensis, 1996 *apud* Veiga Junior and Pinto, 2002).

The copaiba tree, despite its potential for timber due to its large size, has become more widespread because of the oil it exudes from its trunk. This oil has been used by man for numerous applications, but its most traditional use is rooted in Amazonian folk medicine, where it is used for its prophylactic action and to treat various diseases (Pieri *et al.*, 2009).

As a result, the medical aspect, in particular, has prompted ongoing chemical and biochemical studies, seeking to gain a better knowledge of the constituents of this oil and its mechanisms of action against certain human and veterinary pathogens (Santos *et al.*, 2012; Izumi, 2012, Pieri *et al.*, 2010; Santos *et al.*, 2008; Veiga Jr *et al.*, 2007;

Veiga Jr *et al.,* 2006). However its use is not limited to the field of health; it is also used in many other areas, including cosmetics, paints, varnishes, soaps, perfumes and even as a lighting fuel, among others (Pieri *et al.,* 2009).

One of the most striking features of copaiba oil is the variation in chemical composition, not only between different species, but also within the same species. This has contributed to a lack of standardization of the product (Veiga Junior and Pinto, 2002).

The aim of this study was to evaluate the importance of the botanical extractive economy of copaiba oil (*Copaifera* spp.) in the main producing states and municipalities of Brazil, through an analysis of data for the period 1990 to 2010.

## Material and Methods

The work was conducted through an analysis of available data in the databases of the Institute of Applied Economic Research (IPEA) between the years 1991 and 2010, of the Brazilian Institute of Geography and Statistics (IBGE) between the years 1990 and 2010, of and Planning and Economic Secretary of the Amazonas State (SEPLAN) between the years 2002 and 2009, which are available on their websites. The data were processed and calculated using the software program Microsoft Excel®.

## Results and Discussion

The genus *Copaifera*, although cited in the literature as being present throughout the Brazilian territory, is not historically associated with significant levels of oil production, even in the states of the Amazon, and this activity is only recorded as occurring sporadically in a few states of the Northeast.

Looking at the consolidated data from the IBGE–Brazilian Institute of Geography and Statistics for twenty-one years of monitoring, from the last decade of the Twentieth Century (1990) to the first decade of the Twenty-First Century (2010), it is seen that Brazilian production of copaiba oil experienced highs and lows until 1995, but then rose steadily from 1996 to 2010. According to accumulated IBGE data for the period 1990 to 2010, Brazil produced a total of 7,172t, of which 6,986t originated in the North Region (97.4 per cent), 178t in the Midwest region (2.5 per cent), and only 8t in the Northeast region (0.1 per cent), as shown in Figure 13.1 (IBGE, 2012).

Analyzing the extreme years of the 1990/2010 series, Brazilian domestic production of copaiba oil rose from 93t in 1990 to 580t in 2010, a relative increase of approximately 523.7 per cent. Furthermore, the biggest leap in production within this period occurred in 1996, when it rose from 72t in 1995 to 279t, representing an increase of 287.5 per cent. It is also noted that in 2006, domestic production exceeded 500t/year, representing another important achievement in the history of extraction of the oil (Figure 13.2).

A retrospective analysis, based on accumulated IBGE data on the production of Brazilian copaiba oil from 1990 to 2010, by state, reveals that Amazonas state contributed the largest share, with around 90.4 per cent, followed by Para state with 4.9 per cent, Mato Grosso with 2.5 per cent, Rondônia state with 2.0 per cent, and the other states accounting for just 0.2 per cent (Table 1).

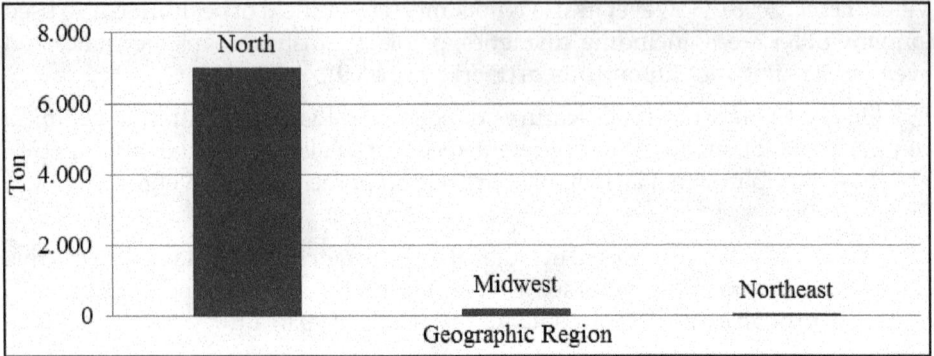

**Figure 13.1**: Brazil–Cumulative production of oil Copaiba by Region–1990-2010.
*Source*: IBGE (2012).

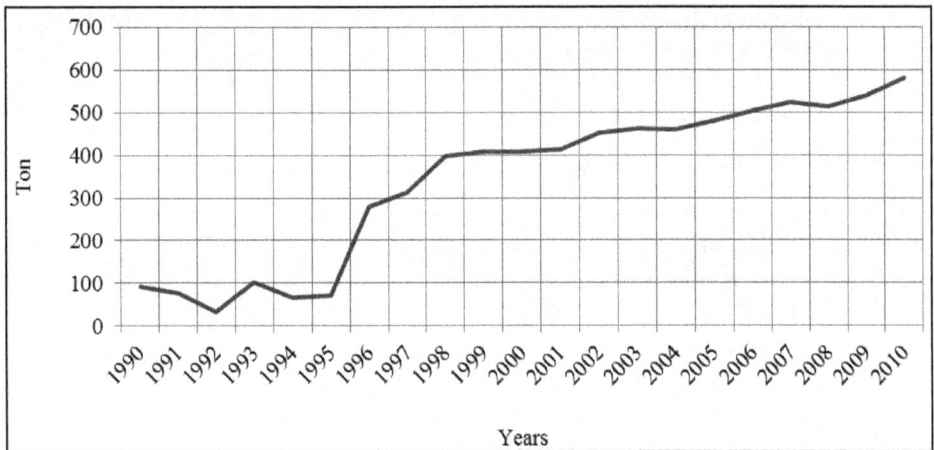

**Figure 13.2**: Brazil–Production of copaiba oil–1990-2010.
*Source*: IBGE (2012).

At the smallest administrative level in Brazil, two municipalities in the state of Amazonas state (Apuí and Novo Aripuan ), of the 5565 municipalities in the country, held the hegemony of this extraction; in the period under consideration, these two accounted for 84.5 per cent of national production. However, another three municipalities, also in the state of Amazonas state (Humaitá, Manicoré and Lábrea) had some extraction that cannot be ignored, though on a smaller scale (Table 13.1).

During the period studied, copaiba oil production in the municipalities of Apuí/ AM and Novo Aripuan /AM was significantly higher. For example, while the cumulative production from 1990 to 2010 in Apuí/AM was 3,129t, the entire state of Pará produced 349t. This shows that the production Apuí/AM was nine times higher than that of the state of Pará. Likewise, compared with Apuí/AM, the state of Rondônia state, has a ratio of 21 to 1, the state of Acre 522 to 1, and the state of Mato Grosso 18 to 1. A similar profile is obtained if we calculate the production of Novo Aripuan / AM, which reached a total of 2.931t (Table 13.1).

**Table 13.1**: Total accumulated production of copaiba oil from 1990 to 2010 in Brazil, according to some geopolitical aspects.

| State | Ton* | National Percentage (per cent) |
|---|---|---|
| BRAZIL | 7171 | 100.00 |
| North Region | 6986 | 97.42 |
| – Amazonas | 6481 | 90.38 |
| – Pará | 349 | 4.87 |
| – Rondônia | 146 | 2.04 |
| – Acre | 6 | 0.08 |
| – Amapá | – | – |
| – Roraima | – | – |
| – Tocantins | – | – |
| Northeast Region | 8 | 0.11 |
| – Piauí | 6 | 0.08 |
| – Bahia | 2 | 0.03 |
| Midwest Region | 178 | 2.48 |
| – Mato Grosso | 178 | 2.48 |
| Southeast Region | – | – |
| South Region | – | – |
| **Main Producing Municipalities** | – | – |
| – Apuí/AM | 3.129 | 43.63 |
| – Novo Aripuanã/AM | 2.931 | 40.87 |
| – Humaitá/AM | 115 | 1.60 |
| – Lábrea/AM | 98 | 1.37 |
| – Manicoré/AM | 56 | 0.78 |

*: Values rounded up.

*Source*: IBGE (2012)

It is noteworthy that since 1996, copaiba oil production in Brazil has been highly concentrated, with relatively high levels in the south of the Amazonas state, where the main producing municipalities (Apuí and Novo Aripuan ) are located. Whether by coincidence or not, these areas are within a context of the expansion national agricultural frontier. Individually this advance in agribusiness, reclaiming land from the Amazon rainforest, does not fully explain the increase in copaiba oil production that has occurred in these municipalities, but it no doubt contributed in some way.

The substantial production of copaiba oil in the municipalities of Apuí and Novo Aripuan has led to their establishment as "Copaiba capitals" within the Amazon, sharing a common location in the state of Amazonas and the Madeira River Valley, one of the main tributaries of the Amazon River. Another common feature of these municipalities is their small population density, with approximately 18,007 inhabitants recorded for Apuí in the 2010 census and 21,451 for Novo Aripuan (IBGE, 2010, *apud* IPEA, 2012). In that same year, 41.2 per cent of the residents of Apuí consisted of the rural population, while in Novo Aripuan this figure was 31.4 per cent, indicating the importance of the local rural economy as a provider of jobs.

The Gross Domestic Product (GDP) of Apuí in the 2009 was approximately R$ 147.0 million (US$ 53.6 million), with agriculture being the main economic activity, accounting for 52.0 per cent of the GDP. In Novo Aripuan the GDP was R$ 85.7 million (US$ 31.2 million), and here, the services sector proved to be primary, contributing 62.0 per cent, while agriculture contributed a share of 27.5 per cent. In both cities, industry was on a smaller scale, contributing less than 9.0 per cent of GDP (SEPLAN/AM, 2011).

The economic contribution of these two municipalities to the total GDP of the Amazonas state was insignificant; 0.3 per cent for Apuí and 0.2 per cent for Novo Aripuan (2009), with GDP per capita of R$ 7,904.00 (US$ 2,900.00) and R$4,522.00 (US$ 1,700.00) respectively, which was below the average for the Amazonas state, of R$ 14,621.00 (US$ 5,300.00) (SEPLAN/AM, 2011).

The economic value of the copaiba oil extraction in Brazil, for the period of this study (1990-2010), is difficult to analyze in monetary terms due to the instability of the national currency during this period, particularly from 1990 to 1994, with hyperinflation, currency trading, and a substantial change in value, skewing the comparability over time.

In an attempt to minimize these influences, and give a historical view of the copaiba oil economy in Brazil in recent times that meets international criteria, the original values of the series, recorded in Cruzeiros and Reais (currencies in Brazil during the study period) by the IBGE were converted to U.S. Dollars, using the exchange rate of the Central Bank of Brazil on December 31 of each year between 1990 and 2010. It should be borne in mind that this procedure incorporates the influences inherent to foreign exchange movements of devaluation or overvaluation of the currency. Nevertheless, it provides an internationally accepted parameter.

In 1990, Brazil produced 93 tons of copaiba oil with a FOB (Free On Board) value equivalent US$ 153,800.00 at that time, giving an average of US$ 1,654.00/ton. From 1992 to 1995 both the production and its value were considerably affected by the economic and political instability in the country, so that 1992 had the lowest average value per ton of oil during the twenty-one year period studied, reaching US$ 110.00. With such low amounts, greater disruption and volatility of production would be expected (Table 13.2).

It was from 1996, after two years of implementation of a new Brazilian currency (Real), that the value of the product began to recover. In that year, the country produced 279t of oil, corresponding to the total value of U$ 514,700.00–an average of US$ 1,845.00/t–compared to the average of 1990 (US$ 1,654.00), showing a nominal gain of 11.5 per cent (Table 13.2). In the period 1996 to 2006, when annual production had risen to 502t, corresponding to the total value of $ 954,200.00, the average income per ton of oil ranged from US$ 837.00 (2002) to US$ 1,901.00 (2006). However, the best period presented by the copaiba oil economy occurred in the last four years of the series, between 2007 and 2010. In 2007, production reached 523t, with a total value of US$ 2.1 million and an average of US$ 4,076.00/t, closing 2010 with 580t of oil extracted, with a total value of US$ 2.9 million and average income US$ 5,082.00/t (Table 13.3). Certainly, this outcome was affected by the exchange rates, and the valuation of the national currency against the dollar

**Table 13.2**: Annual production, production value and average value per ton of extracted copaiba oil in Brazil from 1990 to 2010.

| Year | Production(t) | Value (US$)* | |
|------|---------------|--------------|--------------|
| | | Total | Per ton Average |
| 1990 | 93 | 153,822 | 1654 |
| 1991 | 76 | 31,170 | 410 |
| 1992 | 34 | 3,756 | 110 |
| 1993 | 101 | 41,195 | 408 |
| 1994 | 65 | 72,104 | 1109 |
| 1995 | 72 | 116,195 | 1614 |
| 1996 | 279 | 514,720 | 1845 |
| 1997 | 313 | 590,290 | 1886 |
| 1998 | 398 | 709,854 | 1784 |
| 1999 | 408 | 544,549 | 1335 |
| 2000 | 408 | 531,522 | 1303 |
| 2001 | 414 | 455,094 | 1099 |
| 2002 | 453 | 378,966 | 837 |
| 2003 | 463 | 465,873 | 1006 |
| 2004 | 459 | 574,518 | 1252 |
| 2005 | 479 | 748,709 | 1563 |
| 2006 | 502 | 954,163 | 1901 |
| 2007 | 523 | 2,131,609 | 4076 |
| 2008 | 514 | 1,621,309 | 3154 |
| 2009 | 538 | 1,512,213 | 2811 |
| 2010 | 580 | 2,947,748 | 5082 |

* Calculated using the exchange rates for the sale of U.S. Dollar on 31/12 of each year.

*Source*: IBGE (2012) basic data–authors' calculations

The data on the production value of copaiba oil, according to the IBGE, considering the series in question, show that the best year for the industry in Brazil was 2010, when the total value generated nationally amounted to the equivalent of US$ 2,947,748.00, obtained from 61 producing municipalities: 26 in the state of Amazonas, 22 in the Pará, six in Rondonia, four in Mato Grosso, two in Maranh o and one in Acre. Of the total income generated in 2010, the ten largest producing municipalities accounted for 89.7 per cent, while the remaining 10.3 per cent was distributed among the other 51 municipalities. Thus, the municipalities of Apuí and Novo Aripuan were highlighted as producing the highest amounts, with total values equivalent to US$ 1,153,153.00 and US$ 1,004,204.00 respectively, together accounting for 73.2 per cent of the national production (Table 13.3). This total volume in revenue, in the reality of the many small towns and villages in the Amazon where the copaiba oil is produced, appears to have a certain degree of importance for income generation in the economy, especially in the two larger producing centers (Apuí/AM and Novo Aripuan /AM). However, compared to the value of municipal GDP, it is seen that the value of oil production in 2010 in Apuí/AM, totaling US$ 1.15 million, represented only 2.2 per cent of local GDP in 2009, that is the latest data available. The same applies to Novo Aripuan /AM, where the oil production represents 3.2 per cent of the GDP.

**Table 13.3**: Total and per capita production of copaiba oil extraction and population in ten major producing municipalities in Brazil in 2010.

| Municipality | Value of Production (US$1.00) | National Participation (per cent) | Population (2010) | Value of Oil Production per Capita (US$) |
|---|---|---|---|---|
| Apuí – AM | 1,153,153 | 39.1 | 18,007 | 64.04 |
| Novo Aripuanã – AM | 1,004,204 | 34.1 | 21,451 | 46.81 |
| Manicoré – AM | 145,946 | 5.0 | 47,017 | 3.10 |
| Humaitá – AM | 107,508 | 3.6 | 37,701 | 2.85 |
| Lábrea – AM | 57,658 | 2.0 | 44,227 | 1.30 |
| Altamira – PA | 46,847 | 1.6 | 99,075 | 0.47 |
| Guajará-Mirim – RO | 43,243 | 1.5 | 41,656 | 1.04 |
| Uruará – PA | 33,033 | 1.1 | 44,789 | 0.74 |
| Machadinho do Oeste – RO | 29,429 | 1.0 | 31,135 | 0.95 |
| Óbidos – PA | 23,423 | 0.8 | 49,333 | 0.47 |
| BRASIL | 2,947,748 | 100.0 | – | – |

Source: Raw data IBGE (2012)–authors' calculations.

Taking into account the income per capita of these municipalities that produce copaiba oil, compared to their GDP per capita, we see a similar profile. In Apuí/AM, the largest domestic producer of oil, where production in 2010 reached a value of US$ 1.2 million, with a population of 18.0 thousand people, the average per capita income generated by the copaiba oil economy was US$ 64.0/year, which when compared to the GDP per capita in 2009 (US$ 2,900.00), represents 2.2 per cent. In Novo Aripuan / AM, the second ranked Brazilian municipality for this production, the total amount in 2010 was US$ 1.0 million, with a resident population of 21,451 inhabitants, making an average per capita of US$ 46.8, which compared to the GDP per capita in 2009, (US$ 1,700.00), represents 2.8 per cent.

These parameters were the highest observed for the municipal economic dimension focusing on copaiba oil in Brazil. The real impact of the activity among all

the other producing municipalities, where production was much lower than in Apuí/
AM and Novo Aripuan /AM, is minimal, with places where the average annual
income per capita generated in 2010 did note even amount to one U.S. Dollar, as was
the case with Altamira, Óbidos and Uruará–the largest producers in the state of Pará
state in 2010 – and also with Machadinho do Oeste, the highest producer in the state
of Rondônia, among other observations shown in Table 13.3.

**Table 13.4**: Number, total and average value of some products of vegetable extraction in
Brazil and the largest cities of copaiba oil producers in 2010.

| Municipality | Quantity | Value Total (in thousand R$) | Average Value (in R$)[A] | Average Value (in US$)[B] |
|---|---|---|---|---|
| **Copaiba Oil[C]** | | | | |
| BRAZIL | 580 | 4,908 | 8,462 | 5,082 |
| - Apuí-AM | 240 | 1,920 | 8,000 | 4,805 |
| - Novo Aripuanã-AM | 212 | 1,672 | 7,887 | 4,737 |
| **Chestnut-Amazon** | | | | |
| BRAZIL | 40,357 | 55,194 | 1,368 | 821 |
| - Apuí-AM | 15 | 33 | 2,200 | 1,321 |
| - Novo Aripuanã-AM | 750 | 2,100 | 2,800 | 1,682 |
| **Charcoal[C]** | | | | |
| BRAZIL | 1,502,997 | 650,614 | 433 | 260 |
| - Apuí-AM | – | – | – | – |
| - Novo Aripuanã-AM | 16 | 19 | 1,188 | 713 |
| **Firewood[D]** | | | | |
| BRAZIL | 38,207,117 | 624,293 | 16 | 10 |
| - Apuí-AM | - | – | – | – |
| - Novo Aripuanã-AM | 80,000 | 640 | 8 | 5 |
| **Wood Logs[D]** | | | | |
| BRAZIL | 12,658,209 | 2,156,610 | 170 | 102 |
| - Apuí-AM | 2,400 m³ | 65 | 27 | 16 |
| - Novo Aripuanã-AM | 40,000 m³ | 1,120 | 28 | 17 |

A: Resulting average of the total value divided by quantity; B: Calculations based on the Exchange rate
on 31/12/2010–U.S. $ 1.00 = R $ 1.665; C: Quantity in tons; D: Quantity in cubic meters.

*Source*: IBGE (2012), raw data, authors' calculations.

Comparison of the latest data provided by the IBGE on plant extraction in Brazil,
for the year 2010, with respect to copaiba oil and other products common to
Amazonian biodiversity, including chestnut-amazon, charcoal, firewood and wood
logs, shows that the total financial resources nationally generated by the oil was the
least significant. Brazil had revenues of US$ 2.2 billion from the extraction of wood
logs (US$ 1.3 billion), R$ 650.6 million (US$ 390.8 million) from charcoal, and R$
624.3 million (US$ 374.9 million) from firewood, but just R$ 4.9 million (US$ 2.9

million) from the oil of the copaiba tree.

In the analysis is made to the plan of the main producing municipalities of copaiba oil (Apuí and Novo Aripuan ), it can be seen that the process is reversed, with this product being the main economic activity associated with extraction from plants. In 2010, according to IBGE data, copaiba oil generated R$ 1.9 million (US$ 1.2 million) for the economy in Apuí-AM and R$ 1.7 million (US$ 1.0 million) in Novo Aripuan -AM, while other products (chestnut-amazon, charcoal, firewood and wood logs) fell far short of these figures (Table 13.4).

Advancing a little more in the issue, seeking to obtain a comparable value for the unit of measurement, the predominance is seen, both nationally and locally of the copaiba oil economy on other products. Comparing the average value per ton of copaiba oil with that of a ton of chestnut-amazon, for example, we see a national average R$ 8,500.00 (US$ 5,100.00) and R$ 1,400.00 (US$ 821.00) respectively, *i.e.*, a ratio of 1: 6. In other words, the value obtained from one ton of oil is equal to that obtained from six tons of chestnut-amazon. For charcoal, this ratio increases to 1: 20. The average of the value of one ton of oil was the same of 518 and 50 cubic meters, for firewood and wood logs respectively, as can be inferred from the data shown in Table 13.4.

## Conclusion

The study revealed that the extraction of copaiba oil, for the last twenty-one years of available data (1990-2010), showed an increase in production volumes, especially from 1996 onwards. Regionally, the North proved to be the main producer, but there was also production in the Midwest, and sporadically in the Northeast. The Amazonas state had the highest production, with just two municipalities in the Madeira River Valley accounting for most of this volume. In economic terms, the value generated by the activity still has relatively little significance, accounting for only around 2 per cent of GDP in the two municipalities with the highest production, despite the fact that due to the many potential applications and uses of the oil, there is great potential for this figure to increase. However, in the universe of extraction from the plant biodiversity of the Amazon, compared to the average value generated per mass or volume units of some products, copaiba oil was the one that proved to have the highest economic potential.

## Acknowledgments

The authors thank the National Council for Scientific and Technological Development (CNPq), Center for Research Analysis and Technological Innovation Foundation (FUCAPI) and Federal University of Amazonas (UFAM).

## References

Benchimol, S.I. (2009). Amazônia – Formaç o Social e Cultural. 3.ed. Valer, Manaus.

IBGE – Instituto Brasileiro de Geografia e Estatística. (2007). Censo Agropecuário– 2006. IBGE, Rio de Janeiro.

IBGE–Instituto Brasileiro de Geografia e Estatística. Produç o da Extraç o Vegetal e da Silvicultura – 1990 a 2010.(2012).Banco de Dados Agregados. SIDRA. Available in: www.ibge.gov.br. Accessed on June 19, 2012.

IBGE–Instituto Brasileiro de Geografia e Estatística. Valor da Produç o da Extraç o Vegetal e da Silvicultura – 1990 a 2010.(2012).Banco de Dados Agregados. SIDRA. Available in: www.ibge.gov.br. Accessed on June 19, 2012.

Index Kewensis.(1996).supl. XX. Claredon Press, Oxford.

IPEA–Instituto de Pesquisa Econômica Aplicada. Populaç o Residente – Municípios – Estado do Amazonas – 1991 a 2010.(2012).IPEADATA, Brasília. Available in: www.ipeadata.gov.br. Accessed on June 21, 2012.

Izumi, E., Ueda-Nakamura, T., Veiga Jr, V.F., Pinto, A.C., and Nakamura, C.V.(2012). Terpenes from Demonstrated *in vitro* Antiparasitic and Synergic Activity. *Journal of Medicinal Chemistry*, **55**: 2994-3001.

Pieri, F.A., Mussi, M.C.M., Fiorini, J.E., and Schneedorf, J.M.(2010).Efeitos clínicos e microbiológicos do óleo de copaíba (*Copaifera officinalis*) sobre bactérias formadoras de placa dental em c es. *Arquivo Brasileiro de Medicina Veterinária e Zootecnia*, **62**: 578-585.

Pieri, F.A., Mussi, M.C.M., and Moreira, M.A.S. (2009). Óleo de Copaíba ( *Copaifera* sp.): Histórico, extraç o, aplicaç es industriais e propriedades medicinais. *Revista Brasileira de Plantas Medicinais*, **11**: 465-472.

Santos, A.O., Ueda-Nakamura, T., Dias Filho, B.P., Veiga Jr, V.F., and Nakamura, C.V. (2012). Copaiba Oil: An Alternative to Development of New Drugs against Leishmaniasis. *Evidence-Based Complementary and Alternative Medicine*, **2012**: 1-7.

Santos, A.O., Ueda-Nakamura, T., Dias Filho, B.P., Veiga Jr, V.F., Pinto, A.C., and Nakamura, C.V. (2008). Antimicrobial activity of Brazilian copaiba oils obtained from different species of the *Copaifera* genus. Memórias do Instituto Oswaldo Cruz, **103**: 277-281.

SEPLAN/AM – Secretaria de Estado do Planejamento e do Desenvolvimento Econômico do Estado do Amazonas.(2011).Produto Interno Bruto Municipal – 2002–2009. SEPLAN/AM, Manaus. Available in: www.seplan.am.gov.br. Accessed on June 22, 2012.

Veiga Jr, V.F., Lima, H.C., and Pinto, A.C.(2006).The Essential Oil Composition of *Copaifera trapezifolia* Hayne Leaves. *The Journal of Essential Oil Research*, **18**: 430-431.

Veiga Jr, V.F., and Pinto, A.C.(2002).O gênero *Copaifera* L. *Química Nova*, **25**: 273-286.

Veiga Jr, V.F., Rosas, E.C., Carvalho, M.V. Henriques, M.G.M.O., and Pinto, A.C.(2007). Chemical composition and anti-inflammatory activity of copaiba oils from *Copaifera cearensis* Huber ex Ducke, *Copaifera reticulata* Ducke and *Copaifera multijuga* Hayne. A comparative study. *Journal of Ethnopharmacology*, **112**: 248-254.

Utilisation and Management of Medicinal Plants Vol. 2 (2014)    *Pages* **309–328**
*Editor-in-Chief:* **V.K. Gupta**
*Published by:* **DAYA PUBLISHING HOUSE, NEW DELHI**

# 14

# Use of Timbó (*Derris* and *Deguellia*) to Control Agriculture Pests

M.R. Alecio[1], M. Fazolin[2], P. de A. Oliveira[1], J.L.V. Estrela[2],
R. de C. Andrade Neto[2], S.B. Alves[1] and V.F. Veiga Júnior[1]*

## ABSTRACT

*The plant species known as timbó belongs to various genera (Deguellia, Derris, Thephrosia, Millettia, Serjania, etc.) and several families (Fabaceae, Papilionaceae, Sapindaceae, and Compostae Cariocaraceae, etc.). In South America, these plants are known by various names: timbó, tingui and cunambi (Brazilian Amazon); cube and barbasco (Peru and Colombia), haiari (Guyana), nekoe (Suriname); it is traditionally used by the indigenous people to catch fish. From 1880 to 1940, pest control using extracts and oils from the plant was at its peak, and the timbó species was one of the most important groups for this purpose. After World War II, with the creation of synthetic pesticides, the mills closed and research on the timbó crop ceased. However, due to problems caused by synthetic insecticides to the environment and humans, many scientists have now begun looking into the use of insecticide plants, including timbó, once again, in search of new alternative forms of pest control that are safe, selective, biodegradable, economically sustainable, and applicable to integrated pest control programs. The chemical composition of timbó extracts present wide qualitative and quantitative variations, with rotenone being identified as the main toxic constituent; however its effect depends on the*

1    Universidade Federal do Amazonas – UFAM, Av. Rodrigo Otávio, 6.000, Japiim, 69077-000, Manaus, AM, Brazil. mralecio@yahoo.com.br, valdirveiga@ufam.edu.br, prix_oliver@hotmail.com, suziagro@yahoo.com.br.

2    Embrapa Acre, Rodovia BR-364, km 14, Dom Moacir, 69.908-970, Rio Branco, AC, murilo@cpafac.embrapa.br, joelma@cpafac.embrapa.br, romeu@cpafac.embrapa.br.

*    *Corresponding author*: E-mail: valdir.veiga@pq.cnpq.br

*combined action of other compounds. This chapter presents the results of research on the use of extracts and powdered roots of timbó for pest control in agriculture, emphasizing its biological effects on animals. It also contextualizes the need of more scientific efforts to determine the best combination of compounds present in the plants, in order to increase their effectiveness in pest control.*

*Keywords:* Pest control, Deguelin, Natural insecticide, Rotenoids, Rotenone toxicity.

## Introduction

The use of synthetic insecticides is still the main tactic used for pest management (Busato *et al.*, 2006). These result in a considerable increase in production costs (Grützmacher *et al.*, 2000), and are linked to various problems, such as biological imbalance, high residual levels in foods, poisoning of applicators, and selection of resistant insect populations (Cruz, 1995; Henandez and Vendramim, 1996; Diez–Rodriguez and Omoto, 2001; Yu, 2006; Yu and MCord, 2007).

Given the problems caused by synthetic insecticides to the environment and man that have emerged in recent decades, there is a growing awareness of the need to develop technologies and products to control insect pests that are efficient, safe, selective, biodegradable, economically viable and applicable to integrated pest management programs (Viegas Júnior, 2003).

The use of alternative products presents a promising proposal for the control of insect pest species (Lima *et al.*, 2010), and involves the use of products less harmful to the environment (Vendramim, 1997), such as botanical compounds (Silva *et al.*, 2002; Bogornie and Vendramim, 2003), which theoretically have the advantage that they degrade rapidly in the environment and are more specific than the pesticides (Schmidt *et al.*, 1997; Breuer and De Loof, 1998), reducing the likelihood of generating resistant species (Valladares *et al.*, 1999; Brito *et al.*, 2004).

Plants with insecticide activity can be used in dried powder and extract forms, providing an alternative to the use of synthetic chemical insecticides, with the advantages of increased efficiency, ease of acquisition and application, rapid degradation, low toxicity to applicators, and greater safety for consumers (Faroni *et al.*, 1995; Talukder and Howse, 1995; Oliveira and Vendramim, 1999; Sousa *et al.*, 2005), adapting to the expectations of modern society in the pursuit of healthy foods, as well as programs for integrated pest management (Baldin *et al.*, 2009).

Melo *et al.* (2011) emphasize that the pest control using plants with insecticidal properties is a promising, viable and environmentally correct proposal that has gained increasing importance and attention from various scientific fields, due to their various effects on insects.

The effects of botanical insecticides on insects are variable, and include toxicity, repellant effects, sterility and deformities, modifications in behavior, the development and reduce the feeding (Arnason *et al.*, 1990, Bell *et al.*, 1990, Gusm o *et al.*, 2002).

Plants known as timbó, belonging to the genus *Derris* and *Deguelia*, are widely found in the Amazon (Costa *et al.*, 1986, Lima, 1987; Tozzi, 1998) and due to the toxins they contain (Luitgards-Moura *et al.,* 2002; Pereira and Famadas, 2004; Kotze *et al.,* 2006, Azevedo *et al.,* 2005; Alecio *et al.,* 2010, Correa, 2011; Alecio, 2012), have potential for the development of new insecticides.

There is some misinformation in the literature concerning the toxicity caused by timbó extracts. Several studies suggest rotenone to be the main toxic constituent (Crombie and Whiting, 1998; Costa *et al.,* 1986; Luitgards-Moura *et al.,* 2002; Marinos *et al.,* 2004, Azevedo *et al.,* 2007; Alecio *et al.,* 2010; Lucio *et al.,* 2011), while others point out that this substance has low toxicity when used alone, and relies on the association of other compounds to provide insecticide activity (Correa, 2011; Li *et al.,* 2011; Alecio, 2012). Meanwhile, new components have been isolated from extracts of these plants, and have shown biological effects against arthropods (Hymavathi *et al.,* 2011; Li *et al.,* 2011; Alecio, 2012; Jiang *et al.,* 2012).

Therefore, the objectives of this literature review are to gather information from research on the use of extracts and powders of timbó roots for pest control in agriculture, with emphasis on their biological effects in animals due to the chemical composition of extracts, and to contextualize the need for greater scientific efforts to determine the best combination of the compounds present in these plants, seeking to increase their effectiveness in pest control.

## History and Use of Timbó

The word timbó is of Tupi origin: *ti* meaning juice, and *mbo* meaning snake, hence, snake's juice, poisonous juice, juice that kills (Corbett, 1940). It is a plant that is known for its toxicity, especially against insects and fish, and is widely distributed in the Amazon region, both in primary forest and in cleared areas (Tozzi, 1998).

Plants that go by the name of timbó belong to several genera, including: *Deguelia*, *Derris, Thephrosia, Millettia, Serjania*, etc., and to different families, such as Fabaceae Papilionaceae, Sapindaceae, Compostae and Cariocaceae (Corbett, 1940; Lima, 1987). In South America, these plants are known by various names: in the Brazilian Amazon: timbó, tingui and cunambi; in Peru and Colombia, cube and barbasco, in Guyana, haiari and in Suriname, nekoe (Pires, 1978).

According to Lima (1987) there are many species of timbó, but the most widely used in the Amazon are red timbó, *Derris urucu* (Killip et Smith) Macbride, and white timbó, *Derris nicou* (Killip et Smith) Macbride. These two species belonged to the genus *Deguelia*, and proceeded to the genus *Derris*, in a review by Francis Macbride, who placed them in the family Fabaceae.

Knowledge of the toxicity of timbó to insects is not recent. These plants have been used for pest control and fishing by the Indians and Chinese since the eighth century (Corbett, 1940). According to Lima (1987), the timbó species were already being cultivated and exploited only by the natives long before the discovery of America by Christopher Columbus. In Brazil, indigenous peoples of various regions, particularly the Amazon, use these plants to catch fish, especially on special occasions, when larger amounts of food are needed for their feasts.

In these fishing activities, the natives use the fresh roots of timbó species, which when thrown into the water and stirred, produce a milky liquid with a strong and peculiar smell. Under the action of this juice, even in very diluted form, the fish lose their balance, become stunned, and rise to the surface, where they swim uncontrollably towards the river banks and can be captured easily. In standing water, the mortality of the fish is total (Corbett, 1940; Lima, 1987).

Amaral (2004) identified the use of timbó preparations as the most common fishing method used by the indigenous Ashaninkas and Kaxinawás peoples (Wonderworker Marshall County, State of Acre, Brazil) to catch various fish species of the families Loricariidae and Curimatidae in small rivers and creeks.

Pires (1978) reports that in the Amazon, it is common to find timbó plantations in backyards or in places where there were ancient Indian dwellings, and the inhabitants of this region still use timbó roots for fishing and to kill lice in pets.

From 1880 to 1940, pest control using plant extracts and oils reached its peak. The timbó species was one of the most important for this purpose, and a large amount of research and experiments were performed with extracts of the roots of timbó, by researchers working in several countries (Caminha Filho, 1940; Corbett, 1940).

Due to the wide availability of these plants in the Amazon, and the huge market demand, the Brazilian Amazon began to export timbó roots in 1939, through the port of Belém-PA, to several countries. The United States and France were among the largest importers of this natural defense (Lima, 1947).

At that time, there were several mills operating in Brazil (Manaus-AM and Belém-PA), and the product was exported in the form of powdered timbó roots, which was used for the extraction of rotenone. Rotenone was used in the preparation of different products with widespread uses, such as pest control of crops and household infestation by insects, as well as to combat ectoparasites in domestic animals (Lima, 1987). Even today, timbó powder shipped from the port of Iquitos, in Peru, still passes through the Brazilian ports (Lima, 1947).

After World War II, with the advent of synthetic insecticides, the trade in powdered timbó roots collapsed, the mills closed, and the research activity with the crop ceased altogether (Pires, 1978). However, due to problems caused by synthetic insecticides to the environment and humans, many scientists have begun, in recent decades, to resume the research on insecticide plants seeking new alternatives forms of pest control that are safe, selective, biodegradable, economically feasible and applicable to integrated pest management programs (Viegas Júnior, 2003).

To this end, several scientists have been researching the applicability of timbó for pest control in the last few decades (Costa *et al.*, 1986; Lima, 1987; Nawnot *et al.*, 1987; Maini e Morallo,1993; Lima e Costa, 1998; Mascaro *et al.*, 1998; Tozzi, 1998; Costa *et al.*, 1999a; Costa *et al.*, 1999b; Tokarnia *et al.*, 2000; Fazolin *et al.*, 2002; Luitgards-Moura *et al.*, 2002; Pereira e Famadas, 2004; Alecio *et al.*, 2005; Azevedo *et al.*, 2005; Correa, 2006; Alecio *et al.*, 2010; Alecio, *et al.*, 2011; Correa, 2011; Alecio, 2012).

Costa *et al.* (1986) studied the effect of aqueous extract of timbó roots (*Derris urucu*) in the control of lice (*Haematopinus tuberculatus* Burmeister) in buffaloes and found that when applied by spraying, the extract was effective in combating lice, presenting the equally effective in a range of 0.25 to 2 per cent (m.v$^{-1}$), at least seven days after application. They concluded that the effectiveness of timbó is comparable to the chemicals currently used to combat this pest, with the advantage that it can be grown on the farms, its low cost, and the fact that it is readily available for use.

Maini and Morallo (1993) studied the effect of *D. elliptica* applied by spraying on snails and slugs, and found that the aqueous extract of the roots was toxic at 2,000 ppm, while the stem extract resulted in 30 per cent mortality at 10,000 ppm.

Azevedo *et al*. (2005) tested the effect of various natural products in the control of adults of *Bemisia tabaci* Gennadius biotype B (Hemiptera: Aleyrodidae) in melon (*Cucumis melo* L), under greenhouse and field conditions, and found that the timbó extract was more effective in controlling adults of these insects at the beginning of cultivation, and nymphs at the end of the cycle.

Correa (2006) determined the toxicity of the *Deguelia floribundus* Benth extract in *Toxoptera citricidus* Kirkald (Hemiptera: Sternorrhyncha: Aphidoidea) by the action of contact spray, finding LC$_{50}$ values of 1.75 per cent and 5.75 per cent (v.v$^{-1}$), respectively, for aqueous and alcoholic extracts. These concentrations were considered promising for the control of insect pests.

Alecio *et al*. (2005) evaluated the insecticidal potential of the extract of the roots of *Derris rariflora* Macbride on the maize weevil (*Sitophilus zeamais* Mots) under laboratory conditions, by means of contaminated surface intoxication on filter paper, resulting in LC$_{50}$ of 0.82 µl of extract.cm$^{-2}$.

The biocidal effect of powdered roots of *Deguelia utillis* was evaluated by Marinos *et al*. (2004) as a regulator of the *Anopheles benarroch* mosquito larvae, and it was confirmed that 3.1 g of a water-powder.L$^{-1}$ is sufficient to control 80 per cent and 90 per cent of the insect population at 24 and 48 hours respectively.

Luitgards-Moura *et al*. (2002) determined the effect on the contaminated surface (filter paper) of the alcoholic extract of *D. amazon* against larvae of *Lutzomyia longipalpis* Lutz (Diptera: Psychodidae: Phlebotominae), obtaining a mean lethal concentration (LC$_{50}$) of 21.2 per cent (m.v$^{-1}$).

Pereira and Famadas (2004) conducted laboratory tests to evaluate the efficiency of the alcoholic extract of roots of the *Dahlstedtia pentaphylla* (Taub.) Burk. against the cattle tick of *Boophilus microplus* Canestrini from Vale do Paraíba. The experiments were conducted with larvae at seven to ten days, not fed, and ticks (female engorged). The determined LC$_{50}$ were 1.65 per cent and 2.87 per cent (m.v$^{-1}$), respectively for ticks and larvae.

Costa *et al*. (1999 a) has evaluated the difference between timbó species ( *D. nicou*, *D. urucu* and *D. elliptica*) collected in different regions of the Brazilian Amazon in the control of larvae of *Musca domestica* L. and observed a wide variation between different types of timbó, in terms of their effectiveness in killing insect pests, with mortalities ranging from almost zero to levels that were highly lethal to the insects. According to Costa *et al*. (1999b), this difference in mortality in the individuals tested was due to the different amounts of rotenone contained in the root powders of each plant, as observed by means of correlation analysis, it was significant between the levels of active principle and the ability to control *M. domestica*. The different levels of rotenone found in plants of the same species of timbó originating from different regions may be due to the isolation to which people were subjected during the time of the Quaternary period, forming the so-called "forest refugia" (Costa *et al.*, 1999a).

Gusm o *et al.* (2002) evaluated the larvicidal activity of ethanol extracts of *Derris urucu* (Leguminosae) against the mosquito *Aedes aegypti* (Diptera: Culicidae) and obtained the $LC_{50}$ of 17.6 mg.ml$^{-1}$, with 100 per cent mortality of insects at a concentration of 150 mg.ml$^{-1}$ 24 h after application of the extract. At this concentration, the larvae showed an imperfect peritrophic matrix and damage to the midgut epithelium and shed a large amount of amorphous feces, while the larvae of the control treatment produced no stools during the test period.

Ephron *et al.* (2011) evaluated the selectivity of commercial natural pesticide doses, used in organic production systems by topical application on adults of the predator *Cryptolaemus montrouzieri* (Coleoptera: Coccinellidae) and found no toxicity of the commercial extract of timbó (Rotenat®), which contained at least 5 per cent of rotenone at all the tested doses (1, 3, 6 and 12 µL.mL$^{-1}$).

Rattanapan (2009) studied the mechanism of toxicity of the *Derris elliptica* Benth extract on *Spodoptera exigua* Hubner by the larvae immersion method, and found that the toxicity increased considerably when the substance concentration and exposure time were increased. The $LC_{50}$ were determined as: 69.15 ppm, 51.32 ppm, 46.60 ppm, at 24, 48 and 72 hours, respectively.

Dahlem *et al.* (2009) evaluated the use of alternative products to control *Diabrotica speciosa* Germar and *Sternechus subsignatus* Boheman in the culture of snap bean (*Vigna unguiculata* L. Walp) in an organic system, and found that the commercial extract of timbó (Rotenat®) containing rotenone, was not efficient for the insects.

Alecio *et al.* (2010) evaluated the insecticidal action of the extract of timbó (*D. amazonica* Killip) for adults of *C. arcuatus* Olivier, determining the $LC_{50}$ values as: 15.14 of the extract.mL-1 (by ingestion of infected leaves) and 0.45 µL of extract.cm-2 (contaminated area) and the $LD_{50}$ value as 1.44 µL of the extract per g$^{-1}$ of the insect (by topical application). The authors concluded that the extract is toxic and inhibits the supply of *C. arcuatus* at concentrations of 1 per cent (m.v$^{-1}$).

Alecio *et al.* (2011) found that *Deguelia floribundus* Benth extract caused low levels of adult mortality of *C. tingomarianus* by topical contact and by ingestion of infected leaves, but decreased insect feeding, therefore it is considered as a promising alternative for pest control.

Correa (2011) evaluated the toxicity of extracts of timbó (*Derris* spp.) and purified rotenone against *Tetranychus desertorum* (Acari: Tetranychidae) on pepper leaves, and found $LC_{50}$ values of 2 per cent and 4.4 per cent (m.v$^{-1}$) for ethanol and acetone extracts of *D. rariflora*, respectively, with 13.8 per cent (m.v$^{-1}$) for the ethanol extract of *D. floribundus* and 26.6 per cent (m.v$^{-1}$) for the acetone extract of *D. rariflora*.

## Active Compounds of Timbó

The timbó root extracts contain a wide diversity of constituents (Mendes, 1960; Mors, 1978, Costa *et al.*, 1999b; Lobo *et al.*, 2009; Jiang *et al.*, 2012). The concentration of active principles of the plants can vary according to ecological factors (nutrients, water, light and plant health), and genetic and physiological factors (Brown Júnior, 1988; Castro *et al.,* 2004; Wu *et al.,* 2008; Hymavathi *et al.,* 2011; Jiang *et al.,* 2012).

Rotenone is cited as a major constituent of timbó extract with insecticidal and acaricidal properties. However, it is commonly available in the form of powdered roots of *Deguelia* and *Derris*, containing 1-5 per cent (m.m-1) of active ingredients, or as an extract, which generally contains up to 45 per cent of rotenoids (Fang and Casida, 1999; Coll, 2005). The timbó species with the highest levels of rotenone belong to the genus *Derris* and *Deguelia*, and the highest concentration of this substance is found in the roots of these plants. Their contents can vary from 5 to 13 per cent depending on the species (Costa, 1999a; Lima and Costa, 1998).

Rotenone is a crystalline, odorless, tasteless isoflavonoid, biosynthesized by the secondary metabolite pathway, with low solubility in water and excellent solubility in chloroform, ether, acetone, carbon tetrachloride and in ethylene derivatives. In extracts of timbó, rotenone is always accompanied by other flavonoid constituents, such as deguelin, tephrosin, sumatrol, toxicarol, ellipton and malacol (Figure 14.1) (Mors, 1978). Deguelin and tephrosin also have insecticidal activities (Corbett, 1940; Cravero *et al.,* 1976, Silva *et al.,* 2002), but their toxicities for insects are respectively three and seven times lower than that of rotenone (Lima, 1987).

Most studies with extracts of *Derris* and *Deguelia* for pest control attribute the toxicity of timbó to rotenone (Caminha Filho, 1940; Corbett, 1940; Pinto, 1953; Cravero *et al.,* 1976; Mors, 1978; Pires, 1978; Lima, 1987, Crombie and Whiting, 1998; Luitgards-Moura *et al.,* 2002; Fazolin *et al.,* 2002; Pereira and Famadas, 2004; Azevedo *et al.,* 2005, Correa, 2006; Alecio *et al.,* 2010, Li *et al.,* 2011; Lucio *et al.,* 2011).

Rotenone, present in timbó extracts, causes death of the animal through inhibition of the mitochondrial respiratory chain by blocking the phosphorylation of ADP to ATP. Fish and insects are particularly sensitive (Mascaro *et al.,* 1998). Insects poisoned with rotenone display: decrease in oxygen consumption, reduced respiration and attacks that cause seizures and ultimately lead to paralysis and death by respiratory failure (Silva *et al.,* 2002). According to Gallo *et al.* (2002), rotenone acts primarily as a potent inhibitor of the enzyme NADH-oxide reductase, the respiratory chain, and toxicity in insects is manifested by reduced heart rate, depression of respiratory movements and reduction in oxygen consumption.

Cavalheiro *et al.* (2004) found that the application of 4 mM of rotenone inhibited the respiratory chain complex I of *Candida albicans*, causing a decrease of about 30 per cent in respiratory rate (29 nmol.min$^{-1}$ O$_2$ mg of protein$^{-1}$). Colman-Saizarbitoria *et al.* (2009) observed that rotenone inhibited the respiratory chain complex I (NADH ubiquinone oxidoreductase coupling) of the crustacean *Artemia salina* Leach (Anostraca: Artemiidae) and they considered the mitochondrial inhibitor rotenone to be a classic of the respiratory chain complex I (NADH ubiquinone oxidoreductase coupling).

The cytotoxic effects of various substances on the ATP levels of insect cells (Sf9) grown *"in vitro"* were compared by Saito (2005), who found that rotenone

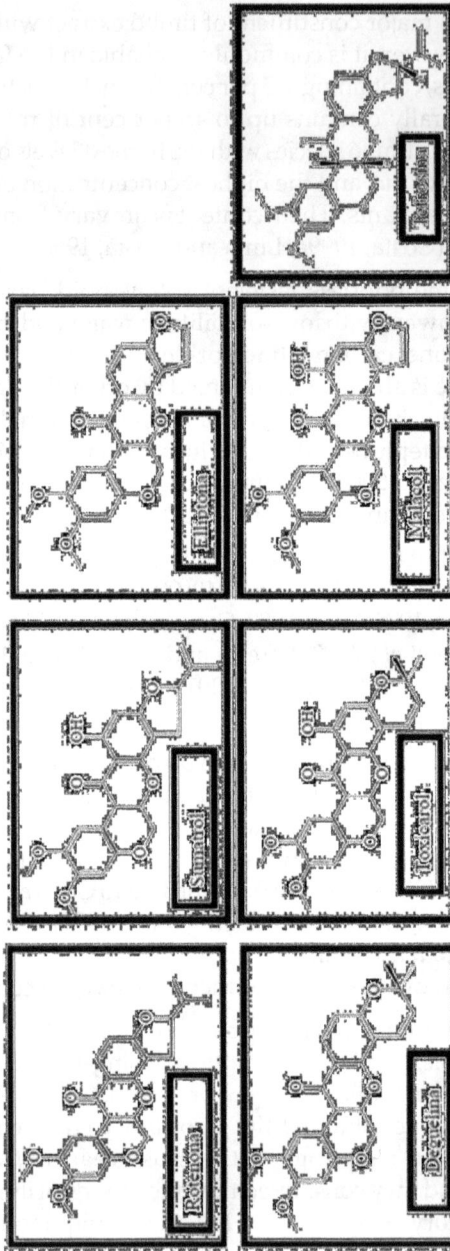

**Figure 14.1:** Rotenoids that can be found in roots of timbó (*Derris* and *Deguelia*).

substantially decreased mitochondrial respiration and reduced ATP content of cells evaluated in culture medium.

Philogene *et al.* (2004) point out that rotenone is highly toxic to insects because it acts on the nervous system and mechanisms of cell respiration. Ducrot (2004) reported that rotenone can exert antifeedant, and can potentially affect biological activity or hormonal changes, causing death of the insects. According to Gosselin (1984) and Silva *et al.* (2002), rotenone is toxic to insects by ingestion and contact, thus, according to Cravero *et al.* (1976), two forms of intoxication are used to control insect pests.

Caminha Filho (1940) points out that rotenone has great value as an insecticide against crop pests, such as coccidia (Coleoptera), scale insects and aphids (Hemiptera), lice (Anoplura), wasps (Hymenoptera), butterflies and moths (Lepidoptera), etc., both as adults and in different periods of development of insects (eggs, larvae, caterpillars and pupae). It is also efficient against ectoparasites affecting domestic animals and humans, such as fleas (Siphonaptera), lice (Anoplura), ticks (Acarina) and grubs (Diptera).

Corbett (1940) found that the concentration of 1,2.10-3 g of rotenone/L in water (m.v-1) was toxic to the acarás fish (*Geophagus brasiliensis*) about half an hour after its application, and death of animals apparently occurred by respiratory paralysis, although the heartbeat continued for some time after the respiratory movements had stopped.

Pinto (1937) successfully used rotenone in powdered root form, in a soapy solution in water, against ticks, but also recommended their use in treatment of lice and fighting ticks and lice.

Saito and Luchini (1998) reported that rotenone is effective in controlling beetles and caterpillars, but its toxicity may be more or less active according to the insect species, and its action may take a little while to appear.

The antifeedant activity of several pesticides was evaluated by Nawnot *et al.* (1987) against several pests of stored products. They concluded that against *Homogyne alpina* L. rotenone obtained from *Derris elliptica* was the more effective substance.

Fazolin *et al.* (2002) obtained satisfactory results when evaluating the insecticide potential of different concentrations of rotenone extracted from *Derris urucu* against *C. tingomarianus* Bechyné in a greenhouse. The authors evaluated the effect on mortality and leaf consumption of insects, and concluded that the concentration of 0.13 per cent (m.v-1) of rotenone was effective in controlling the pest, causing significant mortality and inhibition of feeding in the insects.

Rotenone-based products were used extensively against the Colorado beetle *Leptinotarsa decemlineata* (Coleoptera: Chrysomelidae), a major potato pest in the northern hemisphere (Costa *et al.*, 1997; Cox, 2002). Rotenone has also been used to catch fish during scientific studies, which found evidence of rapid recolonization (around 3-12 days) in tide pools subjected to poisoning with this substance (Lardner *et al.*, 1993; Rosa *et al.*, 1997).

Azevedo *et al.* (2007) evaluated the effectiveness of several natural products to control *Callosobruchus maculatus* (Fab.) in cowpea (*Vigna unguiculata* (L.) Walp stored and found that commercial extract of timbó (Rotenat®), the basis of rotenone, evaluated at a concentration of 7.5 mL.L$^{-1}$, was the most effective form of control of the weevil in stored cowpea.

The oleoresin extract of yam bean (*Pachyrhizus erosus* L. Urban), at concentrations of 0.03 per cent and 0.06 per cent (m.v$^{-1}$), containing 15 mg.L$^{-1}$ of rotenone, was effective in the control of *Acanthoscelides obtectus* Say (Coleoptera: Bruchidae) by contact and ingestion (Lucio *et al.*, 2011).

For Tokarnia *et al.* (2000) rotenone, despite being very toxic to fish, has low toxicity to warm-blooded animals, and experimentally producing plants of this active ingredient (timbó species) has not been shown to be toxic to cattle and other livestock.

There are several factors that highlight the rotenone present in timbó extracts for pest control. It is not phytotoxic (Reynolds, 1989), it is biodegradable, and it is photosensitive when exposed to light, as it degrades within three days (Moreira *et al.*, 2005). When applied in raw form, it decomposes more quickly than nicotine and pyrethrins and is considered practically harmless to humans, due to its low concentration in the mixtures (Mariconi, 1981).

On the other hand, some studies have demonstrated the high toxicity of rotenone to arthropods (Marinos *et al.*, 2004; Azevedo *et al.*, 2007; Guirado *et al.*, 2007; Almeida, 2010; Correa, 2011; Efrom *et al.*, 2011) and other studies have pointed out that rotenone has low toxicity when used alone and that this substance depends on the association of other compounds to show toxic effects (Decker, 1942; Tyler, 1979; Costa *et al.*, 1999a; Costa *et al.*, 1999b; Correa, 2011; Li *et al.*, 2011; Alecio, 2012; Pena, 2012). Meanwhile, new constituents have been isolated and identified from extracts of timbó that have shown biological effects on arthropods (Mors, 1978; Magalh es *et al.*, 2003; Gassa *et al.*, 2005; Wu *et al.*, 2008; Lobo *et al.*, 2009; Babu *et al.*, 2010; Hymavathi *et al.*, 2011; Li *et al.*, 2011; Jiang *et al.*, 2012).

Almeida (2010) found that a defensive botanical trade, with at least 5 per cent (m.m$^{-1}$) of rotenone, showed low efficiency of insecticides on nymphs of *Euphalerus clitoriae* (Hemiptera: Psyllidae), obtaining the highest mortality rate of the insects (16.9 per cent) at the highest concentration measured (0.6 per cent v.v$^{-1}$). A similar effect was obtained by Lima *et al.* (2008) when they studied the efficiency of commercial extract of timbó (Rotenat ®) against larvae of *S. frugiperda* under natural infestation in corn grown in lowland areas. In this study, the concentration of 0.5 per cent (v.v$^{-1}$) timbó extract, with at least 5 per cent of rotenone, showed low efficiency in pest control.

The insecticidal action of several alternative products was evaluated by Guirado *et al.* (2007) for the control of *C. arcuatus* Olivier (Coleoptera: Crhysomelidae) in the sunflower crop under field conditions, confirming that rotenone, used at a concentration of 1 per cent (m.v$^{-1}$), was not effective in controlling the insects. Ephron *et al.* (2011) also evaluated doses of commercial extract of timbó (Rotenat®), containing at least 5 per cent rotenone, and observed no toxicity with topical application of the poison in adults of the predator *Cryptolaemus montrouzieri* (Coleoptera: Coccinellidae).

Correa (2011) found that the extract of the timbó *Derris rariflora* showed high toxicity against mite of the pepper *Tetranychus desertorum* (Acari: Tetranychidae), while the purified rotenone was not toxic to humans, causing only 2.5 per cent mortality in mites at the highest concentration assessed (1 per cent m.v$^{-1}$).

The biological effects of extracts of *D. urucu* and *D. utillis* were studied by Costa *et al.* (1999a) in populations of *Musca domestica* L. The authors found that the species *D. urucu* was more efficient, even with lower levels of rotenone in the roots. These authors hypothesized that other substances in addition to rotenone, could be acting to control the larvae.

Catto *et al.* (2009) evaluated rotenone in the form of *Derris nicou* root extract against *Boophilus microplus*, and found that the product was effective against larvae infestation, but produced an insufficient degree of antiparasitic activity against the tick *in vitro* and in experimentally infested animals. Neither did it show any significant decrease in the tick infestation (engorged) in the test field.

Decker (1942) reports that the crude extract of *D. elliptica* is more efficient than rotenone against insects, and that the insecticidal effect of timbó roots may be related to the joint action of rotenone with other substances present in extracts of *Derris* and *Deguelia*, including the deguelin and toxicarol. Tyler (1979) states that timbó roots may contain rotenone, deguelin, toxicarol or tephrosin, structurally similar compounds that may have insecticidal properties.

Besides rotenone, Mors (1978) has isolated other active ingredients from roots of *Derris urucu*, including a high-powered foam forming saponin, named "derrisídio". This author emphasizes that in timbó, a toxic complex of this saponin appears to have a dispersant action on the rotenone.

Mendes (1960) found significant variation in the chemical composition of 153 plant extracts of timbó of the same species (*Derris nicou* Benth), which were attributed to the different sampling sites and the age range of plants. In this study, the concentrations of the constituents of the extracts ranged from 12.5 per cent to 61.1 per cent (m.v$^{-1}$) for rotenone, from 12.7 per cent and 79 per cent (m.v$^{-1}$) for deguelin and from 2.2 per cent to 64.7 per cent (m.v$^{-1}$) to other components, enabling the author to classify the plants into two groups (A and B). Plants belonging to the group A had ¼ rotenone, $^2/_4$ deguelin and ¼ of other components, while those in group B had $^3/_6$ of rotenone, $^2/_6$ and deguelin and $^1/_6$ of other compounds in the extracts.

Alecio (2012) also observed a wide qualitative and quantitative variation of the major constituents found in extracts of roots timbó (*Deguelia* and *Lonchocarpus*), with a predominance of rotenone, deguelin, tephrosin and hidroxirotenone.

Magalh es *et al.* (2003) found three new constituents (scandenin, methyl robustate and 4',5-dihydroxy-6-3,3-dimethyilallyl methoxyflavone -7) among the flavonoids isolated from *Derris hatschbachii*. These were identified by comparison of the spectroscopic data (NMR, NMR-2D and MS/MS).

From extracts of *Derris laxiflora*, Wu *et al.* (2008) isolated nine compounds which they characterized by spectroscopic analysis. Seven of these were considered new: O-trans-cinnamoylglutinol (1) 22β-hydroxy-12-oleanen-3-one (2), 15α, 16α-epoxyoleanen-12-3-one (3), 29-hydroxy-12-oleanen-3,22-dione (4), 22 β,29-dihyroxy-12-oleanen-3-one (5), 2,3-(methylenedioxy)-4-methoxy-5-methylphenol (8) and 2,3,6-trimethoxy-5-methylphenol (9), as well as isolated from natural sources: 25-cycloarten-3,24-dione (6) and 24î-hydroxy-25-cycloarten-3-one (7).

Lobo *et al.* (2009) isolated and identified five stilbenes from the leaves of "timbó red" (*Derris rufescens* var. *urucu*): metoxilonchocarpene 4-(1) 3,5-dimethoxy-4 '-hydroxy-3'-prenyl-trans stilbene (2), lonchocarpene (3) 3,5-dimethoxy-4 '-O-prenyl-trans-stilbene (4) and pteroestilbene (5). Components 2 and 4 were reported as new natural products, but compound 2 had already been mentioned as a synthetic product.

Three new dihydro-flavonoids; urucuol, denominated A, B and C and dihydroflavonol isotirumalin, were isolated and identified by Lôbo *et al.* (2009) from the ethanol extract of the leaves of *Derris urucu* (Leguminosae). Their structures were elucidated by extensive spectroscopic analysis of the 1D and 2D-NMR, UV, IR and MS data, and comparison with the literature data. The isolated compounds were evaluated for their DPPH scavenging capacity, and showed low antioxidant power when compared to the commercial antioxidant trans-resveratrol.

Babu *et al.* (2010) conducted a phytochemical study of the plant *Derris scandens* Benth (Leguminosae) using spectroscopic methods, especially 1D and 2D NMR and mass spectral analysis, which resulted in the isolation of a new derivative of isoflavones, the scandinone A, as well as 11 known compounds.

The toxicity of nine isolates compounds of *Derris scandens* was evaluated by Hymavathi Benth *et al.* (2011) for four species of insect pests in stored grains (*Callosobruchus chinensis* L., *Sitophilus oryzae* L., *Rhyzopertha dominica* L. and *Tribolium castaneum* H.), using a fumigation bioassay. The sensitivity of individuals to the compounds varied with exposure time, concentration, and insect species, reaching 100 per cent mortality in individuals after 24 h with the compounds osajin, scandinone, sphaerobioside and genistein against all insects tested, while laxifolin and lupalbigenin showed 100 per cent mortality of *T. castaneum* and *R Dominica* 72 h after the start of the bioassays.

The insecticidal action of two compounds (khayasin and 2'S-methylbutanoylproceranolide) isolates of *Xylocarpus moluccensis* Lam was compared to purified rotenone and they was assessed by ingestion of infected leaves on larvae of the coconut beetle *Brontispa longissima* Gestro (Coleoptera: Chrysomelidae), grown in the laboratory. It was observed that the two compounds were more potent than purified rotenone at a concentration of 50 mg.ml-1 (Li *et al.*, 2011).

Jiang *et al.* (2012) isolated and characterized 12 compounds (11 known and one not reported in the literature) from extracts of the aerial parts of *Derris trifoliata* Lour and evaluated their toxicity against *Artemia salina* Leach (Anostraca: Artemiidae). They identified eight constituents with significant toxicity ($LC_{50}$ between 0.06 to 9.95 mg.mL$^{-1}$: rotenone, tephrosin, 12a-hidroxirotenone, deguelin, 6a, 12a-dehidrorotenone, dehidrodeguelin, 7a methyldeguelol-o-and 4'-hydroxy-7-metoxiflavanone), while the isolated novel compound showed low toxicity ($LC_{50}$ = 211.31 mg.mL$^{-1}$) to the crustacean.

Alecio (2012) evaluated the biological activity of extracts from timbó (*Derris sandens* Aubl. and *Deguelia floribundus* Benth) compared to purified rotenone against adults of *Cerotoma tingomarianus* Bechyné (Coleoptera: Chrysomelidae) and *Spodoptera frugiperda* Smith (Lepidoptera: Noctuidae) and verified that: the toxicity of timbó species is related to the chemical composition of the extracts, the form of exposure and the target insect species; the chemical composition of extracts timbó decisively influence the biological toxic effects on insects; rotenone, when used alone, has low toxicity but when associated with other constituents, its toxic effect is enhanced; and greater toxicity of timbó species to larvae of *S. frugiperda* is promoted when the timbó extracts are prepared with approximately $^3/_{10}$ of rotenone, $^3/_{10}$ of deguelin, $^3/_{10}$ of tephrosin, and $^1/_{10}$ hidroxirotenone and smaller amounts of other constituents. These proportions of components were considered to be most suitable for the selection of extracts from timbó, aiming at the development of biotech products to control insect pests in agriculture.

Thus, there is great interest ecological and economic in these active principles as control agents of insect pests in agriculture, or in certain plant/insect relationships (Crombie and Whiting, 1998). Further studies are needed, to define the most appropriate quantities and proportions of the active ingredients of timbó to be used for the control of each species of insect pest.

# References

Alecio, M.R., Alves, S.B., Gonzaga, A.D., Ribeiro, J.D., and Correa, R.S. (2005). Avaliaç o do potencial inseticida *"in vitro"* do extrato aquoso de raízes de timbó (*Derris rariflora*) sobre *Sitophilus zeamais Mots* In: I Jornada Amazonense de Plantas medicinais. Manaus. *Anais da I Jornada Amazonense de Plantas medicinais*, p. 42

Alecio, M.R., Fazolin, M., Coelho Netto, R.A., Catani, V., Estrela, J.L.V., Alves, S.B., Correa, R. da S., Andrade Neto, R. de C., and Gonzaga, A.D. (2010). Aç o inseticida do extrato de *Derris amazonica* Killip para *Cerotoma arcuatus Olivier* (Coleoptera: Chrysomelidae). *Acta Amazonica*, **40(4)**: 719-728.

Alecio, M.R., Fazolin, M., Veiga Júnior, V.F., Estrela, J.L.V., Alves, S.B., Andrade Neto, R. de C., Paiva, F. de F.G., Monteiro, A.F.M., Damaceno, J.E. de O., Cavalcante, A.S. da S., Albuquerque, E.S. de A., andOliveira, P. de A. (2011). Aç o inseticida do extrato de *Deguelia floribundus* Benth para *Cerotoma tingomarianus* Bechyné (Coleoptera: Chrysomelidae). *Anais do V Congresso Brasileiro de Defensivos Agrícolas Naturais*, Embrapa Meio Ambiente. Jaguariúna/SP.

Alecio, M.R. (2012). Atividade Biológica de extratos de timbó (*Derris scandens* Aubl. e *Deguelia floribundus* Benth) sobre *Cerotoma tingomarianus* Bechyné (Coleoptera: Chrysomelidae) e *Spodoptera frugiperda* (J. E. Smith, 1797) (Lepidoptera: Noctuidae). Doctor's Thesis, Universidade Federal do Amazonas/Pós-Graduaç o em Biotecnologia, Manaus, Amazonas, p. 203.

Almeida, M.N. de. (2010). Eficiência de um inseticida botânico no controle de ninfas de *Euphalerus clitoriae* (Hemiptera: Psyllidae). *Revista Controle Biológico (BE-300) On-Line*. 2: Disponível em: http://www.ib.unicamp.br/profs/eco_aplicada. Acesso em: 20 de março de 2012.

Amaral, B.D. (2004). Fishing territoriality and diversity between the ethnic populations Ashaninka and Kaxinawá, Breu river, Brazil/Peru. *Acta Amazonica*, **34(1)**: 75-88.

Arnason, J.T., Philogcne, B.J.R., and Morand, P. (1990). Insecticide of plant origin. Washington, DC, American Chemical Society.

Azevedo, F.R., Guimar es, J.A., Braga Sobrinho, R., and Lima, M.A.A. (2005). Eficiência de produtos naturais para o controle de *Bemisia tabaci* Biótipo B (Hemiptera: Aleyrodidae) em meloeiro. *Arq. Inst. Biol. S o Paulo*, **72(1)**: 73-79.

Azevedo, F.R., Leit o, A.C.L., Lima, M.A.A., and Guimar es, J.A. (2007). Eficiência de produtos naturais no controle de *Callosobruchus maculatus* (Fab.) em feij o caupi (*Vigna unguiculata* (L.) Walp) armazenado. *Revista Ciência Agronômica*, **38(2)**: 182-187.

Babu, T.H., Tiwari, A.K., Rao, V.R.S., Ali, A.Z.; Rao, J.M., and Babu, K.S. (2010). A new prenylated isoflavone from *Derris scandens* Benth. *J. Asian Nat Prod Res.*, **12(7)**: 634-638.

Baldin, E.L.L., Prado, J.P.M., Christovam, R.S., and Dal Pogetto, M.H.F.A. (2009). Uso de pós de origem vegetal no controle de *Acanthoscelides obtectus* Say (Coleoptera: Bruchidae) em gr os de feijoeiro. *BioAssay*, **4(2)**: 1-6.

Bell, A., Fellows, L.E., andSimmonds, M.S.J. (1990). Natural products from plants for the control of insect pests. *In*: *Safer insecticide development and use*. Hodgson, E. and Kuhr, R.J. New York and Basel. Marcel Dekker. pp. 337-383.

Bogornie, P.C., and Vendramim, J.D. (2003). Bioatividade de extratos aquosos de *Trichilia* spp. sobre *Spodoptera frugiperda* (J. E. Smith) (Lepidoptera: Noctuidae) em milho. *Neotropical Entomology*, **32**: 665-669.

Breuer, M., De Loof, A. (1998). Meliaceous plant preparations as potential insecticides for control of the oak precessionary, *Thaumatopoea precessionea* (L.) (Lepidoptera: Thaumetopoeidae). *Med Fac Landbouww Univ Gent.*, **63/2b**: 526-536.

Brito, C.H., Mezzomo, J.A., Batista, J.L., Lima, M.S.B., and Murata, A.T. (2004). Bioatividade de extratos vegetais aquosos sobre *Spodoptera frugiperda* em condiç es de laboratório. Costa Rica. *Manejo Integrado de Plagas y Agroecología*, **71**: 41-45.

Brown Júnior, K.S. (1988). Engenharia ecológica: novas perspectivas de seleç o e manejo de plantas medicinais. *Acta Amazônica*, **18(1)**: 291-303.

Busato, G.R., Grützmacher, A.D; Garcia, M.S., Zotti, M.J.; Nörnberg, S.D., Magalh es, T.R., and Magalh es, J.B. (2006). Susceptibilidade de lagartas dos biótipos milho e arroz de *Spodoptera frugiperda* (J.E. Smith, 1797) (Lepidoptera: Noctuidae) a inseticidas com diferentes modos de aç o. *Ciência Rural*, **36(1)**: 15-20.

Caminha Filho, A. (1940). *Timbó e rotenona: uma riqueza nacional inexplorada*. Serviço de Informaç o Agrícola, Rio de Janeiro.

Castro, H.G., Ferreira,F. A.;, Silva, D. J. H., and Mosquim, P. R. (2004). *Contribuiç o ao estudo das plantas medicinais: metabólitos secundários*. UFV, Viçosa.

Catto, J.B., Bianchin, I., Santurio, J.M., Feijo, G.L.D., Kichel, A.N., and Silva, J.M. (2009). Sistema de pastejo, rotenona e controle de parasitas em bovinos cruzados: efeito no ganho de peso e no parasitismo. *Rev. Bras. Parasitol. Vet.* Jaboticabal, **18(4)**: 37-43.

Cavalheiro, R.A., Fortes, F., Borecký, J., Faustinoni, V.C., and Schreiber, A.Z. (2004). Respiration, oxidative phosphorylation, and uncoupling protein in *Candida albicans*. *Braz J Med Biol Res.*, **37(10)**: 1455-1461.

Coll, J. (2005). Cubé Resin Insecticide: Identification and Biological Activity of 29 Rotenoid Constituents. *J. Agric. Food Chem.*, **53**: 3749–3750.

Colman-Saizarbitoria, T., Montilla, L.; Rodriguez, M., Castillo, A., and Hasegawa, M. (2009). Xymarginatin: a new acetogenin inhibitor of mitochondrial electron transport from *Xylopia emarginata* Mart., Annonaceae. *Brazilian Journal of Pharmacognosy*, **19(4)**: 871-875.

Corbett, C.E. (1940). *Plantas ictiotóxicas: farmacologia da rotenona*. S o Paulo: Faculdade de Medicina da Universidade de S o Paulo. S o Paulo.

Correa, R.S. (2006). Toxicidade de extratos de *Lonchocarpus floribundus* Benth (timbó) sobre *Toxoptera citricida* Kirkald (pulg o preto do citros) (Sternorrhynda: Aphididae). Master's Thesis, Instituto Nacional de Pesquisas da Amazônia/ Fundaç o Universidade do Amazonas, Manaus, Amazonas, p. 71.

Correa, R.S. (2011). Toxicidade de extratos de timbós (*Derris* spp.) sobre *Tetranychus desertorum* (Acari: Tetranychidae) em folhas de piment o. Doctor's Thesis, Universidade Federal do Amazonas/Pós-Graduaç o em Biotecnologia, Manaus, Amazonas, p. 72.

Costa, N.A., Nascimento, C.N.B., Moura Carvalho, L.O.D., and Pimentel, E.S. (1986). *Uso do timbó urucu (Derris urucu) no controle do piolho (Haemotopinus tuberculatus) em bubalinos*. Belém: EMBRAPA-CPATU, Belém. 16 pp. (Boletim de Pesquisa, 78).

Costa, J.P., Belo, M., and Barbosa, J.C. (1997). Efeitos de Espécies de Timbós ( *Derris* spp.: Fabaceae) em Populaç es de *Musca domestica* L. *An. Soc. Entomol. Brasil,* **26(1)**: 163-168.

Costa, J.P.C., Alves, S.M., and Belo, M. (1999a). Teores de rotenona em clones de timbó (*Derris* spp., Fabaceae) de diferentes regi es da Amazônia e os seus efeitos na emergência de imagos em *Musca domestica* L. *Acta Amazônica*, **29(4)**: 563-573.

Costa, J. P. C., Alves, S. M., and Belo, M. (1999b). Diferença entre as espécie de timbó (*Derris* spp., Fabaceae) de diferentes regi es da Amazônia no controle de *Musca domestica* L. *Acta Amazônica*, **29(4)**: 573-583.

Cox, C. (2002). Pyrehrins/Pyrethrum. *Journal of Pesticide Reform,* **22**: 14-20.

Cravero, E.S., Guerra. M. de S., and Silveira, C.P.D. (1976). *Manual de inseticidas e acaricidas:* aspectos toxicológicos. Aimara Ltda. Pelotas.

Crombie, L., and Whiting, A.D. (1998). Biosynthesis in the rotenoid group of natural products: applications of isotope methodology. Oxford. *Phytochemistry*, **49(6)**: 1479-1507.

Cruz, I. (1995). *A lagarta-do-cartucho na cultura do milho*. Sete Lagoas. Embrapa-CNPMS. 45p. (Circular Técnica, 21).

Dahlem, A.R., Possenti, J.C., and Paulus, D. (2009). Produtos alternativos no controle de pragas na cultura do feij o-vagem em sistema orgânico. *XIV Seminário de Iniciaç o Científica e Tecnológica da UTFPR*, p. 4.

Decker, S. (1942). *Inseticidas vegetais*. S o Paulo: Secretaria de Agricultura, Indústrias e Comércio do Estado de S. Paulo. 18pp (Boletim da Agricultura).

Diez-Rodriguez, G.I., and Omoto, C. (2001). Herança da resistência de *Spodoptera frigiperda* (J. E. Smith) (Lepidoptera: Noctuidae) à lambda-cyjalothrin. *Neotropical Entomology*, **30(2)**: 311-316.

Ducrot, P.H. (2004). Contribución de la química al conocimiento de la actividad biopesticida de los productos naturales de origen vegetal. *In: Biopesticidas de origen vegetal*. Ed. Regnault-Roger, C., B.J.R. Philogene y C. Vincent. Ediciones Mundi-Prensa, Madrid, p. 337.

Efrom, C.F.S., Redaelli, L.R., Meirelles, R.N., and Ourique, C.B. (2011). Selectivity of phytosanitary products used in organic farming on adult of *Cryptolaemus montrouzieri* (Coleoptera, Coccinellidae) under laboratory conditions. *Ciências Agrárias. Londrina*, **32(4)**: 1429-1438.

Fang, N., and Casida J.E. (1999). Rotenone, deguelin, Their metabolites, and the rat model of Parkinson's disease. *J. Agric. Food Chem.*, **47**: 2130–2136.

Faroni, L.R.A., Molin, L., Andrade, E.T., andCardoso, E.G. (2005). Utilizaç o de produtos naturais no controle de *Acanthoscelides obtectus* em feij o armazenado. *Rev. Bras. Armaz.*, **20**: 44-48.

Fazolin, M. (2002). Avaliaç o de plantas com potencial inseticida no controle da vaquinha-do-feijoeiro- Rio Branco: EMBRAPA Acre 42pp. (Boletim de pesquisa e desenvolvimento, 37).

Gallo, D., Nakano, O., Silveira Neto, S., Carvalho, R.P.L., Batista, G.C. de, Berti Filho, E., Parra, J.R.P., Zucchi, R.A., Alves, S.B., Vendramim, J.D., Marchini, L.C., Lopes, J.R.S., andOmoto, C. 2002. *Manual de Entomologia Agrícola*. Piracicaba: Fealq. 920 pp. (Biblioteca de Ciências Agrárias Luiz de Queiroz, 10).

Gassa, A., Fukui, M., Sakuma, M., Nishioka, T., and Takahashi, S. (2005). Rotenoids in the Yam bean *Pachyrrhizus erosus*: Possible defense principles against herbivores. *J. Environ. Entomol. Zool.*, **15(4)**: 251-259.

Gosselin, R.E. (1984). *Clinical Toxicology of Commercial Products*. Williams and Wilkins. Baltimore/London.

Guirado, N., Mendes, P.C.D., Ambrosano, E.J., Rossi, F., and Arévalo, R.A. (2007). Controle de *Cerotoma arcuatus* com produtos alternativos na cultura de girassol. *Rev. Bras. Agroecologia*, **2(1)**: 587-590.

Grützmacher, A.D., Lima, J.F.M., Cunha, U.S., Porto, M.P., Martins, José F.S., and Dalmazo, G.O. (2000). Efeito de inseticidas e de tecnologia de aplicaç o no controle da lagarta-do-cartucho na cultura do milho no agroecossistema de várzea. Embrapa CPACT. p.567-573 (Documentos, 70).

Gusm o, D.S., Páscoa, V., Mathias, L., Vieira, I.J.C., Braz-Filho, R., and Lemos, F.J.A. (2002). *Derris* (*Lonchocarpus*) *urucu* (Leguminosae) extract modifies the peritrophic matrix structure of Aedes aegypti (Diptera: Culicidae). *Mem Inst Oswaldo Cruz. Rio de Janeiro.*, **97(3)**: 371-375.

Henandez, C.R., and Vendramim, J.D. (1996). Toxicidad de extractos acuosos de Meliaceae en *Spodoptera frugiperda* (Lepidoptera: Noctuidae). *Manejo Integr. Plagas.*, **42**: 14-22.

Hymavathi, A., Devanand, P., Suresh Babu, K., Sreelatha, T., Pathipati, U.R., and Madhusudana, R, J. (2011). Vapor-phase toxicity of *Derris scandens* Benth-derived constituents against four stored-product pests. *J Agric Food Chem.*, **59(5):** 1653-1657.

Jiang, C., Liu, S., He, W., Luo, X., Zhang, S., Xiao, Z., Qiu, X., and Yin, H. (2012). A New Prenylated Flavanone from *Derris trifoliata* Lour. *Molecules*, **17**: 657-663.

Kotze, A.C., Dobson, R.J., and Chandler, D. (2006). Synergism of rotenone by piperonyl butoxide in *Haemonchus contortus* and *Trichostrongylus colubriformis in vitro*: Potential for drug-synergism through inhibition of nematode oxidative detoxification pathways. *Veterinary Parasitology*, **136**: 275-282.

Lardner, R., Ivantsoff, W.L., and Crowley, E.L.M. (1993). Recolonization by fishes of a rocky intertidal pool following repeated defaunation. *Australian Zool.*, **29(2)**: 85-92.

Li, M.Y., Zhang, J., Feng, G., Satyanandamurty, T., and Wu, J. (2011). Khayasin and 2′S-methylbutanoylproceranolide: Promising candidate insecticides for the control of the coconut leaf beetle, *Brontispa longissima. J. Pestic. Sci.*, **36(1)**: 22–26.

Lima, R.R. (1947). Os timbós da Amazônia brasileira. *Bol. Min. Agric.*, 36: 14-29.

Lima, R.R. (1987). Informaç es sobre duas espécies de timbó: *Derris urucu* (Killip *et al.*, Smith) Macbride e *Derris nicou* (Killip et Smith) Macbride, como plantas inseticidas. Belém: EMBRAPA-CPATU. 23 pp. (Documentos, 42).

Lima, M.P.L., Oliveira, J.V., Gondim Junior, M.G.C., Marques, E.J., and Correia, A.A. (2010). Bioactivity of neem (*Azadirachta indica* A. Juss, 1797) and *Bacillus thuringiensis* sub sp. *aizawai* formulations in larvae of *Spodoptera frugiperda* (J.E. Smith) (Lepidoptera: Noctuidae). *Ciênc. agrotec.*, **34(6)**: 1381-1389.

Lima, R.R., and Costa, J. P. C. da. (1998). *Coleta de plantas de cultura pré-colombiana na Amazônia brasileira*. Belém: EMBRAPA-CPATU. 102pp. (EMBRAPA-CPATU. Documentos, 107).

Lôbo, L.T., Silva, G.A., Ferreira, M., Silva, M.N., Santos, A.S., Arruda, A.C., Guilhon, G.M.S. P., Santos, L.S., Borges, R.S., and Arruda, Mara S.P. (2009). Dihydrofavonols from the leaves of *Derris urucu* (Leguminosae): Structural Elucidation and DPPH Radical-Scavenging Activity. *J. Braz. Chem. Soc.*, **20(6)**: 1082-1088.

Lôbo, L.T., Silva, G.A., Freitas, M.C.C., Souza Filho, A.P.S., Silva, M.N., Arruda, A.C., Guilhon, G.M.S.P., Santos, L.S., Santos, A.S., and Arruda, M.S.P. (2010). Stilbenes from *Deguelia rufescens* var. urucu (Ducke) A.M.G. Azevedo leaves: effects on seed germination and plant growth. *J. Braz. Chem. Soc.,* [online]. **10**: 1838-1844.

Lucio, J.A.R., Goiz, J.M.J., Moya, *E.G.,* Andrés, M.D.F., Hernández, C.R., and Bárcenas, E.A. (2011). Oleorresina de jícama y calidad de semilla de frijol infestada con Acanthoscelides obtectus Say. *Agron. Mesoam.,* **22(1)**: 109-116.

Luitgards-Moura, J.F., Bermudez, *E.G.C.,* Rocha, A.F.I., Tsouris, P., and Rosa-Freitas, M.G. (2002). Preliminary Assays Indicate that *Antonia ovata* (Loganiaceae) and *Derris amazonica* (Papilionaceae), Ichthyotoxic Plants Used for Fishing in Roraima, Brazil, Have an Insecticide Effect on *Lutzomyia longipalpis* (Diptera: Psychodidae: Phlebotominae). *Mem. Inst. Oswaldo Cruz.,* Rio de Janeiro, RJ. **97(5)**: 737-742.

Magalh es, A.F., Tozzi, A.M.G.A., Magalh es, *E.G.,* and Moraes, V.R.S. (2003). New spectral data of some flavonoids from *Deguelia hatschbachii* A.M.G. Azevedo. *J. Braz. Chem. Soc.,* **14(1)**: 133-137.

Maini, P.N., and Morallo, R.B. (1993). Molluscicidial activity of *Derris elliptica* (Farm. Leguminosae). *Philipp. J. Sci.,* **122(1)**: 61-69.

Mascaro, U.C.P., Rodrigues, L.A., Bastos, J.K., Santos, E., and Costa, J.P.C. (1998). Valores de $DL_{50}$ em peixes e no rato tratados com pó de raízes de *Derris* sp. e suas implicaç es ecotoxicológicas. *Pesq. Vet. Bras.,* **18(2)**: 53-56.

Marinos, C., Castro, J., and Nongrados, D. (2004). Biocidal effect del barbasco *Lonchocarpus utilis* (Smith, 1930) as regulator of mosquitoes larvae. *Revista Peruana de Biologia,* **11(1)**: 87-94.

Melo, A.B., Oliveira, S.R., Leite, D.T., Barreto, C.F., and Silva, H.S. (2011). Inseticidas botânicos no controle de pragas de produtos armazenados. *Revista Verde.,* **6(4)**: 1-10.

Mendes, L.O.T. (1960). Seleç o e melhoramento do timbó: II–Estudo de uma populaç o de 153 plantas de timbó macaquinho–*Derris nicou* (Benth.) Timbó improvement: II–Study of a population of 153 plants of timbó macaquinho (*Derris nicou*). *Bragantia* [online], **19**: 273-305.

Moreira, M.D., Picanço, M.C., Silva, E.M., Moreno, S.C., and Martins, J.C. (2005). Uso de inseticidas botânicos no controle de pragas. In: *Controle alternativo de pragas e doenças.* Ed. by Venson, M., Paula Júnior, T.S., Pallini, A. Viçosa: EPAMIG/CTZM. pp. 89-120.

Mors, W. (1978). Plantas ictiotóxicas: aspectos químicos. *Ciênc. Cult.,* **32**: 42.

Nawnot, J., Harmatha, J., and Bloszy, E. (1987). Secondary plant metabolites with antifeeding activity and their effects on some stored product insects. *In: Internacional Working Conference on Stored – Product Protecion,* Donahaye 4. pp. 591-597.

Oliveira, J. V., and Vendramim, J. D.(1999). Repelência de óleos essenciais e pós-vegetais sobre adultos de *Zabrotes subfasciatus* (Boh.) (Coleoptera: Bruchidae) em sementes de feijoeiro. *Anais da Sociedade Entomológica do Brasil,* **28(3)**: 549-555.

Pena, M.R. (2012). Bioatividade de extratos aquosos e orgânicos de diferentes plantas inseticidas sobre a mosca-negra-dos-citros, *Aleurocanthus woglumi* Ashby 1915 (Hemiptera: Aleyrodidae). Doctor's Thesis, Universidade Federal do Amazonas/Pós-Graduaç o em Agronomia Tropical, Manaus, Amazonas, p. 188.

Pereira, J.R., and Famadas, K.M. (2004). Avaliaç o "*in vitro*" da eficiência do extrato da raiz do timbó (*Dahlstedtia pentaphylla*) (leguminosae, papilionoidae, millettiedae) sobre *Boophilus microplus* (Canestrini, 1887) na regi o do vale do Paraíba, S o Paulo, Brasil. *Arq. Inst. Biol.*, **71(4)**: 443-450.

Philogene, C., Regnault, R., andVincent, C. (2004). Productos fitosanitarios insecticidas de origen vegetal: promesas de ayer y de hoy. In: *Biopesticidas de origen vegetal*. Ed. by Regnault-Roger, C., B.J.R. Philogene y C. Vincent. Ediciones Mundi-Prensa, Madrid, p. 337.

Pinto, A.N. (1937). Os timbós e suas aplicaç es e possibilidades. *Chimica e Indústria*, p. 247

Pinto, G.P. (1953). Estudo sobre a composiç o química de das raízes de *Derris urucu* Killip and Smith. *Anais da Associaç o Brasileira de Química*, **4(12)**: 173-179.

Pires, J.M. (1978). Plantas ictiotóxicas – aspectos botânicos. *Ciênc. Cult.*, **32**: 37-41.

Saito, S. (2005). Effects of Pyridalyl on ATP Concentrations in Cultured Sf9 Cells. J. *Pestic. Sci.*, **30(4)**: 403-405.

Saito, M.L., and Luchini, F. (1998). Substâncias obtidas de plantas e a procura por praguicidas eficientes e seguros ao meio ambiente. Jaguriúna: Embrapa-CNPMA. 46pp. (EMBRAPA-CNPMA. Documentos, 12).

Silva, G.A., Lagunes, J.C., and Rodríguez, D. (2002). Insecticidas vegetales: Una vieja-nueva alternativa en el control de plagas. *Revista Manejo Integrado de Plagas* (CATIE), **2(3)**: 21-56

Sousa, A.H., Maracajá, P.B., Silva, R.M.A. Moura, Antonia M.N., and Aandrade, W.G. (2005) Bioactivity of vegetal powders against *Callosobruchus maculatus* (Coleoptera: Bruchidae) in caupi bean and seed physiological analysis. *Revista de Biologia e Ciências da Terra.*, **5(2)**: 1-5.

Rosa, R.S., Rosa, I.L., and Rocha, L.A. (1997). Diversidade da ictiofauna de poças de maré da praia do Cabo Branco, Jo o Pessoa, Paraíba, Brasil. *Rev. Bras. Zool.*, [online]. **14(1)**: 201-212.

Rattanapan, A. (2009). Effects of rotenone from *Derris* crude extract on esterase enzyme mechanism in the beet armyworm, *Spodoptera exiqua* (Hubner). *Commun Agric Appl Biol Sci.*, **74(2)**: 437-44.

Reynolds, J.E.F. (1989). Martindale The Extra Pharmacopoeia. *Pharmaceutical Press.* London. **29**: 1896.

Schmidt, G. H., Ahmed, A. A. I., and Breuer, M. (1997) Effect of Melia azedarach extract on larval development and reproduction parameters of *Spodoptera littoralis* (Boisd.) and *Agrotis ipsilon* (Hufn.) (Lep. Noctuidae). *Anzeiger fur Schadlingskunde Pfanzenschutz Umweltschutz*, Berlin, **70(1)**: 4-12.

Silva, G. A., Lagunes, J. C., and Rodríguez, D. (2002). Insecticidas vegetales: Una vieja-nueva alternativa en el control de plagas. *Revista Manejo Integrado de Plagas* (CATIE), **2(3)**: 21-56

Talukder, F.A, and Howse, P.E. (1995). Evaluation of *Aphanamixis polytachya* as a source of repellents, antifeedants, toxicants and protectants in storage against *Tribollium castaneum* (Herbest). *J. Stored Prod. Res.*, **31**: 55-61.

Tokarnia, C.H., Dobereiner, J.E., and Peixoto, P.V. (2000). Plantas tóxicas do Brasil. Rio de Janeiro.

Tozzi, A.M.G.A. (1998). A identidade do timbó-verdadeiro: *Deguelia utilis* (A.C.Sm.) A.M.G. Azevedo (Leguminosae: Papilionoideae). *Rev. Brasil. Biol.*, **58(3)**: 511-516.

Tyler, V.E., Brady, L.R., and Robbers, J.E. (1979). *Farmacognosia*. Editorial El Ateneo, Buenos Aires.

Valladares, G.R., D. Ferreyra, M.T., Defago, M.C., and Palácios, S. (1999). Effects of *Melia azedarach* on *Triatoma infestans*. *Fitoterapia*, **70**: 421-424.

Vendramim, J.D. (1997). Uso de plantas inseticidas no controle de pragas. *In: ciclo de palestras sobre agricultura orgânica.* S o Paulo. Palestras. Campinas: Fundaç o Cargill, pp. 64-69.

Viegas Júnior C.J. (2003). Terpenos com atividade inseticida: Uma alternativa para o controle químico de insetos. *Quím. Nova.*, **3(26)**: 390-400.

Wu, J.H., Tung, Y.T., Lee, T.H., Chien, S.C., and Kuo, Y.H. (2008). Triterpenoids and aromatics from *Derris laxiflora*. *J. Nat Prod.*, **71**: 1829-32.

Yu, S.J. (2006). Insensitivity of acetylcholinesterase in a field strain of the fall, *Spodoptera frugiperda* (J. E. Smith). *Pesticide Biochemistry and physiology*, **84**: 135-142.

Yu, S.J., and MCord, E.J. (2007). Lack of cross-resistance to indoxacarb in insecticide-resistant *Spodoptera frugiperda* (Lepidoptera: Noctuidae) and *Plutella xylostella* (Lepidoptera: Yponomeutidae). *Pest Management Science*, **63**: 63-67.

Utilisation and Management of Medicinal Plants Vol. 2 (2014)    *Pages* **329–342**
*Editor-in-Chief:* **V.K. Gupta**
*Published by:* **DAYA PUBLISHING HOUSE, NEW DELHI**

# 15

# Essential Oils: Their Chemistry, Extraction Techniques, Quality Control and Utilization

Archana Peshin Raina[1]*

## ABSTRACT

*Essential oils are chemical compounds of odoriferous nature, which are highly volatile, insoluble in water but soluble in organic solvents, obtained from herbs, flowers, woods, seeds including spices by hydro/steam distillation, expression, absorption in to fat or solvent extraction. Essential oils are present in different parts of plants such as roots, stems, leaves, flowers, and seeds. Natural essential oils are complex chemical compounds having terpenes or hydrocarbons in unsaturated straight chains molecule based isoprene ($C_5H_8$) structure. Growing and harvesting conditions are optimized for production of best fragrances. Essential oils should be subjected to both qualitative and quantitative test to know its purity. Essential oils are the volatile oils of aromatic plants used in perfumery, cosmetic and flavour industries. This review will describe complete details of essential oil chemistry, various extraction methods, quality evaluation, storage and uses of essential oil.*

***Keywords***: Absolute, Aromatic plants, Chemistry, Concrete, Distillation, Essential oil, Extraction, Terpenes, Perfumery, Quality.

## Introduction

Nature is an important supplier of raw material to the perfumery industry. Wild plants have always been a natural reservoir of new and more exotic fragrances.

---

1    Germplasm Evaluation Division, National Bureau of Plant Genetic Resources, New Delhi – 110 012

*    *Corresponding author*: E-mail: aprraina@yahoo.co.in

Limitations in the availability of natural essentials in sufficient quantities and steady price rise leads to the gradual replacement of nature by synthetics. Worldwide demand for fragrances is constantly growing. In India, we can grow almost all different types of aromatic plants, moreover our flora is also very rich in essential oil bearing plants. These plants have a large potential of yielding new, diverse and more exciting perfumes. World trade in essential oils continues to increase every year in spite of introduction and consumption of synthetics. The essential oils are used for flavour or perfumes and also in wide variety of finished products from soaps to musk, lipsticks to carbonated beverages, pharmaceuticals specialties to toilet papers and so on. Essential oils are the volatile oils of aromatic plants used in perfumery, cosmetic and flavour industries. They are also used in aromatherapy. They contain mixtures of organic compounds belonging to different classes of compounds such as terpenes, phenols, phenyl propanoids, aliphatic compounds etc. Monoterpenes and their oxygenated derivatives are mostly the class of compounds present in several essential oils. Also some heterocyclic compounds and amino compounds are present as minor constituents of some oils and floral absolutes. Oxygenated monoterpenes are the main odour carriers of the essential oils and are more stable than the terpene hydrocarbons that have a tendency to get oxidized and resinified under the influence of light, air and moisture. Improper storage conditions also cause deterioration in quality and odour of the oils.

## What are Essential Oils?

Essential oils are highly concentrated natural plant extracts; a drop or two can produce significant results. The term essential indicates that the oil carries distinctive scent (essence) of the plant. An entire plant, when distilled, might produce only a single drop of essential oil. That is why their potency is far greater than dried herbs. Pressing or distillation extracts the subtle, volatile liquids (meaning they evaporate quickly) from plants, shrubs, flowers, trees, roots, bushes, and seeds, that make up essential oils. *Essential oils are complex mixtures of odorous and steam-volatile compounds, which are deposited by plants in the subcuticular space of glandular hairs, in cell organelles (oil bodies of Hepaticae), in idioblasts, in excretory cavities and canals or exceptionally in heartwoods.*

## Physico-Chemical Properties

Essential oils are liquids at ambient temperature, but they are also volatile, which differentiates them from "Fixed oils". They are very rarely colored. Their density is generally lower than that of water (the essential oils of clove, cinnamon or vetiver are the exceptions). Essential oils are very different from vegetable oils (also called fatty oils), such as corn oil, olive oil, peanut oil, etc. Pressing nuts or seeds produces fatty oils. They are quite greasy, are not antimicrobial nor help transport oxygen, and will go rancid over time. Essential oils, however, are not greasy nor do they clog the pores like vegetable oils can.

## Why Do Plants Produce Essential Oils?

Essential oils are product of plant's secondary metabolism, which also produces the larger molecular structures that form tannins, steroids, alkaloids, glycosides, bitters, gums and so on. Essential oils are the life-blood of the plant, protecting it from bacterial and viral infections, cleansing breaks in its tissue and delivering oxygen and nutrients into the cells. In essence, they act as the immune system of the plant and give the plant its characteristic flavor and its aroma.

## Essential Oil Localization

Essential oils are natural plant products, which accumulate, in specialized structures such as oil cells, glandular trichomes, and oil or resin ducts. Essential oils are produced by tiny oil glands located at the petals, leaves, stem, root, bark and wood of plants. These oils evaporate very easily at room temperature coming in contact with air releasing a pleasant fragrance. In case of leaves and petals of flowers they are produced in the innermost cell membrane in the parenchymatous tissue and in case of other plant parts it is found in the cytoplasm or separate cell centers. Under normal natural conditions, essential oils are released from the plant slowly to the surroundings. When heated or crushed these glands break releasing the aroma. The formation and accumulation of essential oils in plants have been reviewed by Croteau (1986), Guenther (1972) and Runeckles and Mabry (1973). Chemically, the essential oils are primarily composed of mono- and sesquiterpenes and aromatic polypropanoids synthesized via the mevalonic acid pathway for terpenes and the shikimic acid pathway for aromatic polypropanoids. Essential oils accumulate in all

**Table 15.1**: Classification of some important aromatic plants based on plant parts containing fragrance.

| Plant Part | Examples |
|---|---|
| Plants with scented leaves | *Artemisia dracunculus, Artemisia pallens, Eucalyptus globulus, Eucalyptus citridora, Tagetes patula, Skimmia laureola, Melissa officinalis, Thuja species, Ocimum basilicum, Ocimum sanctum, Pelargonium graveolens, Mentha species, Lavendula officinalis, Rosmarinus officinalis, Cymbopogon spp., Salvia spp.* |
| Plants with scented flowers | *Tagetes patula, Rosa damascena, Jasmine spp., Kaempferia galanga, Matricaria chamomila, Polianthes tuberosa, Crocus sativus, Hypericum perforatum* |
| Plants with aromatic wood | *Cinnamomum camphora, Boswellia serrata* |
| Plants with aromatic bark | *Juglans regia, Cinnamomum species, Boswellia serrata* |
| Plants with aromatic gums | *Boswellia serrata, Commiphora myrrha* |
| Plants with scented underground parts | *Acorus calamus, Allium cepa, Alpinia galanga, Alpinia calcarata, Vetiveria zizanioides, Zingiber officinale, Costus speciosus, Valeriana officinalis, Hedychium spicatum, Nardostachys jatamansi, Curcuma aromatica* |
| Plants with aromatic seeds | *Coriandrum sativum, Carum carvi, Foeniculcm vulgare, Cuminum cyminum, Anethum graveolens, Apium graveolans* |
| Plants containing fruits with scented peel | *Citrus limonum* |
| Plants with all parts scented | *Tagetes erecta, Mentha species, Salvia sclerea, Ocimum species, Artemisia species* |
| Plants with scented fruits and berries | *Citrus aurantium, Piper longum, Zanthoxylum armatum* |

types of vegetable organs (Table 15.1)-flowers (Tuberose, Jasmine), leaves (Citronella, Eucalyptus), bark (Cinnamon), woods (Rosewood, Sandal wood), roots (Vetiver), rhizomes (Turmeric, Ginger), fruits (all spices, anises) and seeds (Nutmeg). They have been known to occur in some 60 plant families but particularly distributed in Labiatae, Rutaceae, Geraniaceae, Umbelliferae, Compositae, Lauraceae, Graminaeae and Leguminosae. All of the organs of a given species may contain an essential oil, but the composition of this oil may vary with the localization.

## Methods of Extraction of Natural Essential Oils

The essential oils from aromatic plants are for the most part volatile and thus, lend themselves to several methods of extraction such as hydrodistillation, water and steam distillation, direct steam distillation, and solvent extraction (ASTA, 1968; Guenther, 1972; Heath, 1981; Sievers, 1928). The specific extraction method employed is dependent upon the plant material to be distilled and the desired end-product. These processes are developed in order to obtain the best aromatic substance to satisfy the needs of the perfumer and of the flavourist.

1. Hydrodistillation
2. Steam distillation
3. Solvent extraction
4. Expression
5. Enfleurage
6. Super-Critical Fluid extraction (SCFE)

### Hydro Distillation (Water Distillation)

The principle of water/hydro distillation is to boil a suspension of an aromatic plant material and water so that its vapours can be condensed. In water distillation the plant material is always in direct contact with water. An extremely important factor is that in stills where the water is boiled by direct contact with the fire, the water present in the still must always be more than enough to last throughout the distillation; otherwise the plant material can overheat and char and can produce off notes.

### Steam Distillation

In this process, steam generated in a boiler is passed through steam pipes into distillation tank, which is filled with harvested aromatic crop. Pressure, temperature and amount of, the steam can be regulated as per requirement, thereby time of distillation can also be varied. The steam converts the essential oil in the aromatic plants into oil vapour and both these pass through delivery pipe into condenser where due to the effect of circulating cold water, steam and oil vapours condense and the resulting mixture of water and essential oil is collected in the receiver or separator. Since the density of most of the essential oil is relatively lesser than water, the oil floats on water and can easily be taken out from the separator through the outlet provided specially for taking out oil. This process is used in large scale production of essential oils from aromatic crops. The equipment (boiler, distillation tank, condenser and separator) needed for this process is expensive and needs technical personnel to run the boiler.

## Solvent Extraction

Solvent extraction has made it possible to create a whole range of oils that could never be produced before because their structures are too delicate to withstand other methods of extraction. Solvents used for extraction include petroleum ether, hexane, toluene, butane, benzene and acetone. Fresh flowers are charged into specially constructed extractor and extracted systematically at room temperature, with a carefully purified solvent, usually petroleum ether. The solvent penetrates the flowers and dissolves the natural flower perfume together with some waxes and albuminous and coloring matter. The solution is subsequently pumped in to an evaporator and concentrated at a low temperature. After the solvent is completely driven off in vacuum, the concentrated flower oil "concrete" is obtained. Compared with distilled oils the solvent extracted flowers oils, therefore more truly represent the natural perfume as originally present in the flowers. Transformation from concrete to absolute is done at low temperature by dissolving concrete in eight to ten times the amount of warm alcohol. The solution is kept for sometimes and clear solution is decanted to remove insoluble waxes. The solution is cooled at –20°C to –25°C overnight, precipitated waxes is filtered off in the cold from the solution, the alcohol is removed by distillation under vacuum at low temperature. The absolute is usually viscous light colored liquid, which closely resembles the fragrance of fresh flower. The final product is called an "absolute". The resinoid is another product derived through solvents or alcohol extraction. The main distinction between resinoids and concretes is that resinoids are produced from plant matter such as tree saps and exudates whereas concretes are extracted from fresh plant matter.

## Cold Expression

Cold expression is referred to as "scarification". This method of extraction is used for citrus fruits, which contain essential oil-bearing pouches in their peels. The peel is shredded, and then mechanically pressed. The resulting emulsion, consisting of essential oil, juice, water and fruit particles, is either filtered or passed through a clarifying centrifuge. The essential oil, which floats on top, is then separated out.

## Enfleurage

Enfleurage is perhaps the oldest method of extracting fragrance from plant. This procedure takes advantages of the liposolubility of the fragrant compounds of plants in fats. It is used for delicate plant that are unable to withstand the high heat of distillation, or for those that continue to exude fragrance after they are picked, such as tuberose and jasmine. Refined lard and tallow are still used in enfleurage today, as they have been for centuries. With their excellent ability to pick up fragrance molecules from flowers, animal fats are the preferred medium for this type of extraction. In traditional enfleurage process, fat is spread thickly on plates of glass and covered with freshly picked flowers. The layers of glass are sealed to retain the scent. The flowers are left undisturbed for 24 to 48 hours, and then replaced with fresh ones. The extraction is achieved by cold diffusion in to the fat, whereas the "digestion" technique is carried out with heat, by immersing plant material in the melted fat also known as "hot enfleurage". This time-comsuming, labour-intensive process is repeated for several weeks until the fat is saturated with the fragrance of the flowers. Then the fat is gently warmed and filtered. The resulting product is called "pomade". Today, it is usually washed with alcohol to remove the fat. The alcohol carried the essential oil and is separated from the fat by chilling. At this stage, the product is called an extrait.

## Super-Critical Fluid Extraction (SCFE)

It is the most recent method of extraction of essential oil from the material of plant origin where the fragrance and flavour ingredients resemble to their source. Supercritical carbon dioxide is used as a solvent in supercritical fluid extraction. It has advantages: it is a natural product, chemically inert, non-flammable, non-toxic, easy to eliminate completely, readily available, selective. Super-critical carbon dioxide has the density of a liquid, low viscosity and diffuses like a gas. It is an excellent solvent for a wide range of natural subtracts. It has the status of safe food grade solvent and can be used for processing of multiple products. In this process, the plant material is placed inside a stainless steel tank, the $CO_2$ is added, and pressure is increased inside the tank. With the high pressure and lower temperatures, the $CO_2$ liquefies and acts as a solvent, extracting the oils and other plant constituents. Within minutes, the pressure is then decreased, and the $CO_2$ returns to its natural gaseous state and evaporates off, leaving a pure extract completely free of solvents. This method has many benefits. Obviously, the low temperatures used the relative ease in which the oils are extracted ensure that none of the constituents will be changed or affected during the process. This is why $CO_2$ extracts are thought to be the truest representation of the essence of the plant. The aroma is much closer to the plant it comes from, the oil yield is higher, and scientists believe it may produce a more potent, therapeutic essential oil. Supercritical $CO_2$ extraction is commercially used, but is less common and beyond the financial means of most processors.

## Storage and Packing of Essential Oil

Storage is an important aspect in essential oil industry. Essential oils are made up of terpeniods, phenolics, acids, sulphur compounds etc. often having unsaturation in their molecules, are likely to undergo oxidation, polymerization, etc., and therefore, care is to be taken for increasing their shelf life. All kinds of essential oils must be free from moisture and metallic impurities. Fresh volatile oils are generally colourless, but on long standing these are affected by light, heat and air. Spoilage occurs due to oxidation and resinification, which causes darkening of oil color. Apart from this polymerization, hydrolysis of esters and interaction of functional groups also cause spoilage. Essential oil should be stored in completely filled, tightly closed containers in cool place, preferably in air conditioned room and protected from light. Geranium, Vetiver, Patchouli and Sandal wood oils can be kept for long time without any spoilage. In fact, oil quality of Patchouli, Vetiver and Geranium improves with age and prolong storage. Prior to storage, oil should be carefully clarified and moisture should be removed.

# Quality Analysis

It is necessary to establish criteria by which identity and purity of essential oil can be judged as well as its origin and adulteration. Essential oil should be subjected to both qualitative and quantitative tests to know its purity. Oils are tested in four stages: First stage is sensory evaluation in which the smell, viscosity, color and clarity of the oil are assessed. The second stage is an odor/smell test, which helps to determine if oil is really what it is claimed to be (Table 15.2).

**Table 15.2**: Classification of odour.

| Notes | Plant |
|---|---|
| Green Notes | Galbanum |
| Citrus Notes | Lemon, Lavender, Rosemary, Bergamot, sweet orange |
| Floral Notes | Jasmine, Rose |
| Aldehydic Notes | Vetiver |
| Lavender Notes | Lavender, Lemon, Clary sage, Bergamot |
| Spicy Notes | Pepper, Ginger, Thyme |
| Woody Notes | Sandal wood, Vetiver, Patchouli, Cedar wood |

## Physico-chemical Constants and Other Instrumental Analysis

In the next stage, physical methods such as specific gravity, optical rotation, refractive index, melting point, solubility and then chemical methods like determination of acid value, ester value and iodine value, carbonyl value, saponification value etc. should be measured. Physico-chemical constants are essential requirements of quality evaluation of essential oils/aroma chemicals. These are compared with those of standard commercial samples, specified by different organizations like- International Organisation for Standardisation (ISO), Essential Oils Association, USA (E.O.A), Bureau of Indian Standards (BIS) and others. If these test results are satisfactory, the oil is then subjected to Gas Chromatography (GC), Mass Spectroscopy (MS) evaluation which results in a separation of the mixture in to the individual components. Identification of the fragrance constituents may be accomplished through ultraviolet, infra-red-NMR and Mass Spectroscopy. Infra red spectroscopy of a chemical mixture gives clean indication of functional group present in it. Nuclear magnetic resonance (NMR) spectroscopy is employed to know about structure of components isolated from essential oils. Thin layer chromatography, column chromatography and HPLC are used for separation of fragrant complex into individual components. Supercritical fluid chromatography (SFC) is also used in fragrance and flavour industry for analysis and quality control. SFC techniques combines desirable characteristics of both gaseous and liquid mobile phases and this technique in fragrance and flavour industry is used for extracting the active components from natural product.

## Factors Affecting Oil Composition

Essential oils from the chemical standpoint are most often extremely complex mixtures whose components in case of any particular oil are relatively constant when viewed qualitatively or quantitatively. The qualitative and quantitative make up of an essential oil is a characteristic, which nevertheless varies between certain limits depending on a number of factors.

☆ Genetic effects

☆ Climatic influence

☆ Type of plant organs from which oils are isolated

☆ Cultural conditions of the plant

☆ Development of plant and plant parts

☆ Post-harvest management of herbs

☆ Method of distillation

# Technology for the Production of Value Added Products from Essential Oils

Value addition of the oils can be achieved by:

1. Rectification,
2. Fractionation,
3. Chemical modification, and
4. Deterpenation.

## Rectification

It is the process of redistillation of an essential oil either with steam or under vaccum. The process is useful to improve the quality of the oil that has deteriorated on long prolonged storage, due to rust formation or due to emulsion formation with water. After rectification, the appearance of the oil will generally be improved as non-volatile and resinified impurities are left behind in the distillation still.

## Fractionation

Fractionation of essential oils is generally done to produce perfumery compounds present as major constituents in the oils. The boiling points of the constituents of the oils at atmospheric pressure are different. By making use of the differences in the boiling points, individual compounds of the oils can be separated. For example, citronellal which is a major constituent of oil of *Eucalyptus citriodora* leaves can be separated by fractional distillation. Fractional distillations are usually carried out under high vacuum which lowers the boiling points of the compounds. This prevents any damage to the compounds by heat. The typical fractional distillation assembly consists mainly a pot in which the oil to be fractionated is placed, fractionating column, condensers, product cooler, receiver and vacuum pump.

## Chemical Modification

Value addition of essential oils can also be achieved by chemical modification of the compounds obtained from essential oils by fractional distillation. Examples of such chemical modification include conversion of the pinenes obtained from turpentine oil to a variety of oxygenated monoterpenes like linalool, terpineol, etc., conversion of citral from lemongrass oil to the ionones.

## Deterpenation

Deterpenation of essential oils is carried out to improve the quality of the oils. If the terpene hydrocarbons which have much inferior perfumery value compared to oxygenated terpenes are removed from the oils, the concentration of oxygenated terpenes will be increased. Terpenes oils have longer shelf life and fetch more price than the crude oils.

# Uses of Essential Oils and their Isolates

Essential oil is a complex mixture of odorous and steam volatile compounds of vegetable origin. The essential oils and their isolates are widely used in number of industries such as perfumery and toiletry, agarbatti, tobacco, soft drinks, pharmaceuticals, paint, food flavouring and aromatherapy etc.

# Essential Oils Chemistry

Chemically, the essential oils are primarily composed of mono- and sesquiterpenes and aromatic polypropanoids synthesized via the mevalonic acid pathway for terpenes and the shikimic acid pathway for aromatic polypropanoids. The basic building block of many essential oils is a five-carbon molecule called an isoprene ($C_5H_8$). Most essential oils are built from isoprene which makes up the terpenoids. When two isoprene units are linked together, they create a monoterpene, when three join they create sesquiterpene, so forth (Table 15.3).

**Table 15.3**: Classification of terpenoids.

| Name of Class | No. of Isoprene Units (No. of Carbon Atoms) | Examples |
|---|---|---|
| Hemiterpenes | 1 (5) | Tiglic acid, angelic acid, isovaleraldehyde |
| Monoterpene | 2 (10) | Pinene, ocimenes, limonene, citral, geraniol, linalool, camphor, menthol |
| Sesquiterpene | 3 (15) | Caryophyllene, germacrene, cadinene, aromadendrene, santalols, farnesol, patchouli alcohol |
| Diterpene | 4 (20) | Ginkolide, sclareol |
| Sesterterpenes (very rare) | 5 (25) | Geranyl farnesol, geranyl nerolidol |
| Triterpenes | 6 (30) | Squalene, jasminol |
| Tetraterpenes | 8 (40) | Carotenes, lycopene |

The terpenes are among the most widespread and chemically diverse groups of natural products. Like all natural products, within this sample classification lies an enormous amount of structural diversity which leads to a wide variety of terpene like (or terpenoid) compounds. The function of terpenes in plants is generally considered to be both ecological and physiological. Many of them inhibit the growth of competing plants (allelopathy). Some are known to be insecticidal; others are found to attract insect pollinators. Eight predominant chemical groups are involved in composition of essential oils and each group contributes a particular therapeutic quality. The main chemical groups found in essential oils are Terpenes, Sesquiterpenes, Esters, Aldehydes, Ketones, Alcohols, Phenols and Oxides (Table 15.4). The aromatic ring structure of essential oil is much more complex than the simpler, linear carbon hydrogen structure of fatty oils. Essential oils also contain sulfur and nitrogen atoms that fatty oils do not have. Chemical analysis using GC and GC/MS can be used to check purity of oil. To fully understand why it is so important to have as pure essential oils as possible, it is necessary to understand the chemical make-up of each essential oil, and the therapeutic effect of each chemical group. This list of natural chemical ingredients and their therapeutic effects illustrates just how important all the natural ingredients are in essential oils. The loss of any of these ingredients, through lack of purity results in essential oils with little or no therapeutic effect.

The chemistry of essential oils is elaborate. An individual oil may have hundreds of ingredients, the principle components being a group of complex substances known as terpenes and their compounds or derivatives. This explains why a single essence has a wide range of therapeutic actions. It is interesting to know the chemical components that nature combines to make up the oils, but it is also humbling to take note of the fact that even with the best human efforts, should you in a laboratory combine all the chemicals in the correct proportions, you would still not have identical oil. Such a copy of oil will not have the same therapeutic effect as the natural and pure essential oil.

## Chemical Variability

Growers and suppliers find it difficult to provide exact information about the quantity of each of the main components contained in an oil. Changes in the soil, the climate and when the crop is harvested mean that no chemical will be present in the same proportions at each distillation. Controlling the growing conditions can reduce these variations. However, since the presence of some constituents can vary from 20 to 70 percent, a sample of each distillation should be analyzed to determine its contents. Readings can be taken by a gas-liquid chromatograph. It must be said, however, that there is no direct and simple connection between the individual constituents of an essential oil and its overall effects. The make-up of oil is intricate; it is more than the sum of its parts. The different chemicals work together to produce a set of effects that often differ from their individual indications. These complications also explain why synthetic oil cannot be a substitute for a true essential oil. An aroma can be reconstituted to make a perfume, or a chemical can be isolated to make a drug, but the therapeutic action of an essential oil owes more than we have yet defined to the complex synergy of its chemical components.

**Table 15.4**: Chemical groups in essential oils and their properties.

| Chemical Component | Examples | Plant Sources | Therapeutic Properties |
|---|---|---|---|
| Terpenes | Limonene, α, β-pinene, camphene, cadinene, cedrene, dipentine, phellandrene, terpinene, sabinene, myrcene | Lemon | Antiseptic, antiviral |
| Sesquiterpenes | Chamazulene, farnesol, β-caryophellene, farnesene, β-bisabolene | Chammoile | Antibacterial, anti-inflammatory |
| Esters | Linalyl acetate, geranyl acetate, bornyl acetate, eugenyl acetate, lavendulyl acetate, citronellyl acetate | Clary sage, lavender | Fungicidal, sedative, fruity aroma, antispasmodic |
| Aldehydes | Citral, citronellal, neral, benzaldehyde, cinnamic aldehyde, cuminic aldehyde, perill aldehyde | Melissa, lemongrass, lemon, eucalyptus and citronella | Antiseptic, sedative, analgesic, Antiinflammatory |
| Ketones | Thujone, pulegone, jasmone, fenchone, camphor, carvone, methone, methyl nonyl ketone, pinocamphone | Mugwort, sage and worm-wood, jasmine and fennel | Respiratory complaints -helps ease congestion and clears mucus |
| Alcohols | Linalool, citronellol, geraniol, borneol, menthol, nerol, terpineol, farnesol, vetiverol, benzyl alcohol, cedrol, Patchouli alcohol, Fenchol, β-Santalol | Rose, Lavender, sandal wood and Geranium | Antiseptic, antiviral, uplifting oils, immune-stimulants, antifungal |
| Phenols | Eugenol, thymol, carvacrol, methyl eugenol, methyl chavicol, anethole, safrole, myristicin, apiol, anethol | Clove, bay, thyme, oregano and savory, fennol | Antibacterial, stimulating, Antiseptic |
| Oxides | Cineole (eucalyptol), linalool oxide, ascaridol, bisabolol oxide, bisabolone oxide | Eucalyptus oil, rosemary, bay, laurel, tea tree and cajeput | Expectorant |

## Chemotypes

Chemotypes are also referred to as 'Chemical breeds", are very common among plants containing essential oils. They refer to plants that are usually the same but produce chemically different essential oils. Many plants have several chemotypes that occur naturally. Thyme (*Thymus vulgaris*) is a good example. This species has seven different chemotypes with either thymol, carvacrol, geraniol, linalool, a-terpineol, or both trans-4- thujanol, cis-8-myrcenol and cineole. Likewise four chemotypes of *Ocimum basilicum* are reported: (1) Methyl chavicol rich, (2) Linalool rich, (3) methyl cinnamate rich, (4) Eugenol rich (Table 15.5)

**Table 15.5**: Chemotypes of sweet basil oil (Guenther, 1949).

| Type of Oil | Principal Constituent |
|---|---|
| European type | Methyl chavicol, linalool (no Camphor) |
| Reunion type | Methyl chavicol, Camphor (no linalool) |
| Methyl cinnamate type | Methyl chavicol, linalool, methyl cinnamate |
| Eugenol type | Eugenol |

## Toxic Effects of Essential Oils

Each essential oil, though having therapeutic properties has some toxic counterparts (Table 15.6). Unwanted effects of essential oils include skin irritation, mucous membrane irritation and photo toxicity and photosensitivity. Some oils are thought to have toxic effects on the brain, liver and kidney.

**Table 15.6**: Some essential oils and possible toxic ingredients.

| Plant Name | Ingredient | Toxic Effect |
|---|---|---|
| Basil (*Ocimum basilicum*) | Methyl eugenol Methyl chavicol | Carcinogenic in mice and rats, may cause nausea, vomiting, sensitization |
| Eucalyptus (*Eucalyptus globulus*) | 1,8-Cineole | Nausea, vomiting, dizziness, skin irritation, photosensitization |
| Lavender(*Lavendula officinalis*) | Linalool Camphor | Sensitizer Dermatitis, Sensitizer |
| Lemon (*Citrus limonum*) | Furocoumarins | Photosensitization, may cause skin cancer |
| Patchouli (*Pogostemon cablin*) | Patchoulenone Cinnamic acid | May reduce appetite Photosensitizer |
| Sandal wood (*Santalum album*) | Borneol | May cause dermatitis, allergy in certain cases |
| Tea tree (*Melaleuca alternifilia*) | Cineole | Nausea, vomiting, dizziness, skin irritation, photosensitizer |
| Thyme (*Thymus vulgaris*) | Carvacrol | Abortification |

Toxic effects are seen when certain threshold concentration are exceeded. This concentration is different for different individuals. Hence we can best use these oils when we know about them in detail, along with their synergic actions.

## Conclusions

The aromatic plants and aroma chemicals contained in them play a vital role in our day-to-day living. All of us use certain spices and condiments, which improve the taste of the food. Some of the essential oils find place in drugs used for cure of diseases. More and more people are using perfumes and perfumed products, which were previously used by affluent people only as these are falling with the reach of more and more. India has to play a dominant role in the production and processing of essential oils. Country's biodiversity coupled with competent scientific force and favourable processing conditions make our country as the natural choice to become a foremost leader in aroma business in the coming years. Hence, it is evident from whole discussion that one cannot be prejudiced that this mode of extracted product will be utilized for this and that for that. In fact in certain cases the essential oil are used as such like rose, jasmine etc. While in case of oil like citronella, basil, lemongrass, palmarosa they need to be fractionated to get citronellol, citronellal, anethole, citral, geraniol etc. In case of spice they are processed by both the mode *i.e.,* steam distillation and solvent extraction to produce essential oil and oleoresin respectively and both of them does have their different kinds of uses in industry. In nutshell it is the use that gives a final verdict for mode of processing not the other way round. India's biodiversity in plants is a boon for the production of many natural isolates, which are needed for blending of natural flavours. The demand for natural flavours is rising as consumers show strong preference in flavour of naturals because of their perception that natural flavours are safe. Natural or herbal cosmetics are another sector, which has a high potential of essential oil. Hence there is a great scope for commercial cultivation of several aromatic crops in India as there is always a demand for new and specific aroma chemicals in market for development of new and exotic flavour and fragrance. Essential oils are natural products of economic importance are indispensable ingredients of cosmetics, perfumes and they impart flavour to many of our foods.

## References

Adams, R. P. (1989). *Identification of Essential Oils by Ion Trap Mass Spectroscopy*. Academic Press, San Deigo, CA., USA.

Ames, G.R. and W.S.A. Matthews. (1969). The distillation of essential oils. *Perf and Essential oil Rec.*, p. 9-18.

ASTA. (1968). *Official analytical methods of the American Spice Trade Association* ASTA, Inc., Englewood Cliffs, NJ.

Chopra, R.N., Nayar, S.L. and Chopra, I. C. (1986). Glossary of Indian Medicinal Plants, Council of Scientific and Industrial Research (CSIR), New Delhi, India.

Croteau, R. (1986). Biochemistry of monoterpenes and sesquiterpenes of the essential oils. Herbs, spices and medicinal plants: *Recent advances in botany, horticulture, and pharmacology*, **1**: 81-135. Oryx Press, Phoenix, AZ.

Davies, N.W. (1990). Gas Chromatographic retention indices of monoterpenes and sesquiterpenes on methyl silicone and Carbowax 20M phases. *J. Chromatography*, **503**: 1-24.

Guenther, E. (1948-1952). *The Essential Oils*, Vol I-IV, Van Nostrand Co. Inc., New York

Guenther, E. (1972). The production of essential oils: methods of distillation, enfleurage, maceration, and extraction with volatile solvents. In: Guenther, E. (ed.). The essential oils. History-origin in plants. production analysis. Vol. **1**: 85-188. Krieger Publ. Co., Malabar, FL.

Guenther, E. (1985). The essential oils. Vol. **3**: 400-433. Krieger Publ. Co., Malabar, FL

Heath, H. B. (1981). *Source book of flavors*. AVI, Westport, CT.

Gupta, R. and Chadha, K.L. (1994). *Recent advances in Horticulture- Medicinal and Aromatic Plants.* Vol. 11 (eds. K.L. Chadha and Rajendra Gupta), MPH Publication, New Delhi.

Kirtikar, K.E. and Basu, B.D. (1935). *Indian Medicinal Plants*, Vol IV, II[nd] edn., M/s. Bishen Singh, Mahendra Pal Singh, New Delhi.

Lawrence, B.M., J.W. Hogg, S. J. Terhune and N. Pichitakul. (1972). Essential oils and their constituents. IX. The oils of *Ocimum sanctum* and *Ocimum basilicum* from Thailand. *Flavor Ind.*, p. 47-49.

Lawrence, B.M. (1979). Commercial production of non-citrus essential oils in North America. *Perf. and Flav.*, **3**: 21-33.

Lawrence, B.M. (1984). A review of the world production of essential oils. *Perf. and Flav.*, **10**: 1-16.

Rajendra Gupta (1993). Status of Medicinal and Aromatic Plants in India. Country Report-2. Proc. Regional Expert Consultation on Breeding and Improvement of Medicinal and Aromatic Plants in Asia. FAO/RAPA-Bangkok, 3 May-4 June, 1993.

Singh, H.P. (2002). Strategies for augmentation of medicinal and aromatic plants wealth. *Indian J. of Arecanut, Spices and Medicinal Plants.* **4**(1): 1-17

Ramaswamy, S. K., Briscese, P., Gargiullo, R.J. and Geldern, T. V. (1988). Sesquiterpene hydrocarbons from mass confusion to orderly line-up. (*In*): *Proc. 10th International Congress of Essential Oils, and Flavours and Fragrance:* pp. 951-980.

Runeckles, V.C. and T.J. Mabry (eds.) (1973). *Terpenoids: structure, biogenesis, and distribution.* Academic Press, New York.

Sievers, A.F. (1928). Methods of extracting volatile oils from plant material and the production of such oils in the United States. USDA Tech. Bull. 16. USDA, Wash., DC.

Simon, J.E., J. Quinn, and R.G. Murray (1990). Basil: A source of essential oils. In: Janick, J. and J.E. Simon (eds.). *Advances in new crops.* Timber Press, Portland, Oreagon, pp 484-489.

Varshney, S.C. (2000). Vision 2005: Essential oil industry of India Presidential address delivered at the National Seminar on "Essential oil beyond 2000" organized by the essential oil association of India at Mussoorie, Sept. 5, 1999. *Indian Perfumer,* **44**(3): 101-118.

Utilisation and Management of Medicinal Plants Vol. 2 (2014)    *Pages* **343–354**
*Editor-in-Chief:* **V.K. Gupta**
*Published by:* **DAYA PUBLISHING HOUSE, NEW DELHI**

# 16

# Extraction Methods and Health Benefits of Nutraceuticals from Medicinal Plants

Dattatreya M. Kadam[1]* and Monika Sharma[1]

## ABSTRACT

*Recent years have witnessed a tremendous resurgence in the interest and use of medicinal plant as pharmaceuticals. Population rise, inadequate supply of drugs, prohibitive cost of treatments, side effects of several allopathic drugs and development of resistance to currently used drugs for infectious diseases have led to increased emphasis on the use of plant materials as a source of medicines for a wide variety of human ailments. In the past few years, many food bioactive constituents have gained huge attention as they have been used in the form of pharmaceutical products exerting beneficial physiological functions. These range of products cannot be truly classified as 'food' and thus a new hybrid term between nutrients and pharmaceuticals, 'nutraceuticals', has been coined to designate them. Nutraceuticals are pure phytomolecules and offer medical or health benefits to the consumer by providing a means for the maintenance of health and well being and protection from diseases. There are several ways in which the nutraceuticals can be extracted from plant materials which include traditional methods such as solvent extraction or soxhlet extraction, hot water extraction etc. The traditional extraction methods have several drawbacks viz., they are time consuming, laborious, have low selectivity and/or low extraction yields. This resulted in increasing demand for new extraction techniques with shortened extraction time, reduced organic solvent consumption, and increased pollution prevention. The emerging extraction methods include supercritical fluid extraction (SFE), microwave assisted extraction, ultrasound assisted extraction etc. The extracted nutraceuticals can be used for the development of functional foods or can be used in the form of pills for the benefit of human health.*

*Keywords*: Extraction methods, Health benefits, Nutraceuticals, Medicinal plants.

---

1    Central Institute of Post-Harvest Engineering and Technology (CIPHET), Ludhiana, Punjab, India

*    *Corresponding author*: E-mail: kadam1k@yahoo.com

# Introduction

Over three-quarters of the world population relies mainly on plants and plant extracts for health care. More than 30 per cent of the entire plant species, at one time or other, were used for medicinal purposes. Among ancient civilizations, India has the rich repository of medicinal plants. The forest in India is the principal repository of large number of medicinal and aromatic plants, which are often used as raw materials for manufacture of drugs and other products. In India, medicinal plants are widely used by all sections of the population with an estimated 7500 species of plant. India's diversity is unmatched due to the presence of 16 different agro-climatic zones, 10 vegetation zones, 25 biotic provinces and 426 biomes (habitats of specific species). Of these, about 15000-20000 plants have good medicinal value. However, only 7000-7500 species are used for their medicinal values by traditional communities. In India, drugs of herbal origin have been used in traditional systems of medicines such as Unani and Ayurveda since ancient times. The Ayurveda system of medicine uses about 700 species, Unani 700, Siddha 600, Amchi 600 and modern medicine around 30 species (Joy *et al.*, 1998).

Modernization of herbal medicine has been a key factor in the widespread acceptance of natural or alternative therapies by the international community. Most of the Ayurvedic medicines are in the form of crude extracts which are a mixture of several ingredients and the active principles when isolated individually fail to give desired activity. This implies that the activity of the extract is the synergistic effect of its various components. So, the standardization and quality control of herbal materials by use of modern science and technology is of great significance.

# Nutraceuticals

Due to risk of toxicity or adverse effects of drugs, consumers are massively turning twoards the food supplements to improve health where pharmaceutical fails. This resulted in a world wide nutraceuticals revolution. Pharmaceuticals and synthetic food additives are relatively new to our genes. Our bodies have had thousands, perhaps millions of years of evolutionary experience from several thousands of phytochemicals of edible sources. Recently a group of products called health products, nutraceuticals or dietary supplements have gained popularity. There are no universally accepted definitions for functional food and nutraceuticals, but generally they are described as foods (fortified with added ingredients or not) with health benefits beyond basic nutrition (Wildman and Kelley, 2007). Stephen DeFelice, MD, founder and chairman of the Foundation for Innovation in Medicine (FIM), Cranford, NJ in 1989, coined the term "nutraceutical" which is an amalgamation of "nutrition" and "pharmaceutical". According to DeFelice, nutraceutical can be defined as, "a food (or part of a food) that provides medical or health benefits, including the prevention and/or treatment of a disease" (Rajat *et al.*, 2012).

Presently over 470 nutraceutical and functional food products are available with documented health benefits. Many of these products which are being promoted to treat various diseases belong to plant origin. Various plants produce secondary compounds as alkaloids to protect themselves from infections and these constituents may be useful in the management of human infection. Many of the phyto medicines are the typical examples. Nutraceutical is a broad term used to describe a product derived from food sources that provides extra health benefits in addition to the basic nutritional value found in foods (Dharti *et al.*, 2010). Ideally, intake of nutrients through food would have been sufficient to prevent curative measures such as pharmaceuticals and traditional medicine to a large extent. But when there is some deficiency in the absence of requisite nutrition through food, an external intervention in the form of nutraceuticals has become imperative.

According to Food Safety and Standards Act (FSSA), 2006, Foods for special dietary use are specially processed or formulated to satisfy particular dietary requirements which exist because of a particular physical or physiological condition or specific diseases and disorders and which are presented as such wherein the composition of these foodstuffs must differ significantly from the Indian Standard (IS) composition of ordinary- foods of comparable nature, if such ordinary foods exist and may contain one or more of the following ingredients, namely:

☆ Plants or botanicals or their parts in the form of powder, concentrate or extract in water, ethyl alcohol or hydro alcoholic extract, single or combination

☆ Minerals or vitamins or proteins or metals or their compounds or amino acids (in amounts not exceeding the Recommended Daily Allowance for Indians) or enzymes (within permissible limits)

☆ Substances from animal origin

☆ Dietary substances for use by human beings to supplement the diet by increasing the total dietary intake

## Methods of Extracting/Isolation of Nutraceuticals from Plants

Extraction and characterization of several active phyto-compounds from medicinal plants have given birth to some high activity profile drugs. The potential natural anticancer drugs like vincristine, vinblastine and taxol can be the best example (Huie, 2002). Amongst the various traditional and conventional extraction techniques, soxhlet extraction has been the most widely used. The conventional methods also include infusion, maceration and decoction. In infusion, the crude form of the drug or the plants parts are kept in hot or cold solvent; mostly water and the liquid solution containing the drug is obtained by maceration. However, in case of decoction, the method suitable for water-soluble, heat-stable constituents; the crude drug is boiled in a specified volume of water for a particular time followed by cooling and straining. The traditional method (maceration), however, is time consuming, requiring timeframes from 2 to 10 days (Cunha *et al.*, 2004; Woisky and Salatino, 1998). Conventional extraction is usually performed at reflux temperature of 90 °C for 2 h. This method which has been used for many decades requires relatively large quantities of solvents. An individual plant may consist of several active phyto-constituents existing in abundance along with certain constituents of low activity profile. Thus, there arises a need for the development and adoption of extraction and analysis techniques with high performance (Smith, 2003). The newer extraction techniques in the herbal drug industry should be aimed at minimizing the extraction time and the cost of solvent consumption. The extraction techniques include solid-phase microextraction, supercritical-fluid extraction, pressurized-liquid extraction, microwave-assisted extraction, solid-phase extraction etc.

## Microwave Assisted Extraction

The electromagnetic energy in the range of microwave has gained special attention to various fields of utilization such as domestic, analytical and biomedical applications. Many organic reactions assisted by microwave heating have the advantages of enhanced selectivity and much improved reaction rate. Other advantages of microwave-assisted reactions are milder reaction conditions, formation of cleaner products with higher yields, minimum wastes, environmental compatibility, low energy consumption, good reproducibility and minimal sample manipulation for extraction process (Asghari *et al.*, 2011). The minute microscopic traces of moisture that occurs in plant cells are the target of heating in microwave treatment. The heating up of this moisture inside the plant cell due to microwave effect, results in evaporation and generates tremendous pressure on the cell wall. Due to this pressure, the cell wall is pushed from inside thereby getting ruptured. Thus the exudation of active constituents from the ruptured cells occurs, which in turn increases the yield of phytoconstituents (Tatke and Jaiswal, 2011).

The microwave- assisted hydro distillation (MAHD) (Lucchesi *et al.*, 2004; Lucchesi *et al.*, 2006) is a new technique for the production of essential oils, solvent free microwave- extraction (SFME). A combination of microwave- heating and dry distillation is also a new green technique developed in recent years. The influence of microwave and $CO_2$- supercritical fluid extraction (SFE) from the extraction of medicinal plants in the presence of moisture has also been reported (Sonnenschein *et al.*, 2002).

This process uses microwave energy to heat solvents in contact with a sample in order to partition some chemical components from the matrix into the solvent. The main classes of polyphenols, phenolic acids and flavonoids, such as flavonols, flavanols, and isoflavones, play a role in the prevention of human pathologies (Tapiero *et al.*, 2002). Extraction using microwaves can result in increased yield in shorter time at the same temperature using less solvent. One must consider the physical parameters while choosing the parameters for microwave extraction including solubility, dielectric constant, and the dissipation factor. Microwaves have been used for the extraction of other biological compounds, such as extraction of essential oils from the leaves of rosemary and peppermint, extraction of glycyrrhizic acid from licorice root and extraction of ergosterol from fungal hyphae and mushrooms (Proestos and Komaitis, 2008). Thus, in view of this, microwave assisted extraction can be performed for isolation/extraction of nutraceuticals or functional components from medicinal plants with slight modification.

## Supercritical Fluid Extraction (SFE)

Supercritical fluids are widely accepted for extraction, purification, re-crystallization, and fractionation operations in many industries. SFE is more efficient than traditional solvent separation methods. Supercritical fluids are selective, thus providing the high purity and good product concentrations. Additionally, it doesn't leave any organic solvent residues in the extract or spent biomass and makes extraction more efficient at modest operating temperatures. Supercritical $CO_2$ is a green solvent which is most commonly used supercritical fluid for various applications. By adjusting the processing pressure and temperature, the gas can act like a liquid solvent, but with selective dissolving powers. In the supercritical fluid phase, extraction/concentration is carried out by simply changing the pressure, which results in a pure product fraction and a clean $CO_2$ gas stream, which can be completely recycled. Supercritical $CO_2$ also serves as a good solvent for the extraction of non-polar compounds such as hydrocarbons (Vilega *et al.*, 1997). To extract polar compounds, some polar supercritical fluids such as Freon-22, nitrous oxide and hexane have been considered. However, their applications are limited due to their unfavorable properties with respect to safety and environmental considerations (Lang and Wai, 2001; Wang and Waller, 2006). SFE has been used for the extraction of alkamides, polyphenolics including chichoric acid, naphthodianthones, hypericin and pseudohypericin, essential oils, oleoresins etc. form various medicinal and aromatic plants (Karale *et al.*, 2011).

## Counter-Current Extraction

In counter-current extraction, wet raw material is pulverized using toothed disc disintegrators to produce fine slurry. In this process, the raw material to be extracted in the form of fine slurry is moved in one direction within a cylindrical extractor where extraction solvent comes in contact with this material. The process is highly efficient, requires less time and poses no risk from high temperature (Handa, 2008).

## Sonication-Assisted Extraction

Sound waves, which have frequencies higher than 20 kHz, are mechanical vibrations in a solid, liquid and gas. Ultrasonic extraction (UE) involves the use of ultrasound with frequencies ranging from 20 kHz to 2000 kHz; this increases the permeability of cell walls and produces cavitation. The benefits of UE are thought to be mainly due to the mechanical effects of acoustic cavitation. Due to the mechanical effects of ultrasound, the extent of solvent penetration into the cellular materials increases which also improves mass transfer. Ultrasound when used for extraction can also disrupt biological cell walls thereby facilitating the release of contents. Therefore, efficient cell disruption and effective mass transfer are cited as two major factors leading to the enhancement of extraction with ultrasonic power (Mason *et al.*, 1996). This method has demonstrated the potential of reducing the extraction times significantly and increasing extraction yields in a number of studies on medicinal plants (Liu and Wang, 2004).

## Accelerated/Pressurized Solvent Extraction

For rapid and effective extraction of analytes from solid matrices such as plant materials, extraction temperature is an important experimental factor, because elevated temperatures could lead to significant improvements in the capacity of extraction solvents to dissolve the analytes, in the rates of mass transport, and in the effectiveness of sample wetting and matrix penetration, all of which lead to overall improvement in the extraction and desorption of analytes from the surface and active sites of solid sample matrices (Huie, 2002). Accelerated solvent extraction (ASE) is a solid–liquid extraction process performed at elevated temperatures, usually between 50 and 200 °C and at pressures between 10 and 15 MPa. Therefore, accelerated solvent extraction is a form of pressurized solvent extraction that is quite similar to SFE. Increased temperature accelerates the extraction kinetics and elevated pressure keeps the solvent in the liquid state, thus achieving safe and rapid extraction. Also, pressure allows the extraction cell to be filled faster and helps to force liquid into the solid matrix (Wang and Weller, 2006). Accelerated solvent extraction is often used for the extraction of high-temperature stable organic pollutants from environmental matrices.

## Solid-Phase Extraction

Solid-phase extraction (SPE) is a simple preparation technique based on the principles used in liquid chromatography, in which the solubility and functional group interactions of sample, solvent, and adsorbent are optimized as they affect sample fractionation and/or concentration. It should, in particular, be noted that SPE is well suited to the treatment of sample matrices with high water content, *e.g.,* extracts of herbal materials (Tekel and Hatrík, 1996). Stobiecki *et al.* (1997) combined $C_{18}$ and an adsorbent containing benzene sulfonic groups as an effective sample-preparation method for profiling quinolizidine alkaloids and phenolic compounds in *Lupinus albus*.

# Health Benefits of Nutraceuticals Obtained from Medicinal Plants

### Antidiabetic Activity

Diabetes is a chronic disease that occurs either when the pancreas does not produce enough insulin or when the body cannot effectively use the insulin it produces (WHO, 2009). Currently 40.9 million people in India are suffering from diabetes (IDF, 2007) and by 2030 there would be 79.44 million diabetics in India alone (WHO, 2007). According to an estimate from the ethnobotanical information, several plant species have anti diabetic potential such as *Momordica charantia, Trigonella foenum greacum* etc. Bael (*Aegle marmelos*) is an important medicinal plant of India. Various parts of this plant such as leaves, fruit and seed possess hypoglycaemic, hypolipidemic and blood pressure lowering property (Vijay *et al.*, 2006). Aegeline 2 present in leaves of Bael have antihyperglycemic activity as it has the ability of lowering the blood glucose levels, decreasing the plasma triglyceride, total cholesterol and free fatty acids (Narender *et al.*, 2007). *Tamarindus indica* Linn. was used as a traditional medicine for the management of diabetes mellitus in human and experimental animals. Treatment of diabetic rats with *Tamarindus indica* seed extract, from one week after diabetes induction, compensated hypoglycemia after 6, 4 and 2 weeks, and increased blood insulin levels (Singh *et al.*, 2012).

In traditional systems of medicine, different parts (leaves, stem, flower, root, seeds and even whole plant) of *Ocimum sanctum* Linn. (Tulsi), a small herb grown throughout India, have been recommended for the treatment of bronchitis, bronchial asthma, malaria, diarrhea, dysentery, skin diseases, arthritis, painful eye diseases, chronic fever, insect bite etc. The *Ocimum sanctum* L. has also been suggested to possess antifertility, anticancer, antidiabetic, antifungal, antimicrobial, hepatoprotective, cardioprotective, antiemetic, antispasmodic, analgesic, adaptogenic and diaphoretic actions (Singh *et al.*, 2012).

## Antioxidant Activity

There are a number of epidemiological studies that have shown inverse correlation between the levels of established antioxidants/phytonutrients present in tissue/blood samples and occurrence of cardiovascular disease, cancer or mortality due to these diseases. Antioxidant-based drugs/formulations for prevention and treatment of complex diseases like atherosclerosis, stroke, diabetes, Alzheimer's disease (AD), Parkinson's disease, cancer, etc. appeared over the past few decades. Various Indian medicinal plants are good sources of antioxidants such as *Aegle marmelos, Aloe vera* (Indian aloe, *Ghritkumari*), *Andrographis paniculata* (Kiryat), *Asparagus racemosus* (*Shatavari*), *Azadirachta indica* (*Neem*), *Bacopa monniera* (*Brahmi*), *Camellia sinensis* (Green tea), *Glycyrrhiza glabra* (*Yashtimadhu*), *Hemidesmus indicus* (Indian *Sarasparilla, Anantamul*) etc. (Devasagayam *et al.*, 2004).

## Antimicrobial Activity

Medicinal plants possess a variety of compounds of known therapeutic properties. The substances that can either inhibit the growth of pathogens or kill them and have no or least toxicity to host cells are considered suitable for developing new antimicrobial drugs. In humans, fungal infections range from superficial to deeply invasive or disseminated, and have increased dramatically in recent years. As a result, antifungal therapy is playing a greater role in health care and various traditional plants are also considered potential antifungal agents. Antimicrobial properties of certain Indian medicinal plants were reported based on folklore information, and a few attempts were made on inhibitory activity against certain pathogenic bacteria and fungi. The antifungal activities of hexane, ethyl acetate and chloroform extracts of 45 medicinal plants were investigated against dermatophytes and opportunistic pathogens. It was found that the ethyl acetate extracts caused maximum inhibition of the fungal strains, followed by hexane and methanol extracts (Duraipandiyan and Ignacimuthu, 2011). In another study conducted by Ahmad and Beg (2001), 45 Indian medicinal plants traditionally used in medicine were examined for their antimicrobial activity against certain drug-resistant bacteria and a yeast *Candida albicans* of clinical origin. Among these, 40 plant extracts showed varied levels of antimicrobial activity against one or more test bacteria. Further, anticandidal activity was detected in 24 plant extracts and broad-spectrum antimicrobial activity was observed in 12 plants. The root extracts of medicinal plant *Heracleum maximum* Bartr. (Umbelliferae) which possess antiviral effects besides antifungal and antibacterial properties (Webster *et al.*, 2006). Also Sambucol, a product isolated from *Sambucus nigra* L., which is effective against various strains of influenza had shown to boost immune responses by secreting inflammatory cytokines (Barak *et al.*, 2001).

**Anti-cancer Activity**

Since 1990 there has been a 22 per cent increase in cancer incidence and mortality with the four most frequent cancers being lung, breast, colorectal, and stomach and the four most deadly cancers being lung, stomach, liver, and colorectal (Parkin *et al.*, 2001). Several classes of dietary compounds have been suggested to reduce the risk of some cancers, especially those of the gut, and there is some evidence that consumption of certain foods leads to a reduction in biomarkers of oxidative damage (Newman *et al.*, 2003). Anticancer agents from plants currently in clinical use can be categorized into four main classes of compounds: vinca (or Catharanthus) alkaloids, epipodophyllotoxins, taxanes, and camptothecins. Vinblastine and vincristine were isolated from *Catharanthus roseus* (L.) (Apocynaceae formerly *Vinca rosea* L.) and have been used clinically for over 40 years (van Der Heijden *et al.*, 2004).

**Other Health Effects**

*Commiphoria mukul* is a small thorny plant indigenous to the India subcontinent (Mesorb *et al.*, 1998). The ole-gum-resin of *C. mukul* is called guggulipid. The yellowish resin produced by the stem of the plant has been widely used in Ayurvedic medicine for more than 2000 years, mainly to treat arthritis and inflammation (Urizar and Moore, 2003; Kimmatkar *et al.*, 2003). The active ingredients in guggulipid are the ketosteroids known as E- and Z guggulsterone (Ding and Staudinger, 2005), are extracted from the resin that is safer and more effective than many cholesterol lowering drugs (Szapary *et al.*, 2003; Asghari *et al.*, 2011).

## Current Scenario and Regulatory Status

The Indian Nutritional market is estimated to be about Rs. 5000 crore; while the global market is growing at a CAGR of 7 per cent, the Indian market has been growing much faster at a CAGR of 18 per cent for the last three years, driven by functional food and beverages categories. However the latent market in India is two to four times the current market size and is between Rs. 10,000 to Rs. 20,000 crore with almost 148 million potential customers. Out of Rs. 5000 crore market size, functional foods share 54 per cent market followed by 32 per cent market share of dietary supplements and 14 per cent share of functional beverages. The Indian nutraceutical market is dominated primarily by pharmaceuticals and FMCG companies with very few pure play nutraceutical companies. Some major companies which are marketing nutraceuticals in India are GlaxoSmithKline consumer healthcare, Dabur India, Cadila Healthcare, EID Parry's, Zandu Pharmaceuticals, Himalaya herbal Healthcare, Amway, Sami labs, Elder pharmaceuticals and Ranbaxy (FICCI, 2009).

The regulatory framework of nutraceuticals in India needs attention from the relevant authorities as less attention is being given to the changing food industry scenario. However, some information regarding nutraceuticals has been given by the Food Safety and Standards act, 2006. Although it specifies the general properties of a nutraceutical, but the traditional medicines have been excluded from it. When talking about nations other than India, United States passed the Watershed legislation in 1994 to regulate the manufacture and marketing of nutraceuticals. This law, known as the Dietary Supplement Health and Education Act (DSHEA), reversed 45 years of increasing FDA regulation of health-related products (Bass and Young, 1996). The passage of the Food and Drug Administration Modernization Act of 1997 (FDAMA) made additional options available to the manufacturers of nutraceuticals. In 1993, the Ministry of Health and Welfare in Japan established a policy of "Foods for Specified Health Uses" (FOSHU) by which health claims of some selected functional foods are legally permitted. In 2001, a new regulatory system, foods with health claims (FHC) with a 'foods with nutrient function claims' (FNFC) system and newly established FOSHU was introduced. In addition, the Govt. changed the existing FOSHU, FNFC and other systems in 2005. Such changes include the new Subsystems of FOSHU such as Standardized FOSHU, Qualified FOSHU and Disease risk reduction claims for FOSHU (Ohama *et al.*, 2006; Gupta *et al.*, 2010).

## Conclusions

Medicinal plants contain a broad range of bioactive compounds. Plant extracts are widely used in the food, pharmaceutical and cosmetics industries. The extraction of nutraceuticals from medicinal plants can be done either by traditional mentods such as maceration, infusion or by emerging methods such as microwave assisted extraction, supercritical fluid extraction. The health benefits offered by the medicinal plants and their extracts are manifold such as antidiabetic activity, antioxidant, antibacterial, antifungal, antiviral, cholesterol lowering action, anti-cancer activity. The Indian nutraceutical market is dominated primarily by pharmaceuticals and FMCG companies with very few pure play nutraceutical companies. The regulatory framework of nutraceuticals in India needs attention from the relevant authorities as there is less attention paid to the changing scenario of the food industry.

## References

Ahmad, I., and Beg, A. Z. (2001). Antimicrobial and phytochemical studies on 45 Indian medicinal plants against multi-drug resistant human pathogens. *Journal of Ethnopharmacology*, **74**: 113–123.

Asghari, J., Ondruschka, B., and Mazaheritehrani, M. (2011). Extraction of bioactive chemical compounds from the medicinal Asian plants by microwave irradiation. *Journal of Medicinal Plants Research*, **5(4)**: 495-506.

Barak, V., Halperin, T., and Kalickman, I. (2001). The effect of Sambucol, a black elderberry- based, natural product, on the production of human cytokines: I. Inflammatory cytokines. *Eur. Cytokine Netw.*, **12 (2)**: 290–296.

Bass, I.S., and Young, A.L. (1996). Dietary Supplement Health and Education Act. The Food and Drug Law Institute, Washington DC.

Cunha, I.B.S., Sawaya, A.C.H.F., Caetano, F.M., Shimizu, M.T., Marcucci, M.C., Drezza, F.T., Povia, G.S., and Carvalho, P.O. (2004). Factors that influence the yield and composition of Brazilian propolis extracts. *J Braz Chem Soc.*, **15**: 964-970.

Devasagayam, T.P.A., Tilak, J.C. Boloor, K.K. Sane, K. S., Ghaskadbi, S.S., and Lele, R.D. (2004). Free Radicals and Antioxidants in Human Health: Current Status and Future Prospects. *JAPI*, **52**: 794-804.

Dharti T.S., Gandhi, S., and Shah, M. (2010). Nutraceuticals-portmanteau of science and nature. *International Journal of Pharmaceutical Sciences Review and Research*, **5(3)**: 33-38.

Ding, X., and Staudingerm, J.L. (2005). The Ratio of Constitutive Androstane Receptor to Pregnane X Receptor Determines the Activity of Guggulsterone against the Cyp2b10 Promoter. *J. Pharmacol. Exp. Therap.*, **314(1)**: 120-127.

Duraipandiyan, V., and Ignacimuthu, S. (2011). Antifungal activity of traditional medicinal plants from Tamil Nadu, India. *Asian Pacific Journal of Tropical Biomedicine*, S204-S215.

FICCI. (2009). Ernst and Young study: Nutraceuticals-Critical supplement for building a healthy India, Health Foods and Dietary Supplements Association conferences, Mumbai Sep. 10, 2009.

Gupta, S., Chauhan, D., Mehla, K., Sood, P., and Nair, A. (2010). An overview of nutraceuticals: current scenario. *Journal of Basic and Clinical Pharmacy*, **1(2)**: 55-62.

Handa, S.S. (2008). An Overview of Extraction Techniques for Medicinal and Aromatic Plants. *In* Extraction Technologies for Medicinal and Aromatic Plants. Eds. Handa, S.S., Khanuja, S. P. S., Longo, G., Rakesh, D.D. Italy, p. 25.

Huie, C.W. (2002). A review of modern sample preparation techniques for the extraction and analysis of medicinal plants. *Anal. Bioanal. Chem.*, **373**: 23-30.

Huie, C.W. (2002). A review of modern sample-preparation techniques for the extraction and analysis of medicinal plants. *Anal Bioanal Chem.*, **373**: 23–30.

Joy, P.P., Thomas, J., Mathew, S., and Skaria, B.P. (2001). Medicinal Plants. Tropical Horticulture Vol. 2. (eds. Bose, T.K., Kabir, J., Das, P. and Joy, P.P.). Naya Prokash, Calcutta, pp. 449-632.

Karale, C. K., Dere, P. J., Honde B. S., Kothule, S., and Kote, A. P. (2011). An overview of supercritical fluid extraction for herbal drugs. *Pharmacologyonline*, **2**: 575-596.

Kimmatkar, N., Thawani, V., Hingorani, L., and Khiyani, R. (2003). Efficacy and tolerability of Boswellia serrata extract in treatment of osteoarthritis of knee- a randomized double blind placebo controlled trial. *Phytomed. Int. J. Phytother. Phytopharm.*, **10(1)**: 3-7.

Lang, Q., and Wai, C.M. (2001). Supercritical fluid extraction in herbal and natural product studies–A practical review. *Talanta*, **53**: 771–782.

Liu, W., and Wang X. (2004). Extraction of flavone analogues from propolis with ultrasound. *Food Sci (China)*, **25**: 35-39.

Lucchesi, M.E., Chemat, F., and Smadja, J. (2004). Solvent free microwave of essential oils from aromatic herbs: comparison with conventional hydro-distillation. *J. Chromatogra.*, **1043**: 323-327.

Lucchesi, M.E., Smadja, J., Bradshaw, S, Louw, W., and Chemat, F. (2007). Solvent free microwave extraction of *Elletaria cardomum* L.; A multivariate study of a new technique for the extraction of essential oil. *J. Food Eng.*, **79**: 1079-1086.

Mason, T.J., Paniwnyk, L. and Lorimer, J.P. (1996). The uses of ultrasound in food technology. *Ultrasonics Sonochemistry*, 3: 253–260.

Mesorb, B., Nesbitt, C., Misra, R. and Pandeyraphy, R.C. (1998). Highperformance liquid chromatographic method for fingerprinting and quantitative determination of E- and Z- guggulsterones in *Commiphora mukul* resin and its products. *J. Chromat. B.*, 720: 189-196.

Narender, T., Shweta, S., Tiwari, P., Papi, R.K., Khaliq, T., Prathipati, P., Puri, A., Srivastava, A., Chander, R., Agarwal, S.C., and Raj, K. (2007). Antihyperglycemic and antidyslipidemic agent from *Aegle marmelos*. *Bioorganic and Medicinal Chemistry Letters*, 17: 1808-1811.

Newman, D.J., Cragg, G.M., and Snader, K.M. (2003). Natural products as sources of new drugs over the period 1981–2002. *Journal of Natural Products*, 66 (7): 1022–1037.

Ohama, H., Ikeda, H., and Moriyama, H. (2006). Health foods and Foods with health claims in Japan. *Toxicology*, 221: 95-111.

Parkin, D.M., Bray, F., Ferlay, J., and Pisani, P. (2001). Estimating the world cancer burden: Globocan 2000. *International Journal of Cancer*, 94 (2): 153–156.

Proestos, C., and Komaitis, M. (2008). Application of microwave-assisted extraction to the fast extraction of plant phenolic compounds. *LWT*, 41: 652–659.

Sarin, R., Sharma, M., Singh, R., and Kumar, S. (2012). Nutraceuticals: A Review. *International Research Journal of Pharmacy*, 3 (4): 95- 99.

Singh, U., Singh, S., and Kochhar, A. (2012). Therapeutic potential of antidiabetic nutraceuticals. *Phytopharmacology*, 2(1): 144-169.

Smith, R.M. (2003). Before the injection- modern methods of sample preparation for Techniques of preparing plant material for chromatographic separation and analysis. *J. Chromatogr. A.*, 1000: 3-27.

Sonnenschein, H., Germanus, I., and Harting, P. (2002). Studied on the effect of microwave radiation on over critical extraction processes. *Chem. Eng. Tech.*, 74: 270-274.

Stobiecki, M., Wojtaszek, P., and Gulewicz, K. (1997). Application of solid phase extraction for profiling quinolizidine alkaloids and phenolic compounds in *Lupinus albus*. *Phytochemical Analysis*, 8(4): 153- 158.

Szapary, P.O., Wolfe, M.L., Bloedon, L.A.T., Cucchiara, A.J., Dermarderosian, A.H., Cirigliano, M.D., and Rader, D.J. (2003). Guggulipid for the treatment of hypercholesterolemia: a randomized controlled trial. *J. Am. Med. Assoc.*, 290: 765-772.

Tapiero, H., Tew, K. D., Nguyen Ba, G., and Mathe, G. (2002). Polyphenols: Do they play a role in the prevention of human pathologies? *Biomedical Pharmacotherapy*, 56: 200–207.

Tatke, P., and Jaiswal, Y. (2011). An Overview of Microwave Assisted Extraction and its Applications in Herbal Drug Research. *Research Journal of Medicinal Plant*, 5 (1): 21-31.

Tekel, J., and Hatrik, S. (1996). Review Pesticide residue analyses in plant material by chromatographic methods: clean-up procedures and selective detectors. *J. Chromat. A.*, **754**: 397-410.

Urizar, N.L., and Moore, D.D. (2003). Gugulipid: A natural cholesterol lowering agent. *Ann. Rev. Nutr.*, **23**: 303-313.

Van Der Heijden, R., Jacobs, D.I., Snoeijer, W., Hallard, D., and Verpoorte, R. (2004). The Catharanthus alkaloids: pharmacognosy and biotechnology. *Current Medicinal Chemistry*, **11 (5):** 607–628.

Vijay, A.K., Thakur, A., and Sinha, K. (2006). The Metabolic Syndrome- Its Prevalence and Association with Coronary Artery Disease in Type 2 Diabetes. *Journal Indian Academy of Clinical Medicine*, **7**: 32-38.

Vilegas, J.H.Y., de Marchi, E., and Lancas, F.M. (1997). Extraction of low-polarity compounds (with emphasis on coumarin and kaurenoic acid) from *Mikania glomerata* ('Guaco') leaves. *Phytochemical Analysis*, **8**: 266–270.

Wang, L., and Weller, C.L. (2006). Recent advances in extraction of nutraceuticals from plants. *Trends in Food Science and Technology*, **17**: 300–312.

Webster, D., Taschereau, P., Lee, T.D., and Jurgens, T. (2006). Immunostimulant properties of *Heracleum maximum* Bartr. *J. Ethnopharmacol.*, **106 (3):** 360– 363.

Wildman, R.E.C., and Kelley, M. (2007). Nutraceuticals and Functional Foods. In: Wildman Robert E.C. Handbook of Nutraceuticals and Functional Foods. Second Edition. New York: CRC Press, pp: 1-9.

Woisky, R.G., and Salatino, A. (1998). Analysis of propolis: some parameters and procedures for chemical quality control. *J Apicult Res.*, **37**: 99-105.

Utilisation and Management of Medicinal Plants Vol. 2 (2014)    *Pages* **355–371**
*Editor-in-Chief:* **V.K. Gupta**
*Published by:* **DAYA PUBLISHING HOUSE, NEW DELHI**

# 17

# Biotechnological Products of Amazonian Fruits

C.V. Lamar o[1], K. Yamaguchi[2], F.E.B. Herculano[2],
M. Campelo[2] and V.F. Veiga Junior[2*]

## ABSTRACT

*The Brazilian Amazon region, due to its vast biodiversity, contains numerous species of fruit-bearing plants that are the object of regional extraction, attracting research studies and generating biotechnology products, especially in the area of popular medicines, foods and cosmetics. Within this scope, this study highlights nine species of Amazonian fruits: açaí (Euterpe oleracea Mart); abiu (Pouteria caimito Ruiz, Pavon); bacuri (Platonia insignis Mart); Amazonian chestnut (Bertholetia excelsa Humb); cupuaçu (Theobroma grandiflorum Willd. ex Spreng. Schum), guarana (Paulinia cupana Kunth); piquiá (Caryocar villosum); pupunha (Bactris gasipaes Kunth) and uxi (Endopleura uchi (Huber), Cuatrec), plus two other species planted in the Amazon in the past, that are now well-distributed throughout the region: acerola cherry, (Malpighia emarginata L.) and palm (Elaeis guineensis).The fruits of these plants are part of the tradition and culture of the peoples of the Amazon, and are present in many of their traditional customs, foods, drinks, pharmaceuticals and cosmetics. The presence of energy-boosting substances, nutrients and other organic elements of importance for human life and health are among the characteristics of these fruits, which have a broad spectrum of uses and applications, as reported in the literature, although this list represents only a small sample of the many species that form part of the biome of the vast Amazon rainforest of Brazil.*

*Keywords:* Brazilian biodiversity, Amazonian fruits, Biotechnology.

---

1    College of Agricultural Sciences, Federal University of Amazonas, Manaus, Brazil.
2    Exact Sciences Institute, Federal University of Amazonas, Manaus, Brazil.
*    *Corresponding author*: E-mail: valdirveiga@ufam.edu.br

# Introduction

Numerous reports in the literature endorse the potential of fruits and vegetables of the Brazilian Amazon rainforest, based on studies of their functional capacity. Many of these fruits and vegetables are rich in bioactive compounds that induce or enhance positive health effects for those who consume them, encouraging ever increasing volumes of research into the presence of nutraceutical compounds, among others (Gorinstein *et al.*, 2011).

In the USA alone, the fresh fruit trade is worth $20 billion/year. Brazil also occupies a position of international prominence in this market, as the third largest producer of tropical fruits in the world (Maia *et al.*, 2009). Nevertheless, many of these Brazilian resources are under exploited, with considerable volumes of waste being generated that could be used as a good nutritional, functional and income source for people and a source of economic profit for industries. This reality needs to change, since the world is increasingly realizing the importance of these fruits for human health, particularly due to their antioxidant and other beneficial properties (Hassimotto *et al.*, 2005; Kuskoski *et al.*, 2005; Roesler *et al.*, 2006; Silva *et al.*, 2007).

Brazil, and especially the Amazon region, has a great biodiversity, another attraction for the many studies that have been carried out, including research into the functional properties of fruits, and their by-products or waste products. Characterization of these plant resources is important for detecting and identifying the bioactive compounds that are so widely appreciated today.

# Material and Methods

We performed a literature review of research published by Brazilian and international groups, using the research databases SCOPUS, ISI, and SCIELO and others Internet search tools.

A summary is given below of the main constituent characteristics and uses of some fruits of the Brazilian Amazon, their main uses among the local populations, and the modest agribusiness that exists in this region based on these fruits.

# Results and Discussion

## AÇAÍ (*Euterpe oleracea* Mart)

The Amazon region has a large number of perennial plants, particularly fruit species. One of these is açaí, which is an important fruit for agroindustrial development in the Amazon region (Rogez, 2000).

Açaí contains lipids, proteins, fibers and anthocyanins. Its consumption in Northern Brazil is generally combined with other regional foods, to produce ice cream, custards, porridges, jams and liqueurs. In the Southeast of Brazil, açaí is consumed differently from in the North region. In the latter region, it is considered a high-energy, nutritionally complete food, and is consumed with cereals, fruits, and fast absorbing carbohydrates, which are added to the acai, to compensate for its lack of simple sugars (Oliveira, 1995; Nogueira, 1995).

The state of Pará is a major exporter of açaí fruit in fresh form. It is also sold as a pasteurized juice, a concentrate, and blended with other fruits such as banana, orange and guaraná. These products are exported mainly to the Southeast and South Brazil, as well as Europe, Asia and North America. The concentrated form of açaí is mainly used in the production of energy drinks (Rogez, 2000).

Açaí powder and dehydrated pulp, which spray-dried, are also used in the production of energy capsules and medicines (Silva *et al.*, 2008).

This fruit is also used in the field of cosmetics, especially its oil. According Rogez (2000), açaí oil, like olive and avocado oil, is rich in monounsaturated and polyunsaturated fatty acids, with contents of 60 per cent and 14 per cent, respectively. The oil can be extracted by the action of solvents, or by the process of enzymatic extraction, which is considered more environmentally sustainable.

The literature still contains contradictory data concerning its fatty acid composition (Christensen, 1991; Da-Silva *et al.*, 1997; Parmentier, 2004). The waste from açaí seeds has been studied to investigate its possible applications, such as in biofuels and in thermochemical processes (Virmond *et al.*, 2012).

## ACEROLA (*Malpighia emarginata* L.)

Acerola (*Malpighia emarginata* L.), a cherry variety, is highly adaptable to various climates and can be found in various regions of the planet, but its commercial cultivation is concentrated in tropical and subtropical regions (Konrad, 2002).

In Brazil, the crop is farmed in the Northeast, North, South and Southeast regions (Ritzinger, 2004). The average productivity of Brazilian acerola orchards is 29.65 tons per hectare per year, equivalent to 59.3 kilograms/plant/year (Agrianual, 2010).

Acerola is sold mainly in the form of frozen pulp and fresh fruit, to consumers who appreciate natural juices. Due to its high vitamin C content, it is considered a high quality product, especially in the field of functional foods. Other acerola products sold on the domestic market are: acerola powder, acerola with vitamin E, pure medicinal vitamin C capsules, jellies and sweets (Manica *et al.*, 2003).

Acerola juice can be used to advantage as an enriching agent in the processing of numerous juices and nectars that are poor in vitamin C (apple, pear, acerola, lime, pineapple and peach) (Ledin, 1958). Nogueira (1991) also mentions the possibility of using acerola juice to enrich the vitamin C content of other juices and nectars.

In this regard, studies have successfully demonstrated the use of acerola in the formulation of blends containing pineapple (Matsuura *et al.*, 2002), papaya, carambola (Matsuura, 2004) and coconut water (Lima *et al.*, 2008).

All these studies conclude that acerola is ideal for mixing in different proportions in the preparation of blends. Furthermore, there was an increase in the amount of vitamin C in the remaining fruit juices, through the addition of varying amounts of acerola juice, to obtain a blend that retains the flavour of the original juices (Matsuura *et al.*, 2002).

Dehydration is also used as a preservation method, preventing deterioration and loss of commercial value, and also as a means of refining the food product, resulting in the introduction of a new product on the market, which usually prompts investments in agricultural production and processing due to the financial benefits derived from transforming the product (Ritter, 1994; Soares *et al.*, 2001).

In this regard, some studies have been carried out on the extraction and dehydration of the pulp by the Foammat method, to obtain a powder that can be used as a food supplement, rich in Vitamin C, in quantities compatible with the minimum recommended daily dose for human consumption. The results showed that the processed product had a high vitamin C content—ten times higher than that of the natural pulp (Soares *et al.*, 2001).

The use of starch in the development of biofilms has the advantage of biodegradability. However, its properties are inferior to those of conventional plastics, forming materials that are brittle and hygroscopic. Many studies have been published using modified starches and starch blends of synthetic polymers, or composites of starch with plant fibers, seeking to improve the mechanical properties of the materials prepared (Souza *et al.*, 2012).

In this context, acerola (*Malpighia emarginata* L.) and its derivatives contain high concentrations of antioxidants, such as carotenoids and phenolic compounds, which can be incorporated into biodegradable films as active compounds. The fruit also contains significant amounts of fibers, especially pectin, starch and cellulose derivatives, and may thus contribute as an additive for enhancing the mechanical and thermal qualities and improving the water barrier of the biofilms, as well as for its antioxidant action resulting from pigments and other compounds (Larrauri *et al.*, 1996; Sanchez-Garcia *et al.*, 2008; Souza *et al.*, 2012).

Studies with acerola have shown that uniform dispersion of these additives at the interface of the matrix results in changes in molecular mobility in relaxation, and consequently, the mechanical and thermal properties of the resulting composite. Molecular compounds with large molecular radii, such as fibers, are particularly interesting because of the greater surface contact, and can promote the desired strengthening (Azizi *et al.*, 2005; Dalmas *et al.*, 2008; Souza *et al.*, 2012).

## ABIU (*Pouteria caimito* Ruiz; Pavon)

Abiu, the fruit of abiu tree, probably originated in Western Amazonia and adjacent areas of Peru, Colombia and Venezuela (Manica, 2000). Although little explored commercially, abiu is widely consumed fruit in the tropics, in its natural form. In Brazil, its use is more widespread in the states of the North, specifically in Acre, Amazonas, Amapá and Pará (Lorenzi, 2000).

Despite its many excellent qualities, the abiu tree has remained largely restricted to Brazil, as a fruit tree grown non-commercially in backyards and orchards. It is found in the wild throughout the Amazon, and is cultivated almost throughout Brazil. There is wide variation in shape, size and quality of fruit, some with firm flesh, others with soft or insipid/mild taste. (Andersen, 1989; LorenziI, 2002).

There is little technological processing applied to abiu and its components. It is used in the local regional cuisine, and is submitted to simple technological processes, such as the production of jellies, ice cream and simple pulp extraction (Gomes, 1980).

The skin is recommended in the popular Amazon pharmacopoeia for combating diarrhea, fever and infections. From the seeds, an oil is extracted that according to regional popular culture, is used to treat skin inflammations (Gomes, 1980; Andersen, 1989; Lorenzi, 2002).

## BACURI (*Platonia insignis* Mart)

Bacuri is a native fruit of the Brazilian Amazon and Guyana, but it also grows in Colombia and Paraguay (Chitarra and Chitarra, 2005). In Brazil, it is mainly found in the State of Pará, but it also grows in Maranh o, MatoGrosso, Goiás and Piauí (Silva and Donato, 1993; Villachica *et al.*, 1996; Aguiar *et al.*, 2008).

This fruit has large, ovoid or subglobose berries weighing between 200 and 1000 g. The flesh is creamy-white, but turns yellow when exposed to air. With a pleasant aroma and flavor, it generally has three seeds, which are also edible (Chitarra and Chitarra, 2001).

The bacuri pulp is widely used in food technology in the form of frozen pulp, jams, and yogurts, and as natural flavoring. The bacuri skin has been studied for its rich fatty acid content, and in particular, its fragranced oil, which is used against dermatitis (Kerolla, 1993, Ferreira *et al.*, 1993; Van Den Berg, 1993). The main techniques used to obtain oil from the skin of bacuri involve pressurized carbon dioxide and the use of solvents (Monteiro *et al.*, 1997).

Some studies have shown the potential of oil extraction from the pulp of bacuri, and also of the production of flour from the pulp bacuri, with positive results in terms of the composition of carotenoids, oleic acid and palmitic acid (Hiane *et al.*, 2003).

## CASTANHA (*Bertholetia excelsa* Humb)

The chestnut (*Bertholletia excelsa* Humb) is the seed of the chestnut tree, which is found mainly in the Amazon region, particularly in the states of Pará, Amazonas, Rondônia and Acre (Freitas *et al.*, 2008).

It is a fruit known for its high biological value, vitamins and minerals and high levels of unsaturated fat (Sun *et al.*, 1987; Felberg *et al.*, 2004; Freitas *et al.*, 2008).

Chestnut is used in the local cuisine, by native peoples, in the preparation of typical dishes of the region. The milk obtained from pressing the nut has the same types of uses as the fruit, but is also used to feed infants. The flour obtained from drying and grinding the nuts is widely used in bakery products like cookies, cakes and breads, as well as to enrich other flours (Camargo *et al.*, 2000; Ferberg *et al.*, 2002).

Food technology applied for industrial purposes produces dried, salted, roasted, or caramelized chestnuts, and it is also used in the production of sweets, and as a component in products such as breakfast cereals, ice cream and chocolate (Vilhena, 2004; Bowles and Demiate, 2006).

Chestnut flour, whether pure or blended with other types of flour, such as peach-palm, provides a good alternative for the preparation of food supplements used in special diets formulated as part of *in vivo* studies. Chestnut has also been used successfully as a source of plant protein in the diet of fish in ponds, and it has been shown that the addition of 30 per cent chestnut flour to the fish feed was able to fatten the fish, without compromising physiological homeostasis or performance in the nursery (Silva *et al.*, 2003; Oliveira, 2005; Tacon *et al.*, 2006; Pereira Jr, 2006; Salze *et al.*, 2010).

## CUPUAÇU (*Theobroma grandiflorum* Willd. ex Spreng. Schum)

Cupuaçu is one of the most popular fruits of the Amazon, with its pleasant taste and odor. The fruit measures 12-15 cm in length and 10-12 cm in diameter, and weighs 1 kg. Of this weight, 30 per cent is pulp and 35 per cent is seeds (Nazaré *et al.*, 1990; Venturieri, 1993; Wolf, 1997).

Various byproducts are obtained from the pulp and seeds of cupuaçu. "Chocolate" powder can be made by grinding the defatted cupuaçu cake, which is then mixed with other ingredients. The raw material for the production of cupuaçu paste consists of the seeds after fermentation, drying and grinding (Medeiros, 1999; Lannes *et al.*, 2002).

Cupuaçu seeds, when subjected to a similar process to that of cocoa (*Theobroma cacao* L.), can develop a similar aroma to chocolate made from cocoa (Nazaré, 2000, Yang *et al.*, 2003; Lopes *et al.*, 2003).

Milling of the defatted cupuaçu cake and the seeds results in a chocolate, known regionally as "cupulate" (cupuaçu + chocolate), while the spray drying process also results in chocolate by-products. The relationship between cocoa and cupuaçu is confirmed by the composition and biochemical properties of the seeds, and the presence of purine alkaloids (methylxanthines) in both (Yang *et al.*, 2003; Reisdorff *et al.*, 2004).

Another product obtained from the cupuaçu seeds is cupuaçu liquor, similar to cocoa liquor, which is defined as a dispersion of cocoa particles surrounded by a continuous fat phase composed of cocoa butter. The cupuaçu liquor can be used in the formulation of products similar to chocolate, cakes, cookies and ice cream. The fat from cupuaçu can be extracted, and this is currently a subject that is being investigated by the food and pharmaceutical industries (Fang, 1995; Cohen, 2003).

## GUARANÁ (*Paulinia cupana* Kunth)

Guaraná, the fruit of the Guaraná plant, is an important and traditional product in the state of Amazonas. It is a genuinely Brazilian plant of great economic and social importance, especially in the Amazon region. Its importance is shown by the high demand for its seeds by the soft drinks and energy drinks industries, both in Brazil and overseas (Suframa, 2003; Tavares and Pereira, 2005).

The most traditional form of marketing is guaraná sticks, a method developed by the Indians of the region, more specifically the city of Maués-AM. In Amazonas and MatoGrosso, there is high demand for this type of product (Suframa, 2003; Tavares and Pereira, 2005).

Guaraná Seeds, after roasting, produce small shells, known as grain "casquilho". These are ground by hand, in a mortar, to decrease the particle size, and then mixed with water to form a thick paste, which is molded into a the form of a small, cylindrical stick. The stick is dehydrated in a prolonged process of curing, resulting in the product widely sold in the region (Suframa, 2003; Tavares and Pereira, 2005).

In the Amazon, guaraná sticks are used as a beverage; the sticks are grated to a fine powder, which is then mixed with water (Suframa, 2003; Tavares and Pereira, 2005).

Guaraná powder is also used in the preparation of drinks, ice creams, creams and other foods. It is a product with higher added value, and is the form most commonly found on the retail market in Manaus-AM (Suframa, 2003; Tavares and Pereira, 2005).

Another biotechnological application of guaraná is in the production of isotonic/energy drinks, targeted specifically at consumers who practice sports. It is also used in the prevention of certain human pathologies (Ashihara, 2003; Mendes and Carlini, 2007).

## Amazonian Palm Oils

Palm oil is a vegetable oil obtained from the pulp of the fruit of the African palm oil, the scientific name of which is *Elaeis guineensis*. It is native to tropical Africa and is also cultivated in Central America, South America and Asia. The State of Pará is the largest producer of this product in Brazil (Silva, 1997; Coelho *et al.*, 2004).

Two types of oil are obtained from the fruit of the oil palm: oil palm (extracted from the pulp) and palm kernel oil (extracted from the kernel). The oil yield represents approximately 22 per cent of the weight of the bunches for palm oil and 3 per cent for palm kernel oil. The main difference between palm oil and palm kernel is the content of palmitic acid and oleic acid (Silva, 1997; Coelho *et al.*, 2004).

The oil is obtained by extraction or pressing, and its color may vary, depending on how it is processed. Due to its high productivity of more than ten tons/hectare per year, palm oil currently occupies second position in the world production of vegetable oils and fats, behind only soybean oil. It is estimated that by 2012, soybean oil production will be superseded by palm oil (Cenbio, 2003).

Palm oil is known internationally for its multiple applications. Due to its low acidity (4-5 per cent) it is widely used in farming, as a component of animal feed. After refining, it is widely used in the manufacture of margarine, biscuits, breads and ice cream (Surré and Ziller, 1969). However, its versatility opens up other prospects for consumer use, and palm oil is now used in the manufacture of soaps, detergents, candles, pharmaceuticals, cosmetics and natural dyes. It is also used in the steel industry, in the manufacture of rolled steel and white iron (Surré and Ziller, 1969).

The palm kernel oil, due to its high quality and high levels of lauric and myristic acids, has similar applications to coconut oil, hence it is used in soaps, detergents, creams, mayonnaise among other things. It is also used to produce chocolate, as a substitute for cocoa butter (Kitamura, 1990).

The clusters are processed to extract the palm and palm kernel oils, and also a series of by-products (fibers, curls nut shells, palm kernel cake and wastewater), all of which have various applications (Kitamura, 1990).

The palm kernel cake has about 13 per cent crude protein and is widely used as a component in food for domestic animals (cattle, poultry, horses and pigs), and as organic fertilizer (Rodrigues Filho *et al.*, 1994).

The nut shell can be used to manufacture brake fiber, or as an alternative source of energy to fuel boilers. The fibers and empty clusters, in turn, may also be used for the same purpose as a fertilizer in the cultivation area itself. The wastewater, after treatment, is also used for this same purpose (Embrapa, 1983).

Fatty acid residues from the refining of palm oil are used as fuel. Besides the production cost compared to other oilseeds, palm allows the manufacture of biodiesel from waste rather than oil, avoiding the formation of byproducts during the production process, such as glycerin, that occurs in other processes for producing biodiesel, such as from soybean, castor and rapeseed (Coelho *et al.*, 2004).

## PIQUIÁ (*Caryocar villosum* Aubl)

Piquiá isdistributed throughout the Amazon, being more highly concentrated in the non-flooded areas. The piquiá tree can reach up to 50 meters in height, with a trunk diameter up to 2.5 meters (Shanley, 2005).

The piquiá tree produces flowers during the dry season, from August to October, and bears fruit from February through April (Lentini, 2005).

The fruits of piquiá are considered as an excellent source of bioactive compounds, as they present high values of total phenols and flavonoids, and have good antioxidant capacity (Krinsky, 1994; Rios *et al.*, 2007; Barreto *et al.*, 2009.;Chisté, Mercadante *et al.*, 2011; Chisté and Mercadante, 2012).

For consumption, the fruit is used mainly in cooking during the preparation of regional dishes, accompanied with rice (Clement, 1993; Bauch and Sieber, 2006).

The piquiá skin produces an oil that is used in foods and chemicals. The oil is used in cooking and in the processing and production of butter (Passos *et al.*, 2003; Grenand *et al.*, 2004). It is also used in soaps and cosmetics. In the latter area, dermatological studies have demonstrated its successful use for skin problems related to the action of fungi, and it also has anti-inflammatory properties (Bauch and Sieber, 2006; Xavier *et al.*, 2011).

## PUPUNHA (*Bactris gasipaes* Kunth)

The origins of the pupunha tree, which produces the pupunha, or peach-palm fruit, are not entirely certain. It grows in the Amazon and Central America, and its domestication is attributed to the indigenous peoples of these regions (Chaimsohn, 2006; Soares, 2011).

The North of Brazil is characterized by an availability of fruits rich in pro-vitamin A, especially peach-palm, a tropical fruit of the palm family. This species represents a potentially rich source of nutritious food, due to its high content of carbohydrates and in particular, bioavailable carotenoids, as well as proteins and lipids (Apgar, 1977; Yuyama *et al.*, 1991; Yuyama and Cozzolino, 1996).

It is a good alternative the production of palm, and can be exploited in organized plantations, owing to its desirable characteristics such as early fruit-bearing, tillering, yield and quality of the palm (Padilha *et al.,* 2003; Soares, 2011). However, a lack of quality control of the palm production process, and predatory extraction of native palms, have so prevented Brazil from becoming the world's largest producer of palm (Sampaio, 2007; Soares, 2011).

The characteristic of pupunha starch makes it ideal for use in flours (use as an addition in food supplements, successfully tested on rats) and alcohol (Yuyama and Cozzolino, 1996; Bianchini *et al.,* 1998).

Pupunha flour has also been tested as a substrate for the production of amylases in rhizobia strains, as it makes the process of obtaining amylases by microbial action less costly, due to its commonly used substrates such as dextrin, fructose and glucose (Stamford, 2001, Strauss, 2001; Buzzini, 2002; Haq, 2002; Oliveira *et al.,* 2007).

The application of pupunha flour together with cassava flour has been studied in the production of third generation extrudates, also known as half products or pellets (Carvalho *et al.,* 2002; Aschieri *et al.,* 2006; Carvalho *et al.,* 2010).

## UXI [*Endopleura uchi* (Huber), Cuatrec]

Uxi originates in the Brazilian Amazon, frequently growing in the Amazon River estuary in the State of Pará, in areas like Bragantina, Guamá and Capim, the west of Marajó Island, and in the small river channels adjacent to it. It is also widely distributed throughout the Amazon Basin (Silva, 2009).

A typically wild species of the upland forest, the Uxi tree can reach 25 to 30 meters in height, 1 meter in diameter or 3 feet in diameter. Its fruit is popularly known as "uxiamarelo" (yellow uxi) or "uxiliso" (smooth uxi) (Shanley, 2005).

The uxi tree blooms from October to November, and bears fruit from February to May, but in some areas near the cities of BelémPará, Viseu and Mosqueiro, trees may bear fruit out of season, from July to August (Shanley, 2005).

There have been few studies on the chemical composition of uxi. Some studies have shown that its pulp has a high amount of fat, predominantly oleic acid, in addition to the carotenoid trans-$\beta$-carotene (Marx *et al.,* 2002; Magalh es *et al.,* 2007).

Uxi is mainly consumed in the form of food by-products, which are submitted to simplified technological process flowcharts. It is also used by the pulp agribusiness, and to make juice and ice cream (Lentini, 2005).

On a smaller scale, in regional folk medicine, the oil is extracted from the fruit skin, and is used to treat sinusitis and constipation (Lentini, 2005). The skin is also used as a tea, with anti-inflammatory and antitumor properties (Corrêa, 1984; Revilla, 2001). Silva (2009) isolated the substance berginina from the skin of this fruit, demonstrating its antimicrobial properties.

Other parts of the uxi that are used include the seeds (to make eco-jewellery and also to produce a mosquito repellent smoke) and the seed powder, which is recommended in ethnomedicine for covering skin blemishes and minimizing allergies (Bauch and Sieber, 2006).

# References

Agrianual (2010). anuário da agricultura brasileira. S o Paulo: FNP Consultoria e Comércio, p. 520.

Aguiar, L.P., Figueiredo, R.W., Alves, R.E., Maia, G.A and Souza, V.A.B. (2008). Caracterizaç o física e físico-química de frutos de diferentes genótipos de bacurizeiro (*PlatoniainsignisMart.*). *Ciência e Tecnologia de Alimentos*, **28:** 423-428.

Andersen, O., and Andersen, V. U. (1989). As frutas silvestres brasileiras. Publicaç es Globo Rural, S o Paulo, 3a ediç o, pp. 37-40.

Ashihara, H., and Crozier, A. (2001). Caffeine: a well-known but little mentioned compound in plant science. *Trends Plant Sci.*, **6(9):** 407-413.

Azizi Samir, M.A.S., Chazeau, L., Alloin, F., Cavaillé, J.Y., Dufresne, A., and Sanchez, J.Y. (2005). POE-based nanocomposite polymer electrolytes reinforced with cellulose whiskers. *Electrochim Acta*, **50:** 3897–3903.

Barreto, G.P.M., Benassi, M.T., and Mercadante, A.Z. (2009). Bioactive compounds from several tropical fruits and correlation by multivariate analysis to free radical scavenger activity.*Journal of the Brazilian Chemical Society*, **20:** 1856-1861.

Ascheri, D.P.R., Andrade, C.T., Carvalho, C.W.P., and Ascheri, J.L.R. (2000). Efeito da extrus o sobre a adsorç o de água de farinhas mistas pré-gelatinizadas de arroz e bagaço de jabuticaba. *Ciência e Tecnologia de Alimentos*, Campinas, **26:** 325-335.

Bauch, S. C., and Sieber, S. S. (1998). Mercado de Produtos N o-Florestais em Belém do Pará. Belém: Imazon, p. 30 p, 2006.

Bianchini, R., and Penteado, M. de V.C. (2008). Carotenóides de piment es amarelos: caracterizaç o e verificaç o de mudanças com o cozimento. Campinas-SP. *Ciência e Tecnologia de Alimentos*, **18(3):** 283-288.

Bowles, S., and Demiate, I. M. (2006). Caracterizaç o físico-química de *Okarae* aplicaç o em p es do tipo francês. *Ciência e Tecnologia de Alimentos*, **26(3):** 652-659.

Buzzini, P., and Martini, A. (2002). Extracellular enzymatic activity profiles in yeast and yeast-like strains isolated from tropical environments. *J. Appl. Microbiol.*, **93(6):** 1020-1025.

Camargo, I. P., Castro, E. M., and Gavilanes, M. L. (2000). Aspectos da anatomia e morfologia de amêndoas e plântulas de castanheira-do-Brasil. *Revista Cerne.*, **6(2):** 11-18.

Carvalho, A.V., Vasconcelos, M.A.M., Silva, P.A., Assis, G.T., and Ascheri, D.P.R. (2010). Caracterizaç o tecnológica de extrusados de terceira geraç o à base de farinhas de mandioca e pupunha. *Ciência e Agrotecnologia*, Lavras, **34(4):** 995-1003.

Carvalho, R.V., Ascheri, J.L.R., and Cal-Vidal, J. (2002). Efeito dos parâmetros de extrus o nas propriedades físicas de pellets de misturas de farinhas de trigo, arroz e banana. *Ciência e Agrotecnologia*, Lavras, **26(5):** 1006-1018.

Chaimsohn, F. P. (2006). Producción y calidaddel palmito al natural, em función de lapoblación, del arreglo de plantas y del tipo de fertilización. CENBIO – Centro Nacional de Referência em Biomassa. Projeto PROVEGAM,205f. Tese(Doutorado em Sistema de Produç o AgrícolaTropical Sustentável)– Universidad de CostaRica, Costa Rica.

Chisté, R.C., and Mercadante, A.Z. (2012). Identification and quantification, by HPLC DAD-MS/MS, of carotenoids and phenolic compounds from the Amazonian fruit *Caryocar villosum*. *Journal of 520 Agricultural and Food Chemistry*. (in press).

Chisté, R.C., Mercadante, A.Z., Gomes, A., Fernandes, E., Lima, J.L.F.C., and Bragagnolo, N. (2011). *In vitro* scavenging capacity of annatto seed extracts against reactive oxygen and nitrogen species. *Food Chemistry*, **127**: 419–426.

Chiatarra, A.B., and Alves, R.E. (2001). Tecnologia de pós-colheita para frutas tropicais. Fortaleza, FRUTAL – SINDIFRUTA, p. 27

Chiatarra, A.B., and Chiatarra, M.I.F. (2005). Pós-colheita de frutos e hortaliças: Glossário. Lavras, UFLA, p. 256.

Christensen, F.M. (1991). Extraction by aqueous enzimatic process. *International News on Fat,Oils and related Materials*, **2(11)**: 984-987.

Clement, C.R. (1993). Piquiá. In: Selected species and strategies to enhance income generation from Amazonian forests. FAO: Rome; 108-114.

Coelho, S.T., Silva, O.C., Velázques, S.M.S.G., Monteiro, M.B., and Silotto, S. E.G.A. (2004). Utilizaç o de óleo de palma "in natura" como combustível em grupos geradores a diesel. Anais I Congresso Internacional em Bioenergia, Campo Grande – MS.

Cohen, K.C. (2003). Estudo do processo de temperagem do chocolate ao leite e de produtos análogos elaborados com liquore gordura de cupuaçu.296p.Tese (Doutorado em Tecnologia de Alimentos) – Faculdadede Engenharia de Alimentos, UNICAMP.

Corrêa, M.P. (1984). Dicionário das Plantas Úteis do Brasil e das Exóticas Cultivadas. Rio de Janeiro, Imprensa Nacional, vol. 6, p. 326.

Da-Silva; R., Franco,R., Célia, M.L., and Gomes, E. (1997). Pectinases, hemicelulases e celulases, aç o, produç o e aplicaç o no processamento de alimentos. *Boletim da SBCTA*, Campinas, **31(2)**: 249-260.

Embrapa. (1983). Dendê: uma nova opç o agrícola. Manaus: EMBRAPA/CNPSD, 22p (Documentos, 14).

Fang, T.N., Tiu, C., Wu, X., and Dong, S. (1995). Rheological behavior of cocoa dispersions. *Journal of Texture Studies*, **26**: 203-215.

Ferberg, I. (2002). Efeito das condiç es de extraç o no rendimento e qualidade do leite de castanha-do-brasil despeliculada. *Boletim do Centro de Pesquisa e Processamento de Alimentos*, **20(1)**: 75-78.

Felberg, I. (2004). Bebida mista de extrato de soja integral e castanha-do-brasil: caracterizaç o físico-química, nutricional e aceitabilidade do consumidor. *Alimentos and Nutriç o*, **15(2):** 163-174.

Ferreira, S.R.S., Meireles, M.A.A., and Cabral, F.A. (1993). Extraction of essential oil of black pepper with liquid carbon dioxide, *J. Food Eng.*, **20:** 121.

Freitas, S. C. (2008). Meta-análise do teor de selênio em castanha-do-brasil. *Brazilian Journal of Food Tecnhology*, **11(1):** 54-62.

Gomes, R. P. (1980). Fruticultura Brasileira. Ed. Nobel. S o Paulo, pp. 80-81.

Goristein, S., Poovarodom, S., Leontowicz, H., Leontowicz, M., Namiesnik, J., Vearasilp, S., Haruenkit, R., Ruamsuke, P., Katrich, E., and Tashma, Z. (2011). Antioxidant properties and bioactive constituents of some rare exotic Thai fruits and comparison with conventional fruits. *In vitro* and *in vivo* studies. *Food Research International*, doi: 10.1016/j.foodres. 2011.10.009.

Grenand, P., Moretti, C., Jacquemin, H., and Prévost, M. F. (2004). Pharmacopées Traditionnelles en Guyane. 2nd ed.; IRD: Marseille, France, pp. 309-319.

Haq, I., Ashraf, S., Omar, S., and Qadeer, M. A. (2002). Biosynthesis of Amyloglucosidase by *Aspergillus niger* using wheat bran as substrate. *Pak. J. Biol. Sci.*, **5(9):** 962-964.

Hassimotto, N. M. A., Genovese, M. I., and Lajolo, F. M. (2005). Antioxidant activity of dietary fruits, vegetables, and commercial frozen fruit pulps. *Journal of Agricultural and Food Chemistry*, **53(8):** 2928–2935.

Kerrola, H. K. (1993). Volatile compounds and odor characteristics of carbon dioxide extracts of coriander (*Coriandrum sativum L.*) fruits.*J. Agric. FoodChem.*, **41(5):** 785-790.

Kitamura, P. C. (1990). Dendê: oferta e demanda no mercado internacional. Belém: EMBRAPA/CPATU, p. 24 (Documentos, 51).

Krinsky, N.I. (1990). The biological properties of carotenoids. *Pure and Applied Chemistry*, **66(5):** 1003-1010.

Kuskoski, E. M., Asuero, A. G., Troncoso, A. M., Mancini-Filho, J., and Fett, R. (2005). Aplicación de diversos métodos químicos para determinar actividad antioxidante enpulpa de frutos. *Ciência e Tecnologia de Alimentos*, **25(4):** 726–732.

Lannes, S. C. S., and Medeiros, M. L. (2002). Formulaç o de "chocolate" de cupuaçu e reologia do produto líquido. *Rev. Bras. Cienc. Farm.*, S o Paulo, **38(4):** 463- 469.

Larrauri, J. A., Rupérez, P., Borroto, B., and Saura-Calixto, F. (1996). Mango peels as a new tropical fibre: preparation and characterization. *Food Science and Technology*, **29(8):** 729-733.

Ledin, R. B. (1956). A comparison of three clones of Barbados Cherry and the importance of improved selections for commercial plantings. *The Proceedings of Florida State Horticultural Society*, Goldenrod, **69:** 293-297.

Lentini, M., Pereira, D., Celentano, D., and Pereira, R. (2005). Fatos Florestais da Amazônia. Belém: Imazon, p. 138.

Lima, A. (2008). Desenvolvimento de bebida mista a base de água de coco e suco de acerola. *Cienc. Tecnol. Aliment.*, Campinas, **28(3)**.

Lopes, A. S., Pezoa-García, N. H., and Vasconcelos, M.A.M. (2003). Avaliaç o das condiç es de torraç o após a fermentaç o de amêndoas de cupuaçu (*Theobroma grandiflorum* Schum) e cacau (*Theobroma cacao* L.). *Brazilian Journal of Food Technology*, Campinas, **6(2):** 309-316.

Lorenzi, H. (2000). Frutas Brasileiras e Exóticas Cultivadas (de consumo in natura). Instituto Plantarum. Nova Odessa – SP, pp. 299-300.

Lorenzi, H. (2002). Árvores Brasileiras: Manual de identificaç o e cultivo de plantasarbóreas nativas do brasil. Nova Odessa, 4ª ediç o, 1: 341.

Magalh es, L.A., Lima, M.P., Marinho, H.A., and Ferreira, A.G.(2007). Identificaç o de bergenina e carotenóides no fruto de uchi (*Endopleurauchi*, Humiriaceae). *Acta Amazônica*, **37(3):** 447- 450.

Maia, G. A., Sousa, P. H. M., Lima, A. S., Carvalho, J. M., and Figueiredo, R. W. (2009). Processamento de frutas tropicais (1st ed.). Fortaleza: Ediç es UFC (Chapter 1).

Manica, I. (2000). Frutas nativas, silvestres e exóticas. Porto Alegre: Cinco Continentes Editora, p. 327.

Manica, I., Icuma, I.M., Fioravanço, J.C., Paiva, J.R. de, Paiva, M.C., and Junqueira, N.T.V. (2003). Acerola: tecnologia de produç o, pós-colheita, congelamento, exportaç o, mercados. Porto Alegre: Cinco continentes, p. 397.

Marx, F., Andrade, E.H.A., Zoghbi, M.G., and Maia, J.G.S. (2002). Studies of edible Amazonian plants. Part 5: Chemical characterization of Amazonian Endopleura uchi fruits. *European Food Research and Technology*, **214(4):** 331-334.

Matsuura, F. (2004). Sensory acceptance of mixed nectar of papaya, passion fruit and acerola. *Sci. agric. (Piracicaba, Braz.)*, Piracicaba, **61(6).**

Matsuura, F., and Rolim, R.B. (2002). Avaliaç o da adiç o de suco de acerola em suco de abacaxi visando à produç o de um "blend" com alto teor de vitamina C. *Rev. Bras. Frutic.*, Jaboticabal, **24(1).**

Medeiros, M. L., Lannes, S. C. S., and Gioielli, L. A.(1999). Gorduras de cacau e de cupuaçu: interaç es físicas. *Rev. Bras. Ciênc. Farm.*, S o Paulo, **35(supl.1):** 114.

Mendes, F.R., and Carlini, E.A. (2007). Brazilian plants as possible adaptogens: an ethnopharmacological survey of books edited in Brazil. *J Ethnopharmacol*, **109:** 493-500.

Monteiro, A.R. (1995). Estudo da cinética de extraç o dos solúveis da casca do fruto bacuri (*Platonia insignis*) com $CO_2$ líquido (Kinetics of soluble solids extraction from bacuri fruit shell with liquid CO2). Tese. UniversidadeEstadual de Campinas, S o Paulo, Brasil.

Nazaré, R. F. R. (2000). Produtos agroindustriais de bacuri, cupuaçu, graviola e açaí, desenvolvidos pela Embrapa Amazônia Oriental. Belém: EMBRAPA Amazônia Oriental, p. 27.

Nazaré, R. F. R., Barbosa, W. C., and Viégas, R. M. F. (1990). Processamento de sementes de cupuaçu para obtenç o de cupulate. *Boletim de Pesquisa EMBRAPA,* Belém, **108**: 38.

Nogueira, C. M. C. (1991). Estudo químico e tecnológico da acerola (*Malpighia glabra* L.). Fortaleza, 117p. Dissertaç o (Mestrado em Ciências), Universidade Federal do Ceará.

Nogueira, O.L., Carvalho, C., Muller, C., Galv o, E., Silva, H., Rodrigues, J., Oliveira, M., Carvalho Neto, J.O., Nascimento, W., and Calvazarra, B. A. (1995). Cultura do açaí. Brasília: Embrapa/Centro de Pesquisa Agroflorestal da Amazônia Oriental (Coleç o plantar, 26), p. 50.

Oliveira, A.M. (2005). Aspectos fisiológicos e bioquímicos do tambaqui alimentados com dietas suplementadas por frutos e sementes de áreas alagáveis. 73f. Dissertaç o (Mestrado em Biologia de Água Doce e Pesca Interior)–Instituto Nacional de Pesquisas da Amazônia. Manaus.

Oliveira, M. (1995). Avaliaç o do modo de reproduç o e de caracteres quantitativos em 20 acessos de açaizeiro (*Euterpe oleracea* Mart. Arecaceae) em Belém-Pa. 146 f. Dissertaç o (Mestrado em Botânica)–Universidade Federal de Pernambuco, Recife.

Oliveira, N.O., Oliveira, L.A., Andrade, J.S., and Chagas-Júnior, A.F. (2007). Produç o de amilase por rizóbios, usando farinha de pupunha como substrato. Campinas-SP. *Ciência e Tecnologia de Alimentos,* **27(1):** 61-66.

Padilha, N. C. C., Oliveira, M. S. P., and Mota, M. G. C. (2003). Estimativa da repetibilidade em caracteres morfológicos e de produç o de palmito em pupunheira (*Bactris gasipaes* Kunth). *Revista Árvore,* **27(4):** 435-442.

Parmentier, M., Guillemin, S., Barbar, R., Linder, M., and Fanni, J. (2004). De nouveaux procédés d'extraction deshuilespourdesproduits finis de haute qualité. *OleagineuxCorpsLipids,* Edinbourg, **11(6):** 377-380.

Passos, X.S., Castro, A.C.M., Pires, J.S.; Garcia, A. C-F., Campos, F.C., Fernandes, O.F.L., Paula, J.R., Ferreira, H.D., Santos, S.C., Ferri, P.H., and Silva, M.D.R.R. (2003). Composition and anti fungal activity of the essential oils of *Caryocar brasiliensis. Pharmaceutical Biology,* **41**: 319–324.

Pereira Jr, G. (2006). Farinha de folha de leucena como fonte de proteína para juvenis de tambaqui (*Colossomamacropomum*). 44f. Dissertaç o (Mestrado emAgricultura no Trópico Úmido)–Instituto Nacional de Pesquisas.

Reisdorff, C. (2004). Comparative study on the proteolytic activities and storage globulins in seeds of The *Obroma grandiflorum* (Willd ex Spreng) Schum and *Theobroma bicolor* Humb Bonpl, in relation to their potencial to generate chocolate-like aroma. *Journal of the Science of Food and Agriculture,* **84(7):** 693-700.

Revilla. J. (2001). Plantas da Amazônia. Oportunidades Econômicas Sustentáveis. Manaus, INPA/SEBRAE, pp. 89-90.

Rios, A.O., Mercadante, A.Z., and Borsarelli, C.D. (2007). Triplet state energy of the carotenoid bixin determined by photoacousticcalorimetry. *Dyesand Pigments*, **74:** 561-565.

Ritter, U.G. (1994). Obtenç o de bebida dietética a partir do suco de acerola (Malpighia glabra L.). Fortaleza, 147p. (Dissertaç o de Mestrado), Universidade Federal do Ceará (UFC).

Ritzinger, R., Ritzinger, C.H.S.P. (2004). Acerola: aspectos gerais da cultura. Cruz das Almas: *Embrapa Mandioca e Fruticultura Tropical* (Boletim Técnico), p. 2.

Rodrigues Filho, J. A., Camar o, A. P., and Guimar es, C. M. C. (1994). Consumo *voluntário e digestibilidade "in vitro" de misturas constituídas parcialmente de subprodutos disponíveis no* Estado *do Pará.* Belém: EMBRAPA/CPATU (Comunicado Técnico, **76**: 5.

Roesler, R., Malta, L. G., Carrasco, L. C., and Pastore, G. (2006). Evaluation of the antioxidant properties of the Brazilian cerrado fruit *Annona crassiflora* (araticum). *Journal of Food Science*, **71(2)**: C102–C107.

Rogez, H. (200). Açaí: preparo, composiç o e melhoramento da conservaç o. Belém: Universidade Federal do Pará.

Salze, G. (2010). Use of soy protein concentrate and novel ingredients in the total elimination of fish meal and fish oil in diets for juvenile cobia, *Rachycentron canadum. Aquaculture*, **298**: 294-299.

Sampaio, L. C. (2007). Análise técnica e econômica da produç o de palmito de pupunha (BactrisgasipaesKunth.) e de palmeira-real (*Archontophoenix alexandrae* Wendl. &Drude). *Revista Floresta e Ambiente*, **14(1)**: 14-24.

Sanchez-Garcia, M. D., Gimenez, E., Lagaron, J. M. (2008). *Carbohydr. Polym*, 235.

Shanley, P. (2005). Frutíferas e Plantas Úteis na Vida Amazônica. Editores: Patrícia Shanley e Gabriel Medina. Belém: CIFOR, Imazon, p. 304.

Silva, S., and Donato, H. (1993). Frutas do Brasil. S o Paulo, Imprensa de Arte e Projetos e Ediç es Artísticas, p. 50.

Silva, O. C. (1997). *Análise do Aproveitamento Econômico e Energético do Óleo dePalma na Guiné Bissau na Perspectiva do Desenvolvimento Sustentável"*,Dissertaç o de Mestrado apresentada ao Instituto de Eletrotécnica e Energia da USP,S o Paulo, SP, Brasil,.

Silva, J.A.M. (2003). Frutos e sementes consumidos pelotambaqui, *Colossomam acrompum* (Cuvier, 1818) incorporadosem raç es: digestibilidade e velocidade de trânsito pelo trato gastrointestinal. *Revista Brasileira de Zootecnia*, **32(6)**: 1815-1824.

Silva, E. M., Souza, J. N. S., Rogez, H., Rees, J. F., and Larondelle, Y. (2007). Antioxidant activities and polyphenolic contents of fifteen selected plant species from the Amazonian region. *Food Chemistry*, **101(3):** 1012–1018.

Silva, S.L., Oliveira, V.G., Yano, T., and Nunomura, R.C.S. (2009). Antimicrobial activity of bergenin from *Endopleura uchi* (Huber) Cuatrec. *Acta Amazônica*, **39(1):** 187-192.

Soares, E.C., Oliveira, G.S.F., Maia, G.A., Monteiro, J.C.S. (2001). Desidrataç o da polpa de acerola (*Malpighia emarginata* D.C.) pelo processo "foam-mat". *Ciênc. Tecnol. Aliment.*, Campinas, **21(2):** ago.

Soares, N.S., Sousa, E.P., Cordeiro, S.A., and Silva, M.L. (2011). Competitividade do palmito de pupunha no Brasil em diferentes sistemas de produç o. *Revista Árvore*, **35(6):** 1287-1997.

Souza, C., Silva, L.T., and Druzian, J.I. (2012). Estudo comparativo da caracterizaç o de filmes biodegradáveis de amido de mandioca contendo polpas de manga e de acerola. *Quím. Nova*, S o Paulo, **35(2).**

Stamford, T. L. M., Stamford, N. P., Coelho, L.C.B.B., and Araujo, J. M. (2001). Production and characterization of a thermostable α-amylase from *Nocardiopsis* sp. endophyte of yam bean. *Bioresour. Technol.*, **76(2):** 137-141.

Strauss, M. L. A., Jolly, N. P., Lambrechis, M. G., and Van Rensburg, P. (2001). Screening for the production of extracellular hydrolytic enzymes by non-Saccharomyces wine yeasts. *J. Appl. Microbiol.*, **91(1):** 182-190.

Sun, S. S. M., Leung, F. W., and Tomic, J. C. (1987). Brazil Nut (*Bertholletia excelsa* H. B. K.) proteins: fractionation, composition, and identification of a sulfur-rich protein. *Journal of Agricultural and Food Chemistry*, **35(2):** 232-235.

Superintendência da Zona Franca de Manaus. (2003). Potencialidades Regionais: estudo da viabilidade econômica, Guaraná. Manaus: SUFRAMA.

Surré, C., Ziller, R. (1969). La palmeira de aceite. Barcelona: Editorial Blume(Coleccion Agricultura Tropical), p. 231.

Tacon, A.G.J. (2006). Use of fishery resources as feed inputs to aquaculture development: Trends and policy implications. Rome: FAO Fisheries(Circularn. 1018), p. 114.

Tavares, A.M., and Pereira, J.C.R. (2005). Cultura do guaranazeiro no Amazonas. 4 Ed. Manaus – AM. Embrapa Amazônia Ocidental.

Van Den Berg, M.E. (1993). Plantas Medicinais na Amazônia: contribuiç o ao seu conhecimento sistemático (Medicine Plants from the Amazon), CNPq (Conselho Nacional de Pesquisa), Museu Emilio Goeldi, Belém.

Venturieri, G. A. (1993). Cupuaçu: a espécie, sua cultura, usos e process amento. Clube do cupu, Belém, p.108.

Vilhena, M. R. (2004). Ciência, tecnologia e desenvolvimento na economia da castanha-do-brasil. 120 p.Dissertaç o (Mestrado em Política Científica e Tecnológica) – Universidade Estadual de Campinas – UNICAMP, Campinas.

Villachica, H., Carvalho, J.E.U., Muller, C.H., Diaz, S.C., and Almanza, M. Frutales (1996). y hortaliças promossoras de la Amazônia. Lima, Tratado de Cooperación Amazônica. Secretaria Pró-Tempore (Publicaciones, 44), p. 152-156.

Virmond, E., Sena, R.F., Albrecht, W., Althoff, C.A., Regina, F.P.M., and Moreira, H.J.J.(2012). Characterisation of agroindustrial solid residues as bio-fuels and potential application in thermochemical processes. *Waste Management*.

Wolf, M. A.(1997). Accumulation of biomass and nutrients in the aboveground organs of four local tree species in monoculture and polyculture systems in central Amazonia. German "Diplom"-thesis [unpubl.]. Technische Universität Braunschweig.

Xavier, W.K.S., Junior Medeiros, B., Lima, C.S., Favacho, H.A., Andrade, E.H.A., Araújo, R.N.M., Santos, L.S., and Carvalho, J.C.T. (2011). Topical anti-inflammatory action of *Caryocar villosum* oil (Aubl) Pers. *Journal of Applied Pharmaceutical Science*, **01:** 62-67, 2011.

Yang, H. (2003). New bioactive polyphenols from the *Obroma grandiflorum* ("Cupuaçu"). *Journalof Natural Products*, **66(11):** 1501-1504.

Yuyama, L.K.O., and Cozzolino, S.M.F. (1996). Efeito dasuplementaç o com pupunha como fonte de vitamina A em dieta: estudo em ratos. *Revista de Saúde Pública*, S o Paulo, **30(1).**

Yuyama, L.K.O., Fávaro, R.M.D., Yuyama, K., and Vannucchi, H. (1991). Bioavailability of vitamin A from peach palm (*Bactris gasipaes* H.B.K.) and mango (*Mangifera indica* L.) in rats. *Nutrition Research*, **11:** 1167-1175.

Utilisation and Management of Medicinal Plants Vol. 2 (2014)     *Pages* **373–385**
*Editor-in-Chief:* **V.K. Gupta**
*Published by:* **DAYA PUBLISHING HOUSE, NEW DELHI**

# 18

# Medicinal and Aromatic Plants: Diversity, Conservation and Future Prospects

Archana Peshin Raina[1]*

## ABSTRACT

*The demand for medicinal and aromatic plants is on increase in pharmaceutical industry in addition to their uses as spices, condiments and perfume; on the other hand due to crude deforestation by the tribal's, the natural sources are continuously depleting and so the medicinal plants. There is an urgent need to conserve the available species of medicinal and aromatic species in the forests and also to start their cultivation on commercial scale to meet domestic and global demands. India can have prominent place in the world market by increasing production through adoption of scientific cultivation practices, by value addition through improved processing and marketing management. A need-based strategy including adoption of improved production packages, planning and proper marketing strategies for buy back arrangements should be chalked out to meet the market requirements of plant based crude drugs. Fortunately, India has huge potential for the production of medicinal and aromatic crops, which should be fully exploited.*

**Keywords:** Conservation, Diversity, *In-situ, Ex-situ*, Medicinal Plants.

## Introduction

India with most varied and diverse soil and agro-climate conditions has one of the world's richest heritages for the production of different medicinal and aromatic plants across diverse habitats with rich biodiversity. Over 18000 species of flowering plant available in India, about 8000 species are of medicinal value in Ayurveda,

---

1   Germplasm Evaluation Division, National Bureau of Plant Genetic Resources, New Delhi – 110 012, India

*   *Corresponding author*: E-mail: aprraina@yahoo.co.in

Siddha, Unani, and Homeopathy and also in modern medicines. Over 200 of these species are being currently used in preparation of various formulations by the pharmacies. In the world, over 30000 medicines are prepared from plants and their derived material.

Medicinal plants are a precious natural resource, both from the perspective of their use in traditional medicine as well as providing natural ingredients for the manufacture of modern pharmaceuticals (Lambert *et al.*, 1997; Balick and Mendelsohn, 1992; FAO, 1997). They play a crucial role in providing new remedies for existing and new diseases. Medicinal plants being natural, non-narcotics having no side effect offer a wide range of safe, cost effective, preventive and curative therapies which are useful in achieving the goal of health for all. The use of medicinal and aromatic plants is as old as the human civilization. India has a glorious tradition of health care system based on plants. The world over, 80 per cent of the population derives its primary health care from medicinal plants.

The demand for medicinal and aromatic plants is increasing in both developing and developed countries. The increase in demand of medicines of Ayurveda, Siddha, Unani and Homeopathy both for domestic market and for export has created a great problem of availability of medicinal plants in sufficient quantities in their primitive source *viz.* forests. A large number of plant species are yet to be screened for active compounds. This suggests that the importance of medicinal plants be expected to grow further. It is, therefore, vital that existing stocks are protected and conserved. There is now broad consensus that cultivation offers the best prospect for conserving many medicinal plants currently found in the wild. In addition to maintaining or expanding supply, cultivation is seen as one type of short term solution for conservation of medicinal plants as it increases the supply for industries. A World Bank commentary has observed that "while commercial cultivation of medicinal plants is taking place miniscule scale, this activity is poised for 'dramatic growth ' in the coming decade" and favours organic and mixed cropping to ensure 'good agricultural practices'.

Majority of the medicinal and aromatic plants are still collected from wild for preparation of herbal drugs and perfumes/cosmetics. Presently, 90 per cent herbal material is harvested from forests without applying scientific management practices for their future growth. The habitat destruction of medicinal plants from forest is also causing the problem of environmental degradation in addition to the extinction of many species of medicinal and aromatic plants from the country. Rapid population growth and rising popularity of herbal drugs and natural essential oils have brought in to focus the acute scarcity in availability of some of the plants due to indiscriminate and unregulated collection, habitat destruction through expanding agricultural lands, deforestation and urbanization. Fall in supply of good quality, genuine raw material has resulted in price rise and deterioration in the quality of formulations. With growing demand and use of medicinal and aromatic plants, the important species have to be introduced in to commercial agriculture. Small farmers on marginal lands are generally cultivating medicinal and aromatic plants with low input in resources. Besides this, information on several aspects of their agricultural productivity is also not easily available to the farmers.

Medicinal plants provide raw material for use in all indigenous systems of medicine in India *viz.* Ayurveda, Unani, Siddha and Tibetan Medicine. According to the World Health Organization, 80 per cent of the population in developing countries relies on traditional medicine, mostly in the form of plant drugs for their healthcare needs. There are estimated to be around 25,000 effective plant based formulations available in Indian medicine. It is estimated that there are over 7800 medicinal drug manufacturing units in India, which consume about 2000 tonnes of herbs annually. The international market for medicinal plant-related trade is to the tune of US$ 60 billion having a growth rate of 7 per cent per annum. The annual export of medicinal plants from India is valued at Rs. 1200 million.

## Biodiversity

India is a treasure chest of biodiversity, which hosts a large variety of plants and has been identified as one of the eight important "Vavilorian" centers of origin and crop diversity. Although its total land area is only 2.4 per cent of the total geographical area of the world, the country accounts for 8 per cent of the total global biodiversity with an estimated 49000 species of plants of which 4900 are endemic. The ecosystems or the Himalayas, the Khasi and Minor hills of Northeastern India, the Vindhya and Satpura ranges of the northern peninsular India, and the Western Ghats contain nearly 90 per cent of the country's higher plant species and are therefore of special importance to traditional medicine. Although a good proportion of species of Medicinal Plants do occur throughout the country, peninsular Indian forests and the Western Ghats are highly significant with respect to varietal richness. India extending downwards from Gujrat, Madhya Pradesh and Southern Bihar was once dominated by a continuum of tropical forests, namely: thorn forests, dry deciduous forests, moist deciduous forests, dry evergreen forests, wet evergreen forests and semi-evergreen forests. The complexity with respect to soils topography and climate has created an exceptional variety of biomass and specialized within this region.

Biological diversity or biodiversity, "The library of life" is the variety of all the genes, species and ecosystems that are found on our planet. It embraces microorganisms, plants and animal wild life and the water, land and air in which they live and interact. The richness of biodiversity forms the basis of human sustenance. It comprises every form of life, from the finest microbe to the mightiest beast and the ecosystems of which they are a part. The "library of life" is on fire, about 100 species are lost every day, and most of them vanish unknown for no more than 1.7 million have yet been identified. Only the minutest proportions of plant and animal species have yet been tested for their usefulness to mankind. Out of estimated 2.65 lac species of plants only 7000 have ever been cultivated for food. And even the most insignificant species plays a crucial role in the ecosystem to which it belongs simply do not know what we are throwing away and loosing forever (Khan, 2005). Most of medicines irrespective of the therapies, to which they belong, are the plant derivatives. The shrubs and herbs, which are the first victims of the depletion of biodiversity, constitute major part of the vegetation, the derivatives of which are useful for our health.

India has been designated as one of the 12 mega diversity states in the world with estimated 16000 vascular plants, 5000 endemic species distributed in to different bio-geographic zones. It is estimated that 10-15 per cent of the Indian flowering plants are under various degrees of threat, 25 per cent will become rare by the turn of the century unless proper conservation measures are not taken (Roy, 2003). Medicinal plants constitute a major part of the biodiversity in India. It is estimated that around 8000 plant species have medicinal properties. The heritage of medicinal plants used in India has an ancient history dating back to the pre-vedic culture. Even today, it is estimated that at least 70 per cent of the country's population rely on herbal medicines for primary health care and many other make use of such treatments in conjugation with other forms of medical therapy. The estimates concentrate mainly on well-documented systems, such as, Ayurveda, Siddha and Unani as well as Homeopathy and Allopathy (Singh, 2004). There is rising demand of Indian herbal medicinal products in the international market. This all is resulting in the heavy exploitation of medicinal plants from wild. The scarcity, in wild, has resulted in the thrust on cultivation of around 70 species much in use. Cultivation of most of the medicinal species is still a dream because of the non-availability of standard practices of cultivation and other much important uses of cultivable land. Moreover, the very existence of medicinal plant species in the wild is necessary for the conservation of our biodiversity.

## Causes of Threat

The threat to the existence of medicinal plants in our biodiversity is, mainly because of the following causes:

1. **Over exploitation**: The trends of using medicinal herbs for curing most of the diseases instead of the synthetic preparations, all over world, has resulted in over exploitation of medicinal plants in India right from the Himalayas to the Western Ghats and the back waters of Kerala. It is estimated that 95 per cent need of the pharmaceutical industry is met through indiscriminate collection from wild. Among endangered plant species, medicinal herbs account for almost 1/3 of the species mentioned in the red data book.

2. **Unscientific exploitation**: Because of unscientific exploitation of medicinal plants in wild, the re-growth, replacement and the realistic substitution are hampered. Medicinal plants like *Atropa acuminata* (Solanaceae), *Balanophora dioica* (Balanophoraceae), *Commiphora nightii* (Burseraceae) and *Withania somnifera* (Solanaceae) are glaring examples being met with such threat (Jain and Sastry, 1980).

3. **Environmental degradation:** Natural calamities like floods, cyclones, typhoons, shifting sand dunes, earthquakes and various types of soil erosions adversely effect the natural flora. Besides, natural calamities the man made factors like sewage waste, garbage dumps, harmful effluents of different industries also cause damage to the medicinal plants like other vegetation. The disturbing composition of different harmful gases in the environment is also affecting the efforts of conserving medicinal plant

diversity. Unscientific use of chemical fertilizers and pesticides have also resulted in destruction of many medicinal plant species.

4. **Biotic pressure**: The increasing human and cattle population is exerting immense pressure over the habitat of medicinal plants in many ways. The grasslands, wastelands and other uncultivated lands outside the government owned forests were good habitats for number of medicinal plant species but with increasing human population such land are were transformed in to cultivable lands or they are under various other commercial uses. In this way a number of medicinal plant species lost their home.

## Need for Conservation of Medicinal Plants

Besides the role of medicinal plants in the conservation of biodiversity, the economic scenario relating to the medicinal plants should also be the basis of for formulating policies and adopting practices for conserving this great heritage of nature. According to 1994 UNDP report, the annual value of medicinal plants derived from developing countries is approximately 32 billion US dollars. There are 47 major modern pharmaceutical plant based drug already in the world market and the predicted 328 drugs, yet to be discovered, have a market potential of 147 billion US dollars. The sale, in the very first year, of the anti cancer drug, Taxol from *Taxus spp.* has been more than 2000 billion US dollars. Phyto-remedies and health food over the country sales in USA is expected to touch 2 billion US dollars according to American Botanical Council. The value of global trade of the medicinal plant product has been put over US$ 75 billion per year and is growing @ 12.5 per cent annually. Of the total value of trade, about 20 billion US dollar over the counter (OTC) drugs, US$ 25 billion for prescription drugs and remaining US$ 30 billion for nutritional supplements.

Bulk of the raw material (90 per cent) is produced in Asia, Africa and Latin America and some in Europe and USA (10 per cent). About 60 per cent of the total material is imported and processed in USA, Canada, UK, Australia, Germany, France, Italy, Switzerland and Japan and about 50 per cent of that is used there and the rest is exported to the raw material producing countries to be sold at higher rates. About 90 per cent of the marketed material are collected from wild resources and the rest from cultivations in China, India, USA, Germany, France, Italy and Eastern Europe.

The trade in medicinal plants in India is estimated to the tune of Rs. 675 crore per year. Of India's total turnover of Rs. 3100 crore of Ayurvedic and Herbal products, major OTC products constitute around RS. 1700 crore, Ayurvedic ethical formulations constitute around Rs. 850 crore and Ayurvedic classical formulations constitute remaining Rs. 550 crore. It is estimated that the global market for herbal drugs is nearly Rs. 800 crore per year. The export of medicinal plant produces from India is growing faster.

Conservation estimates put the economic value of medicinal plants related trade over US$ 90 billion. Demand and trade in medicinal plants species globally indicates a steep upward trend and world trade in medicinal plants and related products is expected to rise to US$ 5 trillion by 2050 A.D (Sharma, 2003).

# Conservation of Medicinal Plants: Strategies and Priorities

Several national and international agencies have formulated appropriate policies and strategies for the conservation of medicinal plants. The world conservation strategy defines conservation as "the management of human use of the biodiversity so that it may yield the greatest sustainable benefit to present generation while maintaining its definition invokes two complimentary components "conservation" and "sustainability". The primary goals of biodiversity conservation as envisaged in the World Conservation Strategy can be summarized as follows:

1. Maintenance of essential ecological processes and life support systems on which human survival and economic activities depend,

2. Preservation of species and genetic diversity, and

3. Sustainable use of species and ecosystems, which support millions of rural communities as well as major industries.

Medicinal plants potential renewable natural resources. Therefore, the conservation and sustainable utilization of medicinal plants must necessarily involve a long term, integrated scientifically oriented action program. This should involve the pertinent aspects of protection, preservation, maintenance, exploitation, conservation and sustainable utilization. A holistic and systematic approach-envisaging interaction between social, economic and ecological systems will be more desirable. The most widely accepted scientific technologies of biodiversity conservation are the *in-situ* and *ex-situ* methods (Sharma, 2003).

## *In-situ* Conservation

All the plants, of which the medicinal properties have been recognized, are the gifts of nature to us. Most of them are found in the forests. So, their *in-situ* conservation is necessary and cannot be replaced by other mode, *i.e., ex-situ* conservation or any other technique or practice. The need of the hour is to focus our attention towards the ecological and commercial wealth of these plants and not only to assess the productivity of our forests in terms of products obtained from tree species providing valuable timber, firewood, resin, gum etc. The success of our management systems, techniques and practices will have to be evaluated not in terms of the enhancement in the productivity of species but the enrichment of diversity of our flora (of course including medicinal plants) and fauna. Thus our approach should be "eco-centric" rather than species-centered. For scientific impetus to evolve *in-situ* conservation techniques we shall have to study extensively and intensively the areas, which are comparatively rich or even less depleted in terms of medicinal plant diversity and shall have to identify the factors prevailing therein. The results of such surveys and studies should be implemented to the areas, which are in peril. The thumb rule can be the least disturbance to the forest areas and leaving them to the nature for enrichment The exploitation of different parts of medicinal plants will also have to be checked till the area is replenished to the extent that it can be used for sustained scientific exploitation.

## *Ex-situ* Conservation

To conserve the germplasm outside their natural habitat is known as *ex-situ* conservation. In fact *in-situ* and *ex-situ* strategies are complementary approaches. *Ex-situ* conservation facilitates to conserve a species of high importance in controlled conditions, its reintroduction in the wild and an insight in to the basic biology of the species to work out new strategies for its conservation. Although this mode of conservation can not substitute *in-situ* conservation but this can acts as a supplement. Conservation of medicinal plants outside natural habitat by cultivating and maintaining plants in botanical gardens, parks, other suitable sites, and through long term preservation of plant propagules in gene banks (seed bank, pollen bank, DNA libraries, etc.) and in plant tissue culture repositories and by cryopreservation need encouragement. It can be implemented through the various techniques:

**Herbal or Botanical garden** maintained by ancient herbal doctors, healers, sages and the Royal families supported the conservation effort where beauty of display was fully justified with the medicinal aspects. Even today, one can find many ashrams in and around Haridwar, Rishikesh (Uttarakhand) and Himalayan region where medicinal plants have been maintained since ages. More recently Ministry of Agriculture under its Horticulture Division has established 16 Herbal gardens all over India which are responsible for maintaining about 150 medicinal plants including 40 tree species, 35 shrubs, 31 herbs and 35 endangered species (Gupta and Chadha, 1994). Apart from producing true seeds for utilization, these gardens will be utilized to educate the public through short-term training programmes. The major botanical gardens in India are Tropical Botanical Garden and Research Institute, Thiruvanthapurm (Kerala); Medicinal and Aromatic Plant Garden and Herbarium, Pune; Lal Bagh Botanical Garden, Bangalore; Royal Botanical Garden, Calcutta; Lloyd Botanical Garden, Darjeeling.

**Field Gene Bank** ensures that a plant material along with its wild related species is conserved and available for breeding, reintroduction, research and other purposes. These are particularly appropriate for long lived perennial trees and shrubs which can not be adequately conserved in the wild and which may take decades to produce seeds. The IPGRI has designated 23-field gene bank for (Global Biodiversity) crops, at either global or regional levels. India, there are many field gene bank for a large number of medicinal plants. A few of them are being maintained by several governmental organization like National Bureau of Plant Genetic Resources, New Delhi; Council for Scientific and Industrial Research, New Delhi; Central Institute for Medicinal and Aromatic Plants, Lucknow; or M.S. Swaminathan Research Foundation, Madras.

*In vitro* **Repository** or *In vitro* (literally in glass) conservation is another important from of strategy for conserving flora with vegetative propagation, recalcitrant seeds or in plant where seed formation is poor. Theoretically, cultures can be stored indefinitely using cryo-genic techniques, which would reduce labor requirement. Species like *Malus, Ribes, Vaccinium, Rubus* and *Pyrus communis* has been stored using this technique. In India many research laboratories, and institutes, including NGO's have been involved in developing protocols for micropropagation of various medicinal plant. Concerted efforts of *in vitro* conservation of this group of crops with the establishment of India's first National Facility of Plant Tissue culture Repository at NBPGR in 1986. The main objective of *in-vitro* methodologies is to reduce the need of sub culturing. The strategies include growth under normal cultural conditions as culture. Though abound with the risk of contamination, it may be useful particularly for developing countries. In slow growth, low temperature incubation alone or in combination with media modification (addition of osmotic or growth retarding agents) prolongs shelf life of cultures. For long term conservation, cryo-genic storage is the only method of choice.

*In-vitro* **cryopreservation** as a promising tool for long term germplasm conservation, cryopreservation of medicinal and aromatic plants is still at experimental stage. In several medicinal and aromatic plants, meristems/shoot tips, cell suspensions, protoplast, somatic embryos and pollen have been studied from preservation perspective.

**DNA banks** or conserving the extracted form of DNA is a very recent approach. The basis of this approach is the DNA sequences in the genome of the accession are the main source of the genes required in the various breeding and crop improvement programmes.

This strategy can be of high relevance in case of the endangered, endemic and threatened forms of the medicinal plants, especially when the threat is from patenting issues.

**Seed gene banks** are the most effective and economic forms of the *ex-situ* conservation for the plant species that produce orthodox type of seed *i.e.*, seeds where viability is not lost if dried to moisture content of 5-7 per cent. They occupy less space by virtue of small in size and by wide genetic variability as each seed represent a unique genetic constitution. The success of this strategy needs careful monitoring and testing of seed viability. Seed viability in medium term storage (0-5°C and 35 per cent RH) can be 5-25 years and in long term storage (-10°C to -20°C) the seed remains viable up to 100 years. In India, there are three gene banks for medicinal and aromatic plants *viz.*, Tropical Botanical Garden and Research Institute, Thiruvananthapuram; Central Institute of Medicinal and Aromatic Plants, Lucknow; and National Bureau of Plant Genetic Resources, New Delhi. All the three gene banks are actively engaged in collection and conservation of these resources in the Indian sub-continent.

For conservation of medicinal plants, in either way, the role of extension and educational programmes is of much importance. The people involved in the conservation works will have to establish a good rapport with the villagers, traders and consumers, so that, everyone who has a direct or indirect relation with medicinal plants may understand the need for conservation and his/her role in achieving the goal.

The medicinal plants should be domesticated and brought under cultivation to maintain constant supply of quality materials and thus reduce the pressure on the wild populations. The process of domestication of a species involves characterizations of its reproductive biology to decide on the method of propagation-seed or propagule and definition of then area in which to cultivate based on the soil and climate characteristic of the plant's natural habitat. The genetic resources of the species are screened for its adaptability and yield and quality of the material that will make the end product, to identify suitable genotypes for cultivation. Soil type to be used, irrigation and fertilizer amounts and application schedule and sowing and harvesting time are standardized, diseases and pest problems of standing crop and harvesting material are solved to keep the product safe for consumption. Initiation of plant breeding programmes will ensure that in future the crop will be high yielding, resistant to pests and economically beneficial to the cultivators.

## Major Constrains in the Field of Medicinal and Aromatic Plant Cultivation

1. **Conservation and cultivation of endangered and exportable plants** : *In-situ* and *Ex-situ* conservation, cultivation of endangered and exportable species with good cultivation practices and identification of marketing export channels are the need of the hour.

2. **Lack of the knowledge of seed biology**: In fact, cultivation of medicinal and aromatic plants is not so easy. It is rather a challenging task and very little is known about the seed biology of medicinal and aromatic plants. For maintaining purity and production of genuine planting material, it is essential to obtain knowledge of seed biology of medicinal and aromatic plants.

3. **Lack of scientific cropping systems**: Scientific cropping system of medicinal and aromatic plants is becoming integrated with agricultural and forestry for optimum use of land and water resources and maintain eco-friendly environment.

4. **Lack of knowledge of modern cytogenetic and breeding tools** : Sometimes plant selected from wild population may be suitable for the cultivation of and there is no immediate necessity for any improvement programmes for improvement of plant species become necessary for its commercial cultivation. Fundamental investigation on breeding behaviour, floral biology, cytology, genetics, chemistry, biosynsthesis of active principles etc. are some essential pre-requisites before undertaking any planned hybridization programs. For this purpose, the breeder should have the proper knowledge of floral morphology, flowering time, maturity of stigma, compatibility of pollens, fertilization, seed-set etc.

5. **Knowledge of molecular investigation coupled with breeding and genetic engineering**: Molecular investigation coupled with breeding and genetic engineering approaches should be pursued to improve specific medicinal and aromatic plants for higher productivity of their important medicinal and aromatic components.

6. **Lack of knowledge of post harvest operations** : Post harvest operation and processing of plants for the extraction of chemical and preparation is another important area, the knowledge on these activities is essential and new methods are to be developed.

7. **Edaphic climate factors**: For cultivation, knowledge on climate, soil factor, plant associations etc. should be available and essential so that most suitable location of cultivation can be made for harvesting potential yield with high returns.

8. **Lack of proper agro-technology**: Many government organizations, public sectors and private institutions and non-government organization have developed standardized practices for the cultivation and propagation of the medicinal and aromatic plant species. However, the farmers still hesitate to practise these packages. Some institutions charge high fees for package developed by them, which are beyond the pockets of common farmers. Reliable and standardized technology package has been a bottleneck in popularization of commercial cultivation of important medicinal and aromatic plants.

9. **Lack of genuine planting material and high costs**: Lack of genuine planting material is also one of the major hurdles in the cultivation of medicinal and aromatic plants. There are some organizations, which are having planting material for large-scale cultivation but prices of planting material are very high. So, there is urgent need to develop nursery and multiplication centers in every state and in every district with in state so that farmers may get genuine planting material in time with reasonable prices near to their farm.

10. **Lack of market potential and systems**: Although demand of medicinal and aromatic plants is too high but there is lack of authentic data for demand and supply. Market is unorganized and does not have any regulatory body/ systems. Furthermore, there is fluctuation in prices of produce of medicinal and aromatic plants and crude drugs. In the present system, farmers are not getting the right price for their produce. Middleman and agents are being benefited necessary to check illegal collection and trading and encourage cultivation of these species.

11. **Lack of awareness in organic farming**: Organic farming is an important practise gaining wide acceptance world over demand (particularly in developed countries) for organically grown crops is rapidly on the increase. Farmers have to be trained in all aspects of organic farming of medicinal plants and herbs including obtaining certification from association that do the monitoring from cultivation to final harvesting. Organic farming which is labor intensive gives the developed countries the comparative advantage to be competitive medicinal and aromatic plants should be grown in organic farming. Pesticides, insecticides and chemical fertilizers should not be used.

## Strategies for Promotion of Cultivation of Medicinal and Aromatic Plants

For successful cultivation of medicinal and aromatic plants, the following key points may be given due consideration.

1. Identifying of crops and their varieties.
2. Suitability of crops species to specific locations and edaphic conditions.

3. Optimum agro technique in relation to:
   - Crop geometry
   - Soil- water- fertility and
   - Plant protection management
4. Economic returns
5. Market accessibility

# Future Thrust

1. **Farmer's awareness program**: The farmer should be educated about the cultivation of important medicinal plants. They should be encouraged and informed about the importance and benefits when compared to the traditional crops.

2. **Bank assistance:** Interested farmers of medicinal plants cultivation should be provided with liberal bank loans with which they can come forward and produce quality material which helps individual and the country by reducing pressure on the forests and by way of getting more foreign exchange.

3. **Subsidies:** Sufficient amount of subsidies should be provided to boost up the production of medicinal and aromatic plants by National State Government as well as National Medicinal Plants Board.

4. **Crop selection**: The medicinal plants are seasonal, annual, biennial and perennial. Some plants are for hedges, bio-diesel producing and food colour yielding. Before taking up for cultivation, farmers should know the cultivation, before harvesting, post harvesting techniques by undergoing training, reading literature, seeing demonstration plots and cultivated lands and interacting with the progressive cultivators who have already taken up these crops for cultivation.

5. **Seeds and sapling**: Genuine seed materials are to be provided with reasonable price. Tissue culture and gene banks are advisable in the government and private sector. Farmers should be warned about the people who exploit the situation and safe their sale material for high costs by promising them the high returns.

6. **Organic farming**: It is a production system of agriculture, which avoids the use of synthetically compounded fertilizers and chemical pesticides. The chemical fertilizers and pesticides may increase the quantity of production, but cause soil degradation and pollution. The resulted produce produces health hazards. The organic manuring will reduce the pest incidence and improves the quality. The major aims of organic agriculture bare, production of quality medicinal plant products, which contain no chemical residues.

7. **Marketing**: Timely marketing for a beneficial rate of his produce helps the farmer. There should be a medicinal plant market in the state level, at National level and International levels. To attract the marketers, sufficiently large quantities of medicinal plants produce is needed.The linkages

between the producers and whole sale marketers are to be established. There are private traditional medical practitioners, pharmaceutical agencies/industries in the state, which should be coordinated to purchase the produce in an organized way from the farmers. To create market awareness and help the farmers, regional buyers and sellers meets should be organized regularly.

## Conclusions

It can be very safely concluded that conservation of medicinal plants shall remain a dream without conserving the biodiversity. Our forests, which are the best pictures of biodiversity, need management with the objective of enriching the plant diversity rather than with the species centered approach. *Ex-situ* conservation of medicinal plants should be encouraged for the role of a supplement to the *in-situ* conservation. It should not be taken as substitute for the *in-situ* conservation. The creation of awareness among all the persons, who are directly or indirectly concerned with the medicinal plants, is the need of the hour.

There is urgent need for promotion of scientific cultivation of MAPs through best management practices to maximize crop yields. A well organized markets ensuring proper regulation to provide remunerative price to the growers should be ensured. Efforts are needed for market development, strengthening of market intelligence, price support system and linkages/tie-up of the marketing system with pharmacies and export markets so that producers share in the final consumer's price increases and price spread gap is reduced. Financial institutions should expand support for cultivation of medicinal and aromatic plants. Research initiatives are needed to increase productivity and reduce the cost of production and make their cultivation economical and attractive so as to complete in the global markets. Market intervention and price support policies are needed in order to meet the increasing demand of internal and export markets.

In order to harness the potential of rich heritage of medicinal and aromatic plants available in Indian sub-continent, a holistic approach is required to regulate and manage these valuable resources in scientific manner. In this context, survey inventorization, collection, characterization, evaluation, conservation and utilization of these genetic resources along with the documentation of associated indigenous knowledge occupies a unique place in the field of PGR management. The search for new genes is an endless exercise and various national institutions are engaged in this pious task. The thrust area being the plants used in organized sectors of industry as well as in the Indian System of Medicine from the core of collection. After agro-morphological and chemical evaluation the promising germplasm will be harnessed for commercial production of raw material as well as to find out the newer source of drugs and aroma chemicals.

## References

Balick, M.J., and R. Mendelsohn, (1992). Assessing the economic value of traditional medicines from tropical forests. *Conservation Biology*, **6**: 128-129.

Gupta R., and Chadha, K.L. (1994). Recent advances in Horticulture- Medicinal and Aromatic Plants, Vol.-**11** eds. K.L Chadha and Rajendra Gupta, MPH Publication, New Delhi.

FAO, (1997). Medicinal Plants for Forest Conservation and Health Care. Global Initiative for Traditional Systems (GIFTS) of Health. Food and Agriculture Organization of the United Nations (FAO), Non-Wood Forest Products No. **11**, Rome.

Jakhar, M.L., Kakralya, B.L., Singh, S.J., and Singh Karan. (2004). Enhancing the Export Potential of Medicinal Plants through Biodiversity Conservation and Development under Multi-Adversity Environment. Medicinal Plants-Utilization and Conservation. Avishkar Publishers, Distributors, Jaipur, India.

Jain, S. K., and Sastry, A. R. K. (1980). Threatened Plants of India. New Delhi: Department of science and Technology.

Khan, T. I. (2005). Biodiversity: Concepts and Need of Conservation and Sustainable Department. Avishkar Publishers, Distributers, Jaipur, India.

Lambert, J., Srivastava, J., and Vietmeyer N. (1997). Medicinal Plants: Rescuing a Global Heritage.World Bank Technical Paper No. 355. The World Bank,Washington D.C.

Rajendra Gupta. (1993). Status of Medicinal and Aromatic Plants in India. Country Report-2. Prod. Regional Expert Consultation on Breeding and Improvement of Medicinal and Aromatic Plants in Asia. FAO/RAPA-Bankok, 3 May–4 June.

Roy, R. K. (2003). Conservation of Plant Genetic Diversity in India: The role of botanic gardens in the new millennium. *Journal of Non-Timber Forest Products*, **10** (1/2): 61-64.

Sharma, Ravindra. (2003). Medicinal Plants of India–An Encyclopedia. Daya Publishing House, Tri Nagar, Delhi.

Singh, B.P. (2004). Germplasm Introduction, Exchange, Collection/Evaluation and Conservation of Medicinal and Aromatic plants- There Export Potential. Medicinal Plants-Utilisation and Conservation. Avishkar Publishers, Distributors, Jaipur, India.

Utilisation and Management of Medicinal Plants Vol. 2 (2014)    Pages 387–402
Editor-in-Chief: **V.K. Gupta**
Published by: **DAYA PUBLISHING HOUSE, NEW DELHI**

# 19

# Waste from the Commercial Production of Edible Pulp Tucum (*Astrocaryum aculeatum*, Meyer) in the Urban Area of Manaus/Amazonas-Brazil

Francisco Elno Bezerra Herculano[1], Klenicy Kazumy de Lima Yamaguchi[2], and Valdir Florêncio da Veiga Jr[3]*

## ABSTRACT

*In Manaus, the capital of the State of Amazonas, the sale of the pulp (mesocarp) of the fruit tucuma-do-Amazonas (Astrocaryum aculeatum Meyer) is sold at open air and indoor public markets around the city. The commercial activity of extracting this pulp produces a significant amount of waste (peels and seeds) that are the focus of this work, which aims to estimate the mean masses (weight) of pulp yield, losses in the process, and the consequent technical coefficient of production of raw materials. The method used was probabilistic searching by proportions, with tolerance of 5 per cent and a confidence interval of 80 per cent, in a sample of 143 ripe, fresh fruits. The fruits were commercially processed by manually removing the pulp, a task done by local market stall holders working at the largest fresh produce market*

1   MSc em economia do desenvolvimento regional pela Universidade Federal do Amazonas, doutorando em gest o da biotecnologia pela Universidade Federal do Amazonas, pesquisador do Núcleo de Estudos e Pesquisas em Inovaç o da Fundaç o Centro de Análise, Pesquisa e Inovaç o Tecnológica.

2   MSc em química orgânica pela Universidade Federal do Amazonas, doutoranda em química orgânica.

3   Pós-doutor em química orgânica pela Universidade Federal do Amazonas, doutor pela Universidade Federal do Rio de Janeiro, professor e pesquisador da Universidade Federal do Amazonas.

*   *Corresponding author*: E-mail: valdirveiga@ufam.edu.br

*in the city, during May 2012. The results showed low levels of productivity of the resource; the average mass of pulp (mesocarp) produced for sale was only 17.5 per cent of the total weight of the fruit, resulting in a technical coefficient of 5.73 kilograms per kilogram of fruit pulp. The waste generated represented an average of 82.55 per cent, including 20.59 per cent of peel (epicarp and perianth) and 61.96 per cent of seed or core (endocarp and endosperm). This waste has potential use in the production of various products that can add value, increase productivity, and generate employment for families, but it is being thrown into the garbage, resulting in environmental, social and economic losses for the urban population of the central Amazon region of Brazil.*

*Keywords*: Edible pulp, Tucum , *Astrocaryum aculeatum* Meyer.

---

# Introduction

The city of Manaus, capital of the Brazilian State of Amazonas, located in the heart of the rainforest, at the center of the Amazon region, close to the confluence of the Negro and the Amazon Rivers, converges naturally to the theme of biodiversity.

Other aspects, however, can also lead to the convergence of the location, particularly with regard to its ethnocentric situation. Interestingly, despite being born under the aegis of the European colonial power in the New World, much of its cultural content still remains. The very name of the city alludes to the roots of one of the native ethnic groups that previously dominated much of the Vale Rionegrino, or Negro River Valley- the Manaos, led by Ajuricaba (IBGE, 2012).

Today, the capital, is a major population center of the North of Brazil, with a size that has given it prominence both nationally and internationally. By 2009, the resident population of Manaus had reached 1,738,641 inhabitants, demographically overtaking other major cities of the world, including Vienna in Austria, Kuala Lumpur in Malaysia, Bhopal in India and Washington DC in the USA (United Nations, 2011, pp. 307-360).

For an urban conglomeration like Manaus–which in 1970 had a population of 311.6 thousand inhabitants–to reach nearly two million residents in approximately fifty years later, indicates immigration on a massive scale. The main reason for this is strong regional economic leverage provide by the creation of the "Zona Franca de Manaus", a free-trade zone that provides fiscal incentives to industry. This expansion has influenced, and is influenced by the various habits and customs of the city, particularly its gastronomic, cultural and social customs.

Consumption of the tucum fruit (*Astrocaryum aculeatum*, Meyer), a custom originating with the indigenous peoples of the Amazon, has thrived despite the many new foods brought by the immigrants, attracted by the economic expansion of the city in recent years. In fact, its consumption has significantly increased. Today, tucum pulp is sold throughout the city. It is used in a sandwich, known locally as a "X-caboquinho", consisting of bread, cheese curds (hence the "X", which in the Portuguese pronunciation, sounds similar to "cheese"), and tucum pulp. Sometimes its name is shortened to "BCT"–bread, cheese and tucum –and it has become something of a calling card of the city. Other traditional foods of the region also contain tucum pulp, such as tapioca and manioc flour, pizza, bread, and ice cream.

The process of extracting the edible tucum pulp, as it is currently sold in Manaus is entirely manual, resulting in extremely low utilization of this natural resource of the Amazon. The high proportions of waste generated are disposed of in the general garbage, as was observed in this study, although small amounts of seeds are used in the creation of regional handicrafts, which are sold in some parts of the city. Thus, the waste generated in the extraction of the pulp could be used for other purposes, thereby reducing the amount of waste generated in the city, adding value to the fruit, and generating income for many more families along the production chain.

The importance of tucum waste has been the object of various studies focusing on its use in the manufacture of products such as biofuels, cosmetics and animal feed. It is therefore important to assess the amount of waste resulting from the extraction operations of tucum pulp sold in Manaus. The aim of study work is to estimate the average masses of waste and the pulp yield from tucum sold in Manaus, seeking to gain a better understanding of the facts and stimulate further studies on the subject, forming a knowledge base that will result in social, economic, and environmental benefits.

## Referential Theoretical

The abundance of living things in the Amazon environment, whether plants, animals or microorganisms, constitutes a massive biodiversity on a global scale. Studies commissioned by the Brazilian government, through its Ministry of Environment, estimates the resource potential of plant biodiversity of the Amazon to be 33,000 higher plant species, of which at least 10,000 contain active ingredients for use in medicine, cosmetics, or biological pest control (MMA, 2008), as confirmed by the INPA–National Institute of Amazon Research (2010).

Some plants of the Amazon already have been analyzed, and have been proven to be rich in proteins, fiber, fat, vitamins, minerals, pigments flavonólicos, carotenoids and other important physical-chemical components. Some of these plants are historically present in the secular culture of the Amazonian peoples, in various different forms, being used in foods, beverages, medicines, cosmetics, dyes and for other purposes. A short list of plants used for such purposes includes: assai (*Euterpe oleracea* Mar.); buriti (*Mauritia flexuosa* L.); pataua (*Oenocarpus bataua* Mart); pupunha (*Bactris gasipaes* Kunth); piquia (*Caryocar villosum* Aubl); castanha-de-cutia (*Couepia edulis* Prance); Amazon nut (*Bertholletia excels* Humb); castanha pendula (*Couepia longipendula* Pilger); castanha sapucaia (*Lecythis pisonis* Miers.); cupuassu (*Theobroma grandiflorum* Shumann); copaiba (*Copaifera* ssp.); andiroba (*Carapa guianensis* Aubl); rosewood (*Aniba rosaeodora* Ducke); ucuúba (*Virola surinamensis* Warb); cumaru (*Coumarouna odorata* Wild); sacaca (*Croton cajucara* Benth) and others (Mota, 1946; Campos, 1951; Altman, 1956 *apud* Clay *et al.*, 2000; Iaderoza *et al.*, 1992; Ozela *et al.*, 1997; Bobbio *et al.*, 2000; apud Menezes *et al.*, 2008).

More specific evaluations, such as studies to produce a detailed breakdown of the composition, and chemical analyses, have concluded that the pulp of the tucum fruit is a potential source of lipid, a precursor of vitamin A, as well as containing fiber, vitamin E, unsaturated fatty acids and tocopherols (Lorenzi *et al.,* 2004; Yuyama *et al.,* 2005; Rodrigues *et al.,* 2010).

Reinforcing this argument, the literature places a strong emphasis on the chemical composition of the mesocarp of tucum , which contains high amounts of carotenoids (Lima *et al.,* 2011). These are recommended as natural food colorants and in the prevention of certain degenerative and cardiovascular diseases (Santos, 2004; Sentanin and Rodriguez-Amaya, 2007; Uenojo *et al.,* 2007).

Nevertheless, in general, studies focusing on the tucum -de-amazons (*Astrocaryum aculeatum* Meyer) are scarce, and those that do exist focus mainly on the basic understanding of the physical, chemical or botanical characteristics. Studies on the commercial, social or economic exploitation of the fruit are very rare, despite its widespread gastronomic use.

The tucum fruit is widely sold in open air and indoor produce markets in the city of Manaus and surrounding region, where it is mainly sold in the form of fresh pulp.

The fruit can take several forms, small, medium or large. The vernacular language also provides a variety of classifications, the most common of which is tucum -do-amazons (*A. aculeatum*). In Manaus, the "arara" and "rajado" varieties are the commonly exploited, while the varieties growing in Eastern Amazônia, particularly in some towns and cities in the neighboring state of Pará, are commonly known as "tucum bab o" or "tucum -do-pará" (*A.vulgare*), and are hardly sold in Manaus.

The technical name of the fruit found in the literature is also diverse. For Leit o (2008, p.15), the synonymy of the tucum -do-amazons or tucum -açu (*A. aculeatum*) comprises 11 variations: *A. tucuma* Mart., *A. aureum* Griseb. And H.Wendl. In H.R.Grisebach, *A.candescens* Barb.Rodr., *A. Princeps* Barb. Rodr., *A. Jucuma* Linden, *A. princeps* var. *Aurantiacum* Barb. Rodr., *A. princeps* var. *Flavum* Barb. Rodr., *A. princeps* var. *Sulphureum* Barb. Rodr., *A. princeps* var. *Vitellinum* Barb. Rodr., *A. Manaoense* Barb. Rodr., and *A. Macrocarpum* Huber.

Ferreira *et al.* (2008) also have a broad understanding of the species for the genus *Astrocaryum,* highlighting 11 denominations: *A. aculeatum* Meyer., *A. segregatum* Dr., *A. princeps* Bard., *A. giganteum* Bar., *A. tucum* Mart., *A. acaule* Mart., *A. cantesis* Mart., *A. chonta* Mart., *A.leisphota* Bard., *A. undata* Mart and *A. vulgare* Mart, the latter found mainly in the states of Pará and Amapá (Brazil).

Moussa and Kahn (1997), however, reported that the tucum palm that grows in the Brazilian Amazon is restricted to just two species: *Astrocaryum aculeatum* Meyer and *Astrocaryum vulgare* Mart. The authors also describe the first species as growing predominantly in central Amazon (Brazil, Bolivia, and Venezuela). It is commonly found in forests and secondary forests in the vicinity of Manaus, where the fruit is widely appreciated by people of all social classes. The *vulgare* species, on the other hand, is concentrated mainly in the eastern part of the Amazon (Suriname, French Guiana and Brazil–Pará), supporting the argument that the tucum consumed in Manaus belongs to the *A. aculeatum* Meyer species, despite the many different denominations given to it. This is, therefore, the species investigated in this study, despite the inconsistent morphologies found.

There have been very few studies focusing on the consumption of tucum pulp in Manaus, despite its prominence in the daily diet of the local population, and the visually noticeable result – discarded peels (epicarp + perianth) and seeds (endocarp + endosperm), generating high volumes of waste, particularly of the pulp, which was the motivation for this work. This waste is abandoned in dumps or landfills, and can cause to problems of an aesthetic, environmental, and economic nature, with potential harm to the environment and to man (Martins and Farias, 2002; Sena and Nunes, 2006).The debate on the recovery of waste from fruit can be found in the literature, highlighting the potential of oil endosperm (kernel) of tucum for biofuel production, as an ingredient for cosmetics, and in the manufacture animal feed (Castro *et al.*, 2007 apud Pig, 2008, p.18; Barbosa *et al.*, 2009, p.372; Figliuolo and Silva, 2009)

## Materials and Methods

The methodology consisted of a study of probability sampling by proportions, carried out from 8 to 11 May 2012, at the market stall of a vendor who had worked in the region for nine years, selling fresh, local produce (tucum , cupuaçu, passion fruit, cashew, pineapple and other fruits) and extracting the pulp (mesocarp) from tucum -do-amazonas (*Astrocaryum aculeatum* Meyer). The tucum pulp produced was sent to the local market of Manaus–small restaurants and snack bars–or sold directly, to customers on the street. The waste from the fruit was disposed of in the garbage and collected by the city's garbage collection service.

The research was conducted with the vendor of the largest stall, out of the five that extracted/sold tucum pulp inside the Colonel Jorge Teixeira de Oliveira market, popularly known as the "Feira da Manaus Moderna" (Manaus modern market). This Market, situated on the banks of the Negro River, in the old center of Manaus, occupies a strategic location, polarizing the retail trade of fresh regional produce in Manaus. Close to it is a bustling regional hub, with river boats sailing to and from the interior of the Amazon region, carrying cargo, passengers, and much of the region's production of fish, fruits and vegetables, attracting a considerable flow of vendors and consumers.

Bearing in mind the aim of this research, which is to find out as much as possible about the tucum waste that is generated in the commercial extraction of fruit pulp in the markets of the city of Manaus, and in order to estimate the rates of exploitation and waste produced by the activity, the work was carried out at the location where the product is actually extracted and sold. Thus, the tucum pulp sampled was extracted by the stallholder himself (who, like everyone else, did this manually using a small knife). Therefore, our findings do indicate the potential yield of the fruit pulp, but only the actual yields recorded. To determine the potential yield, this would involve, for example, experiments to measure the maximum pulp (mesocarp) content of the fruit, minimizing the portion adhered to the peel and seed that is regarded as waste. In this study, the focus was exclusively on the quantitative estimate of the amounts of pulp waste resulting from the manual extraction process, as practiced by sellers in the city.

To calculate the sample size, a pilot test was initially conducted with one hundred units of ripe fresh fruits of tucum -do-amazonas, drawn randomly from a bag containing 55 kg of the fruits, reported as coming from the rural area of Rio Preto da Eva/AM. When the bags of fruit arrived at the stall, they were removed for processing and sale of the pulp.

The total mass of 100 fruits in the pilot test was 5,300 g, with a simple arithmetic mean of 53 g/ripe fresh fruit. This led to an estimated 19 ripe fresh fruits per kg, and a total of 1,045 fruit per 55 kg bag being processed. Based on this, we estimated the universe of population (batch of fruit bag with 55 kg) in 1.045 fruit, where the sampling error was ± 5 per cent at a confidence interval of 80 per cent, resulting in a sample size of 143 fruits (Tagliacarne, 1986, p. 173).

The 143 sample fruits were identified, weighed (per unit and total), then manually peeled and pulped by the market vendor. All the pulp removed, and the waste generated, were separated and weighed: peel (remains adhering to the mesocarp + to the epicarp + to the perianth) and seeds (remains adhering to the mesocarp + endocarp + endosperm). For convenience, the terms "peel" and "seeds" are adopted in this work, to describe all the waste content detailed above.

All the weights were determined using a Gural, EGI model digital electronic scale, with 5g divisions. The data were entered and processed in a microcomputer, and calculations and charts were performed using the software program Microsoft Excel.

## Results and Discussion

The results found in the research, for ripe, fresh fruit, were segmented into three groups, comprising: 1–Indicators of mass (weight) of fruit, 2–Indicators of mass yield (weight) of fruit and 3–Other estimates. It should be noted that the quantitative findings, in relation to the universe of the population, must take into account the lower and upper limits of the range of tolerances inherent to the adopted sampling error of ± 5 per cent, with an 80 per cent Confidence Interval (CI) of the sample.

### Indicator Mass (Weight) of Fruit

The survey revealed that the total weight of the 143 sampled fruits was 7965 g, resulting in a simple arithmetic average of 55.70 g/fruit. However, considering the sampling error, the average of the fruit can be estimated between a lower limit of 52.9 g/fruit and an upper limit of 58.5 g/fruit. Thus, the standard deviation found in 10.32 g is between 9.8 g/fruit and 10.8 g/fruit. The largest fruit found had a mass of 75g. However, in the universe of the population that has an 80 per cent confidence interval, it is possible to find maximum values of between 71 g and 79 g. The minimum value, in terms of unit weight of fruit recorded, was 25 g, but can be considered between 24 g and 26 g. The mode and median, both for the amount of 60 g/fruit and for the estimated effect, can also be considered as being between 57.0 g/fruit and 63 g/fruit (Table 19.1).

**Table 19.1**: Indicators mass (weight) of fruit.

| Item | Average Value Found | Tolerance Interval (± 5 per cent) | |
|---|---|---|---|
| | | Lower Limit | Upper Limit |
| Total mass of fruit (grams) | 7965 | – | – |
| Mean (grams/fruit) | 55.7 | 52.9 | 58.5 |
| High (grams/fruit) | 75.0 | 71 | 79 |
| Low (grams/fruit) | 25.0 | 24 | 26 |
| Standard Deviation (grams/fruit) | 10.3 | 9.8 | 10.8 |
| Mode (grams/fruit) | 60.0 | 57,0 | 63.0 |
| Median (grams/fruit) | 60.0 | 57.0 | 63.0 |

The frequency distribution of the masses (weights) in sample units of the fruits showed a sliding scale with 11 at intervals of five grams each. The fruits with the minimum mass (25 g) represented only 0.70 per cent while the highest (75 g) represented 2.80 per cent. Fruit weighing 45 g to 65 g represented 76.92 per cent of the sample (Table 19.2).

**Table 19.2**: Simple frequency distribution, cumulative, absolute and relative masses of the fruits of the sample unit.

| Unit Weight of Fruit (g) | Single Frequency | | Cumulative Frequency | |
|---|---|---|---|---|
| | Absolute | Relative (per cent) | Absolute | Relative (per cent) |
| 25 | 1 | 0.70 | 1 | 0.70 |
| 30 | 2 | 1.40 | 3 | 2.10 |
| 35 | 2 | 1.40 | 5 | 3.50 |
| 40 | 14 | 9.79 | 19 | 13.29 |
| 45 | 14 | 9.79 | 33 | 23.08 |
| 50 | 19 | 13.29 | 52 | 36.36 |
| 55 | 14 | 9.79 | 66 | 46.15 |
| 60 | 39 | 27.27 | 105 | 73.43 |
| 65 | 24 | 16.78 | 129 | 90.21 |
| 70 | 10 | 6.99 | 139 | 97.20 |
| 75 | 4 | 2.80 | 143 | 100.00 |
| Total | 143 | 100.00 | – | – |

Separately, the largest group was fruits with a weight of 60 g, representing 27.27 per cent of the sample, so this was the median and modal mass of the sample. The frequency distribution of the relative percentages of the mass unit are shown in Graph 19.1.

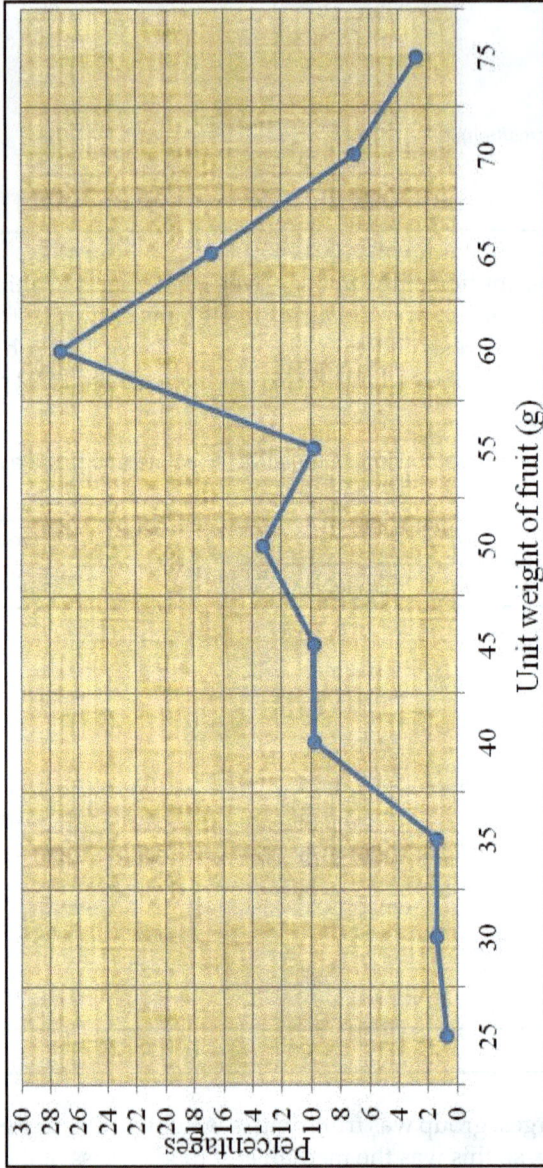

**Graph 19.1**: The frequency distribution of the masses of fruit unit (Percentage x grams).

## Performance Indicators by Mass (Weight) of Fruit

Before determining the yield of the fruit in mass (weight), for the pulp extracted and the waste generated (+ seed husks), one must take into account factors such as the subjective efficiency of the person extracting the pulp, who uses entirely manual processes and technology. A greater or lesser amount of pulp may be left adhered to the peel and seeds, and other factors inherent to the genetic and edaphoclimatic variables can also lead to lack of standardization of the amounts of pulp.

The yield indicators, in mass (weight), demonstrated that from 7965 g of ripe fruit 1390 g of pulp (mesocarp) was extracted, the only part of the fruit currently used by the vendors, resulting in a rate of return of about 17.45 per cent. Taking into account the sampling error of the research ( ± 5 per cent), the weight of the pulp extracted may be between 1321 and 1460 g, so that the yield may vary in the range from a minimum of 12.45 per cent to a maximum of 22.45 per cent as shown in Table 19.3. The average weight per fruit pulp found in the research was 9.72 g, however, assuming the sample error, it is between 9.23 g and 10.21 g (Table 19.3).

**Table 19.3**: Indicators of yield, in mass (weight) of fruit.

| Item | Average Value Found | Tolerance Interval ( ± 5 per cent) | |
|---|---|---|---|
| | | Lower Limit | Upper Limit |
| Pulp (mesocarp) | | | |
| – Weight of pulp removed (g) | 1390 | 1321 | 1460 |
| – Yield of the mass of pulp (per cent) | 17.45 | 12.45 | 22.45 |
| – Average weight of fruit pulp per unit (g) | 9.72 | 9.23 | 10.21 |
| Peels (pericarp + mesocarp+ remains of perianth) | | | |
| – Total weight (g) | 1640 | 1558 | 1722 |
| – Mass shells on the total weight of the fruit (per cent) | 20.59 | 15.59 | 25.59 |
| – Average weight of peel per unit of fruit (g) | 11.47 | 10.90 | 12.04 |
| Seed (endocarp + mesocarp + remains of endosperm) | | | |
| – Weightofseeds (g) | 4935 | 4688 | 5182 |
| – Percentage of total weight of fruit (per cent) | 61.96 | 56.96 | 66.96 |
| – Average weight of seed per unit of fruit (g) | 34.51 | 32.78 | 36.24 |
| Total Waste–Waste (seed+ peels) | | | |
| – Weight of bark (g) | 1640 | 1558 | 1722 |
| – Weightofseeds (g) | 4935 | 4688 | 5182 |
| – Total weight of waste (g) | 6575 | 6246 | 6904 |
| – Percentage of waste for the total fruit weight | 82.55 | 77.55 | 87.55 |

The peels produced in the commercial pulp production process resulted in a total mass of 1640 g, varying in range of tolerances between 1558 g and 1722 g. Thus, the levels of waste inherent to the peels were 20.59 per cent on average, but may be between 15.59 per cent and 25.59 per cent. The mass of peels generated per fruit unit presented with an average 10.90 g, but may range between 10.90 g and 12.04 g, due of the systematic error inherent to the sampling.

The same observation made for the seeds reveals that the total of 7965 g of fruit, after extracting the peels and pulp, resulted in 4935 g of seeds, accepting, due to sampling error, values of between 4688 g-5182 g. Thus, the average rate of seed mass for the total weight of fruits was 61.96 per cent, but may range from 56.96 per cent to 66.96 per cent–Table 19.3.

In these circumstances, the 7965 g of the fruit generated a total mass of waste (peels + seeds) sent to the garbage of around 6575 g, or 82.55 per cent of the resource. Applying the sampling error of the research (± 5 per cent), this total discarded mass ranges between a minimum of 6.246 g and a maximum of 6.904 g, resulting in waste levels of between 77.55 per cent and 87.55 per cent–Table 19.3. Moreover, it was observed that the total mass of waste generated was 6575 g, of about 25 per cent was peel and 75 per cent was seeds.

## Other Estimatives

The results obtained in the research, considering the mean values and sampling error, allowed other estimates, as an offshoot of the work, in order to stimulate new thinking. In this scope, it was observed that the ratio of the mass of peel/pulp is 1.18. This means that the production process adopted results around 18 per cent more peel than pulp, but taking into account the tolerances of the sampling error, this ratio may be from 1.12 to 1.24. In this case, it is possible that mechanizing the production process could result in gains in productivity, decreasing the amount of pulp adhering to the peel, and therefore minimizing the loss of pulp, without compromising the quality of the pulp.

The ratio of the masses of seed/pulp, however, is much greater than for the peel/pulp. On average, the mass of seeds was 3.55-fold greater than that of pulp, *i.e.* 255 per cent higher. This may vary, within a margin of error, from a minimum of 3.37 to a maximum of 3.73. This means that for every 1000 g of pulp, 3550 g of seeds are sent to non-selective, general waste as was found in this study. Here again, it is also estimated that mechanization process could result in gains in pulp yield, while minimizing the portion attached to the endocarp and in particular, reducing the manual labor time required for the operation.

However, the improvements regarded as having the highest impact on increasing the productivity of this Amazonian resource are innovations in the use of the seeds, not only for use in the creation of seed jewellery, known as "biojóias", but also for use in several other products could be made using only endocarp, *e.g.*, bio scrub for cosmetics, activated carbon, and organic fertilizer, among others.

The endosperm, the inner part of the seed where the biological core is situated, capable of generating a new biological plant, is also sent to garbage, but could be used in many different ways, such as in biofuel production, natural cosmetics and food.

According to data obtained from SEBRAE–Brazilian Service of Support for Micro and Small Enterprises (2012), the FOB price of tucum nut oil practiced in the sale of the oil to the cosmetic industries, by a cooperative of extractors of oils biodiversity of the Amazon (andiroba, tucum and others) in the municipality of Manaquiri/AM, reached a share price of $ 20.00/kg, in May 2012, equivalent to U.S. $ 10.00/kg. Under these conditions and in monetary terms, one kilogram of this oil is enough to acquire, in Manaus, about 6.9 liters of gasoline or 8.7 gallons of automotive ethanol fuel. This comparison suggests greater economic attractiveness in selling the oil to the cosmetics industry than for processing into biodiesel, because even tests on the maximum yield of oil as biodiesel reached the 95 per cent (Barbosa *et al.*, 2009, p. 375).

Alternatively, a less economical but most environmentally friendly hypothesis; if no portion of the waste is industrialized, at least, the healthy seeds could simply be used for new plant seedlings, for replacement/maintenance of the species in the environment, gene banks, and breeding, among other purposes, instead of just adding more garbage to the city.

Considering the low yield of pulp due in the procedure – an average of 17.45 per cent–this being the only use made of the fruit, it is seen that the raw material is indeed underutilized; the technical production coefficient (fruit/pulp) found was around 5.73, *i.e.*, 5730 g of fruit is needed to generate 1000 g of pulp. Allowing for sampling error, this ratio varies between 5.44 and 6.02. In practical terms, just one more bag of fruit, of the type purchased by the merchants/extractors – at an average FOB price of $ 70.00, equivalent to US$ 35.00 at the time of the survey in the port of Manaus, with waste of around 82.55 per cent, may mean proportionally the same as almost R$ 58.00/bag or US$ 29.00/bag being thrown into the garbage. Of course cost this is passed on to the consumer in the price of the pulp.

**Table 19.4**: Other estimatives.

| Discrimination | Average Value Found | Tolerance Interval ( ± 5 per cent) | |
|---|---|---|---|
| | | Lower Limit | Upper Limit |
| Average ratio of the weight peel/pulp | 1.18 | 1.12 | 1.24 |
| Average ratio of seed weight/pulp | 3.55 | 3.37 | 3.73 |
| Weight average ratio fruit/pulp (technical coefficient of raw material) | 5.73 | 5.44 | 6.02 |
| Average number of fruits per kilogram (unit) | 18 | 17 | 19 |
| Average number of fruits per kilogram of pulp (unit) | 103 | 97 | 108 |
| Average number of fruits/50 bag (unit) | 900 | 855 | 945 |
| Average yield of pulp per 50 kg bag of fruit (kg) | 8.73 | 8.29 | 9.16 |

Although not the focus of the study, other relevant costs for the extractor were observed, although there are were no formal records of these; for example, the cost of transporting the bags of fruits from the boat to the vendor's stall–the "carreto ", or hand-held cart, which is R $ 3.00 a bag, and the loss of surplus pulp that is produced but not sold, and which quickly deteriorates with poor refrigeration.

At the time of this study, the price of retail kilogram of pulp was between R$ 20.00 and R$ 35.00 (U$ 10.00 and U$ 17.50) and a 50 kg bag of fruit yielded around 8.73 kg of pulp, but with wide variation (Table 19.4). It is possible that in these conditions, and without effective management, the activity would have more difficulty surviving, either from a business point of view, or from and environmental perspective.

Finally, all the results found in this study, when compared to those described in the literature, although rare, on research carried out at other times, with different goals and procedures, particularly in terms of the subjectivity of the manual pulp extraction, were, as expected, convergent in some respects but divergent in others.

One very important piece of information for this research is the average mass of fruit pulp (pulp yield), for which the mean value was 17.4 per cent. But if this figure is observed within the range of tolerances from 12.4 per cent to 22.4 per cent, it is compatible with the findings of Moussa and Kahn (1997, p.108) who report 21.9 per cent, Leit o (2008, p.23) who reports 19.4 per cent and Figliuolo and Silva (2009) who reports 21.2 per cent. In theory, even with these differences, the results, both in this study and in the literature, point to low levels of utilization of a resource that has much more to offer, even the parts currently considered as waste.

The average unit weight of the fruit was 55.7 g, with a range of 52.9 g to 58.5 g, which is close to the findings of Schroth *et al.* (2004) apud Leit o (2008, p. 23) and Figliuolo and Silva (2009, p.sn). Other parameters of the literature and this research are summarized in Table 19.5.

In relation to the endosperm, studies by Moussa and Kahn (1997, p. 108) show that the average weight of the kernel (endosperm) of tucum corresponds to 22.4 per cent of the weight of the fruit, and can yield, depending upon the extraction method used (mechanical or chemical), 40 per cent to 50 per cent in weight of oil rich in fatty acids of short and medium carbon chains, making it a very suitable raw material for biodiesel production or for use in cosmetics (Figliuolo and Silva, 2009, p.sn).

Therefore, considering that on average, each kilogram of tucum pulp commercially extracted in Manaus uses 5.730 g of raw fruit (the technical coefficient of pulp production of found in this study) and that 22.4 per cent of this weight corresponds to the kernel (Moussa and Kahn, 1997, p. 108) a ratio of 1.284 g of endocarp/1 kg of pulp can be estimated. With the oil extraction yield of this material ranging from 40 per cent to 50 per cent by weight (Figliuolo and Silva, 2009, p.sn), this would result in 514 g to 642 g of oil per kilogram of pulp removed, a resource that is currently being wasted and converted into garbage.

Although readily available in public markets, and sold by street vendors throughout the city, tucum pulp is not yet sold in supermarkets, shopping centers or similar (Figure 19.1).

Under the management of the City Council of Manaus, there are two itinerant markets, 34 permanent markets, and 8 public markets, totaling 44 establishments, not including informal trade, and the 7084 street vendors registered by the municipal government (Sempab, 2012) who sell various types of products, including tucum pulp. Therefore, it can be deduced that a significant amount of waste from tucum

**Table 19.5**: Data from the literature on ripe fresh fruits tucumã-do-amazonas (*A. aculeatum* Meyer) and findings of the research.

| Item | Sources | | | | | | | |
|---|---|---|---|---|---|---|---|---|
| | 1 | 2 | 3 | 4 | 5 | 6 | 7 | 7* |
| Samplesize (fruit) | – | – | 717 | – | – | – | 143 | – |
| Sampling error | – | – | – | – | – | – | ± 5 per cent | – |
| ConfidenceInterval | – | – | – | – | – | – | 80 per cent | – |
| Fruitmass | | | | | | | | |
| – Mean (g) | 47.4 | 52.6 | 47.5 | 70/75 | – | 57.3 | 55.7 | 52.9/58.5 |
| – Highest (g) | 72.5 | 90.1 | 76.4 | – | – | – | 75 | 71/79 |
| – Minimum (g) | 19.8 | 18.6 | 25.9 | – | – | – | 25 | 24/26 |
| – Standard deviation (g) | 14.7 | – | – | – | – | – | 10.3 | 9.8/10.8 |
| Mass ofepicarp (peel) | | | | | | | | |
| – Mean (g) | 7.7 | – | 8.6 | – | – | – | 11.5 | 10.9/12.0 |
| Mass mesocarp (pulp) | | | | | | | | |
| – Mean (g) | 9.1 | – | 10.3 | – | – | – | 9.7 | 9.2/10.2 |
| – Percentage/weightoffruit (per cent) | 19.4 | – | 21.9 | – | – | 21.2 | 17.4 | 12.4/22.4 |
| Mass ofpyrene (seed) | | | | | | | | |
| – Mean (g) | 30.0 | – | – | – | 34.8/53.3 | – | 34.5 | 32.8/36.2 |

Sources: 1: Leitão (2008); 2: Schroth *et al.*, 2004, cited by Leitão (2008); 3: Moussa and Kahn (1997); 4: Cavalcante (1996) cited by Leitão (2008); 5: Mendonça (1996) cited by Leitão (2008); 6: Figliuolo and Silva (2009); 7: Authors of this paper; 7*: Range tolerances of the results obtained by the authors of this work.

**Figure 19.1**: Aspects of the extraction and sale of tucumã pulp in Manaus
*Photo*: Herculano, F.E.B.

fruit is produced in Manaus, placing an additional burden on the garbage collection services in the city. In the year 2006 alone, according to the IBGE, the sale of tucum in the State of Amazonas as a whole reached 2790 tons (IBGE, 2007). If Manaus, which concentrated 51 per cent of the state population in 2007 (Ibge, 2007 apud Ipea, 2012) consumed equal proportions of the fruits, or, 1422.9 tons/year, corresponding to an average of 118.58 tons/month or 3.95 tons/day, confirming the significant mass of waste from the fruit, even though much of it could be reutilized.

## Conclusions

The activity of commercial extraction of the pulp (mesocarp) of tucum (*A. aculeatum* Meyer), which is so common in the public markets in the city of Manaus, results in considerable amounts of waste, reflecting the under-utilization of a rich natural resource of the Amazon rainforest.

This waste- with just a few notable exceptions, such as its use in handicrafts – goes to the general garbage, but it has potential use for various social and economic purposes that could result in improved productivity and income, as well as helping to reduce the volume of urban waste produced by the city.

## Acknowledgments

The authorsthankstoFUCAPI, UFAMandSrs. Raimundo Cortez e Ivan Tavares (feirantes)

## References

Aragón, L.E.(2002). Há futuro para o desenvolvimento sustentável na amazônia ? In: O futuro da Amazônia: dilemas, oportunidades e desafios no limiar do Século

XXI. Melho A.F. (org). Editora da Universidade Federal do Pará, Belém. Available at http://www.desenvolvimento.gov.br. Accessed on 14/03/2011.

Barbosa, B.S. *et al.* (2009). Aproveitamento do óleo das amêndoas de tucum do amazonas naproduç o de biodiesel. *Acta Amazônica*, **39(2):** 371-378. Available at: http://acta.inpa.gov.br. Accessed on: 23/05/2012.

Ferreira, E.S., *et al.* (2008). Caracterizaç o físico-química do fruto e do óleo extraído de tucum (*Astrocaryum vulgare* Mart). *Brazilian Journal of Food and Nutrition*, **19(4):** 427-433.

Figliuolo, R., and Silva, J.D. (2009). Cadeia Produtiva Sustentável e Integral do Tucum do Amazonas: do lixo à produç o de cosméticos e biodiesel. 32nd Annual Meeting of the Sociedade Brasileira de Química (Brazilian Chemical Society) held in Fortaleza-CE, 30/5 on 02/06/2009. Sociedade Brasileira de Química– (SBQ), CDrom Resumos. Available at: http://sec.sbq.org.br/cdrom/32ra. Accessedon: 23/05/2012.

IBGE (2012). Instituto Brasileiro de Geografia e Estatística–Cidades, Manaus – AM, Histórico. Available at: on 16/05/2012.

IBGE (2006). Instituto Brasileiro de Geografia e Estatística – Censo Agropecuário– IBGE, Rio de Janeiro, 2007.

INPA (2011). Instituto Nacional De Pesquisas Da Amazônia. Amazônia tem 10 mil plantas com potencial econômico. Available at: http://www.inpa.gov.br Accessed on 24/10/2011.

IPEA (2010). Instituto de Pesquisa Econômica Aplicada. Populaç o Residente – Municípios – Estado do Amazonas – 1991 a 2010. IPEADATA. Available at: www.ipeadata.gov.br. Accessed on12/06/2012.

Leit o, A.M. (2008). Caracterizaç o morfológica e físico-química de frutos e sementes de *Astrocaryum aculeatum* Meyer (Arecaceae), de uma floresta secundária. Thesis presented to the Integrated Postgraduate Program in Tropical Biology and natural resources. Instituto Nacional de Pesquisas da Amazônia (National Research Instituteof Amazônia–INPA) and Universidade Federal do Amazonas (UFAM), Manaus.

Lorenzi, H. *et al.*(2004). Palmeiras Brasileiras e Exóticas Cultivadas. Nova Odessa, Editora Plantarum.

Lima, A.L.S. *et al.* (2011). Aplicaç o de baixas doses de radiaç o ionizante nofruto brasileiro tucum (*Astrocarium vulgare* Mart.). *Acta Amazônica*, **41(2):** 377-382.

Martins, C. R., and Farias, R. M. (2002). Produç o de alimentos x desperdício: tipos, causas e como reduzir perdas na produç o agrícola. *Revista da Faculdade de Zootecnia, Veterinária e Agronomia*. **9(1):** 83-93.

Menezes, E.M.S., *et al.* (2008). Valor nutricional da polpa de açaí (Euterpe oleracea Mart) liofilizada. *Acta Amazônica*, **38(2).**

MMA (2008). Ministério do Meio Ambiente. Plano Amazônia Sustentável: diretrizes para o desenvolvimento sustentável da Amazônia Brasileira. Presidência da República–MMA. Brasília.

Moussa, F., and Kahn, F. (1997). Uso y potencial económico de dos palmas, Astrocaryumaculeatum Meyer y A. vulgareMartius, enlaAmazoníabrasile a. Institut Français de Recherche Scientifique pour le Développement en Coopération (ORSTOM), pp. 101-116. Brasilia, D.F. – Brazil.

Rodrigues, A. M. C.; Darnet, S., and Silva, M. (2010). Fattyacid profiles and tocopherol contents of buriti (*Mauritia flexuosa*), patawa (*Oenocarpus bataua*), tucuma (*Astrocaryum vulgare*), mari (*Poraqueibap araensis*) and inaja (*Maximiliana maripa*) fruits. *J. Braz. Chem. Soc.*, **21(10)**.

Santos, R. I. (2004). Metabolismo básico e origem dos metabólitos secundários. In: C.M.O. Sim es *et al.* (eds.). Farmacognosia da planta ao medicamento. Universidade UFRGS/Ed. da UFSC, 5th edition rev., Porto Alegre, chapter 16, Florianópolis.

SEBRAE/AM (2012). Serviço Brasileiro de Apoio às Micro e Pequenas Empresas – Amazonas. Unidade de Atendimento Coletivo Agronegócios. Information by email: fatima@am.sebrae.com.br.

SEMPAB (2012). Secretaria Municipal de Produç o e Abastecimento da Prefeitura de Manaus. Mercados e Feiras–Comércio Informal.Available at: http://sempab.manaus.am.gov.br. Accessed on: 01/065/2012.

Sena, R. F., and Nunes, M. L. (2006). Utilizaç o de resíduos agroindustriais no processamento de raç es para carcinicultura. *Revista Brasileira de Saúde e Produç o Animal*, **7(2):** 94-102.

Sentanin, M. A., and Rodriguez-Amaya, D. B. (2007). Teores de carotenóides em mam o e pêssego determinados por cromatografia líquida de alta eficiência. *Ciências e Tecnologia de Alimentos*, **27:** 13-19.

Tagliacarne, G. (1986). Pesquisa de mercado: técnica e prática. Trad. Maria de Lourdes Rosa da Silva, 2ª. ed., Atlas, S o Paulo.

Uenojo, M. (2007). Maróstica Junior, M.R.; Pastore, G.M.; Carotenóides: Propriedades, aplicaç es e biotransformaç o para formaç o de compostos de aroma. *Química Nova*, **30(3):** 616-622.

United Nations, (2011). Demography Yearbook 2009 – 2010. Sixty-first issue. United Nations, New York.

Yuyama, L. K. O. *et al.* (2005). Polpa e casca de tucum (*Astrocaryum aculeatum* Meyer): quais os constituintes nutricionais? *Nutrire: Revista Soc. Bras. Alim. Nutr.*, **30(Supl.):** 225.

Utilisation and Management of Medicinal Plants Vol. 2 (2014)    *Pages* **403–414**
*Editor-in-Chief:* **V.K. Gupta**
*Published by:* **DAYA PUBLISHING HOUSE, NEW DELHI**

# 20

# Pharmacognostic Evaluation and Antibacterial Activity of *Prosopis cineraria* Leaf

Mahendra S. Khyade[1]* and Nityanand P. Vaikos[2]

## ABSTRACT

*The pharmacognosy and antibacterial activity of Prosopis cineraria (L.) Druce leaf, one of the important medicinal plants in the indigenous systems of medicine in India has been studied in detail. The present work was carried out in terms of macroscopic and microscopic structure of the organ including histochemistry, physiochemical analysis and preliminary phytochemical testing. The antibacterial activity was also screened using different solvent extracts of leaf by agar well diffusion method. The leaf microscopic characters like stomatal frequency, stomatal index, vein islet number and vein termination number including some other features were determined. Physiochemical constants like percentage of moisture content, total ash, acid insoluble ash, chloroform soluble extractive, ethanol soluble extractive and water-soluble extractive were determined. The phytochemical analysis reveals the presence of alkaloids, tannin, saponin, steroids, flavonoids, leucoanthocyanins and phenolics. Antibacterial activity of leaves was studied using different solvent system against various gram positive, gram-negative bacteria. The results of antibacterial activity of P. cineraria against tested strains of bacteria examined in the current study and their potency were qualitatively assessed by the presence or absence of inhibition zones and zone of diameter.*

1    Post Graduate Department of Botany, Sangamner Nagarpalika Arts, D.J. Malpani Commerce and B.N. Sarda Science College, Sangamner – 422 605, M.S., India

2    Department of Botany, Dr. B.A.M.University, Aurangabad – 431 004, M.S., India
     [Presently at 15B Sonchafa, Mahavirnagar, Osmanpura Aurangabad]

*    *Corresponding author*: E-mail: mskhyade@rediffmail.com, drvaikos@gmail.com

*Pharmacognostical studies and phytochemical screening can serve as a valuable source of information and provide suitable standards to determine the quality of this plant material in future investigations or applications. Moreover, the leaf extract can be effective antibiotics, both in controlling Gram positive and Gram-negative human pathogens.*

**Keywords:** Prosopis cineraria, Leaf, Pharmacognosy, Antibacterial activity.

## Introduction

*Prosopis cineraria* (L.) Druce belongs to the family Mimosaceae is known as Khejra in hindi, Shami in sanskrit (Yoganarasimhan, 2000), Soundad in Marathi (Naik, 1998) and has been advocated as one of the important medical plants in India. It is thorny, irregularly branched, multipurpose tree species of arid areas and commonly found in dry and arid regions of north-western India, southern India, Pakistan, Afghanistan, Iran and Arabia (Kirtikar and Basu, 1984). It has been extensively used in indigenous system of medicine as folk remedy for various ailments (Kirtikar and Basu, 1984; Duke, 1983) like leprosy, dysentery, bronchitis, asthma, leucoderma, piles, muscular tremors and wandering of the mind. It is also known to possess anthelmintic, antibacterial, antifungal, antiviral, anticancer and several other pharmacological properties. Leaf paste of *P. cineraria* is applied on boils and blisters, including mouth ulcers in livestock and leaf infusion on open sores on the skin (Chopra *et al.*, 1956; 1958; Nadkarni, 1976).The smoke of the leaves is considered good for eye troubles (Rastogi and Mehrotra, 1995). The leaves and pods are consumed by livestock and are beneficial forage. Jewers *et al.* (1974; 1976) have studied the phytochemicals in the leaves of *P. cineraria* and reported alkaloid namely spicigerine; steroids namely campesterol, cholesterol, sitosterol, stigmasterol; alcohols namely octacosanol and triacontan-1-ol; and alkane hentriacontane.

The phytochemical studies on the leaves of *Prosopis cineraria* resulted in isolation of methyl docosanoate, diisopropyl-9,10–dihydroxyicosane-1,20-dioate, tricosan-1-ol and 7,24-tirucalladien-3-one. While diisopropyl-10, 11-dihydroxyicosane-1,20-dioate is a hitherto unreported compound, methyl docosanoate, tricosan-1-ol and 7,24-tirucalladien-3-one are being reported for the rst time from *P. cineraria* (Malik, 2007). A comparative anatomical study of the leaflets of *P. cineraria* and *P. juliflora* is reported (Stellaa *et al.*, 2010). Also phytochemical and antibacterial study carried out (Sivanarayan and Suriyavathana, 2010).

In spite of the numerous medicinal uses attributed to this plant, there are no pharmacognostical reports on the leaf of this plant. Hence, the present investigation is an attempt in this direction and includes morphological and anatomical evaluation, determination of physico-chemical constants and the preliminary phytochemical analysis and antibacterial screening of the different extracts of *P. cineraria*

## Materials and Methods

### Plant Material

The leaves of *P. cineraria* were collected from the well grown healthy plants inhabiting from the local area of Aurangabad, Maharashtra (India). It was identified, confirmed and authenticated by comparison with and authentic specimen. A voucher specimen of the plants deposited in the departmental herbarium for the future reference. The samples were cut suitably and removed from the plant and thoroughly washed with water to remove the adherent impurities and dried in shade at room temperature for 15 days. The fresh leaf was used for the study of macroscopic and microscopic characters.

## Pharmacognosy

### Macroscopical Examinations

For morphological observations, fresh young leaves were used, the macro-morphological features of the plant parts (leaves) were observed under magnifying lens and simple microscope (Tyler *et al.*, 1977).

### Microscopical Examinations

Fresh leaves of the plant were studied transversely and longitudinally, using surface preparations and sections. The different parts of leaf like lamina and midrib were studied. For microscopic studies, the leaves were cut and removed from the plant and fixed in FAA (Formalin 5 ml + Acetic acid 5 ml + 70 per cent Ethanol 90 ml). After 24 hours of fixing, the epidermal peels and transections of leaf and petiole were taken by free hand. The sections were stained in safranin (1 per cent), light green (1 per cent) and mounted in DPX after the customary dehydration. Some hand sections were also examined in glycerin. Microphotographs of leaf were taken by using Jenaval and Mirax Laborec Cameras affixed to microscope.

### Quantitative Microscopy

The leaf epidermal studies were carried out on fresh specimens. Peels were removed mechanically using some chemicals. They were stained in 1 per cent safranin mounted in glycerin and made semi-permanent by ringing with DPX solution. Stomatal index (SI) and stomatal frequency was calculated (Salisbury, 1927; 1932). The vein islet number and vein termination number the leaf were determined according to the reported method (Levin, 1929).

### Histological Colour Reactions

The histochemical colour reactions were performed by the standard methods (Johanson, 1940; Guerin *et al.*, 1971).

### Physio-chemical Characters

Physiochemical values such as the moisture content, percentage of total ash, acid insoluble ash and acid soluble ash; extractive values like chloroform-soluble extractives, alcohol-soluble extractives and water-soluble extractives were calculated according to the methods described in the Indian pharmacopoeia (Anonymous, 1966; 1985).

## Phytochemical Evaluation

Phytochemical studies such as qualitative were done from the shade-dried powdered material. For qualitative phytochemicals, standard prescribed methods were used (Smolenski *et al.*, 1972; Dan *et al.*, 1978; Harborne, 1984; Karumi *et al.*, 2004; Edeoga *et al.*, 2005).

## Antibacterial Evaluation

### Preparation of Extracts

The dried plant material was pulverized into fine powder using a grinder (mixer). About 50 gm of powdered material was extracted in soxhlet extraction apparatus with 250 ml of each of the following solvents; Petroleum ether, chloroform, Acetone and Methanol (Vogel, 1988). The extracts obtained with each solvent were filtered through Whatman filter paper No. 1 and the respected solvents were evaporated (at 40°C) with the help of heating mantle. The sticky greenish-brown substances were obtained and stored in refrigerator and were suspended in dimethyl sulphoxide (DMSO) for prior to use (Beyer and Walter, 1997).

### Tested Microorganisms

Various cultures of human pathogenic, gram positive and gram negative bacteria were used. These are *Staphylococcus aureus, Bacillus megaterium, Bacillus subtilis, Escherichia coli, Salmonella typhi, Pseudomonas aeruginosa, Corynobacterium glutamicum* and *Klebsella planticola*. The cultures were obtained from Microbial Type culture Collection (MTCC), IMTEC, Chandigarh, India. The microorganisms were repeatedly sub cultured in order to obtain pure isolates. A loop full test organism was inoculated on nutrient broth and incubated for 24 h at 37 ± 1°C and maintained in sterile condition.

### Screening for Antibacterial Properties

Antibacterial activities of plant extracts were tested by Agar well diffusion method (Kavanagh, 1972). The culture plates were prepared by pouring 20 ml of sterile nutrient agar.1 ml inoculum suspension was spread uniformly over the agar medium using sterile glass rod to get uniform distribution of bacteria. A sterile cork borer (8 mm) was used to make wells in each plate for extracts. These plates were labeled and 100 µl of each plant extracts (at concentration of 100 mg/ml) was added aseptically into the well. Then the plates were incubated for 24 h at 37°C during which the activity was evidenced by the presence of zone of inhibition surrounding the well. Each test was repeated three times and the antibacterial activity was expressed as the of diameter of the inhibition zones (mm) produced by the plant extracts when compared to the controls.

## Results and Discussion

Macroscopically, the leaves are bipinnate, main rachis glabrous or puberlous, pinnae, usually 2-pairs, opposite, 2-7cm long, often with round leaflets 7-12 pairs, subsessile, oblong obliquely round and mucronate at the apex and very unequal sided. The upper side much smaller reticulatly veined, grey in colour, glabrous, base round and very oblique (Figure 20.1A), the thorns are straight with a conical base and sparsely distributed on the internodes.

The shape of rachis in cross section is circular. The epidermal cells are thick walled, unequal in size and tightly arranged. It is followed by parenchymatous

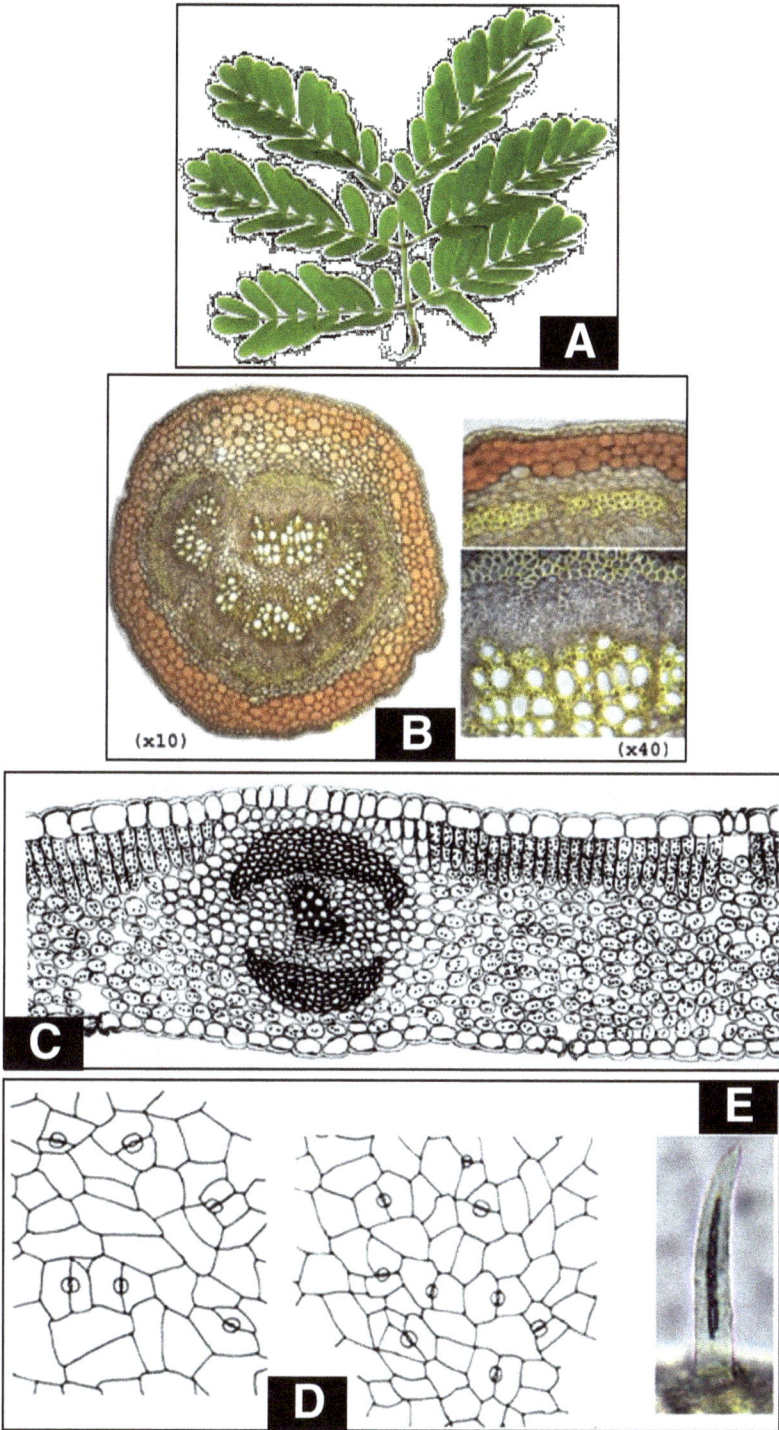

**Figure 20.1:** A: Bipinnate leaves; B: T. S. of Rachis with enlarged part; C: T.S. of Lamina; D: Epidermal structures (adaxial and abaxial epidermis); E: Unicellular trichome.

cortex. The vascular tissue comprises of 3-5 bicollateral vascular bundles which occur in a more or less circular ring. Some are interconnected to each other. The vascular bundles are surrounded by sclerenchymatous tissue.

The leaflet is dorsiventral and amphistomatic. The upper epidermal cells are larger and unequal in size with thick outer wall and cuticle. The lower epidermal cells are also unequal in size smaller and thin walled. Stomata occur on both the surfaces. However they are more on lower side. The mesophyll is made up of palisade and spongy tissue. Palisade is 2-3 layered and spongy tissue is of compactly arranged cells. The tannin content is of common occurrence. The vascular bundles are many, bicollateral which extends upwards into the mesophyll tissue. Adaxial epidermal cells are large, polygonal, thick walled with straight anticlinal walls. Abaxial epidermis is smaller celled and thin walled. Cell walls are straight. Stomata are paracytic, rarely anisocytic. and more in number in the lower surface. Trichomes are one celled (Figures 20.1B-E).

The leaf microscopic characters like stomatal frequency, stomatal index, vein islet number and vein termination number were determined (Figure 20.2). Stomatal index in abaxial surface was 21.57 μm and 20.9 μm in adaxial side. The average vein islet numbers are 91.76/sq.mm and vein termination number is 116/sq.mm. The quantitative determination of some pharmacognostic parameters are useful for setting standards for crude drugs. The vein islet and vein termination numbers and other parameters determined in the quantitative microscopy, are relatively constant for plants and can be used to differentiate closely related species. Stellaa *et al.* (2010) compare some microscopic differences between the leaves of two species in *Prosopis*.

The histochemical colour reactions were carried out on transverse section of the fresh leaf (Table 20.1). The results indicated that presence of lignin, starch, fats, alkaloids, saponins, tannins, and flavonoids. Histochemical localization of certain important compounds enables to get a preliminary idea of type of compounds and their accumulation in the plant tissues. Based on this study, one can choose the organ or tissue where the required compounds are located.

**Table 20.1**: Histochemical tests.

| Reagents | Constituents | Colour | Histological Zone | Degree of Intensity |
|---|---|---|---|---|
| Anilline $SO_4 + H_2SO_4$ | Lignin | Yellow | Scl, Xy. | +++ |
| Weak Iodine solution | Starch | Blue | Co. | ++ |
| Sudan III/IV | Fats | Red/Pink | Co. | + |
| Dragendroffs reagent | Alkaloids | Turbidity brown | Co. | + |
| Ba $(OH)_2 + K_2Cr_2O_7 + CaCl_2$ | Saponins | Yellow | Co. | ++ |
| $FeCl_3$ | Tannins | Blue green | Co. | ++ |
| Vanillin + HCl | Flavonoids | Yellow | Co. | ++ |
| $AgNO_3 + H_2O_2$ | Ca. crystals | — | — | — |

$: Scl.: Sclerenchyma; Xy.: Xylem; Co.: Cortex.

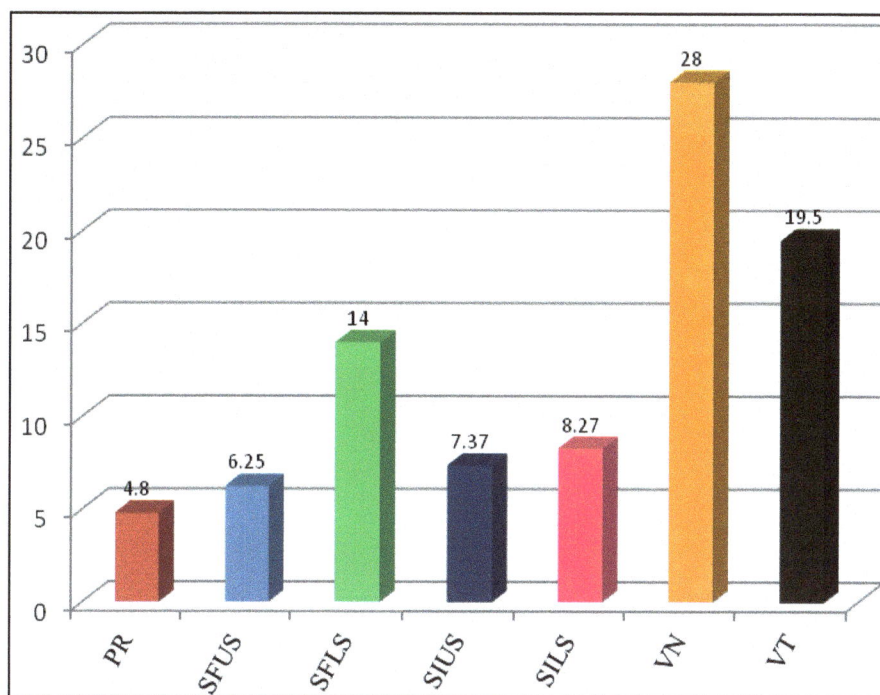

**Figure 20.2**: Quantitative microscopy of *P. cineraria* leaf.
PR: Palisade ratio; SFUS: Stomatal frequency upper surface; SFLS: Stomatal frequency lower surface; SIUS: Stomatal index upper surface; SILS: Stomatal index lower surface; VN: Vein islet number; Vein termination number.

Physico-chemical constants like percentage of moisture content, total ash, acid insoluble ash, chloroform soluble extractive, ethanol soluble extractive and water soluble extractive were determined and depicted in Figure 20.3. The physical constants evaluation of the drug is an important parameter for in detecting adulteration or improper handling of the drugs. The moisture content of the drug is not high (5.5 per cent), thus it discourages bacteria, fungi or yeast growth, as the general requirement for moisture content in crude drug is not more than 14 per cent. Equally important in the evaluation of crude drugs, is the ash value and acid insoluble ash value determination. The total ash is particularly important in the evaluation of purity of drugs, *i.e.* the presence of or absence of foreign inorganic matter such as metallic salts or silica. Thus it appears that the plant *P. cineraria* is useful in the traditional medicine.

Presence or absence of certain important compounds in an extract is determined by colour reactions of the compounds with specific chemicals (Table 20.2). Various tests have been conducted qualitatively to find out the presence or absence of bioactive compounds. Different chemical compounds such as alkaloids, tannin, saponin, steroids, flavonoids, leucoanthocyanins and phenols are detected in the leaves of *P. cineraria* which could made the plant useful for treating different ailments as having a potential of providing useful drugs of human use.

**Table 20.2**: Qualitative chemical analysis.

| Chemical Constituents | Observations |
|---|---|
| Acubins/Iridoids | — |
| Alkaloids | |
| a) Dragendorff's reagent | ++ |
| b) Mayers reagent | + |
| c) Wagner's reagent | + |
| Anthraquinone | — |
| Cardiac glycoside | — |
| Coumarins | — |
| Flavonoids | +++ |
| Leucoanthocyanins | ++ |
| Phlobatannin | — |
| Simple phenolics | +++ |
| Steriods | ++ |
| Saponins | ++ |
| Tannins | +++ |
| Terpenoid | — |

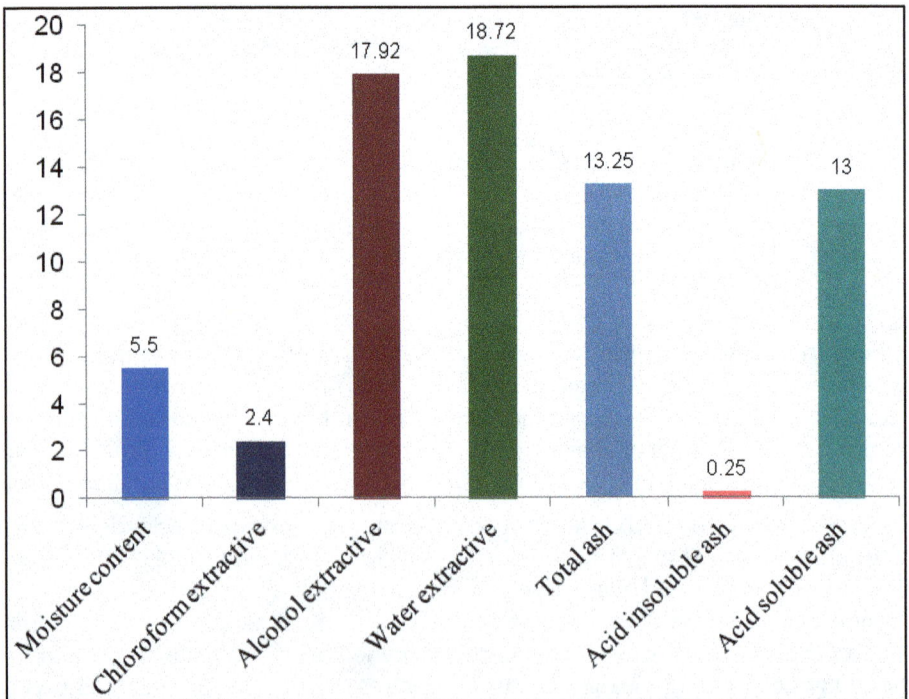

**Figure 20.3**: Physio-chemical parameters.

**Table 20.3**: Antibacterial efficacy of different solvent extracts of *Prosopis cineraria* leaf.

| Organisms | Gram +/- | Dose (mg/ml) | Petroleum Ether | Chloroform | Acetone | Methanol | DMSO | Ampicillin (40 mg/ml) |
|---|---|---|---|---|---|---|---|---|
| Staphylococcus aureus | + | A | 0 | 14 | 13 | 15 | 0 | 23 |
| Bacillus subtilis | + | A | 0 | 16 | 14 | 13 | 0 | 21 |
| Bacillus megaterium | + | A | 0 | 13 | 14 | 14 | 0 | 25 |
| Micrococeus luteus MTCC 106 | + | A | 0 | 17 | 15 | 12 | 0 | 30 |
| Escherichia coli | – | A | 0 | 15 | 13 | 11 | 0 | 17 |
| Salmonella typhi | – | A | 0 | 12 | 9 | 10 | 0 | 19 |
| Pseudomonas aeruginosa MTCC2488 | – | A | 0 | 10 | 10 | 13 | 0 | 16 |
| Klebsella planticola | – | A | 0 | 0 | 0 | 0 | 0 | 21 |

$: A-100 mg/ml (100 µl/well); Antibiotic-40mg/ml (100µl/well); 0 -no inhibition; Figures are diameter of zone of inhibition.

The results of antibacterial activity of *P. cineraria* against bacteria examined in the current study and their potency were qualitatively assessed by the presence or absence of inhibition zones and zone diameter. Different solvent extracts tested at 100μl concentration against eight important pathogenic bacteria are presented in Table 20.3. Among four solvent extracts *viz.* petroleum ether, chloroform, acetone and methanol tested against all the strains of bacteria; chloroform and acetone extracts recorded considerable antibacterial activity followed by methanol extracts against all the test pathogens. Antibacterial activity was not observed in petroleum ether extracts against all the strains of bacteria. Among the extracts tested, chloroform extracts showed broader spectrum of activity, being active to both Gram-positive and Gram-negative organisms compared to acetone and methanol, while petroleum ether showed negative inhibition. The chloroform extract at 100 mg/ml for example, 17 mm was recorded as diameter zone of inhibition against *M. luteus*. This was followed by 16 mm *B. subtilis*, 15mm *E. coli*, 14 mm *S. aureus*, 13 mm *B. megaterium*, 12 mm *S. typhi* and *P. aeruginosa* respectively. Whereas at the same concentration the acetone extracts exerted highest activity against *M. luteus* with diameter 15 mm followed by 14 mm *B.subtilis*, *B. megaterium*, 13 mm *E. coli*, *S.aureus* and 10 mm *P. aeruginosa*. The least activity 9 mm against *S. typhi* at 100 mg/ml was recorded. Activities of the various extracts were comparable to those of standard antibacterial agent ampicillin and DMSO as control. The differences in the observed activities of the various extracts may be due to varying degree of solubility of the active constituents in the four solvents used. It has been documented that different solvents have diverse solubility capacities for different phytochemical constituents (Marjorie, 1999).

Herbal drugs are derived from heterogeneous sources leading to variations. This makes the standardization of herbal medicines all the more important as erroneous results can cause variations in pharmacological and phytochemical studies. Pharmacognostical studies and phytochemical screening can serve as a valuable source of information and provide suitable standards to determine the quality of plant material in future investigations and applications. The pharmacognostic characters and phytochemical values reported in this paper could be used as a diagnostic tool for the standardization of this medicinal plant. Since *Prosopis cineraria* leaf is known for its various medicinal properties hence the present study may be useful to supplement information in respect to its identification, authentication and standardization, since no such data is available for the same.

## Acknowledgements

The authors wish to thank Head of Botany Department, Dr. Babasaheb Ambedkar Marathwada University, Aurangabad for providing the necessary laboratory facilities.

## References

Anonymous. (1966). Indian Pharmacopoeia. Govt. of India, Ministry of Health, Controller of Publications, New Delhi.

Anonymous. (1985). Indian Pharmacopoeia. Govt. of India, Ministry of Health, Controller of Publications, New Delhi.

Beyer, H., and Walter, W. (1997). Organic chemistry. Albion Publishing. Ltd, Chichester.

Chopra, R.N., Chopra, I.C., Handa, K.L. and Kapur, L.D. (1958). Indigenous Drugs of India. UN Dhur and Sons Pvt. Ltd, Calcutta.

Chopra, R.N., Nayar, S.L., and Chopra, I.C. (1956). Glossary of Indian Medicinal Plants. Council for Scientific and Industrial Research, New Delhi.

Dan, S.S., Mondal, N.R., and Dan, S. (1978). Phytochemical screening of some plants of Indian botanical garden. *Bulletin of Botanical Survey of India*, **20**: 117–23.

Duke, J.A. (1983) Handbook of Energy Crops. http://www.hort.purdue.edu/newcrop/duke energy/Prosopis cineraria.html.

Edeoga, H.O., Okwu, D.E., and Mbacbie, B.O. (2005). Phytochemical constituents of some Nigerian medicinal plants. *African Journal of Biotechnology*, **4**: 685–88.

Guerin, H.P., Delaveau, P.G., and Paris, R.R. (1971). Localizations of histochimiques II: Procedes simples de localization de pigments flavoniques. Applications on quelques phanerogames, *Bulletin de La Societe Botanique de France*, **118**: 29–36.

Harborne, J.B. (1984). Phytochemical Methods. Chapman Hall, London.

Jewers, K., Nagler, M.J., Zirvi, K.A., and Amir, F. (1976). Lipids, sterols, and a piperidine alkaloid from *Prosopis spicigera* leaves. *Phytochemistry*, **15**: 238-240.

Jewers, K., Nagler, M.J., Zirvi, K.A., Amir, F., and Cottee, F.H. (1974). Spicigerine, a new alkaloid from *Prosopis spicigera*. *Pahlavi Medical Journal*, **5**: 1-13.

Johansen, D.A. (1940). Plant Microtechnique. Tata McGraw Hill Publishing Company Ltd., New Delhi.

Karumi, Y., Oneyili, P.A., and Ogugbuaja, V.O. (2004). Identification of active principles of *M. balsamia* (Balsam apple) leaf extract. *Journal of Medical Sciences*, **4**: 179–82.

Kavanagh, F. (1972). Analytical microbiology. Academic Press, New York.

Kirtikar, K.R., and Basu, B.D. (1984). Indian Medicinal Plants. Lalit Mohan Publication, Allahabad.

Levin, F.A. (1929). The taxonomic value of the vein islet areas based upon a study of the genera *Berosma, Cassia, Erythroxylon* and *Digitalis*. *Journal of Pharmaacy and Pharmacology*, **2**: 17-43.

Malik, A., and Kalidhar, S.B. (2007). Phytochemical examination of *Prosopis cineraria* L. (Druce) Leaves. *Indian Journal of Pharmaceutical Sciences*, **69**: 576-578.

Marjorie, M.C. (1999). Plant products as antimicrobial agents. *Clinical Microbiology Review*. **12**: 564-582.

Nadkarni, K.M. (1976). Indian Materia Medica. Popular Prakashan, Bombay.

Naik, V.N. (1998). Flora of Marathwada Vol. I, Amrut Prakashan, Aurangabad [M.S.], India.

Rastogi, R.P., and Mehrotra, B.N. (1995). Compendium of Indian Medicinal plants: A CDRI Series. Publication and information Directorate, New Delhi.

Salisbury, E.J. (1927). On the causes of ecological significance of stomatal frequency with special reference to wood land flora. London. _Philosophical Transactions of the Royal Society_, **216:** 1-65.

Salisbury, E.J. (1932).The interpretation of soil climate and the use of stomatal frequency as an interesting index of water relation to the plant. _Beih Bot Zeni-ralb;_ **49:** 408-20.

Sivanarayan, V., and Suriyavathana, M. (2010). Preliminary studies on phytochemicals and antimicrobial activity of _Delonix elata_ and _Prosopis cineraria_. _International Journal of Current Research_, **8:** 066-069.

Smolenski, S.J., Silinis, H., and Farnsworth, N.R. (1972). Alkaloid screening I. _Lloydia_, **35:** 1–34.

Stellaa, R., Narayanan, N., Deattu, N., and Ravi Nargis N.R. (2010). Comparative anatomical features of _Prosopis cineraria_ (L.) Druce and _Prosopis juliflora_ (Sw.) DC (Mimosaceae). _International Journal of Green Pharmacy_, **4:** 275-280.

Tyler, V., Brady, L., and Robbers, J. (1977). Pharmacognosy. K. M. Varghese Company, India.

Vogel, A.I. (1988). Elementary practical organic chemistry. Orient Longman Limited, New Delhi.

Yoganarasimhan, S.N. (2000). Medicinal Plants of India, Tamil Nadu. Vedams Books (P) Ltd., Bangalore.

Utilisation and Management of Medicinal Plants Vol. 2 (2014)   *Pages* **415–419**
*Editor-in-Chief:* **V.K. Gupta**
*Published by:* **DAYA PUBLISHING HOUSE, NEW DELHI**

# 21

# Antibacterial Effect of Rhizobial Exo-Polysaccharide Against some Multidrug Resistance Human Pathogens

Rabindranath Bhattacharyya[1], Sandip Das[2],
Souryadeep Mukherjee[3] and Abhijit Dey[4]*

## ABSTRACT

*Polysaccharides are known for their biological efficacy which includes antimicrobial properties. The crude exo-polysaccharide extracted by phenol-sulfuric acid method from Rhizobium sp. isolated from the root nodule of a leguminous plant, Sesbania cannabina was investigated for its antibacterial properties against a number of non pathogenic bacteria such as Escherichia coli, Bacillus subtilis, Staphylococcus aureus, Micrococcus luteus, Pseudomonas aeruginosa and multidrug resistant human pathogenic bacteria such as Enterobacteriaceae bacterium IK1_01, Shigella dysentery, Enterobacter cloacae and Serratia marcescens. Antibiotic susceptibility tests were carried out to investigate the multi-drug resistance patterns of the strains all of which have exhibited multi-drug resistance against commercial antibiotics. Minimum Inhibitory Concentration (MIC) was measured accordingly. The crude polysaccharide showed antibacterial activity against some of the strains assayed by agar-*

1   Associate Professor, Microbiology Laboratory, Department of Botany, Presidency University, 86/1, College Street, Kolkata – 700 073, West Bengal, India.

2   Research Scholar, Department of Botany, Presidency University, 86/1, College Street, Kolkata – 700 073, West Bengal, India.

3   Assistant Professor, Department of Zoology and Molecular Biology and Genetics, Presidency University, 86/1, College Street, Kolkata – 700 073, West Bengal, India.

4   Assistant Professor, Department of Botany, Presidency University, 86/1, College Street, Kolkata – 700 073, West Bengal, India.

*   *Corresponding author*: E-mail: abhijitbio25@yahoo.com

*diffusion method. The authors suggest a possible exploitation of microbe-derived novel products against the pathogenic microorganisms. Further purification of the crude polysaccharide may lead to the discovery of pure compounds with pharmacological potentials.*

*Keywords:* Exo-polysaccharide, Multidrug resistance (MDR), Minimum inhibitory concentration (MIC).

## Introduction

Various natural products are known to possess pharmacological efficacy such as antimicrobial activity (von Nussbaum *et al.*, 2006) which is manifested in terms of antibacterial, antifungal (Han *et al.*, 2007) and antiplasmodial (Laurent and Pietra, 2006) properties. Antimicrobial properties of such preparations are shown either by the crude product or by isolated biomolecules. Interestingly, antibacterial activity has been reported from the active compounds and crude extracts from certain bacteria (Ripa *et al.*, 2010). These findings may open up an exciting aspects of bacterial derived product as biocontrolling agent. In the context of development of resistance while using synthetic drugs and rapidly increasing tendency of using natural products as therapeutic agents, discovery of new natural products with pharmacological activities represent one of the major research trends these days.

In the present investigation, the authors have tried to find out antibacterial activity, if any, of the crude exo-polysaccharide extracted by phenol-sulfuric acid method from *Rhizobium* sp. isolated from the root nodule of a leguminous plant, *Sesbania cannabina*. This exo-polysaccharide was investigated for its antibacterial properties against a number of non pathogenic bacteria such as *Escherichia coli, Bacillus subtilis, Staphylococcus aureus, Micrococcus luteus, Pseudomonas aeruginosa* and multidrug resistance human pathogenic bacteria such as *Enterobacteriaceae bacterium IK1_01, Shigella dysentery, Enterobacter cloacae* and *Serratia marcescens*.

## Materials and Methods

### Material

Fresh healthy and matured pink coloured root nodules were selected for isolation of *Rhizobium* bacteria. After selecting the root carefully, these were washed with distilled water and surface sterilized with 0.1 per cent $HgCl_2$ and 70 per cent ethanol for 5 minutes and 3 minute respectively. The nodules were washed 3 times with sterile distilled water and crushed between 2 sterile glass slide and the fluid coming out of the crushed nodules were streaked aseptically on Yeast Extract Mannitol (YEM) agar plate. Dilution plating was followed for making individual colonies and these were selected and designated as s1-s15. Bacterial colonies were white, opaque and gummy. Purified culture were kept in YEM agar slants and maintained by subculturing once in a month. After proper growth the slants were stored at 5°C-10°C in a refrigerator.

## Bacterial Strains

Five non pathogenic bacterial strains such as *Escherichia coli, Bacillus subtilis, Staphylococcus aureus, Pseudomonas aeruginosa* and *Micrococcus luteus* were used in the experimentation. Another four human pathogenic bacteria *viz. Enterobacteriaceae bacterium IK1_01, Shigella dysentery, Enterobacter cloacae* and *Serratia marcescens* were used in the study. All were cultured on nutrient agar (Himedia, India). Grown bacteria were taken out from the medium, and kept in nutrient agar stab at room temperature and also in nutrient broth with 10 per cent glycerol (Merck, Germany) at -20°C for further testing.

## EPS Extraction

The 24 hrs culture of the strains was centrifuged at 10,000 rpm and 3 volumes of acetone were added to the cell free supernatant to precipitate polysaccharides. The polysaccharides were collected by centrifugation, dissolved in minimum volumes of distilled water and re-precipitated by 3 volumes of acetone, centrifuged and re-suspended in distilled water and used for EPS production.

## Colorimetric Estimation of EPS

The above cell supernatant was used for the EPS determination by phenol-sulfuric acid method (Dubois *et al.*, 1956), *i.e.* 1ml of EPS solution was taken and 1 ml of 5 per cent aqueous solution of phenol was added to it in a test-tube. Then 5 ml of concentrated sulfuric acid was added to it and shaken well. It was kept for 10 minutes at room temperature. After that it was kept in water bath set at 25°C -30°C for 20-25 minutes. The mixture develops brown colour. The brown colour thus developed was read at 540 nm. The amount of EPS was calculated, after spectrophotometric estimation, from a standard curve prepared by glucose.

## Antimicrobial Susceptibility Testing

Agar-diffusion assay depicted by Reeves, 1989 was used to determine the antibacterial activity of the extract. Bacteria were grown in Mueller Hinton Broth (MHB) under shaking condition for 4 h at 37 °C and after incubating, 1 ml of culture was spread on Mueller Hinton Agar (MHA) plate. Wells were made by using sterile 6mm cork borer in the inoculated MHA plate. The wells were filled with 200 µl of the crude exo-polysaccharide extract. The concentrations of extract employed were 1, 10, 25, 50, 100 and 250 mg/ml. Tetracycline (150 µg/ml, 200µl) was used as positive control. Zone diameter was measured after 24 h incubation at 37 °C.

## Determination of Minimum Inhibitory Concentration (MIC)

For the determination of MIC of EPS against *E. coli*, a loopful of bacteria from a 24 hours old slant was transferred to 20ml nutrient broth incubated on rotary shaker (150 rpm) for 6 hours at 37 ± 1°C. A dilution series of EPS were prepared aseptically in nutrient broth tubes. The dilutions were 0, 10, 25, 50, 100 and 250 µg/ml. These tubes were inoculated with 0.1 ml of the 6 hours old liquid culture. The final volume of each tube was adjusted to 10 ml with the medium. These were incubated at 37 ± 1°C for 24 hours and optical density (OD) was measured at 650 nm.

# Results and Discussion

## *In-vitro* Antibacterial Activity of Crude Rhizobial Exo-Polysaccharide Extract

Strains of nine bacterial cells, *Escherichia coli, Bacillus subtilis, Staphylococcus aureus, Micrococcus luteus, Pseudomonas aeruginosa* and multidrug resistance human pathogenic bacteria such as *Enterobacteriaceae bacterium IK1_01, Shigella dysentery, Enterobacter cloacae* and *Serratia marcescens* were tested to evaluate the antibacterial activity of exo-polysaccharide extracted by phenol-sulfuric acid method from *Rhizobium* sp. isolated from the root nodule of a leguminous plant, *Sesbania cannabina*. However, the crude Rhizobial exo-polysaccharide extract had shown insignificant inhibitory activities against most of the bacterial strains except *E. coli*. Table 21.1 represents the antimicrobial activity of rhizobial polysaccharide. Table 21.2 depicts the determination of minimum inhibitory concentration (MIC) of EPS against *E. coli*.

**Table 21.1**: Antimicrobial activity of rhizobial polysaccharide.

| Concen-tration (µg/ml) | Diameter of Inhibition Zone (mm) | | | | | | | | |
|---|---|---|---|---|---|---|---|---|---|
| | *E. coli* | *B. subtilis* | *M. luteus* | *S. aureus* | *P. aeru-ginosa* | *E. bact-erium* | *S. dyse-ntery* | *E. cloacae* | *S. marce-scens* |
| 1 | – | – | – | – | – | – | – | – | – |
| 10 | – | – | – | – | – | – | – | – | – |
| 50 | 1.5 | – | – | – | – | – | – | – | – |
| 100 | 2.5 | – | – | – | 1 | – | 0.5 | – | – |
| 250 | 2.5 | – | – | 0.5 | 1 | – | 1.0 | – | 0.5 |

**Table 21.2**: Determination of Minimum Inhibitory Concentration (MIC) of EPS against *E. coli*.

| Concentration of EPS (µg/ml) | O.D. at 650 nm |
|---|---|
| 0 | 0.35 |
| 1 | 0.30 |
| 10 | 0.15 |
| 50 | 0 |
| 100 | 0 |
| 250 | 0 |

In the present investigation, the crude Rhizobial exo-polysaccharide extract has shown insignificant antibacterial activity against *Bacillus subtilis, Staphylococcus aureus, Micrococcus luteus, Enterobacteriaceae bacterium IK1_01, Shigella dysentery, Enterobacter cloacae* and *Serratia marcescens*. Only *E. coli* was found to be significantly inhibited by the extract. From the above observation, it might be speculated that inability of the extract to inhibit most of the bacterial strains may be attributed to its crude form. Further purification of the crude exo-polysaccharide may yield some active principles with enhanced pharmacological properties. Antimicrobial activity of crude extracts and purified compounds has been assayed (Rojas *et al.*, 1992; Yasunaka *et al.*, 2005) and enhanced or novel activities of the purified forms have been detected.

## Conclusion

In the modern era of multi-drug resistance pathogenesis and the ever changing pathogenecity of the microbes, the present venture was an attempt to explore the bio-controlling ability of a microbe-derived product against some of the potent human pathogenic microbes. Although, the results did not show significant antimicrobial properties, further investigation could lead to purification of natural compounds with enhanced biological activity.

## References

Dubois, M., Gilles, K.A., Hamilton, J.K., Rebers, P.A., and Smith, F. (1956). Colorimetric method for determination of sugars and related substances. *Anal Chem.*, **28(3):** 350-356.

Han, L., Zheng, D., Huang, X.S., Yu, S.S., and Liang, X.T. (2007). [Natural products in clinical trials: antibacterial and antifungal agents]. [Article in Chinese], *Yao Xue Xue Bao.*, **42(3):** 236-244.

Laurent, D., and Pietra, F. (2006). Antiplasmodial marine natural products in the perspective of current chemotherapy and prevention of malaria: a review. *Mar Biotechnol* (NY)., **8(5):** 433-447.

Reeves, D.S. (1989). Antibiotic assays. In: Medical Bacteriology: A Practical Approach. Hawkey, P.M. and D.A. Lewis, (Eds.). IRL Press, Oxford, pp. 195-221.

Ripa, F.A., Nikkon, F., Rahman, B.M., and Khondkar, P. (2010). *In vitro* antibacterial activity of bioactive metabolite and crude extract from a new *Streptomyces* sp. *Streptomyces rajshahiensis. Int J PharmTech Res.*, **2(1):** 644-648.

Rojas, A., Hernandez, L., Pereda-Miranda, R., and Mata, R. (1992). Screening for antimicrobial activity of crude drug extracts and pure natural products from Mexican medicinal plants. *J Ethnopharmacol.*, **35(3):** 275-283.

von Nussbaum, F., Brands, M., Hinzen, B., Weigand, S., and Häbich, D. (2006). Antibacterial natural products in medicinal chemistry—exodus or revival? *Angew Chem Int Ed Engl.*, **45(31):** 5072-5129.

Yasunaka, K,. Abe, F., Nagayama, A., Okabe, H,. Lozada-Pérez, L., López-Villafranco, E., Mu iz, E.E., Aguilar, A., and Reyes-Chilpa, R. (2005). Antibacterial activity of crude extracts from Mexican medicinal plants and purified coumarins and xanthones. *J Ethnopharmacol.*, **97(2):** 293-299.

Utilisation and Management of Medicinal Plants Vol. 2 (2014)    *Pages* **421–429**
*Editor-in-Chief:* **V.K. Gupta**
*Published by:* **DAYA PUBLISHING HOUSE, NEW DELHI**

# 22

# Bioassay-Guided Fractionation for the Isolation of the Antibacterial Compounds from *Loranthus* Species of the Mistletoe Plant

S. E. Ukwueze[1]* and P.O. Osadebe[2]

## ABSTRACT

*The present study was aimed at the isolation and characterization of antibacterial compounds contained in solvent extracts of leaf of the African mistletoe species, Loranthus micranthus, using biological assay as a guide. The crude methanol fraction (CMFM) of powdered mistletoe leaves harvested from Persea americana (host-tree) was tested against some standard bacteria using the agar-well diffusion method. The phytochemical and bioassay results obtained were utilized to carry out further fractionation of the crude extracts. These fractions (MFM, EFM, AFM and CFM) were also tested against the bacteria. Results indicate that CMFM and MFM (defatted methanol fraction) had appreciable activity against the microorganisms. When compared with a standard antibiotic {Ceftriaxone}, however, these activities were found to be statistically non-significant {P > 0.05}. Antibacterial screening of the solvent fractions showed a significant {P<0.001} loss of activity against all the test organisms in comparison to that of CMFM. When the fractions were compared with one another AFM (acetone-soluble fraction) showed the best antibacterial activity, but, its column fractionation yielded sub-fractions with no observable antibacterial activity. All the extracts/ fractions that showed antibacterial activity had higher zones of inhibition against B. subtilis and Pseudomonas than against Staph and E. coli. Phytochemical results equally followed the*

1   Department of Pharmaceutical and Medicinal Chemistry, Faculty of Pharmaceutical Sciences, University of Port Harcourt, Port Harcourt, Nigeria

2   Department of Pharmaceutical and Medicinal Chemistry, Faculty of Pharmaceutical Sciences, University of Nigeria, Nsukka, Nigeria.

*   *Corresponding author*: E-mail: stanley.ukwueze@uniport.edu.ng

same trend in their distribution across the crude and the solvent extracts/fractions. CMFM (and to an extent MFM) showed a preponderance of such constituents like alkaloids, flavonoids, terpenoids and tannins when compared to the other fractions. Also, AFM showed the presence of tannins, alkaloids and terpenoids which were only randomly distributed in the other fractions. These experimental data show that the observed antibacterial activity in mistletoe might be as a result of some interactions among these plant constituents rather than that of any one in isolation. We therefore recommend the use of the crude aqueous or alcoholic extract of mistletoe leaves in the management of non-complicated community acquired bacterial infections.

*Keywords:* Antibacterial, Bioassay-guided, Ceftriaxone, Fractionation, *Loranthus mistletoe*.

# Introduction

The mistletoe plant is one of the oldest known herbs with vast folkloric usage as a medicinal plant in many countries and regions of the world. It is an obligate parasite that depends partly on its host to obtain water and minerals but can carry out photosynthesis (Griggs, 1991). The mistletoe is an evergreen semi-parasite that can grow in most parts of the globe and has different families and species that are well known worldwide. The most common ones are: the European mistletoe (*Viscum album* L.); American mistletoe (*Phoradendron flavescens*); Australian/Argentine mistletoe (*Ligaria cuneifolia* R and T), African mistletoe, etc.

*Loranthus micranthus* L. is the Eastern Nigerian specie of the commonly known African Mistletoe from the family Loranthaceae. This specie has been used traditionally by the people in that region as an anti-cancer, anti-diabetic, anti-hypertensive, and indeed as an 'all-purpose herb' (Kafaru, 1993). Many of these folkloric uses have already been investigated (Obatomi *et al.*, 1996; Osadebe and Ukwueze, 2004; Ukwueze, 2008; Osadebe and Omeje, 2009; Obatomi *et al.*, 1994). Studies have, however, suggested that several factors play important roles in the phytochemical composition and pharmacological activities of the mistletoe plant. Such factors include: the host, specie of mistletoe used, season of harvest, etc. (Osadebe *et al.*, 2004; Osadebe and Ukwueze, 2004; Osadebe *et al.*, 2008; Schink and Bussing, 1997; Wagner *et al.*, 1996)

Several works have been carried out in the past to verify the folkloric use of the African mistletoe in the management of microbial infections. Earlier studies by the authors on the crude plant powder and some of its solvent fractions have established some significant antibacterial properties by *L. micranthus*, though with negligible anti-fungal activity (Osadebe and Ukwueze, 2004; Osadebe *et al.*, 2008; Ukwueze and Osadebe, 2012). The optimal harvesting season as well the host tree of choice for mistletoe as an antimicrobial agent has equally been documented (Osadebe and Ukwueze, 2004; Osadebe *et al.*, 2008).

Hence, the present work is an effort geared towards the bioassay-guided fractionation of the leaf extracts of *Loranthus micranthus* with a view to isolating and characterizing the constituents responsible for the antibacterial properties observed in the plant.

# Materials and Methods

## Plant Materials

The leaves of *Loranthus micranthus L.* were collected in mid June at Nsukka, South-Eastern Nigeria, from the stem of *Persea americana* and authenticated by Mr. A. Ozioko, a taxonomist with the Bio-resources Development and Conservation Project (BDCP), Nsukka, Nigeria. Voucher specimen was deposited in the herbarium of the Faculty of Pharmaceutical Sciences, University of Nigeria, Nsukka. The leaves were air-dried at room temperature to a constant weight, pulverized and passed through a 1mm sieve. The powdered material was stored in an air-tight container and kept in a refrigerator.

## Microorganisms Used

The studies were performed with *Staphylococus aureus (ATTC 25923)*, *Pseudomonas aeruginosa (ATTC 27833)* and *Escherichia coli (ATTC 35219)* obtained from the Nigerian Institute of Medical Research (NIMR),Yaba, Nigeria. The clinical isolate of *Bacillus subtilis* was used. All the microorganisms were grown in nutrient broth (*Biotec, Suffolk, UK*) at 37°C and maintained on nutrient agar (*Biotec, Suffolk, UK*) slants at 4°C. The standardized cultures of the organisms were used throughout the experiment.

## Solvents and Reagents

*N*-Hexane (*Sigma-Aldrich*, South Africa), Chloroform (*Sigma-Aldrich*, South Africa), Ethyl acetate (*BDH*, England), Methanol (*Sigma-Aldrich*, South Africa), Acetone (*BDH*, England), DMSO (*Sigma-Aldrich*, USA), Acetic acid (*Hopkins and Williams*, England), Cutter mill (*Manesty*, England), Electronic Balance (*Sartorius*, Germany) and Weighing balance (*Ohans*, USA). Other reagents used were freshly prepared in the laboratory according to official specifications.

## Extraction and Fractionation of Plant Materials

A portion of the plant material (800g) was defatted with n-hexane (to yield HFM extract) and the dried Marc extracted in a soxhlet extractor with absolute methanol. The methanol soluble extract (MFM) was further fractionated by treating it in succession with chloroform, ethyl acetate and acetone to yield the chloroform (CFM), ethyl acetate (EFM) and acetone (AFM) soluble fractions. Fractions with the best antibacterial activity were subjected to column chromatographic separation with a view to isolating the active constituents. Another portion (300 mg) of the crude powder from the plant was macerated in 90 per cent methanol for 48 hours and the filtrate concentrated *in vacuo* to yield the crude methanol extract, CMFM.

## Antibacterial Activity Testing

Each of the fractions was screened for antibacterial activity using the microorganisms provided. The inhibition zone diameters (IZDs) of the extracts and reference antibiotics (Ceftriaxone – *MAY and BAKER PLC*, Nigeria) were determined by agar-well diffusion method (Lovian, 1980). 0.1 ml of the standardized broth culture of the appropriate microorganism was introduced into a sterile Petridish and 20 ml of molten agar added. The content was mixed thoroughly and allowed to solidify. Four 2-fold serial dilutions of the extracts (including the fractions/sub-fractions) and reference antibiotics were obtained from their stock solutions (20 mg/ml in DMSO). Four quadrants were marked on each petri-dish and a cup (6 mm) was bored on each quadrant using a sterile cork borer. Two drops of each dilution were placed in each cup using Pasteur pipettes, allowed to diffuse for about an hour and then incubated at 37°C. This procedure was repeated for each of the microorganisms. The IZDs were recorded after 24 hours of incubation.

## Phytochemical Tests

Phytochemical tests were carried out to detect the presence of steroids, alkaloids, tannins, glycosides, reducing sugars, flavonoids and saponins. These were carried out according to the procedures outlined by Harbourne (1998).

## Statistical Analysis

All the data obtained were analyzed by GraphPad Prism® (Model 5) using two-way ANOVA and subjected to Bonferroni post-tests to compare replicate means. The statistical results were presented as mean ± SEM. Differences between means were considered significant at $P<0.05$.

# Results and Discussion

## Antibacterial Activity Results

The results from the antibacterial screening of the extracts and solvent fractions of the leaves of mistletoe plant showed that they all had some level of activity against the test bacteria (Table 22.1). When compared to the control, however, all the test materials showed statistically lower values (up to $P<0.001$) suggestive of a generally weak antibacterial action. Among the extracts/fractions, there was a conspicuous and progressive loss of antibacterial activity with the initial methanol extracts (CMFM and MFM) having the highest activities while the other solvent fractions showed diverse but significantly lower values (Figure 22.1). For the fractions, that of acetone exhibited the highest level of antibacterial activity. The activity trend is therefore CMFM > MFM > AFM > EFM > CFM > HFM.

**Table 22.1**: Table of Mean IZD (mm) +/- SEM for *L. micranthus* extracts/fractions.

| Extracts{20mg/ml} | Staph | Bacillus | Pseudo | E. coli |
|---|---|---|---|---|
| CMFM | 16.00 ± 0.58*** | 18.00 ± 0.58 | 18.00 ± 0.58*** | 10.00 ± 1.16*** |
| HFM | 1.98 ± 0.88*** | 3.82 ± 1.20*** | 2.10 ± 0.33*** | 0.00 ± 0.00*** |
| MFM | 9.67 ± 1.20*** | 13.00 ± 2.31*** | 10.00 ± 2.08*** | 5.67 ± 1.20*** |
| EFM | 4.67 ± 1.20*** | 6.67 ± 1.20*** | 6.67 ± 0.33*** | 6.33 ± 0.88*** |
| AFM | 7.33 ± 1.20*** | 9.67 ± 1.20*** | 7.67 ± 0.88*** | 5.00 ± 1.15*** |
| CFM | 2.67 ± 0.33*** | 5.67 ± 0.67*** | 3.00 ± 0.58*** | 1.33 ± 0.33*** |
| Control (10mg/ml) | 26.00 ± 0.58 | 20.00 ± 0.58 | 33.67 ± 2.03 | 27.00 ± 1.15 |

CMFM: Crude Methanolic Extract; HFM: Hexane extract; MFM: Defatted Methanol extract; EFM: Ethyl acetate soluble fraction; AFM: Acetone soluble fraction; CFM: Chloroform soluble fraction; Control: Ceftriaxone.

*: $P<0.05$; **: $P<0.01$ and ***: $P<0.001$ significantly lower when compared with Control.

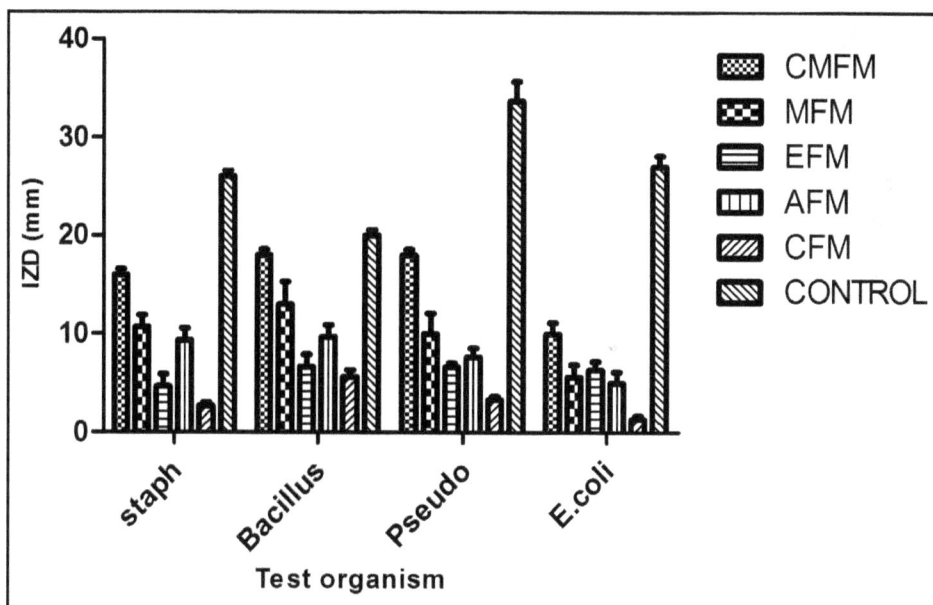

**Figure 22.1**: Graph of Mean IZD +/- SEM of the solvent fractions of *Loranthus micranthus* against bacteria.

The antibacterial activity results equally showed some significant variations in the level of activity exhibited by the test materials against the organisms screened. All the extracts/fractions displayed the highest activity against *bacillus* with little or no activity against *E. coli*. The activity of the extracts against *pseudomonas* and *staphylococcus*, though intermediate, is of great interest here bearing in mind the level of resistance displayed by these organisms towards conventional antibiotics.

## Column Chromatography of AFM

Further fractionation of the acetone soluble fraction (AFM) yielded sub-fractions with no observable zone of inhibition (Figure 22.2). This implies a total loss of antibacterial activity at this stage leading to the suspension of the isolation process, having followed, since the start of the experiment, the outcome of the antibacterial assay as the only guide for the procedure.

## Phytochemical Screening Results

The phytochemical screening results followed the same trend as above with CMFM and MFM showing the presence of most of the constituents screened (Table 22.2). CFM and HFM only showed traces of terpenoids with an abundance of phenolic compounds. AFM with a slightly higher activity than the other fractions tested positive for such constituents like tannins, alkaloids, terpenoids and saponins, while the presence of tannins and flavonoids in EFM might have contributed to its improved antibacterial activity over CFM and HFM. These deductions are in tandem with several documented evidence about the medicinal potentials of these plant secondary metabolites. Many authors have demonstrated the antibacterial activity of these phytochemicals (Hufford *et al.*, 1974; Karou *et al.*, 2006; Takhi *et al.*, 2011). Also, many

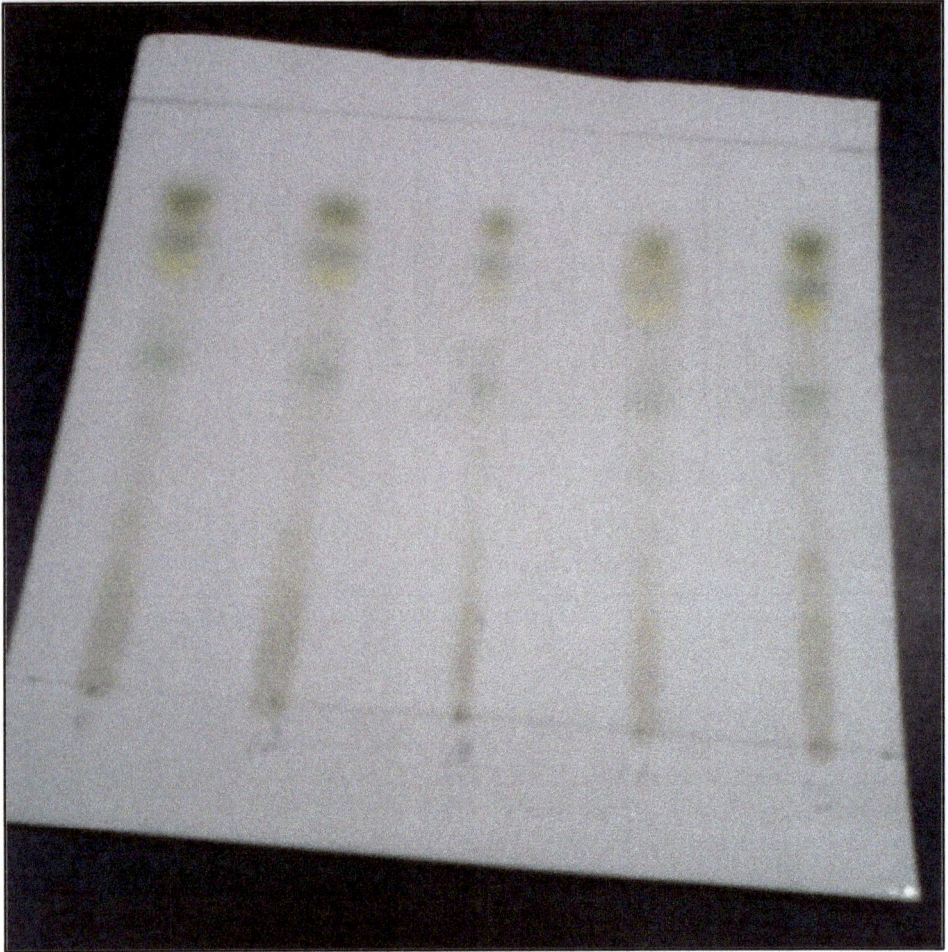

**Figure 22.2**: Analytical Thin-Layer Chromatogram of some of the sub-fractions obtained from *AFM*.

**Table 22.2**: Results of phytochemical tests on the solvent fractions of *Loranthus micranthus* leaves.

| Fractions | Tannins | Flavonoids | Alkaloids | Reducing Sugars | Terpenoids | Saponins | Glycosides |
|-----------|---------|------------|-----------|-----------------|------------|----------|------------|
| CMFM | + | ++ | ++ | + | + | + | + |
| HFM | – | – | – | – | + | – | – |
| MFM | + | + | + | – | + | + | + |
| EFM | + | ++ | – | – | – | – | – |
| AFM | + | – | + | – | + | + | – |
| CFM | – | – | – | – | + | – | – |

researches devoted to substances extracted from plants have established that such metabolites like essential oils, terpenoids, flavonoids, alkaloids, etc significantly inhibit the growth of bacteria (*Staphylococcus aureus, Klebsiella pneumoniae, Enterococcus fecalis, Escherichia coli, Staphylococcus epidermidis*,. etc) and fungi (Hammer *et al.*, 1999; Takhi *et al.*, 2011; Amaral *et al.*, 1998; Barrey *et al.*, 1997; Ahmed *et al.*, 1993; Habtemariam *et al.*, 1993). Indeed, metabolites like flavonoids and terpenoids are known to be synthesized by plants in response to microbial infection, and thus have been found *in vitro* to be effective antimicrobial substances against a wide array of microorganisms (Dixon *et al.*, 1983; Cowan, 1999; Himejima *et al.*, 1992).

An analysis of these IZD and phytochemical results therefore suggest that the antibacterial activity observed in Loranthus might have arisen as a result of a number of the phytoconstituents present in the plant. This followed from the above results which clearly showed that no singular fraction/constituent could be said to be solely responsible for the antibacterial action of the plant. Among these constituents, however, tannins, flavonoids, terpenoids and alkaloids appear to have the greatest impact on the activity under review (Osadebe and Ukwueze, 2004; Osadebe *et al.*, 2008; Ukwueze, 2008). Though the complete isolation of the implicated compounds were not concluded under this investigation due to a very significant loss in bioactivity, we can confidently opine that the observed antibacterial activity in *Loranthus* is as result of synergy among such constituents like tannins, flavonoids, terpenoids, alkaloids and/or saponins. According to reports, the main reasons behind apparent loss or total absence of any biological activity include the amount at which a particular potential metabolite is produced, synergistic or antagonistic relationship among the molecules when the crude extracts are tested for the biological activities; and, along with these main reasons, the geographical location of plants, seasonal variation in particular constituents, difference in accumulation of secondary metabolites in different parts of the same plant or time of collection are the variables that can significantly alter the yield or efficacy of a particular metabolite from a potential plant specie (Khurram, 2010; Ncube *et al.*, 2008; Ukwueze and Osadebe, 2004; Wagner *et al.*, 1996).

It follows therefore, using bioassay as a guide, that the utilization, in isolation, of any of the constituents present in *Loranthus* for the management of microbial infections might not only lead to treatment failure but may also result in the development of resistance among the pathogenic organisms being treated. We therefore recommend the use of the crude aqueous or alcoholic extract of mistletoe leaves in the management of non-complicated community acquired bacterial infections.

## Acknowledgements

The authors wish to acknowledge the financial support by the John D. and Catherine T. MacArthur Foundation towards the studies through which this research work was accomplished.

## References

Ahmed, A. A., Mahmoud, A. A., Williams, H. J., Scott, A. I., Reibenspies, J. H., and Mabry, T. J. (1993). New sesquiterpene α-methylene lactones from the Egyptian plant *Jasonia candicans*. *J. Nat. Prod.*, **56**: 1276-1280.

Amaral, J. A., Ekins, A., Richards, S. R., and Knowles, R. (1998). Effect of selected monoterpenes on methane oxidation, denitrification, and aerobic metabolism by bacteria in pure culture. *Appl. Environ. Microbiol.*, **64**: 520-525.

Barre, J. T., Bowden, B. F., Coll, J. C., Jesus, J., Fuente, V. E., Janairo, G. C., and Ragasa, C. Y. (1997). A bioactive triterpene from *Lantana camara*. *Phytochemistry*, **45**: 321-324.

Cowan, M. M. (1999). Plant products as antimicrobial agents. *Clinical Microbiology Reviews*, **12** (4): 564-582

Dixon, R. A., Dey, P. M., and C. J. Lamb. (1983). Phytoalexins: enzymology and molecular biology. *Adv. Enzymol.*, **55**: 1-69.

Griggs, P. (1991) Mistletoe, myth, magic and medicine. *The Biochemist* **13**: 3-4.

Habtemariam, S., Gray, A.I., and Waterman, P.G. (1993). A new antibacterial sesquiterpene from *Premna oligotricha*. *J. Nat. Prod.*, **56**: 140-143.

Hammer, K.A., Carson, C.F., and Riley, T.V. (1999). Antimicrobial activity of essential oils and other plant extracts. *Journal of Applied Microbiology*, **86**: 985-990.

Harborne, J.B., 1998. Phytochemical Methods: A Guide to Modern Techniques of Plant Analysis, 3rd ed. Chapman and Hall, London.

Himejima, M., Hobson, K.R., Otsuka, T., Wood, D.L., and Kubo, I. (1992). Antimicrobial terpenes from oleoresin of ponderosapine tree *Pinus ponderosa*: a defense mechanism against microbial invasion. *J. Chem. Ecol.*, **18**: 1809-1818.

Hufford, C.D., Funderburk, M.J., Morgan, J. M., and Robertson, L.W. (1974). Two antimicrobial alkaloids from heartwood of *Liriodendron tulipifera* L. *Journal of Pharmaceutical Sciences*, **64** (5): 789-792.

Kafaru, E. (1993) Herbal remedies, *The Guardian*, Thursday June 3, 1993, p. 24.

Karou, D., Savadogo, A., Canini, A., Yameogo, S., Montesano, C., Simpore, J., Colizzi, V., and Traore, A.S. (2006). Antibacterial activity of alkaloids from *Sida acuta*. *African Journal of Biotechnology*, **5** (2): 195-200.

Khurram, M. (2010). Studies on the isolation and characterization of secondary metabolites from *Dodonaea viscosa* and *Quercus baloot* and their potentials as antibacterial agents. Retrieved from prr.hec.gov.pk/Thesis/533S.pdf. (Assessed on 22/05/2012).

Ncube, N.S., Afolayan, A.J., and Okoh, A.I. (2008). Assessment techniques of antimicrobial properties of natural compounds of plant origin: current methods and future trends. *African Journal of Biotechnology*, **7** (12): 1797-1806.

Obatomi, D.K., Aina, V.O., and Temple, V.J. (1996). Effect of African Mistletoe on blood pressure in spontaneously hypertensive rats. *International Journal of pharmacognosy*, **34** (2): 124-127.

Obatomi, D.K., Bikomo, E.O., and Temple, V.J. (1994) Anti-diabetic properties of African Mistletoe in streptozotocin-induced diabetic rats. *Journal of ethnopharmacology*, **43**: 13-17.7.

Osadebe, P.O., and Omeje, E.O. (2009). Main immunomodulatory constituents of Eastern Nigerian mistletoe, *Loranthus micranthus* Linn. *Asian Pacific Journal of Tropical Medicine,* **2**(4): 11-18.

Osadebe, P.O., and Ukwueze, S.E. (2004). A comparative study of the phytochemical and antimicrobial properties of the Eastern Nigeria species of African Mistletoe (*Loranthus micranthus*) sourced from different host trees. *Journal of Biological Research and Biotechnology,* **2**(1): 18-23.

Osadebe, P.O., Dieke C.A., and Okoye, F.B.C. (2008). A study of the seasonal variation in the antimicrobial constituents of the leaves of the leaves of *Loranthus micranthus* sourced from *Persea americana. Research Journal of Medicinal Plant,* **2** (1): 48-52.

Osadebe, P.O., Okide, G.B., and Akabogu I.C. (2004). Study on anti-diabetic activities of crude methanolic extracts of *Loranthus micranthus* (Linn.) sourced from five different host trees. *J. Ethnopharmacol.,* **95**: 133-138.

Schink, M., and Bussing A. (1997). Mistletoe therapy for human cancer: the role of the natural killer cells. *Anti-cancer drugs,* **8** (1): 47-51.

Ukwueze, S.E. (2008). An evaluation of the antibacterial activities of leaf extracts from *Loranthus micranthus* L, Parasitic on *Persea americana. Scientia Africana,* **7**(1): 51-55.

Ukwueze, S.E., and Osadebe P.O. (2012). Determination of anti-fungal properties of the African mistletoe species: *Loranthus* micranthus L. *International Journal of Pharma and Bio Sciences,* **3** (1): 454-458.

Wagner, M.L., Teresa, F., Elida, A., and Rafeal, A.G. (1996). Micromolecular and macromolecular comparison of Argentina Mistletoe (*Ligaria Cuneifolia*). *Acta Farmacentica Babaerense,* **15** (2): 99 – 105.

Utilisation and Management of Medicinal Plants Vol. 2 (2014)    *Pages* **431–445**
*Editor-in-Chief:* **V.K. Gupta**
*Published by:* **DAYA PUBLISHING HOUSE, NEW DELHI**

# 23

# Processing of *Aloe vera* Leaf

## C.T. Ramchandra[1]* and P. Srinivasa Rao[2]

## ABSTRACT

*Proper scientific investigations on Aloe vera have gained more attention over the last decade due to its reputable, medicinal, pharmaceutical and food properties. Some publications have appeared in reputable scientific journals that have made appreciable contributions to the discovery of the functions and utilizations of Aloe vera lacking processing of leaf gel. Present processing techniques aims at producing best quality aloe products but end aloe products contain very little or virtually no active ingredients. Hence, appropriate processing techniques should be employed during processing in order to extend the use of aloe vera gel. Further research needs to be done to unravel the myth surrounding the biological activity and the exploitation of aloe constituents.*

***Keywords:*** Cold process, Qmatrix process, Whole leaf process, Desiccant air dehydration, Time temperature, Sanitation process.

## Introduction

*Aloe vera* (*Aloe barbadensis* Miller) is a perennial plant of liliacea family with turgid green leaves joined at the stem in a rosette pattern. *Aloe vera* leaves are formed by a thick epidermis (skin) covered with cuticle surrounding the mesophyll, which can be differentiated into chlorenchyma cells and thinner walled cells forming the parenchyma (fillet). The parenchyma cells contain a transparent mucilaginous jelly

1   Assistant Professor, Department of Processing and Food Engineering, College of Agricultural Engineering, University of Agricultural Sciences, Raichur – 584 102, Karnataka, India

2   Associate Professor, Agricultural and Food Engineering Department, Indian Institute of Technology, Kharagpur, India

*   *Corresponding author*: E-mail: ramachandract@gmail.com

which is referred to as *Aloe vera* gel. Potential use of aloe products often involves some type of processing, *e.g.* heating, dehydration and grinding. Processing may cause irreversible modifications to the polysaccharides, affecting their original structure which may promote important changes in the proposed physiological and pharmaceutical properties of these constituents.

Processing of *Aloe vera* gel derived from the leaf pulp of the plant, has become a big industry worldwide due to the application in the food industry. It has been utilized as a resource of functional food, especially for the preparation of health drinks which contain *Aloe vera* gel and which have no laxative effects. It is also used in other food products, for example, milk, ice cream confectionery and so on. However, *Aloe vera* gel juice was not very popular due to their laxative effect and majority of them contained absolutely no active mucilaginous polysaccharides or acemannan. Although colour changes have little relation to the therapeutic effectiveness of stabilized gel, they are rarely acceptable psychologically to the user. The colour change is totally unacceptable in some products. It therefore becomes imperative that a simple but efficient processing technique needs to be developed, especially in the aloe beverage industry, to improve product quality, to preserve and maintain almost all of the bioactive chemical entities naturally present in the plant during processing.

The production process of aloe products involve crushing, grinding or pressing of the entire leaf of the *Aloe vera* plant to produce an *Aloe vera* juice, followed by various steps of filtration and stabilization of the juice. The resulting solution is then incorporated in or mixed with other solutions or agents to produce a pharmaceutical, cosmetic or food product. In the food industry, *Aloe vera* has been utilized as a resource of functional food, especially for the preparation of health food drinks and other beverages, including tea.

The amount of *Aloe vera* that finds its application in the pharmaceutical industry in not negligible as far as the manufacturing of topical ointments, gel preparations, tablets and capsules are concerned. *Aloe vera* gel also finds its application in the cosmetic and toiletry industries, where it is used as a base for the preparation of creams, lotions, soaps, shampoos and facial cleaners. Unfortunately, because of improper processing procedures, many of these so-called aloe products contain, very little or virtually no active ingredients, namely, mucopolysaccharides.

In view of the known wide spectrum of biological activities possessed by the leaves of the *Aloe vera* plant and its wide spread use, it has become imperative that the leaf be processed with the aim of retaining essential bioactive components. The review aims to provide a succinct resume of information regarding Aloe vara to serve as a reference for further investigations about this potential ingredient, to develop an effective method for processing of *Aloe vera* leaf, in the process, preserve and maintain almost all of the bioactive chemical entities naturally present in the *Aloe vera* leaf. The analysis deals with biological activity of leaf gel, gel stabilization technique, heat treatment of leaf gel, processing methodologies like cold-process, whole leaf process Qmatrix process, activealoe process, desiccant air dehydration, total process *Aloe vera* and Time Temperature and Sanitation (TSS) process.

## Biological Activity of *Aloe vera*

The controversy over the identity of the active substance(s) in *Aloe vera* has not been settled. Also, various mechanisms have been proposed for the alleged healing properties of *Aloe vera*. Since no single definitive active ingredient has been found, it is commonly suggested that there may be some synergitic action between the polysaccharide base and other components (Leung, 1978). According to Mackee (1938), vitamin D was the healing agent, but Row and Parks (1941) reported the absence of vitamin D. Morton (1961) suggested a theory stating the seeming efficacy of aloe pulp may be attributed to its high water content, *i.e.*, 96 per cent +, providing a means of making water available for injured tissue without sealing it off from the air. This recovery would explain the instant soothing effect of *Aloe vera* gel has on burns, but would not account for the long term effect of healing.

The action of *Aloe vera* is simply due to its moisturizing and emollient effects, hence, its use in cosmetics. Various researchers reported that the effective components for wound healing may be tannic acid (Freytag, 1954) and a type of polysaccharide (Kameyama, 1979). Other researchers have also reported anti-inflammatory effects of complex polysaccharides, glycoproteines and sulfated polysaccharides. However, there are many examples in the literature indicating that polysaccharides can exhibit pharmacological and physiological activities without help from other components. It is therefore, logical that the mucilaginous gel of *Aloe vera* plant, which is essentially a polysaccharide, holds secrete to *Aloe vera*'s medicinal properties. Many researchers such as Collins and Collins (1935), Fine and Brown (1938) and Crew (1939) have attributed pain-relieving properties to *Aloe vera* gel.

It is virtually impossible to prevent contamination by the leaf exudates during commercial extraction of *Aloe vera* gel. It is also believed that the intact leaves anthraquinones and their derivatives may diffuse into the gel from the bundle sheath cells; this possibly supports the conclusion of Row *et al.* (1941) who states that the healing agent is passed from the rind into gel on standing. Davis (1997) using the conductor-orchestra concept, explains the relationship that exists among over 200 biologically active compounds within *Aloe vera*. One of these molecules, a polysaccharide and acts as the conductor that leads a symphony composed of 200+ biologically active compounds. Davis concluded that, as the conductor, the polysaccharide modulates the biological activity between the surrounding orchestra molecules to work synergistically. In view of these findings, it has seen presumptuous for any scientific research to consider or even to postulate that any one substance is responsible for the biological activity seen in *Aloe vera* gel. Unfortunately, it is not easy to differentiate between a good quality product and one that has been adulterated. Although price can be a guide-the more expensive the *Aloe vera*, the better the product-this does not always apply. In the end, the key to judging *Aloe vera* is by results.

The things that happens to make aloe products less desirable or cause it to become virtually non beneficial are stem from the harvesting of the leaves, processing and distribution of leaves. The freshly removed leaves must go directly into production or must be appropriately refrigerated to prevent a loss of biological activity, principally through the degradative decomposition of the gel matrix. The value of aloe further diminishes if the processing procedure applies too much heat for too long a time.

Extended heating renders the product free from bacterial contamination but effectively destroys aloe's mucopolysaccharide and consequently reduces its efficacy (http://www.aloecorp.com). For therapeutic purposes, the most efficacious *Aloe vera* is that derived from whole-leaf aloe and cold-processed. Aloe is not just aloe because the manufacturer says so. To assure that an aloe product at a price worth paying and to achieve the desired results, it is recommended to look for International Aloe Science Council (IASC) certification seal on literature and packaging. Another way to ascertain whether an *Aloe vera* product has a high healing capacity is to find out the number of mucopolysaccharides (MPS) present. This is sometimes included on the labeling. The highest therapeutic value is found in product containing between 10,000 and 20,000 MPS per liter.

## Gel Stabilization Technique

*Aloe vera* gel is the mucilaginous jelly obtained from parenchyma cells of the *Aloe vera* plant. When exposed to air, the gel rapidly oxidizes, decomposes and looses much of its biological activities. Different researchers have described different processing techniques of the gel with regards to its sterilization and stabilization, *i.e.*, cold processing or heat treatment. However, the fundamental principle underlying these processing techniques remains almost the same. Regardless of the relative quality of the plant, the best results are obtained when leaves are processed immediately after harvesting. This is because degradative decomposition of the gel matrix begins due to natural enzymatic reactions, as well as the growth of bacteria, due to the presence of oxygen.

The entire process involves washing the freshly harvested *Aloe vera* leaves in a suitable bactericide, followed by processing of the leaves to mechanically separate the gel matrix from the outer cortex. The separation of the gel from the leaf could be facilitated by the addition of cellulose dissolving compounds, *e.g.*, cellulose. Thus, the aloe liquid obtained is treated with activated carbon to decolourize the liquid and remove aloin and anthraquinones, which have laxative effects. This is especially so if the stabilized gel is to be used as a drink formulation for internal use. The resultant liquid is then subjected to various steps of filtration, sterilization and stabilization. The stabilized liquid, thus, obtained could be concentrated to reduce the amount of water or, alternatively, almost all of the water removed to yield a powder. In cold processing technique, the entire processing steps are accomplished without the application of heat. Coats (1979) reported the use of enzymes, like glucose oxidase and catalase, to inhibit the growth of aerobic organisms within *Aloe vera* gel and, thereby, sterilize it. Other sterilization steps reported in the cold processing includes exposing the gel to ultraviolet light, followed by a micron filtration.

In the heat treatment processing, sterilization is achieved by subjecting the aloe liquid obtained from the activated carbon treatment to pasteurization at high temperature. Aloecorp (http://www.aloecorp.com) has reported the biological activity of *Aloe vera* gel essentially remains intact when the gel is heated at 65°C for periods less than 15 min. Extended periods or higher temperatures have resulted in greatly reduced activity levels. They, however, suggested that the best method of pasteurization is HTST (High Temperature Shot Time), followed by flash cooling to 5°C or below. In all these processing techniques, stabilization can be achieved by the addition of preservatives and other additives. The use of sodium benzoate, potassium sorbate, citric acid, vitamin E in synergism and the resultant efficacy, has been reported.

## Heat Treatment of Gel

Xiu *et al.* (2006) conducted research on the gel juice from *Aloe vera* to investigate the effects of heat treatment on bioactive substances including polysaccharide and barbaloin. The effect of methanol solvent on compositional variations of barbaloin was also taken into consideration. Results show that the polysaccharide from *Aloe vera* exhibited a maximal stability at 70°C decreasing either at higher or lower temperatures. Heating promoted a remarkable decrease in barbaloin content depending on temperature and time, more affected than polysaccharide of the gel juice from *Aloe vera*. Barbaloin is unstable when dissolved in methanol resulting in the transformation into a series of unidentified compounds, in addition to aloe emodin with the period of storage at 4°C in refrigerator.

The effect of air-drying temperature (from 30 to 80 °C) on dehydration curves and functional properties (water retention capacity, WRC; swelling, SW; fat adsorption capacity, FAC) of *Aloe vera* cubes has been investigated by Simal *et al.* (2000). A diffusion model taking into account sample shrinkage has been proposed and solved by using a finite difference method. The effective diffusivities estimated with the proposed model varied with the air-drying temperature according to the Arrhenius law except for the experiment carried out at 80°C, where case-hardening took place. Simulation of *Aloe vera* drying curves by using the model was accurate (percentage of explained variance (per cent var): 99.7 ± 0.1 per cent). Furthermore, drying kinetics of *Aloe vera* cubes of different sizes to those used to develop the model could be satisfactorily predicted (per cent var: 99.5 ± 0.2 per cent). The three studied functional properties exhibited a maximum when drying temperature was 40°C decreasing either at higher or lower temperatures. Physico-chemical modifications promoted by heat treatment and dehydration at different temperatures (30-80 °C) on acemannan, a bioactive polysaccharide from *Aloe vera* parenchyma were evaluated by Antoni *et al.* (2003) Modification of acemannan, a storage polysaccharide, was particularly significant when dehydration was performed above 60°C. Heating promoted marked changes in the average molecular weight of the bioactive polysaccharide, increasing from 45 kDa, in fresh aloe, to 75 kDa, for samples dehydrated at 70 and 80 °C respectively.

The importance of physico-chemical modifications detected in dehydrated *Aloe vera* parenchyma depends on temperature used during the drying process. Regarding the chemical composition, the bioactive polysaccharide acemannan underwent similar losses of mannosyl residues when dehydration was performed between 30 and 60°C, above the latter temperature, losses increased significantly. The physico-chemical alterations of the main type of polysaccharide may have important implications on the physiological activities attributed to the *Aloe vera* plant.

# Processing of *Aloe vera* Leaves

Basic methods of processing *Aloe vera* leaves are

☆ Traditional hand filleted Aloe processing

☆ Whole leaf *Aloe vera* processing

☆ Total process *Aloe vera* processing

## Traditional Hand Filleted *Aloe vera*

In order to avoid contamination of internal fillet with the yellow sap, the traditional hand-filleting method of processing Aloe leaves was developed. In this method, the lower 1 inch of the leaf base (the white part attached to the large rosette stem of the plant), the tapering point (2-4 inch) of the leaf top and the short, sharp spines located along the leaf margins are removed by a sharp knife, then the knife, is introduced into the mucilage layer below the green rind avoiding the vascular bundles and the top rind is removed. The bottom rind is similarly removed and the rind parts, to which a significant amount of mucilage remains attached, are discarded. Another portion of the mucilage layer accumulated on the top of the filleting table. This is of critical concern because the highest concentration of potentially beneficial Aloe constituents are found in this mucilage, as this layer represents the constituents synthesized by the vascular bundle cells empowered by energy developed in the green (chlorophyll- containing) rind cells through sun induced photosynthesis.

The materials of the mucilage layer, subsequent to their synthesis, are distributed to the storage cells (cellulose-reinforced hexagons) of the fillet, a process that is accompanied by dilution owing to the water (the major fillet constituent), which is stored in the fillet cells. The fillet consists of more than 99 per cent water. The fillet is washed again ensuring that there is no possibility of bacterial contamination, after which, the fillet is inserted into the pulper. The pulper has a refrigerated system that reduces the temperature of the resulting juice for optimum conversion, when the holding tank is full; it is left for 24 h to decant. Each tank is scientifically analyzed and certified, which takes approximately 170 h. The way the inner gel is extracted from the leaf is very important. As mentioned above, the latex portion of the leaf is located between the rind and the inner gel. The gel should be removed from the leaf without disrupting this area so that little or no latex (aloin) gets into the gel. If latex does get into the gel, it makes the gel very bitter. This bitter taste can be distinguished from the vegetable taste of the inner gel with little experience. Just because Aloe juice is bitter, it does not mean that it contains 100 per cent pure Aloe juice from the inner fillet. If the gel is extracted by mechanical methods, the Latex can mix with the inner gel resulting in a loss in purity. Only by hand filleting the leaf it is able to cleanly separate the gel from the rind. The gel is then ground to a liquid and the pulp is removed. All this is performed at the farm, so only freshest leaves are processed. The hand filleting method is very labour intensive. Owing to this fact, machines have been designed and employed which attempt to simulate the hand filleted techniques, but generally the product contains higher amounts of the anthraquinones laxatives than the traditional hand filleted approach.

## Whole Leaf *Aloe vera* Processing

This whole leaf process employed in the making of aloe juice allows the cellulose (skin) to be dissolved, as well as measurable amounts of aloin is to be removed. This total procedure is done entirely by a cold process treatment. Maximum efficiency is thus assured, resulting in a product rich in polysaccharides. In this process, the base and tip are removed as previously delineated and then the leaf is cut into sections and ground into particulate slurry.

The method for producing whole leaf *Aloe vera* begins by placing the whole leaf in a Fitz Mill grinding unit that pulverizes the entire leaf into a soup-like structure (Figure 23.1). The material is then treated with, special chemical products which break down the hexagonal structure of the fillet releasing the constituents, by means of a series of coarse and screening filters, or passage through a juice press, the rind particles are removed, the expressed juice is then passed through various filtering columns which remove the undesirable laxative agents. This liquid is then pumped into large, stainless steel holding tanks that have been thoroughly cleaned and sanitized.

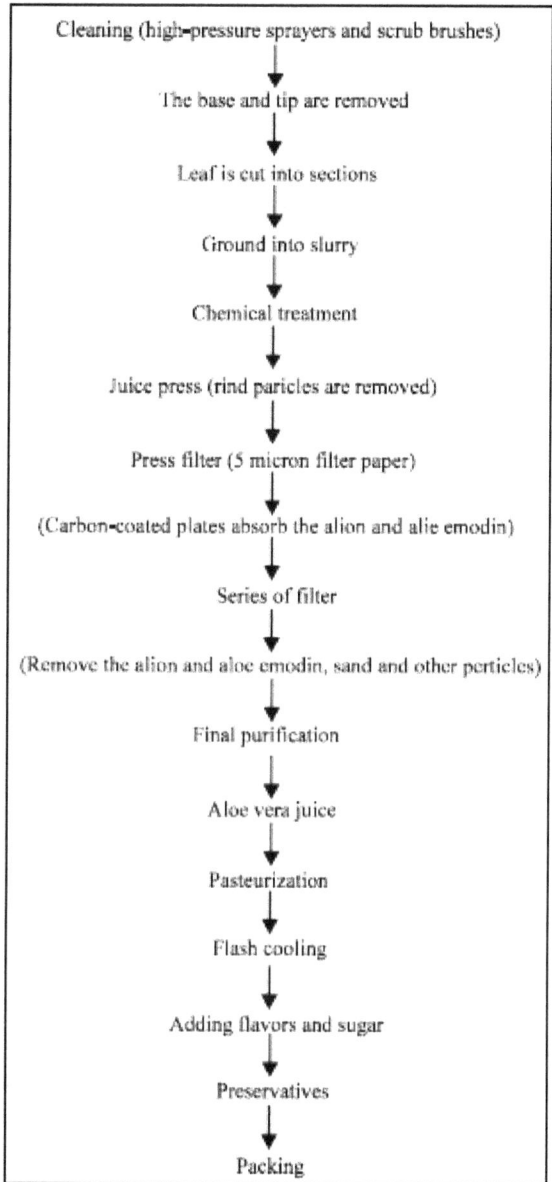

Cleaning (high-pressure sprayers and scrub brushes)
↓
The base and tip are removed
↓
Leaf is cut into sections
↓
Ground into slurry
↓
Chemical treatment
↓
Juice press (rind paricles are removed)
↓
Press filter (5 micron filter paper)
↓
(Carbon-coated plates absorb the alion and alie emodin)
↓
Series of filter
↓
(Remove the alion and aloe emodin, sand and other perticles)
↓
Final purification
↓
Aloe vera juice
↓
Pasteurization
↓
Flash cooling
↓
Adding flavors and sugar
↓
Preservatives
↓
Packing

**Figure 23.1**: Process flow diagram for whole leaf *Aloe vera* processing.

Once the tank is filled, it is hooked-up to a depulping extractor. This machine removes the large pieces of pulp and leaves that the initial grinding process developed. The result is the separation of the *Aloe vera* liquid and the pulp, which consists of the particles of Aloe leaf that have been ground and the naturally occurring pulp in the Aloe gel. The second phase of processing consists of passing the Aloe liquid through a series of filters that remove the aloin and aloe emodin (bitter-tasting, harsh laxatives) as well as any microscopic traces of leaves, sand or other particles. A press filter is used during this phase. First, the press filter is attached to the storage tank containing the pre-filtered Aloe liquid. The press filter's carbon-coated plates absorb the aloin and aloe emodin that is a byproduct of grinding the whole leaf. The Aloe liquid is continually passed through the filter press until 99 per cent of the aloin and aloe emodin are removed. This filtered product is then placed in a second holding tank. At this point, a press filter containing five micron filter paper is attached to this holding tank. The Aloe liquid is passed through this filter medium until it shows no signs of residue. Cold filtration processing is then done as final purification procedure before the Aloe liquid is ready for stabilization. This process, performed properly, can produce a constituent-rich juice (generally containing three times or more constituents than hand filleted juice), which should be virtually free from the laxative anthraquinones, this process was developed in the 1980's.

## Total Process *Aloe vera* Processing

In this new revolutionary approach, The Aloe leaves are hand filleted by the traditional, old fashioned, labour intensive method. Then the green rinds and the mucilage layer from the tabletop are processed by a newly developed propriety methodology. A combination of the products produced by these two procedures produces an aloe product called Total Process Aloe, which contains an enviably high concentration of desirable constituents, which are virtually free from undesirable laxative anthraquinones. The traditional Hand-Fillet methodology, coupled with the newly developed proprietary handling of the refuse of the traditional methods (green rinds and tabletop mucilage) and a geographical area where aloe plants thrive have been combined in achieving the superior quality of Total Process Aloe. Total Process Aloe contains considerably higher concentrations of total solids, calcium, magnesium and malic acid, the major parameters of quality utilized and recommended by the International Aloe Science Council (IASC) for certification.

## Major Unit Operations in Processing of *Aloe vera* Leaf Gel

### Reception of Raw Materials

The *Aloe vera* leaves after harvesting were preferably transported in refrigerated vans from the field to the processing place. The leaves should be sound, undamaged, mold/rot free and matured (3-4 years) in order to keep all the active ingredients in full concentration Lawless and Allan (2000). However, the composition of these active ingredients are subtly affected by seasonal, climatic and soil variations. One important factor that must be considered is the handling/treatment of the leaves after its harvesting because the decomposition of the gel matrix occurs on cutting due to natural enzymatic reactions and the activity of bacteria that are normally present on the leaves. This degradative process can adversely affect the quality of the end product. Therefore, there is a need to carefully work towards refrigerating the freshly removed leaves within 4-6 h or get the raw material directly into production. Some information regarding the quality of a batch of Aloe leaves can be obtained by visual inspection (Figure 23.2).

## Filleting Operation

The losses of biological activity appeared to be the result of enzymatic activity after the aloe leaf was removed from the plant. In fact, it was shown that the aloe gel, once extracted from the leaf, had greater stability than the gel left in the leaf. In order to avoid the decomposition of the biological activity, the filleting operation must be completed within 36 h of harvesting the leaves. In the other hand, the anthraquinone was one important factor leading to nonenzymatic browning in aloe gel product.

## Grinding/Homogenization

The major steps in this process include crushing or grinding. The aloe gel fillets should be crushed and homogenized using a commercial high speed tissue crusher at room temperature (25°C). Due to the reaction of enzymatic browning, the longer the crushing/grinding time, the higher the browning index in *Aloe vera* gel juice (Liu, 2001). Therefore, crushing or grinding should be shortened within 10-20 min in order to avoid the enzymatic browning reaction of *Aloe vera* gel.

## Addition of Pectolytic Enzyme

Enzymatic treatment of *Aloe vera* gel for a long duration prior to processing is detrimental to biologically active compound such as polysaccharide which is the single most important constituent in aloe. Many researches have been done on the polysaccharides (Gowda *et al.*, 1980; Waller *et al.*, 1978; Yagi *et al.*, 1982). It has been reported that the enzyme treatment at 50°C and within 20 min did not induce the loss of biological activity of polysaccharide in *Aloe vera* gel (Maughan, 1984).

Reception of raw material (Aloe vera leaves)
↓
Washing operation
↓
Filleting operation
↓
Grinding/Homogerization
↓
Enzyme addition
↓
Filtation
↓
Un-pasteurized juice
↓
addition of vitamin C and citric acid
↓
Deration
↓
Pasteurization
↓
Falsh cooling
↓
Packaging
↓
Storage

**Figure 23.2**: Processing flow diagram of single-strength *Aloe vera* gel juice production.

## Filtration

This operation influences on the stability of *Aloe vera* gel juice. For example, the product showed the sedimentation of particles as the filtration operation lost its control.

## Addition of Vitamin C and Citric Acid

The unpasteurized aloe gel juice was fortified with vitamin C and citric acid to avoid browning reaction, to improve the flavor of *Aloe vera* gel juice and to stabilize the juice (Eison-Perchonok *et al.*, 1982; Kacem *et al.*, 1987; Kennedy *et al.*, 1992; Tramell

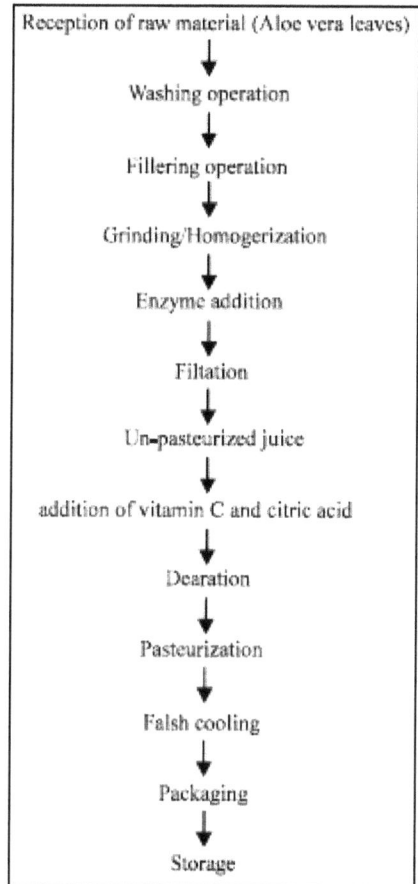

*et al.*, 1986). The pH of aloe gel juice was adjusted between 3.0 and 3.5 by adding citric acid to improve the flavour of *Aloe vera* gel juice.

## Dearation

The aim of dearation step is to avoid the oxidation of ascorbic acid (Chan *et al.*, 1986), which eventually improves the shelf life of the *Aloe vera* gel juice.

## Pasteurization

Like the process of other vegetable juice, this step may affect the taste, appearance and the content of biological activity of aloe gel product. HTST treatment (at 85-95 °C for 1-2 min) is an effective method to avoid the bad flavour and the loss of biological activity of the *Aloe vera* gel (Eshun, 2003).

## Flash Cooling

After pasteurization, the juice is flash cooled to 5°C or below within 10-15 sec. This is a crucial step to preserve biological activity of the *Aloe vera* gel (Eshun, 2003).

## Storage

Relative humidity and temperature are two most important environmental parameters that affect product quality. Those two parameters can also affect the amount of the volatile substances of the juice absorbed by the packaging material (Hernadez and Giacin, 1998) and consequently, affect the shelf-life of the product (Hirose *et al.*, 1988; Sadler and Braddock, 1990).

# Time Temperature and Sanitation (TTS) Process

The stages of this innovative process technology is discussed hereunder

## Timing of Leaf Process

Leaves show losses of biological activity beginning at 6 h following the harvest when the leaves are stored at ambient temperatures. A decrease in activity is also evident when the leaves are stored refrigerated, even though the rate of activity loss is greatly reduced. The losses of activity appear to be result of enzymatic activity after the leaf is removed from the plant. In fact it has been shown that the gel, once extracted from the leaf, has greater stability than gel which is left in the leaf. This means that shipping of leaves, even at refrigerated temperatures, will result in loss of biological activity. The overall timing of TTS production phase are extremely critical. The processing must be completed within 36 h of harvesting of leaves.

## Leaf Harvesting and Handling

Biological activity is also due to the microbial decay of the gel. The first exposure of the inner gel to microbes is when the leaves are harvested from the plant. Leaves in which the base is not intact and sealed will greatly increase the microbial counts in the finished product. To prevent contamination of the gel, the leaves are handled carefully and soaked in a food grade sanitizer which effectively reduces the microbial count in the leaf exterior to acceptable levels.

## Flash Cooling

As a crucial step to preserve biological activity, the gel should be cooled below 5°C in 10 to 15 sec following the gel extraction. Rapid cooling leads to enzymatic and microbial deterioration of the gel, but also aids in reducing the microbial counts in the product.

## Pasteurization

Biological activity remains active when the gel is heated at 65°C for periods of less than 15 min. Extended periods or higher temperatures will results in greatly reduced activity levels. The best method of pasteurization is HTST (High Temperature Short Time), which expose the gel to elevated temperatures for periods of 1 to 3 min. Once heated the gel is flash cooled to 5°C or below.

## Concentration

The gel obtained using the pasteurization and flash cooling methods can be concentrated under vacuum without the loss of biological activity. The concentration operation must be conducted under 125 mm mercury vacuum at temperature below 50°C and must not exceed 2 min. Higher vacuum and temperature will cause activity loss, as will extend concentration times.

## Freeze or Spray Drying

The concentrated product can then be freeze-dried at temperature between -45 and 30°C or can be spray dried with product temperature below 60°C without the loss in biological activity.

# Desiccant Dehydration Process

This system employs a low-tech procedure used for many years to dehydrate foods. The pure intact aloe fillets are first washed so that the first remaining aloin is removed. Then they are placed into a desiccant dehydration chamber where desired level of relative humidity and temperatures are maintained. Here the desiccant air is passed over the fillets to dry them. They come out of the chamber looking a little like a loofah sponge. This material is then ground to powder and packed. By using this several important objectives are achieved. There is no concentration of the aloe gel. There by eliminating one step of the process. When the aloe is gently dried in the natural fillet form, the macromolecules do not break down like they do with mechanical pressing. The result is when the powder is re-hydrated, it comes back to its natural slippery form it had inside the leaf. It is generally believed that these delicate macromolecules are responsible for many *Aloe veras*' proteins. Because there is no need to pre-treat or pre-concentrate the aloe, there are no residual preservatives present in the final powder.

# Qmatrix Process (Aloecorp)

Qmatrix drying is 4th generation dehydration technology, which also includes microwave and radio frequency drying. Microwave and radio frequency drying are not appropriate for aloe as they can deacetylate aloe polysaccharides and denature proteins. For high quality foods, freeze-drying is traditionally used but it is relatively expensive (up to 10 times that of forced air dryers) and is limited to relatively small throughputs. Spray drying can be used for large throughput but the quality of the resultant product is inferior to that produced by freeze-drying due to volatile losses and heat damage. The Qmatrix process is a novel proprietary method of dehydration in enabling the dehydration of aloe while maintaining its integrity with respect to flavour, colour and nutrients. It is comparable to freeze drying in quality aspects but without the high operation costs (http://www. aloecorp.com).

Advantages of this process:

☆ Unique in the Aloe industry
☆ Exclusive to Aloecorp
☆ Gentle low temperature/short time drying
☆ Superior sensory attributes retained (Academic study)
☆ Superior retention of nutrients and bioactivity (Academic study)
☆ Atmospheric pressure (no vacuum)
☆ Energy efficient (Green)
☆ Environmentally friendly (Green)
☆ Superior solubility characteristics
☆ New proprietary products due to versatility

## Solubility Analysis of Qmatrix Process

Spray dried aloe gel powder solubility was compared with Qmatrix processed powder. Equal quantities of powders were added at the same time to the same volume of room temperature water. Spray dried aloe clumps and floats whereas Qmatrix processed powder immediately disperse and settle to the bottom of the vessel. After 15 sec of gentle stirring the Qmatrix processed material is completely in solution while spray dried powder is still clumped on the surface.

# Active Aloe Process

Activealoe is *Aloe vera* manufactured by a patented process, developed using bioactivity guided research. The unique characteristics are as follows:

☆ Polysaccharide guarantee of not less than 10 per cent by weight solids
☆ Controlled digestion of polysaccharides to enhance bioactivity
☆ Rapid processing to prevent breakdown of bioactive components
☆ Extensively tested and proven biologically active

## Development of Activealoe Process

Univera Pharmaceuticals investigated the role molecular weight played in the biological activity of aloe polysaccharides in order to develop a processing method that would retain and enhance the biological activity of native aloe resulting in the patented processing methods now used exclusively by Aloecorp.

# Conclusions

A review on processing of *Aloe vera* leaf gel has revealed *Aloe vera* as a highly potential functional and valuable ingredient that exhibits relatively impressive biological functions of great interest in cosmetic, pharmaceutical and food industries. It also revealed the present processing technologies *viz.*, gel stabilization technique, biological activity of aloe leaf gel and the effect of heat treatment on various constituents of gel. The process technologies like desiccant dehydration of aloe cubes, Qmatrix process, low temperature short time heat treatment process, activealoe process, Time Temperature and Sanitation Process, Total Process *Aloe vera* are the potential innovative process technologies.

# References

Antoni, F., Pablo, G., Susana, S., and Carmen, R. (2003). Effect of heat treatment and dehydration on bioactive polysaccharide acemannan and cell wall polymers from *Aloe barbadensis* Miller. *Carbohydrate Polymers*, **51**: 397-405.

Chan, H.T., and Cavaletto, C.G. (1986). Effects of dearation and storage temperature on quality of aseptically packaged guava puree. *J. Food Sci.*, **51**: 165-168.

Coats, B.C. (1979). Hypoallergenic stabilized *Aloe vera* gel. US patent 4: 178,372.

Collins, C.E. and C. Collins, (1935). Roentgen dermatics treated with fresh aloe leaf. *Am. J.Roentgenol.*, **33**: 396-397.

Crewe, J.E. (1939). Aloe in the treatment of burns and scalds. *Minnesota Med.*, **22:** 538-539.

Davis, R.H. (1997). *Aloe vera*-A Scientific Approach. Vantage Press Inc., New York, USA, pp. 290-306.

Eison-Perchonok, M.H., and Downes, T.W. (1982). Kinetics of ascorbic acid oxidation as a function of dissolved oxygen concentration and temperature. *J. Food Sci.*, **47**: 765-767, 773.

Eshun, K. (2003). Studies on *Aloe vera* gel: Its application in beverage preparation and quality assessment. Thesis submitted to Food Science and Technology School of Southern Yangtze University in partial fulfillment of the requirements for the Degree of Master of Science.

Fine, A.F., and Brown, S. (1938). Cultivation and clinical application of *Aloe vera* leaf. *Radiol.*, **31: ** 735-736.

Freytag, A. (1954). Suggested role of traumatic acid in Aloe wound healing. *Pharmzie*, **9: ** 705.

Gowda, D., Neelisiddaiah, B., and Anjaneyalo, Y. (1980). Structural studies of polysaccharides from Aloe saponaria and Aloe vanbalenii. *Carbohydrate Res.*, **83**: 402-405.

He, Q., Liu, C., and Zhang, T. (2002). Study on noenzymatic browning of aloe products and its inhibition methods. *Food Sci. (Chenses)*, **23(10):** 53-56.

Hernadez, R.J., and Giacin, J.R. (1998). Factors affecting permeation, sorption and migration processes in package-product systems. In: Food storage stability, Boca Raton, CRC Press, pp. 269-329.

Hirose, K., Harte, B., Giacin, J.R., Miltz, J., and Stine, C. (1988). Sorption of d-limonene by ealant films and effects on mechanical properties. In: Food and Packaging Interactions; ACS Symposium Series (vol. 365).

http://www.aloecorp.com.

Kacem, B., Mathews, R.F., Grandall, P.G., and Cornell, J.A. (1987). Non-enzymatic browning in aseptic packaged orange juice and orange drinks. Effect of amino acids, dearation and anaerobic storage. *J. Food Sci.*, **52**: 1665-1667, 1672.

Kameyama, S. (1979). Wound healing composition from *Aloe arborescens* extracts. Jap. Patent, 7856995.

Kennedy, F.C., Rivera, Z.S., Lloyd, L.L., Warner, F.P., and Jumel, K. (1992). L-ascorbic acid stability in aseptically processed orange juice in tetra brick cartons and the effect of oxygen. *Food Chem.*, **45**: 327-331.

Lawless, J., and Allan, J. (2000). *Aloe vera*–Natural Wonder Cure. Harper Collins Publishers. London.

Leung, A.Y. (1978). *Aloe vera* in cosmetics. *Excelsa*, **8**: 65-68.

Liu, C. (2001). Study on preservatives in the aloe gel juice system. *J. Wuxi University Light Ind. (Chenses)*, **20(5)**: 480-484.

Mackee, G.M. (1938). X-ray and Radium in the Treatment of Diseases of the Skin. Lea and Febiger (Eds.). Philadelphia, PA, pp. 319-320.

Maughan, R.G. (1984). Method to increase colour fastness of stabilized *Aloe vera*. US Patent.

Mortan, J.F. (1961). Folk uses and commercial exploitation of Aloe leaf pulp. *Econ. Bot.*, **15**: 311-319.

Robert, H.D. (1997). *Aloe vera*: A Scientific Approach. Vantage Press, Inc. New York.

Row, T.D., and Parks, L.M. (1941). Phytochemical study of *Aloe vera* leaf. *J. Am. Pharm. Assoc.*, **30**: 262-266.

Row, T.D., Lovell, B.K., and Parks, L.M. (1941). Further observations on the use of *Aloe vera* leaf in the treatment of third-degree X-ray reactions. *J. Am. Pharm. Assoc.*, **30**: 266-269.

Sadler, G.D., and Braddock, R.J. (1990). Oxygen permeability of low density polyethylene as a function of limonene absorption. An approach to modelling flavour (Scalping). *J. Food Sci.*, **55**: 587-590.

Simal, S., Femenia, A., Llull, P., and Rossell, C. (2000). Dehydration of *Aloe vera*: Simulation of drying curves and evaluation of functional properties. *J. Food Eng.*, **43**: 109-114.

Tramell, D.J., Dalsis, D.E., and Malone, C.T. (1986). Effect of oxygen on taste, ascorbic acid loss and browning for HTST-Pasteurized, single-strength orange juice. *J. Food Sci.*, **51**: 1021-1023.

Waller, G.R., Mangiafica, S., and Ritchey, C.R. (1978). A chemical investigation of *Aloe barbadensis* Miller. *Proceedings of the Oklahoma Academy of Science*, **58:** 69-76.

Xiu, L.C., Changhai, W., Yongmei, F., and Zhaopu, L. (2006). Effect of heat treatment and dehydration on bioactive polysaccharide acemannan and cell wall polymers from *Aloe barbadensis* Miller. *J. Food Eng.* (**75**) **2**: 245-251.

Yagi, A., Shibata, S., Nishioka, I., Iwadre, S., and Ishida, Y. (1982). Cardiac stimulant action of constituents of *Aloe saponaria. J. Pharm. Sci.*, **71**: 739-741.

Utilisation and Management of Medicinal Plants Vol. 2 (2014)   *Pages* **447–460**
*Editor-in-Chief:* **V.K. Gupta**
*Published by:* **DAYA PUBLISHING HOUSE, NEW DELHI**

# 24

# A Study of Photo Damage of Hair and its Protection Using *Aloe vera* as a Natural Agent

Farhat S. Daud[1]* and Sheela B. Kulkarni[2]

## ABSTRACT

*Exposure to sunlight has both beneficial and harmful effects on human body depending on the length, frequency, intensity of radiation and sensitivity of the individual concerned. Moderate exposure has many beneficial effects like increase in blood circulation, increase in formation of hemoglobin, increase in vitamin D absorption etc.*

*However, it also has several adverse effects (short term and long term) like sunburns, eczema, wrinkles, ageing and skin cancer. Since historical time's societies have been aware about the harmful effects of sunlight on skin and sunscreens have been used in skin care since a long time. But the harmful effects of Sun or the UV light on hair have been realized only a few years ago. Hence the concept of a sunscreen or sun protectant for hair is relatively new.*

*Hair also needs protection from the harmful sunrays or UV rays as these cause morphological, physical and chemical damage to hair which is manifested in terms of cuticle damage, breakage, loss of strength, loss in color, texture etc.*

*There are some synthetic sunscreen agents available in the market for use on hair for sun protection however the use of a natural sunscreen agent for sun protection is still at a nascent stage.*

1   Department of Cosmetic Technology, L.A.D. and Smt. R.P.College for Women, Seminary Hills, Nagpur – 440 006, Maharashtra, India

*   *Corresponding author*: E-mail: fairydaud@yahoo.com

The objective of this study was to study the effect of UV radiation on hair damage and evaluate Aloe vera (Aloe barbadensis Mill.; Family Liliaceae) known for its use in moisturizing, conditioning and skin protecting activity as a natural sun screening agent for hair through cuticle damage studies.

**Keywords:** *Aloe Vera,* Hair, Photo protection, Sunscreen, UV radiation.

# Introduction

## Sun and Hair Damage

The sun's ultraviolet light is made up of UVC, UVB and UVA Radiation. The UVC (shortest rays < 290 nm) does not reach the earth as the stratosphere absorbs it. The UVB (280–320 nm k/a short burning rays), about 4-6 per cent and UVA (320-400 nm k/a long tanning rays), about 6-7 per cent of the total terrestrial UVR are responsible for damage to skin and hair; the rest of spectrum bounces off (Campbell and Alexandra, 1996).

On skin, it is the UVA which is more harmful as it goes deep into the dermis; however skin has a natural protective mechanism via melanin generation (Mitsui, 1997).

Besides, skin remains mostly covered by clothing which again acts as a protectant (Campbell and Alexandra, 1996). However hair remains relatively unprotected and bare. In addition, being black in colour, it's a good absorber of light and more prone to its effects. The damage to the hair fiber is mostly by both but more by the UVB light (290-320 nm).

Sunlight causes many undesirable effects on hair like local production of photo induced free radicals that destroy natural and artificial hair color rupture of cystine bonds and reduction in hair strength making them look dull and lifeless (Wood, 2005; Signori, 2004).

Thus, human hair undergoes changes in morphological, chemical, mechanical and cosmetic properties when exposed to sunlight (UV radiation). The damage is both physical and chemical (Garcia *et al.*, 1998) and can be described as:

## Physical Damage

☆ Alteration or elimination of cuticle cells

☆ Roughening of hair surface causing difficulty in combing or brushing

☆ Loss of mechanical resistance and increase in porosity

☆ Swelling and alkaline solubility.

## Chemical Damage is Seen as

☆ Breakage of disulphide bond and decomposition of tryptophan

☆ Photo oxidation of cysteine, cholesterol and fatty acids

☆ Photo bleaching of melanin or artificial hair colour.

All this leads to:

☆ Difficulty in combing;

☆ Increased dryness, roughness, brittleness;

☆ Loss of elasticity;

☆ Fading of artificial and natural hair colour (Patel and Neil, 1997).

Most of the damage is expressed in terms of cuticle damage as it 'most at risk' being the outermost component and the first to be exposed to environmental and various chemical and physical stresses (Garcia *et al.*, 1998) (Figures 24.1 and 24.2).

**Figure 24.1**: Damaged hair cuticle.
(www.scott.k12.va.us/rita/photo_tinted.gif).

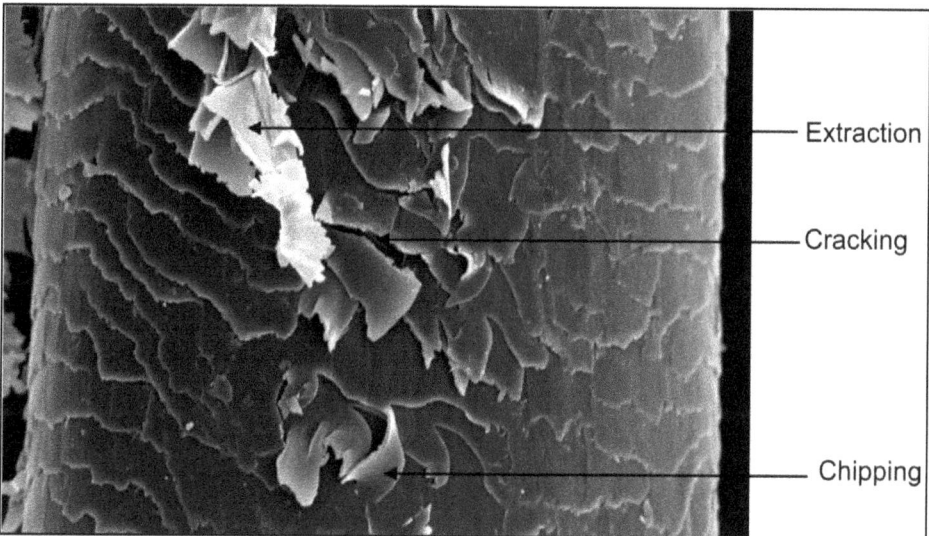

**Figure 24.2**: Different types of damage to hair cuticle (Daud and Kulkarni, 2008).

Thus it is clear that hair needs UV protection and is most beneficial when done at an early stage. This protection can be achieved by using a 'Sunscreen'- an agent (organic or inorganic) which protects by 'absorbing' or 'scattering' sun's radiation in the desired range.

For an agent to be effective on hair it must possess certain specific characteristics:

☆ Be substantive to hair fiber (bond to hair at a minimum, concentration)

☆ Possess photo filtering ability in the UVB range (Peter, 1997)

☆ Form a thin layer on hair fiber to prevent hair from becoming limp and heavy (Cosmetics and Toiletries, 1996).

## Why *Aloe vera*

### *Aloe vera* (Figure 24.3)

'Aloe' derived from Arabic word *'alloch'* means a shining bitter substance and amongst different species 'Vera' means *'true'* while 'barbadensis' refers to *habitat* of plant (Kokate *et al.*, 2003; www.internethealthlibrary.com/plant-remedies/Aloe-Vera).

*Aloe vera* or *Aloe barbadensis* Mill. also referred to as Curcuao aloe is a coarse-looking perennial, drought-resisting, succulent plant belonging to the Lily (Liliaceae) family (Dagnea, 2000; www.internethealthlibrary.com/Plant-Remedies/AloeVera.htm; Naik *et al.*, 1980).

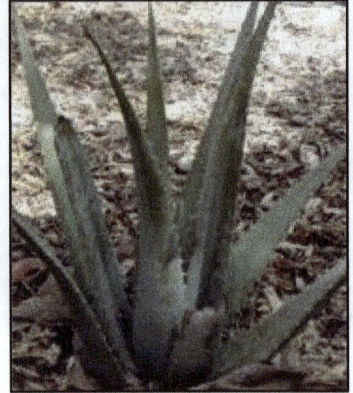

**Figure 24.3**: *Aloe vera* plant.

It is not a part of the cactus family as many believe from the rosette like long spiked leaves. Its use as a purgative is (Cheney, 1920) well known since years and because of aloe gel's reputation as a folk remedy for burns and wounds, it is grown even in homes (Chow *et al.*, 2005; Swami Prakashananda Ayurveda Research Centre, 1992).

Over the years the plant has been known by a number of names such as 'the wand of heaven', heaven's blessing' and 'the silent healer' (Dagnea, 2000; www.internethealthlibrary.com/Plant-Remedies/AloeVera.htm; Naik *et al.*, 1980).

| | | |
|---|---|---|
| *Botanical Name* | – | *Aloe barbadensis* Mill. |
| *Family* | – | Liliaceae |
| *Common Names* | – | *Aloe vera*, Aloe indica, Curacao Aloe, Barbados Aloe, Lily of the desert (Kokate *et al.*, 2003) |

### Other Names in India

Indian Aloe

| | | |
|---|---|---|
| *Sanskrit* | – | Ghrita Kumari |
| *Gujarati* | – | Kunwar |
| *Hindi* | – | Musabhar, Ghikanwar |
| *Marathi* | – | Korphad |
| *Oriya* | – | Kumari, Mushabaro |
| *Tamil* | – | Bhottu-katrazhae |
| *Telugu* | – | Kalabanda (www.globalherbalsupplies.com/herb-information/aloe-vera.htm, *Aloe vera*; Dagnea *et al.*, 2000) |
| *Part Used* | – | Leaves, fresh leaf gel and juice, latex, dried juice of leaves and pulp (Dagnea *et al.*, 2000; Swami Prakashananda Ayurveda Research Centre, 1992) |

## Geographical Source

*Aloe barbadensis* Mill. is the native of North Africa, Mediterranean region of Southern Europe and Canary Islands but is spread to east and West Indies, India, China and other countries. It is grown in Cape colony, Zanzibar and Islands of Socotra. It is also cultivated in Caribbean islands and Europe. There are 2 to 3 easily recognizable varieties in India (North West Himalaya region), but their exact delimitation is not clear (Rajpal, 2000; Kokate *et al.*, 2003; Ross, 1996; Chopra *et al.*, 1982).

The *Aloe vera* variety *Chinensis baker* is common in Maharashtra, Karnataka, Tamil Nadu, Kerala Andhra Pradesh and Madhya Pradesh. The *Aloe vera* variety *Littoralis koehing ex Baker* is found on the beach shingle in Tamil Nadu right up to Rameshwaram. Another variety, which thrives on the Saurashtra coast, is *Jaffarbad Aloes*. This has also been called as *Aloe variegata* Linn (Rajpal, 2000; CSIR, 1985).

## Botanical Description (Figure 24.4)

The plants are characterized by stem less large, thick, leaves which are fleshy in rosette, sessile densely crowded with horny prickles on the margins, strong spine at the apex and convex below. Leaves are about 35-50 cm long, 8-12 cm broad, and 1.5-2.5 cm thick and tapering to a blunt point. Surface is pale green with irregular white blotches, flowers in racemes, bright yellow, tubular stamens frequently projected beyond the perianth tube, fruits loculicidal capsule. It flowers during September-December and harvested during and after monsoon (Rajpal, 2000; Kokate *et al.*, 2003; CSIR 1985; Swami Prakashananda Ayurveda Research Centre, 1992).

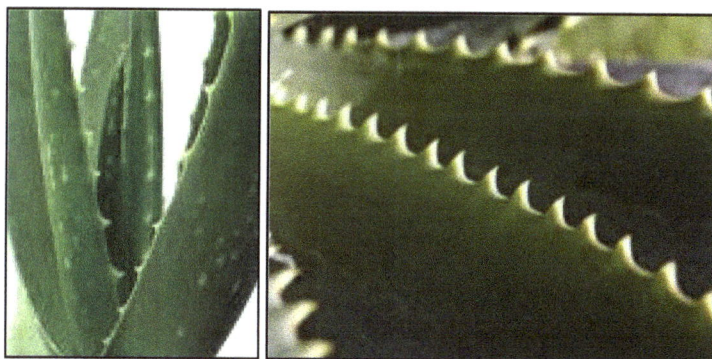

Figure 24.4: *Aloe vera* leaves (stem less) (www.forpeaceofmind.com.au/./ aloe_vera.jpgcontent.answers.com/./230px- aloe_vera_leaf.jpg; http://en.wikipedia.org/wiki/ File: Aloe_vera_leaf.jpg)

## Main Products of *Aloe vera* Plant

'Aloe' or 'Aloe juice' (dried) which is the solid residue obtained by evaporating the latex obtained from pericyclic cells beneath the skin. The bitter yellow latex is mainly composed of Anthraquinone glycosides, free anthraquinones and resinous matter. It has medicinal value and purgative action.

☆ Aloe gel and juice which is clear, thin, gelatinous material obtained by crushing the mucilaginous cells found in inner tissue of the leaf. The gel is the product used most frequently in cosmetic and health food industries. It is generally devoid of Anthraquinone glycosides. The gel contains a polysaccharide known as Glucomannan, which contributes mostly to the emollient effect of the gel. Fresh juice may contain anthraquinones if the sap is mixed with it along with all other components of gel (Luta and McAnalley, 2005; Warrier, *et al.*, 1994; Behl *et al.*, 1993).

## Chemical Constituents of *Aloe barbadensis* Miller

*Aloe vera* is 99-99.5 per cent water, with an average pH of 4.5. The remaining solid material *i.e.* 1-0.5 per cent contains over 75 different ingredients including vitamins, minerals, enzymes, sugars, anthraquinones or phenolic compounds, chromones, flavanoids, lignin, saponins, sterols, amino acids and salicylic acid. These are compiled below. (Atherton, www.positivehealth/atherton.htm; Dagnea *et al.*, 2000)

### Major

☆ Hydroxy Anthraquinone and Anthraquinone derivatives *viz.* aloin (=barbaloin, a mixture of aloin A and B), aloe-emodin and 7-hydroxyaloin isomers

### Minor

☆ Chromone derivatives (Aloe resins)

☆ Polysachharide and Monosachharide (glucomannan)

☆ Coumarins, Pyrans and Pyrones

☆ Flavanoids

☆ Sterols

Indian aloes contain aloinosides as major constituents with traces of aloin. (Indian Herbal Pharmacopea, 1998)

## *Aloe-Vera* and UV Absorption

The UV absorption of various Aloe species and aloin was examined by Proserpi (1976) and found that both Aloe extracts and aloin have spectrophotometric peaks at about 297nm. *i.e.*, UVB range which damages the hair most.

The aloin spectrogram has a second peak at 360 nm. The author concluded that cosmetics containing one to two percent of aloe extract should give effective sunburn protection due to the substances contained within Aloe performing selective screening of UV radiation, by absorbing mainly in the erythemogenic rays.

Bader *et al.* (1981) examined dry extracts of Aloe (containing anthraquinones) for their ability to absorb light in the UVB range. Extraction was done with a 50 per cent water ethyl alcohol solution. The aloe extracts had maximum absorption around

294nm. It has also been reported that Aloin in extract blocks 20 per cent to 30 per cent of sun's ultraviolet rays hence acts as a sunscreen (Rajpal, 2000).

Exposure to UV radiation from the sun suppresses immune system cell functioning in skin; Acetylated polysaccharides (acemannan) found in aloe gel prevent this photo suppression (particularly UVB induced) of the skin's immune cells (epidermal LC) (Jones, 2004; Fox 2003).

On the basis of this data *Aloe vera* was selected as the natural agent for the studies in a shampoo base which is the most used hair care product.

## Material and Methods

### Procurement of *Aloe vera*

*Aloe vera* leaves were obtained from a Aloe farm and the plant was authenticated at the Botany Department, Nagpur University Campus, Nagpur with an Authentication number 9029.

### Preparation of *Aloe vera* Fresh Juice

Leaves were plucked from the base of the plant and immediately cleaned with water. They were cut transversely from the base with a knife, the skin was then scrapped off and the inner gel was removed with a spoon and collected in a container. This gel was then pulverized in a home blender and sieved to obtain fresh juice.

This was then preserved with Sodium benzoate 0.1 per cent, Citric acid 0.1 per cent, Sodium sulphite 0.04 per cent. (Benes, 2004).

### Preparation of Shampoo with Aloe juice (Table 24.1)

Two batches of basic Cream Shampoo with 2 per cent and 10 per cent as minimum and maximum levels of *Aloe vera* juice were prepared as follows. They were coded as 'X' and 'Y' respectively.

### Method of Manufacture

☆ All the ingredients were weighed in the specified quantity.

☆ Phase A and B were heated to 70 degree and mixed at this temp. with constant stirring.

☆ Phase C was added at 45 degree C and mixed properly.

☆ The pH of the shampoo was noted and Phase D was added accordingly to adjust pH.

☆ The batches were allowed to stand for 24 hrs for maturation and consistency development.

### Selection of Hair Sample

Dry hairs were selected for the study. Hair samples were collected from single subject *i.e.*, ladies of age group between 25-40 years. The samples were about 15 cm in length.

*Note*–The physiological and clinical conditions of the subjects were ignored.

**Table 24.1**: Shampoo with *Aloe vera*.

| Formula | Ingredients | Quantities in Per cent | |
|---|---|---|---|
| | | 'X' | 'Y' |
| Phase A | SLES | 28.0 | 28.0 |
| | CMA | 2.0 | 2.0 |
| | CDA | 1.0 | 1.0 |
| | CB | 2.0 | 2.0 |
| | EGMS | 1.5 | 1.5 |
| | Propyl Paraben | 0.15 | 0.15 |
| Phase B | Methyl Paraben | 0.15 | 0.15 |
| | Distilled Water | upto 100.0 | upto100.0 |
| Phase C | Aloe juice | **2.0** | **10.0** |
| | Perfume | q.s | q.s |
| Phase D | Lactic Acid | To adjust pH 4.5-6.0 | To adjust pH 4.5-6.0 |

SLES: Sodium Lauryl Ether Sulphate; CMA: Cocomonoethanolamide; CDA: Cocodiethanolamide; CB: Cocobetaine; EGMS: Etheleneglycol mono stearate.

## Pre-Treatment of Hair Samples

Hair samples were soaked in 20 ml of 10 per cent SLS solution for one minute washed under running tap water for another minute and were then air dried overnight (Croda Inc, 1999). These samples were divided into 4 sets of 1 g each.

## UV Source and Length of Study

Osram Ultra vitalux 230V-E27/ES, 300W with a Quartz burner and a tungsten filament was fixed in a fabricated UV chamber. The UV lamp was calibrated using digital Lux meter. The flux obtained was 6.173 J/cm$^2$ (Lowe *et al.*, 1997).

The exposure period was a total 4 hours to UV radiation in UV chamber under standard humidity and temperature conditions (Petter, 1997). The distance fixed for exposure of hair samples was 20 cm from the UV lamp.

## Instrument Used

### Scanning Electron Microscope (SEM)

The SEM is a microscope that uses electrons rather than light to form an image. There are many advantages to using the SEM instead of a light microscope.

## Method to Determine Hair Damage/Protection

To study the effect of UV radiation on hair and to evaluate the protective efficacy of a natural sunscreen, the parameter used was '**Degree and Extent of Cuticle Damage**' as on exposure to UV radiation the cuticle undergoes a process of chipping, extraction and erosion, causing dryness of the hair and greater susceptibility to further damaging action, which may involve large segments being ripped from the hair (Monteiro *et al.*, 2003).

In healthy hair the edges of cuticle cells are smooth, scales flat and patterns of the cuticle are regular while in damaged hair the degree of damage is observed as:

☆ *In slightly damaged hair:* part of cuticle edge is either peeling off or lost.

☆ *In damaged hair:* cuticle edge is missing in some parts, peeling and loss progresses to another layer. This type of hair has no luster due to random scattering of reflected light and hair does not feel smooth.

☆ *In badly damaged hair:* cuticle is almost completely missing and the cortex is exposed. This type of hair splits and breaks easily (Mitsui, 1997) (Figure 24.5).

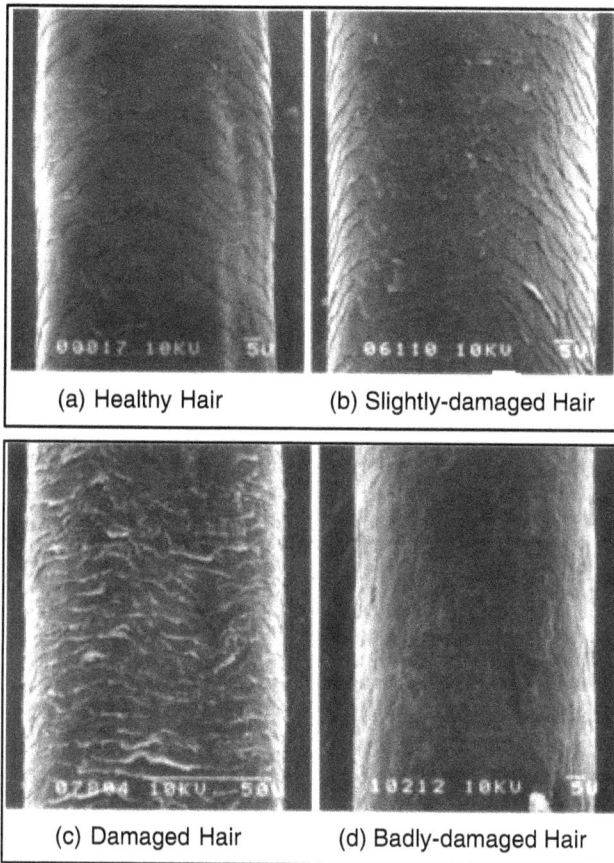

(a) Healthy Hair    (b) Slightly-damaged Hair

(c) Damaged Hair    (d) Badly-damaged Hair

**Figure 24.5**: Degrees of hair damage (Mitsui, 1997).

These effects can be determined by observing the cuticle topography before and after irradiation in a Scanning Electron Microscope (Monteiro *et al.*, 2003).

The assessment of 'Degree and Extent of Cuticle Damage' and protection by *Aloe vera* was done through following parameters where hair were categorized as (Monteiro *et al.*, 2003).

☆ *Comparatively undamaged/Well Protected (Hh/Wp)*: The edges of the cuticle cells are smooth, scales flat and patterns of cuticle regular.

☆ *Slightly Damaged Hair/Considerably Protected (SDh/Cp)*: Part of cuticle edge is either peeling off or is lost.

☆ *Considerably Damaged Hair/Slightly Protected (CDh/Sp)*: Cuticle edge is missing in some parts, the peeling and loss progresses to another layer. This type of hair has no luster due to random scattering of reflected light and hair does not feel smooth.

☆ *Badly Damaged Hair/Least Protected (BDh/Lp)*: The cuticle is almost completely missing and the cortex is exposed. This type of hair splits and breaks easily.

## Procedure for Evaluation of Photo Protecting Efficacy of *Aloe vera*

1. *Treatment of Hair sample with Shampoo 'X' and 'Y'*: Two sets of pre-treated hair samples were treated with shampoo 'X' and 'Y' separately by messaging the product for 10 seconds.

   They were then washed under running water for 1 minute and dried at R.T.

   The strands were then mounted on two glass slides by pasting the two ends with a cellophane tape.

   They were then exposed to artificial UV radiation for 4 hrs.

   One set of hair was left untreated for comparison but was exposed to UV radiation for 4 hours and was called 'Reference'.

   One set was left untreated and unexposed and was called 'Virgin' (Table 24.2).

**Table 24.2**: Hair samples.

| Type of Hair Samples |
| --- |
| ☆ Set 1 –treated with Sample X and Exposed |
| ☆ Set 2- treated with Sample Y and Exposed |
| ☆ Set 3 –Reference- untreated but Exposed |
| ☆ Set 4 – Virgin- untreated and unexposed |

2. All the 4 samples *i.e.*, X, Y, Reference and Virgin were then observed under Scanning Electron Microscope for changes in cuticle layer and photomicrographs were obtained which were assessed for their degree of cuticle damage/protection (Figure 24.6).

The OBSERVATIONS were noted in Table 24.3.

## Results and Interpretation

*Aloe Vera* which is widely used in skin and hair care for its moisturizing and conditioning ability since years was selected for the study as the photo protecting agent on the basis of recent reports stating that *Aloe vera* can block UV radiation by 25-30 per cent (Rajpal, 2002).

**Table 24.3**: Comparative degree of damage/protection.

| Sample No. | Condition of Hair Cuticle | | | |
|---|---|---|---|---|
| | Badly Damaged/ Least Protected | Considerably Damaged/ Slightly Protected | Slightly Damaged/ Considerably Protected | Comparatively Undamaged/ Well Protected |
| Virgin | | * | | |
| Reference | * | | | |
| Sample x | | | * | |
| Sample Y | | | | * |

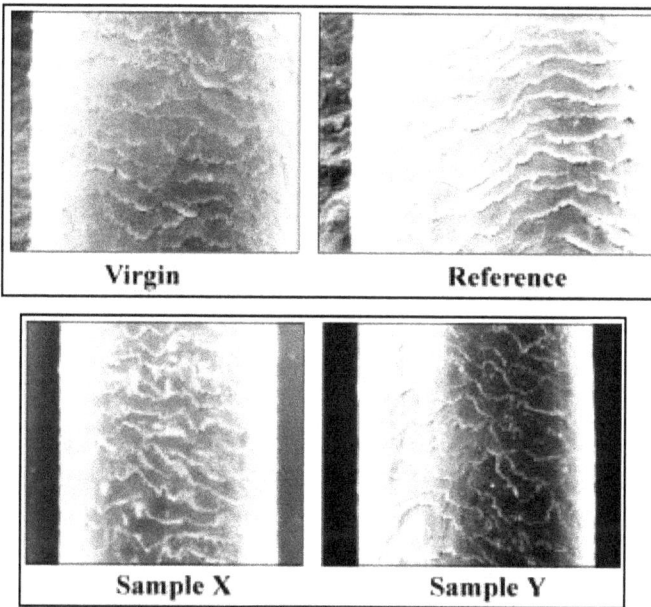

**Figure 24.6**: SEM pictures showing degree of protection of AV shampoos X and Y.

An evaluation was done to study its efficacy as a protectant on dry hair using cuticle damage as the parameter. The results of the study showed that hair treated with freshly extracted *Aloe vera* juice in shampoo base offered good protection by keeping the cuticles scales well aligned and smooth after exposure to UV radiation. The result can be compiled as:

1. Reference was more damaged than Virgin which confirms the theory that hair when unprotected and is exposed to UV radiation Shows UV damage.

2. *Aloe vera* showed better activity at 10 per cent concentration that 2 per cent concentration as Sample Y containing 10 per cent Aloe juice was undamaged compared to X made with 2 per cent.

# Conclusion

Hair has a definite chemical and physical structure. Hence inspite of the fact that hair shaft is a dead entity, the above study and results indicate that hair exposed to UV radiation shows damage detrimental to its health, thus justifying the need for photo protection. Above all, this damage may be irreversible; suggesting protection as the best solution.

Though several synthetic sunscreens for hair photo protection have been launched in the market today, the discovery of a natural counterpart is yet not well established. In the above study, *Aloe vera* which was selected as natural sunscreen agent showed good UV protecting ability on hair. Thus it can also be concluded from the above study that *Aloe vera* acts as a UV protecting agent on hair and can be used as a natural sunscreening agent in hair care products.

# References

Atherton, P. *Aloe vera* Myth or Medicine? M.B.Ch.B., D.Obst. R.C.O.G., M.R.C.G.; www.positivehealth\atherton.htm.

Behl, P. N., Arora, R. B., Srivastava, G., and Malhotra, S. C. (1993). Herbs used in Dermatological Therapy. CBS Publishers and Distributors, 22-23.

Benes, D. M. (2004). Final Report of the Cosmetic Ingredient Review Expert Panel. Cosmetic Ingredient Review, Washington. 3-4.

Campbell and Alexandra. (1996). The Body shop, Suntan Special; Futura publications; A division of Macdonald and Co. (Publishers) ltd., London, p. 4.

Campbell and Alexandra. (1996). The Body shop, Suntan Special; Futura publications; A division of Macdonald and Co. (Publishers) ltd., London, p. 29.

Chopra, R. N., Handa, K. L., and Kapur, L. D. (1982). Chopras Indegenous Drugs of India, 2nd edn. (Revised). Academic Publishers, Calcutta. 61-62.

Chow, J. T-N., Williamson, D. A., Yaes, K. M., and Goux, W. J. (2005). Chemical characterization of immunomodulating polysaccharides of *Aloe vera* L. *Carbohydrate Research*, **340:** (6): 1132.

Cosmetics and Toiletries. (1996). Vol. (111) : 4. Allured publishing Corporation, U.S.A. 57.

Dagnea, E., Bisrata, D., Viljoena, A., and Wykb, B-E. V. (2000).Chemistry of *Aloe* species. *Current Organic Chemistry*, **4:** 1055.

Dagnea, E., Bisrata, D., Viljoena, A., and Wykb, B-E. V. (2000).Chemistry of *Aloe* species. *Current Organic Chemistry*, **4:** 1057.

Dagnea, E., Bisrata, D., Viljoena, A., and Wykb, B-E. V. (2000).Chemistry of *Aloe* species. *Current Organic Chemistry*, **4:** 1058.

Daud, F.S., and Kulkarni, S. B. (2008). A study on sunscreen (Photoprotection) for hair, *Doctoral Research Project*, RTM Nagpur University, p. 17.

Fox, C. (2003). From *Aloe vera* Gel to Stabilized Vitamin C. *Cosmet. Toil.* **118**(3): 22.

Garcia, Gamez, M., and Challoner, N.I. (1998). Evaluation of the Formation and Repair of Hair Cuticle Damage. Croda Colloids Limited, England. (Paper Presented At Image 98, Mumbai, Dec. 5-7, 1998), 1.

http://en.wikipedia.org/wiki/File: Aloe_vera_leaf.jpg

Incroquat UV-283. (1999). DS-127R-5, Croda Inc., New Jersey. **21**: 5.

Indian Herbal Pharmacopeia. (1998).A Joint Publication of Regional Research Laboratory and Indian Drug Manufacturers Association, Vol.1, 13.

Jones, K. (2004). Aloe science: A brief history of skin care with Aloe. *Cosmet. Toil.,* **119**(12): 71-72.

Kokate, C. K., Purohit, A. P., and Gokhale, S. B. (2003). Pharmacognosy. 23ʳᵈ edn, Nirali Prakashan, 186.

Lowe, N. J., Shaath, N. A., and Pathak, M. A. (1997). Sunscreens-Development Evaluation And Regulatory Aspect. 2ⁿᵈ end, Cosmetic Science and Technology series, Marcel Dekker Inc., New York, pp. 506.

Luta, G., and McAnalley, B.H. (2005). *Aloe vera*: Chemical composition and methods used to determine its presence in commercial products. *Glycoscience and Nutrition,* **6(4):**

Mitsui, T. (1997) New Cosmetic Science. Published by Elseiver Science BV. 61-63.

Monteiro, V. F., Pinheiro, A. S., Leite, E. R., Agnell,i J. A. M., Pereira-Da-Silva E., and Longo, M.A. (2003). UV radiation: aggressive agent to the hair- AFM, a new methodology of evaluation. *J. Cosmet. Sci.,* **54**(3): 275.

Naik, V. N., and Associates (1980). Flora of Marathwada. Amrut Prakashan, Vol. II, 858.

Patel, D., and Neil, T. (1997). A new sun filters for hair products. *Inside Cosmetics,* 17-18.

Peter, J. P. (1997). A new substantive photo filter for hair care. *Soaps, Perfumery and Cosmetics,* 29-33.

Petter, P. J. (1997). A new substantive photo filter for hair care. *Soaps, Perfumery and Cosmetics.* **70**(3): 31.

Rajpal, V. (2002). Standardization of Botanicals–Testing and Extraction methods of Medicinal Herbs. Eastern publishers, New Delhi. Vol. 1, 11.

Rajpal, V. (2002). Standardization of Botanicals–Testing and Extraction methods of Medicinal Herbs. Eastern publishers, New Delhi. Vol. 1, 22.

Ross, I. A. (1999). Medicinal Plants of the World. Humana Press, New Jersey. 66.

Selected Medicinal Plants of India. (1992). Swami Prakashananda Ayurveda Research Centre, Mumbai. Tata Press. 28.

Selected Medicinal Plants of India. (1992). Swami Prakashananda Ayurveda Research Centre, Mumbai. Tata Press. 27.

Signori, V. (2004). Review of the current understanding of the effect of ultraviolet and visible radiation on hair structure and options for photo protection. *J. Cosmet. Sci.*, **55** (1): 109.

The Wealth of India. (1985). A dictionary of Indian Raw materials and Industrial products, Raw materials. CSIR, New Delhi. Vol. I: A (Revised); 192.

Warrier, P. K., Nambiar, V. P., and Ramankutty, C. (1994). Indian Medicinal Plants. Vol.1, Orient Longman. 103.

Wealth of India. (1985). A dictionary of Indian Raw materials and Industrial products; Raw materials. Vol.-I: A (Revised). CSIR, New Delhi. 191.

Wood, J. (2005). Screen savers. *Soaps, Perfumery and Cosmetics.* **76**(94): 38.

www.forpeaceofmind.com.au/./aloe_vera.jpgcontent.answers.com/./230px-aloe_vera_leaf.jpg

www.globalherbalsupplies.com/herb-information/aloe-vera.htm,Aloe.

www.internethealthlibrary.com/plant-remedies/Aloe-Vera; Page 2.

www.internethealthlibrary.com/Plant-Remedies/AloeVera.htm

www.scott.k12.va.us/rita/photo_tinted.gif

Utilisation and Management of Medicinal Plants Vol. 2 (2014)     *Pages* **461–469**
*Editor-in-Chief:* V.K. Gupta
*Published by:* DAYA PUBLISHING HOUSE, NEW DELHI

# 25

# Anti-Proliferative Potential and Induction of Apoptosis from *Catharanthus roseus*

Madhulika Bhagat[1]* and S.K. Singh[2]

## ABSTRACT

*In the present investigation we explore the antiproliferative properties and mechanisim of cell death against various human cancer cell lines in vitro. The madagascar prewinkle (Catharanthus roseus [L] G. Don) is traditionally well known medicinal plant used for the treatment of various diseases including cancer. Three extracts were prepared by the whole plant of C. roseus viz., alcoholic, hydro-alcoholic and aqueous extracts and anti-proliferative activity was measure by using SRB assay and induction of apoptosis was measured by using commet assay and Dna fragmentation was also performed. The significant growth inhibition by crude extracts of the plant was observed against all the human cancer cell line viz., A-549 (lung), HEP-2 (liver), KB (oral) and COLO-205(colon) respectively. Growth inhibition was observed in dose dependent manner by all the extracts against all the four cancer cell lines. Alcoholic extract was found to be more potential than the other hydro-alcoholic and aqueous extracts with percent growth inhibition of 69, 79 and 88 against A549 (lung), 65, 78 and 87 against HEP-2 (liver), 46, 65 and 67 against KB (oral) and 65, 76 and 83 against COLO-205 (colon) cancer cell lines at 10, 30 and 100 µg/ml respectively. Single strand DNA break was observed by the commet assay performed against cancer cell line and also DNA fragmentation was observed against cell line this confirm induction of apoptosis by alcoholic extract of the C. roseus. Therefore, alcoholic extract of Cathranthus roseus can be further explored for new bioactive compounds for chemotherapeutic drug.*

*Keywords:* Cathranthus roseus, Anti-proliferation, Apotosis, Sulforohadamine B assay, Commet assay.

---

1   School of Biotechnology, University of Jammu, Jammu – 180 006, J&K, India
2   Cancer Pharmacology Division, CSIR–Indian Institute of Integrative Medicine, Canal Road, Jammu – 180 001, J&K, India
*   *Corresponding author*: E-mail: madhulikasbt@gmail.com

# Introduction

Cancer is a major public health problem worldwide with millions of new cancer patients diagnosed each year and many deaths resulting from this disease. The limited success of clinical therapies including radiation, chemotherapy, immunomodulation and surgery in treating cancer, as evident by the high morbidity and mortality rates, indicates that there is an imperative need of new cancer management. Chemoprevention involves the use of pharmacological, dietary bio-factors, phytochemicals and even whole plant extracts to prevent, arrest, or reverse the cellular and molecular processes of carcinogenesis due to its multiple intervention strategies (Mehta *et al.*, 2010).

Medicinal plants are potential source of drugs, used to treat various serious diseases. The World Health Organization (WHO) estimates that almost 75 per cent of the world's population employs plant-based traditional remedies (Liu *et al.*, 2008). Over 60 per cent of the currently used anticancer chemotherapeutics are derived in one way or another from plants (Cragg and Newman, 2009; Tan *et al.*, 2006). Natural phytochemicals derived from medicinal plants have gained significant recognition in the potential management of several human clinical conditions, including cancer (Soobrattee *et al.*, 2006; Desai *et al.*, 2008; Guilford and Pezzuto, 2008). A number of new chemopreventive agents are being identified based on the ability to modulate one or more specific molecular events. The discovery of effective herbs and elucidation of their underlying mechanisms could lead to the development of an alternative and complementary method for cancer prevention and/or treatment. The Indian sub-continent has great botanical diversity and widespread use of traditional medicine practice known as ayurvedic medicine; however, only a relatively small number of these plants have been subjected to accepted scientific evaluation for their potential anticancer effects (Krishnaswamy, 2008).

*Catharanthus roseus* (L.) G. Don (Apocyanaceae) is known with various names (Madagascar periwinkle; *Vinca rosea*; *Lochnera rosea*) in India and all over the world. This short-lived perennial with dark green and glossy leaves is native to Madagascar. This plant is frequently mentioned in Ayurveda (an ancient Indian literature) and has traditionally been used to treat various diseases. Plant pacifies vitiated pitta, kapha, diabetes, hypertension, leukemia and malignancy, the root is stomachic, tonic and used internally to treat excess menstrual bleeding and vaginal discharges and externally to treat nosebleeds and piles. *V. minor* is used in homeopathy (Vinca) for haemorrhages. The related plant *Catharanthus roseus* is the source of two alkaloids, vincristine and vinblastine, used in the treatment of leukaemia and Hodgkin's disease in orthodox medicine (Ozgen *et al.*, 2003). The plant has more than 70 types of alkaloids (mostly monoterpene indole alklaloids), and some are known to be effective in treating various types of cancers including breast and lung cancer, uterine cancer, melanomas, and Hodgkin's and non-Hodgkin's lymphoma (EL-Sayed and Cordell, 1981; Ueda *et al.*, 2002). The anticancer drugs vincristine and vinblastine are synthesized from alkaloids of *Catharanthus roseus*. The plant is also known for its antihypertensive and antispasmodic properties. This plant have possesses known antibacterial, antimicrobial, antifungal, antioxidant, anticancer and antiviral activates (Marcone *et al.*, 1997; Prajakta and Ghosh, 2010). The present study is an attempt to assess antiproliferative properties and to understand the mechanisim of cell death by the whole plant of *Catharanthus roseus*.

# Materials and Methods

## Collection of Plant Material

Whole plant of *Catharanthus roseus* were collected locally from Jammu in the month of December and were authenticated at source by the taxonomist of the institute.

## Preparation of Plant Extracts

For the alcoholic extract, dried powdered plant material was extracted in soxhlet extractor with 95 per cent ethanol, and then concentrated to dryness under reduced pressure. Hydro-alcoholic extract was prepared by percolating another lot of dried ground plant material with 50 per cent ethanol and then concentrating it to dryness under reduced pressure. For the preparation of aqueous extract, the dried powdered plant material was heated with distilled water on steam bath for 2 hour, the supernatant was decanted and filtered through celite powder and the process was repeated four times, pooled extract was concentrated on rotavapour and dried in a lyophilizer, extract was obtained.

## Anti-Proliferative Activity

### Human Cancer Cell Lines

The human cancer cell lines *viz.*, A-549 (lung), HEP-2 (liver), KB (oral), COLO-205 (colon) were obtained from National Center for Cell Science, Pune, India and cultured in RPMI/MEM medium (pH 7.4), supplemented with FCS 10 per cent, pencillin 100 units/ml, streptomycin 100 µg/ml and glutamine 2 mM. Positive control 5-Flurouracil was prepared in distilled water and then diluted in gentamycin medium to obtain desired concentrations of $1 \times 10^{-5}$ M.

Test material was subjected to *in vitro* anticancer activity against human cancer cell lines (Monks *et al.*, 1991). For the assay (in brief), the cells were grown in tissue culture flasks in growth medium at 37°C in an atmosphere of 5 per cent $CO_2$ and 90 per cent relative humidity in a $CO_2$ incubator (Hera Cell; Heraeus; Asheville, NCI, USA). The cells at subconfluent stage were harvested from the flask by treatment with trypsin (0.05 per cent trypsin in PBS containing 0.02 per cent EDTA) and suspended in growth medium. Cells with more than 97 per cent viability (trypan blue exclusion) were used for determination of cytotoxicity. An aliquot of 100µl of cells ($10^5$cells/ml) was transferred to a well of 96-well tissue culture plate. The cells were allowed to grow for 24 h. Test material was then added to the wells and cells were further allowed to grow for another 48 h.

The anti-proliferative SRB assay (Shekhan *et al.*, 1990) was performed to assess growth inhibition which estimates cell number indirectly by staining total cellular protein with the dye SRB. In brief, the cell growth was stopped by gently layering 50µl of 50 per cent (ice cold) trichloroacetic acid on the top of growth medium in all the wells. The plates were incubated at 4°C for an hour to fix the cells attached to the bottom of the wells. Liquid of all the wells was then gently pipetted out and discarded.

The plates were washed five times with distilled water and were air-dried. SRB 100µl (0.4 per cent in 1 per cent acetic acid) was added to each well and the plates were incubated at room temperature for 30 min. The unbound SRB was quickly removed by washing the cells five times with 1 per cent acetic acid. Plates were air-dried, tris buffer (100µl, 0.01M, pH 10.4) was added to all the wells to solubilise the dye. Plates were gently stirred for 5 min on a mechanical stirrer. The optical density was recorded on ELISA reader at 540 nm. Suitable blanks and positive controls were also included. Each test was done in triplicate and the values reported herein are mean values of three experiments.

## COMET Assay (Single Cell Gel Electrophoresis, SCGE)

Drug-induced DNA damage was analyzed using the comet assay with modifications (Singh *et al.*, 1988). Cell pellets (treated with 0, 10, 30 and 100 µg/ml of chloroform fraction for 72 hr) were collected by centrifugation and re-suspended with 200 µl PBS and 800 µl of 1 per cent low melting point (LMP) agarose. The mixture was then pipetted onto a frosted glass microscope slide pre-coated with a layer of 1.0 per cent normal melting point agarose, prepared in PBS, covered with cover slips, and incubated at 4 °C for 10 min. After the LMP agarose solidified, the cover slips were gently removed, then 0.8 per cent LMP agarose pre-coated cover slips were added and the slides were allowed to solidify at 4 °C for 10 min. After 10 min, the cover slips were removed and the cells were lysed in high salt solution (2.5 M NaCl, 10 mM Tris–HCl, 100 mM EDTA, pH 10, with 1 per cent Triton and 10 per cent dimethyl sulfoxide added fresh) for one hour. The slides were then placed in a horizontal electrophoresis unit containing fresh buffer (1 mM EDTA, 300 mM NaOH pH 13) and incubated for 20 min to allow unwinding of DNA. Electrophoresis was then conducted in freshly prepared electrophoresis buffer (pH 13) for 20 min. at 25 V and 300 mA (0.8 V/cm) at 4 °C. Subsequently, the slides were gently washed with neutralization solution (0.4 M Tris–HCl, pH 7.5) for 20 min and stained with 20 µl ethidium bromide (15 µg/ml). Damaged cells were characterized by an extensive DNA fragmentation which allowed 90 per cent of the DNA to migrate during electrophoresis, forming the comet tail. Stained nucleoids were scored visually using a fluorescence microscope equipped with a digital camera.

## DNA Fragmentation Assay

Fragmentation of chromatin to units of single or multiple nucleosomes that form the nucleosomal DNA ladder in agarose gel is an established hallmark of programmed cell death or apoptosis. The cells grown to about 70 per cent confluence and treated with 5 per cent aqueous plant extracts for 48h and subjected to processing for DNA isolation and fragmentation assay as previously described (Shukla and Gupta, 2004; Srivastava and Gupta, 2007). The bands were visualized under an UV transilluminator, followed by digital photography.

## Statistical Analysis

Data were expressed as means ± SD Significant differences ($p<0.05$) between means of controls and treated cells were analyzed by ANOVA analysis.

# Results and Discussion

Herbal plants have been the basis for nearly all medicinal therapies since ancient times and the relevance of *Catharanthus roseus* is well documented for the treatment of various human diseases in ayurveda.

The anti-proliferative effect of whole plant of *Catharanthus roseus* was determined after 48 hours of incubation of the 95 per cent alcoholic, hydro-alcoholic and aqueous extracts against human cancer cell lines *viz.*, A-549 (lung), HEP-2 (liver), KB (oral), COLO-205 (colon) at the concentrations of 10, 30 and 100 µg/ml respectively. Growth inhibition was observed in a dose dependent manner in all the cell lines by all the extracts (Figure 25.1). Data expressed as mean ± S.E., unless otherwise indicated. Among all the three extract maximum growth inhibition was observed by alcoholic extract followed by hydro alcoholic extract and then by the aqueous extract. Also, significant growth inhibition was observed against A-549 (lung) followed by HEP-2 (liver), COLO-205 (colon) and KB (oral) cancer cell lines. Percent growth inhibition

**Figure 25.1:** *In vitro* cytotoxicity of root part of *Catharanthus roseus* against human cancer cell line. Data are mean±S.D. (*n* = 6-wells) and representative of three similar experiments.

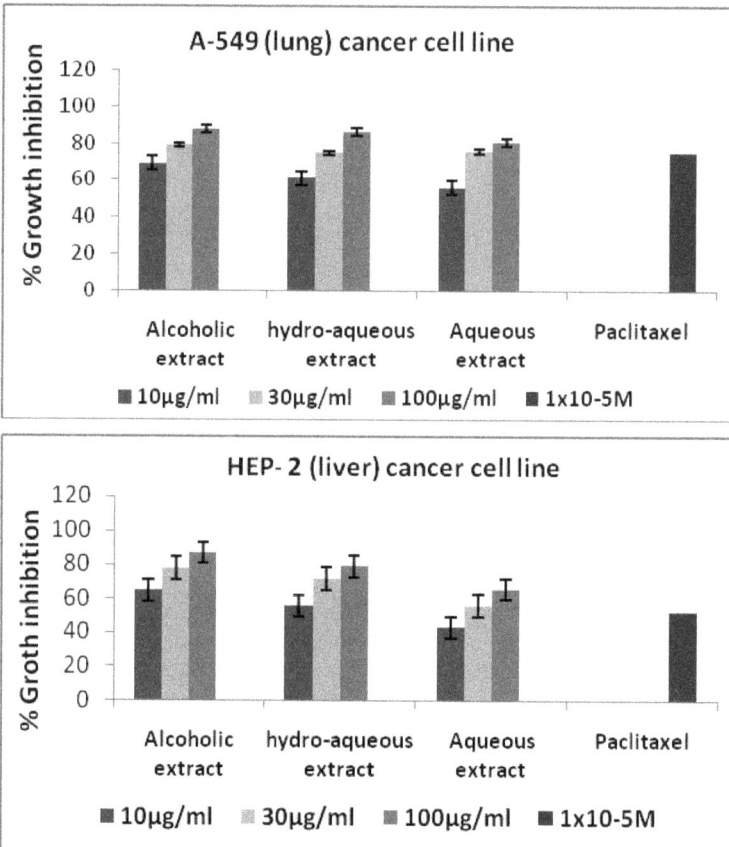

*Contd...*

**Figure 25.1**–*Contd...*

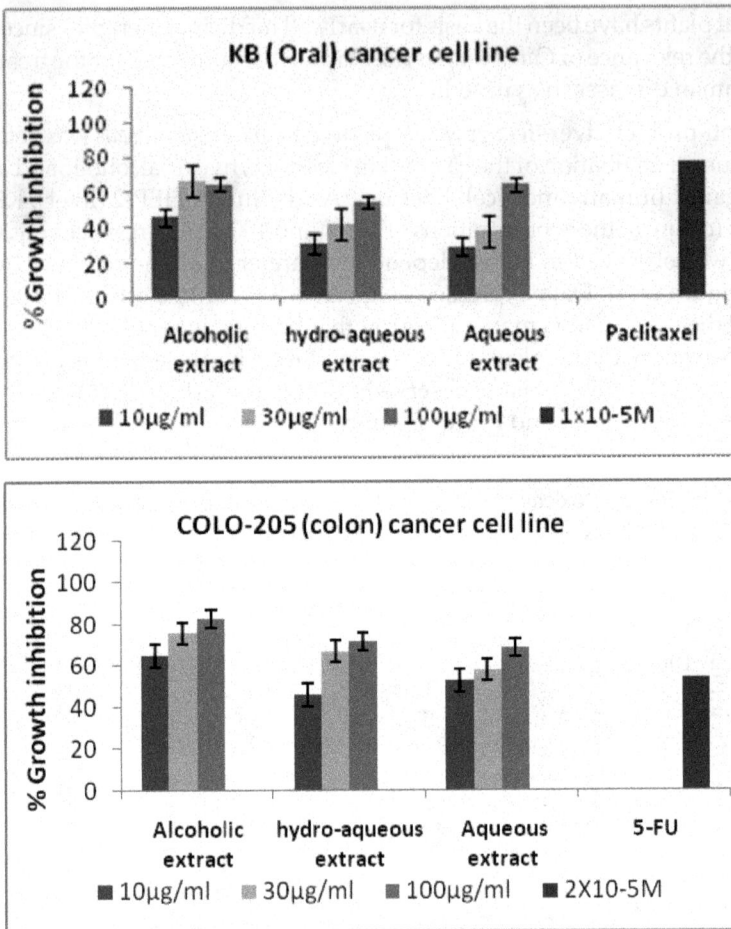

observed for alcoholic extract was 69, 79 and 88 against A549 (lung), 65, 78 and 87 against HEP-2 (liver), 46, 65 and 67 against KB (oral) and 65, 76 and 83 against COLO-205 (colon) cancer cell lines at 10, 30 and 100 µg/ml respectively. Whereas, hydro-alcoholic extract showed percent growth inhibition of 61, 75 and 86 against A-549 (lung), 56, 72 and 79 against HEP-2(liver), 31, 42 and 54 against KB (oral) and 46, 67 and 72 against COLO-205 (colon) cancer cell lines at 10, 30 and 100 µg/ml respectively. Aqueous extract showed significant percent growth inhibition of 56, 76 and 81 against A-549 (lung) and for rest of the cell lines percent growth inhibition was less than 69 percent (Figure 25.1). Comparing the data it was observed that maximum percent growth inhibition was observed against lung cancer cell line (A-549) and lowest antiproliferation against oral (KB) cancer cell line. Further to observe the mode of cell death the active extract *i.e.*, alcoholic extract was also evaluated for the assessment of DNA damage of the cancer cells by using comet assay A-549 cells.

The comet assay is a sensitive method used to monitor single strand (ss) DNA breaks at the single-cell level. Nuclei with condensed chromatin and apoptotic bodies, which are typical characteristics of apoptosis, were observed in A-549 cells incubated with alcoholic extract of *C.roseus,* and the number of apoptotic cells increased as the concentration of alcoholic extract of *C.roseus* was increased (Figure 25.2). So, the DNA damage induced by the alcoholic extract was also dose-dependent.

Current studies on development of effective cancer preventive approaches have focused mainly on the utilization of natural bioactive agents that can induce selective apoptosis in cancer cells (Mukherjee *et al.,* 2001). Alcoholic extracts have been screened for anticancer properties because traditional practitioners believed that mostly the polar compounds were responsible for the claimed anticancer potential. In this study all the extracts of *C. roseus* showed antiproliferative activity against four human cancer cell lines but among all the three extracts alcoholic extract showed significant activity. The activity of the alcoholic extract may due the presence of the various phytochemical such as Indole alkaloids, including majdine, majordine, akuammigine,

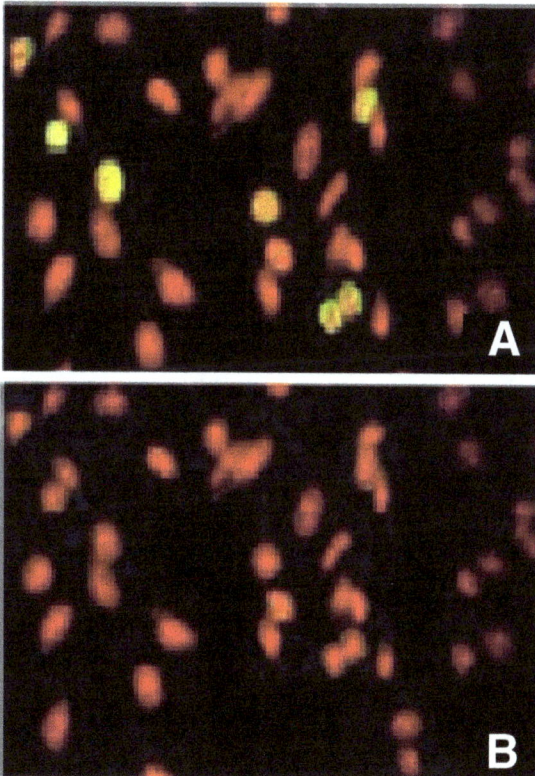

**Figure 25.2:** Morphological changes of HCT 15 cells after treatment with alcoholic extract of *Catharanthus roseus* for 48h followed by DAPI staining. (A) cells treated with 30 μg/ml alcoholic extract of *Catharanthus roseus* and (B) Control cells treated with 0.1 per cent DMSO. Light yellow colour cells indicate apoptotic bodies of nuclear fragmentation. Magnification x 100.

reserpine, sarpagine, serpentine, and vincamajine, tannins. As more than 200 monoterpenoid indole alkaloids are produced by *Catharanthus roseus* (Madagascar periwinkle), some of them are of pharmaceutical uses. The aerial parts of the plant are known to contain maximum amount of alkaloids. In this study the alcoholic extract of *Catharanthus roseus* was found to possess antiproliferative activity against human cancer cell lines and among them significant activity was demonstrated against A549 (lung) cancer cell lines, the extract also demonstrated induction of apoptosis. Since apoptosis is regarded as a new target in discovery of anticancer drugs, these results confirm the potential of *Catharanthus roseus* as an agent of chemotherapeutic activity against human lung cancer cells. Further investigation is under progress to elaborate this possibility.

## Acknowledgements

Authors are grateful to National Centre for Cell Science, Pune (India). We are also thankful to Dr. A.K Saxena's for their help and support to use the facilities of their Lab.

## Refrences

Cragg, G.M., and Newman, D.J.(2009). Nature: a vital source of leads for anticancer drug development. *Phytochemistry Review*, 8: 313-331.

Desai, A.G., Qazi, G.N., and Ganju, R.K., *et al.* (2008). Medicinal plants and cancer chemoprevention. *Current Drug Metabolism*, 9: 581-591.

El-Sayed, A., and Cordell, G.A. (1981). Catharanthamine: A new antitumor bisindole alkaloid from *Catharanthus roseus*. *Journal of Natural Product*, 44: 289-293.

Guilford, J.M., and Pezzuto, J.M. (2008). Natural products as inhibitors of carcinogenesis. *Expert Opinion of Investigation Drugs*, 17: 1341-52.

Krishnaswamy, K. (2008). Traditional Indian spices and their health significance. *Asia Pacific Journal of Clinical Nutrition*, 17: 265-8.

Liu, Y., and Wang, M.W. (2008). Botanical drugs: Challenges and opportunities. Contribution to Linnaeus Memorial Symposium 2007. *Life Sci*, 82: 445-449.

Marcone, A., Ragozzino, E., and Seemuller. (1997). Dodder transmission of alder yellows phytoplasma to the experimental host *Catharanthus roseus* (periwinkle*)*. *Forest Pathology*, 27: 347–350.

Mehta, R.G., Murillo, G., Naithani, R., and Peng, X. (2010). Cancer chemoprevention by natural products: how far have we come? *Pharmaceutical Research*, 27: 950-61.

Monks, A., Scudiero, D., Skehan, P., Shoemaker, R., Paull, K., Vistica, D., Hose, C., Langley, J., Cronise, P., Vaigro–Wolff, A., Gray–Goodrich, M., Campbell, H., Mayo, J., and Boyd, M. (1991). Feasibility of a high-flux anticancer drug screen using a diverse panel of cultured human tumor cell lines. *Journal of National Cancer Institute*, 83, 757-766.

Mukherjee,A.K., Basu, S., Sarkar, N., and Ghosh, A.C. (2001). Advances in cancer therapy with plant-based natural products. *Current Medicinal Chemistry*, **8:** 1467–1486.

Ozgen, U., Turkoz, Y., Stout, M., Ozugurlu, F., Pelik, F., Bulut, Y., *et al.* (2003). Degradation of vincristine by myeloperoxidase and hypochlorous acid in children with acute lymphoblastic leukemia. *Leukemia Research*, **27**: 1109-13.

Prajakta, J.P., and Ghosh, J.S. (2010). Antimicrobial Activity of *Catharanthus roseus* – A Detailed Study *British Journal of Pharmacology and Toxicology*, **1**: 40-44.

Shukla, S., and Gupta, S. (2004). Molecular mechanisms for apigenin-induced cell-cycle arrest and apoptosis of hormone refractory human prostate carcinoma DU145 cells. *Molecular Carcinogenesis*, **39**: 114-26.

Singh, N.P., McCoy, M.T., Tice, R.R., and Schneider, E.L. (1988). A simple technique for quantitation of low levels of DNA damage in individual cells. *Experimental Cell Research*, **175**: 184-191.

Skehan, P., Storeng, R., Scudiero, D., Monks, A., McMohan, J., Vistica, D., Warren, J.T., Bokesh, H., Kenny, S., and Boyd, M. (1990). New colorimetric cytotoxicity assay for anticancer drug screening. *Journal of National Cancer Institute*, **82**, 1107-1112.

Soobrattee, M.A., Bahorun, T., and Aruoma, O.I. (2006). Chemopreventive actions of polyphenolic compounds in cancer. *Biofactors*, **27**: 19-35.

Srivastava, J.K., and Gupta, S. (2007). Antiproliferative and apoptotic effects of chamomile extract in various human cancer cells. *The Journal of Agricultural and Food Chemistry*, **55**: 9470-8.

Tan, G., Gyllenhaal, C., and Sorjarto, D.D. (2006). Biodiversity as a source of anticancer drugs. *Curr Drug Targets*, **7**: 265-277.

Ueda, J.Y., Tezuka, Y., Banskota, A.H., Le Tran, Q., and Tran, Q.K. (2002). Antiproliferative activity of Vietnamese medicinal plants. *Biology of Pharmaceutical Bulletin*, **25**: 753-760.

Utilisation and Management of Medicinal Plants Vol. 2 (2014)   *Pages* **471–487**
*Editor-in-Chief:* **V.K. Gupta**
*Published by:* **DAYA PUBLISHING HOUSE, NEW DELHI**

# 26

# Indian Gooseberry (*Emblica officinalis*): Pharmacognosy Review

Madhulika Bhagat[1]*

## ABSTRACT

*Wonder berry, Emblica officinalis Gaertn., commonly known as Indian gooseberry or amla, is used in Ayurveda as a potent Rasaayana. The traditional system of medicine used almost all of its parts i.e. roots, leaves, stems but mostly it is known for astonishing properties of fruits. The fruit is used either alone or in combination with other plants to treat many ailments such as common cold and fever, as a diuretic, laxative, liver tonic, refrigerant, stomachic, restorative, alterative, antipyretic, anti-inflammatory, hair tonic, to prevent peptic ulcer and dyspepsia, and as a digestive. The present review gives an account of updated information on its phytochemical and pharmacological properties. The fruit is rich in quercetin, phyllaemblic compounds, gallic acid, tannins, flavonoids, pectin and vitamin C and also contains various polyphenolic compounds. A wide range of phytochemical components including terpenoids, alkaloids, flavonoids and tannins have been shown to possess useful biological activities. In addition, experimental studies have shown that some of its phytochemicals such as gallic acid, ellagic acid, pyrogallol, some non-sesquiterpenoids, corilagin, geraniin, elaeocarpusin, and prodelphinidins B1 and B2 also possess anti-neoplastic effects. Preclinical studies have shown that amla possesses anti-pyretic, analgesic, cardioprotective, gastroprotective, anti-hypercholesterolemia, wound healing, hepatoprotective, chemopreventive, free radical scavenging, antioxidant, anti-inflammatory, anti-mutagenic and immunomodulatory properties. In view of its reported pharmacological properties and relative safety, it can be said that E. officinalis is a source of potential therapeutically useful products.*

*Keywords*: *Emblica officinalis*, Triphala, Phytochemical constitutes, Pharmacological uses.

---

1   School of Biotechnology, University of Jammu, Jammu – 180 006, J&K, India
*   *Corresponding author*: E-mail: madhulikasbt@gmail.com

# Introduction

Medicinal plants have been used for thousands of years in herbal preparations of the Indian traditional health care system (Ayurveda) named Rasayana. *Emblica officinalis* (Amla) has an important position in Ayurveda- an Indian indigenous system of medicine. According to belief in ancient Indian mythology, it is the first tree to be created in the universe. It is a deciduous tree which belongs to family Euphorbiaceae. It is also commonly known as Amla in Hindi, *Phyllanthus emblica* or Indian gooseberry. The species is native to India and also grows in tropical and sub-tropical regions. The fruits of Indian gooseberry are pale yellowish, fleshy, globose (Figure 26.1) and are widely used in the Aryuveda and are believed to increase defense against diseases.

## Classification

| | | |
|---|---|---|
| *Kingdom* | : | Plantae |
| *Division* | : | Angiospermae |
| *Class* | : | Dicotyledonae |
| *Order* | : | Geraniales |
| *Family* | : | Euphorbiaceae |
| *Genus* | : | Emblica |
| *Species* | : | *officinalis* Geartn. |

**Figure 26.1:** *Emblica officinalis* (Amla).

## Botanical Description

A small to medium sized deciduous tree, 8-18 meters height with thin light grey bark exfoliating in small thin irregular flakes, leaves are simple, sub-sessile, closely set along the branchlets, light green having the appearance of pinnate leaves; flowers are greenish yellow, in axillary fascicles, unisexual, males numerous on short slender pedicels, females few, subsessile, ovary 3-celled; fruits globose, fleshy, pale yellow with six obscure vertical furrows enclosing six trigonous seeds in 2-seeded 3 crustaceous cocci (Nemmani *et al.*, 2002).

## Geographical Distribution

It is commonly found in tropical, sub-tropical, the sea-coast districts and on hill slopes upto 200 meters, also cultivated in plains and in heights of Kashmir and also found in Burma, it is abundant in deciduous forests of Madhya Pradesh (Chaudhuri, 2004; Sai *et al.*, 2002; Bhattacharya *et al.*, 1999).

## Planting

Amla is generally propagated through seeds, but seed propagated trees bear inferior quality fruits and have a long gestation period. Shield budding is done on one year old seedlings with buds collected from superior strains yielding big size fruits. Older trees of inferior types can be rejuvenated and easily changed into superior type by top working. The pits of $1m^3$ are prepared during May-June at a distance of 4.5 m spacing and should be left for 15-20 days exposed to sunlight. Each pit should be filled with surface soil mixed with 15 kg farm yard manure and one kg of super phosphate before planting the grafted seedling.

## Description

Tree; leaves alternate, bifarious, pinnate, flower -'bearing; leaflets numerous, alternate, linear-obtuse, entire; petioles striated, round; calyx 6-parted; flowers in the male very numerous in the axils of the lower leaflets, and round the common petiole below the leaflets; in the female few, solitary, sessile, mixed with some males in the most exterior floriferous axils; stigmas 3; drupe globular, fleshy, smooth, 6-striated; nut obvate-triangular, 3-celled; seeds 2 in each cell; flowers small, greenish yellow. Flower during October (Treadway, 1994).

## Photochemical constituents

Amla contains calcium, phosphorous, iron, carotene, thiamine, riboflavin, and niacin. The seeds of Indian gooseberry contain fixed oil, phosphatides and essential oil. The fruits, leaves and bark are rich in tannins. The roots contain ellagic acid and lupeol and bark contains leucodelphinidin. The seeds yield a fixed oil (16 per cent) which is brownish-yellow in colour. It has the following fatty acids: linolenic (8.8 per cent), linoleic (44.0 per cent), oleic (28.4 per cent), stearic (2.15 per cent), palmitic (3.0 per cent) and myristic (1.0 per cent). Ethanol soluble fraction contains free sugars, D-glucose, D-fructose, D-myo-inositol. The acidic water soluble fraction contains pectin with D-galacturonic acid, D-arabinosyl, D- rhamnosyl, D-xylosyl, D-glucosyl, D-mannosyl and D-galactosyl residues (Bhattacharya *et al.*, 1999). The low molecular weight hydrolyzable tannins (<1,000), namely Emblica nin A and Emblica nin B, along with pedunculagin and punigluconin are the key ingredients in Emblica (Kim *et al.*, 2005). It contains 3-6-di-o-galloyl-glucose (Fruit), 3-6-di-o-galloyl-glucose (Shoot), Alanine (Fruit), Amlaic-acid (Leaf), Arginine (Fruit), Ascorbic-acid (Fruit), Ascorbic-acid (Plant), Ash (Fruit), Aspartic-acid (Fruit), Astragalin (Leaf), β-carotene (Fruit), β-sitosterol (Bark, Seed Oil, Tissue Culture, Shoot), Boron (Fruit), Calcium (Fruit), Carbohydrates (Fruit), Chebulagic acid (Fruit), Chebulagic acid (Shoot), Chebulaginic acid (Fruit), Chebulic acid (Fruit), Chibulinic acid (Fruit), Chibulinic acid (Shoot), Chloride (Fruit), Copper (Fruit), Corilagic acid (Fruit), Corilagin (Fruit, Shoot), Cystine (Fruit), D-fructose (Fruit), D-glucose (Fruit), Ellagic acid (Fruit, Shoot, Root, Pericarp and Leaf), Emblicol (Fruit, Pericarp), Ethyl gallate (Fruit), Fat (Fruit and Seed), Fibre (Fruit), Gallic acid (Fruit, Shoot, Pericarp), Gallic acid ethyl ester (Fruit, Tissue Culture), Gallo-tannin (Leaf), Gibberellin-a-1 (Fruit), Gibberellin-a-3 (Fruit), Gibberellin-a-4 (Fruit), Gibberellin-a-7 (Fruit), Gibberellin-a-9 (Fruit), Glucogallin (Fruit), Glucogallin (Shoot), Glucose (Fruit), Glutamic acid (Fruit), Glycine (Fruit), Histidine (Fruit), Iron (Fruit), Isoleucine (Fruit), Kaempferol (Leaf), Kaempferol-3-o-glucoside (Leaf), Leucine (Fruit), Leucodelphinidin (Bark), Linoleic acid (Seed, Seed Oil, Linolenic acid (Seed, Seed Oil), Lupenone (Plant), Lupeol (Bark, Root, Shoot), Lysine (Fruit), Magnesium (Fruit), Manganese (Fruit), Methionine (Fruit), Myo-inositol (Fruit), Myristic acid (Fruit, Seed Oil), Niacin (Fruit), Nitrogen (Fruit), Oleic acid (Seed, Seed Oil), Palmitic acid (Seed, Seed Oil), Pectin (Fruit), Phenylalanine (Fruit), Phosphorus (Fruit), Phyllantidine (Fruit, Tissue Culture, Leaf), Phyllantine (Fruit, Leaf, Tissue Culture), Phyllemblic acid (Fruit, Pericarp), Phyllemblin (Fruit), Phyllemblinic acid (Fruit), Polysaccharide (Fruit), Potassium (Fruit), Proline (Fruit), Protein (Fruit), Quercetin (Tissue Culture), Riboflavin (Fruit), Rutin (Fruit, Leaf), Selenium (Fruit), Serine (Fruit), Silica (Fruit), Sodium (Fruit), Starch (Fruit), Stearic acid (Seed, Seed Oil), Sucrose (Fruit), Sulfur (Fruit), Tannin (Bark, Fruit, Twig, Leaf), Terchebin (Fruit), Thiamin (Fruit), Threonine (Fruit), Trigalloyl glucose (Fruit), Tryptophan (Fruit), Tyrosine (Fruit), Valine (Fruit), Water (Fruit), Zeatin (Fruit), Zeatin nucleotide (Fruit), Zeatin riboside (Fruit), Zinc (Fruit) (Srikumar *et al.*, 2007).

## Active Principle

Tannins, Gallic acid and Pyrogallol (Veena *et al.*, 2006)

## Ethno-botanical Uses

In traditional Indian medicine, dried and fresh fruits of the plant are used. All parts of the plant are used in various Ayurvedic/Unani medicine (*Jawarish amla*) herbal preparations, including the fruit, seed, leaves, root, bark and flowers. According to Ayurveda, amla fruit is sour and astringent in taste, with sweet, bitter and pungent secondary tastes. Its qualities are light and dry, the post-digestive effect is sweet, and its energy is cooling. It may be used as rejuvenative to promote longevity, and traditionally to enhance digestion, treat constipation, reduce fever, purify the blood, reduce cough, alleviate asthma, strengthen the heart, benefit the eyes, stimulate hair growth, enliven the body, and enhance intellect. They are useful in vitiated conditions of tridosha, diabetes, dyspepsia, colic, flatulence, hyperacidity, peptic ulcer, erysipelas, skin diseases, leprosy, haematogenesis, inflammations, anemia, emaciation, hepatopathy, jaundice, strangury, diarrhoea, dysentery, hemorrhages, leucorrhoea,

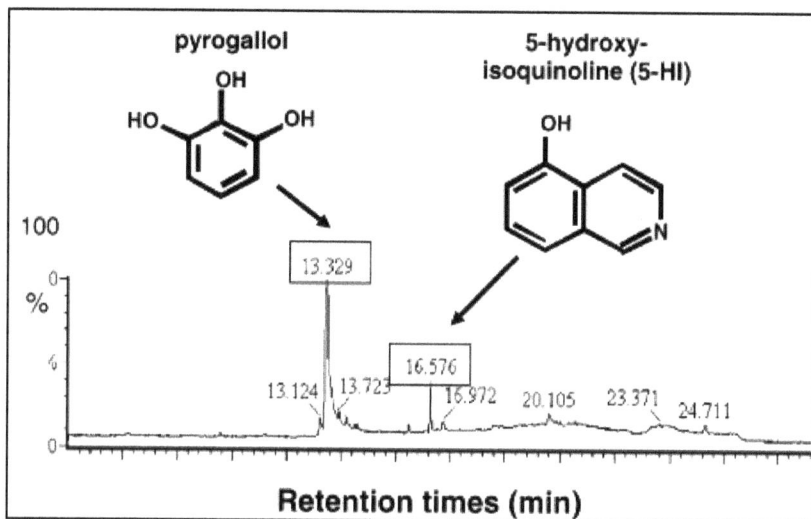

**Figure 26.2**: Gas chromatography/mass spectrometry analysis of the n-butanol fraction of *Emblica officinalis* extracts.

menorrhagia, cardiac disorders, intermittent fevers and greyness of hair (Thakur, 1989; Saeed and Tariq, 2007; Jayaweera, 1980; Chaudhuri *et al.*, 2003; Tasduq *et al.*, 2005; Biswas *et al.*, 2001).

In Ayurvedic polyherbal formulations, Indian gooseberry is a common constituent, and most notably is the primary ingredient in an ancient herbal *rasayana* called *Chyawanprash*.http://en.wikipedia.org/wiki/Indian_gooseberry - cite_note-dharm-12 This formula, which contains 43 herbal ingredients as well as clarified butter, sesame oil, sugar cane juice, and honey, was first mentioned in the Charaka Samhita as a premier rejuvenative compound. Another very important and popularly used Ayurvedic herbal rasayana churna is Triphala churna consisting of equal parts of three myrobalans, taken without seed: Amalaki (*Phyllanthus emblica*), Bibhitaki (*Terminalia bellirica*), and Haritaki (*Terminalia chebula*). Triphala is a mild laxative, which cleanses and tonifies the gastro-intestinal tract, including a blood cleanser. The herb also has a high nutritional value, including high levels of vitamin C (McIntyre, 2005; Jagetia, 2004).

# Pharmacology and Clinical Studies

## Antioxidant and Anti-ageing Effect

The use of amla as an antioxidant has been examined by a number of researchers (Vani, 1997; Bhattacharya *et al.*, 1999; Golechha *et al.*, 2012). The studies showed that Amla preparations contained high levels of the free-radical scavenger, superoxide dimutase (SOD), in the experimental subjects (Treadway, 1994). *Emblica officinalis* (Eo) reduced UV-induced erythema and showed free-radical quenching ability, chelating ability to iron and copper as well as MMP-1 and MMP-3 inhibitory activity (Chaudhuri, 2003). In another study, amla was studied against the cold stress-induced alterations in the behavioural and biochemical abnormalities. Triphala was administered orally about 1g/kg/animal body weight for 48 days. It significantly prevented cold stress-induced behavioral and biochemical abnormalities in albino rats (Dhanalakshmi *et al.*, 2007). The hydrolysable tannins emblicanin A (2,3-di-*o*-galloyl-4,6-(*S*)-hexahydroxydiphenoyl- 2-keto-glucono-d-lactone) and emblicanin B (2,3,4,6-bis-(*S*)-exahydroxydiphenoyl-2-keto-glucono-d-lactone) proved as very strong antioxidant. These two emblicanins A and B also preserve erythrocytes against oxidative stress induced by asbestos, generator of superoxide radical. Emblicanin A oxidates when put in contact with asbestos becoming emblicanin B and together they have a stronger protective action to erythrocytes than vitamin C. Moreover they improve the efficacy of vitamin C in reducing dihydroascorbic acid to ascorbic acid. The same recycling process has been observed in the rutin-vitamin C combination (Scartezzini and Speroni, 2002).

## Immunomodulation Effect

*Emblica officinalis* had shown to modulate the immune system and the inflammatory response. Immunomodulatory activity of Triphala (an herbal formulation containing fruits of *Emblica officinalis*, *Terminalia chebula* and *Terminalia belerica* in equal proportions) was reported by Srikumar and his co-worker in albino rats. They showed that Triphala stimulates the neutrophil functions in the immunized rats and stress induced suppression in the neutrophil functions (Srikumar *et al.*, 2005). Another ayurvedic polyherbal formulation Immu-21, containing extracts of an *Emblica officinalis*, *Ocimum sanctum*, *Withania somnifera* and *Tinospora cordifolia* showed immunomodulatory response in mice. Pretreatment with Immu-21 selectively elevated the proliferation of splenic leukocyte to B cell mitogen, LPS and cytotoxic activity against K 562 cells in mice (Nemmani *et al.*, 2002). The immunosuppressive effects of Cr on lymphocyte proliferation and restoration of the IL-2 and gamma-IFN production was reported (Sai *et al.*, 2002).

## Antipyretic, Analgesic and Anti-inflammatory effect

Extracts of *Emblica officinalis* leaves and fruits possess potent anti-pyretic, analgesic as well as anti-inflammatory activity (Mythilypriya *et al.*, 2007; Gupta *et al.*, 2013; Asmawi *et al.*, 1993; Ihantola-Vormisto *et al.*, 1997; Muthuraman *et al.*, 2011). A single oral dose of ethanolic extract and aqueous extract (500 mg/kg, i.p.) showed significant reduction in hyperthermia in rats induced by brewer's yeast. Both of these extracts elicited pronounced inhibitory effect on acetic acid-induced writhing response in mice in the analgesic test (Perianyagam *et al.*, 2004). Yet in other studies, fruit extract was found to be an effective anticoagulant and anti-inflammatory agent as it potentially and significantly reduced lipopolysaccharide (LPS)-induced tissue factor expression and von Willebrand factor release in human umbilical vein endothelial cells (HUVEC), it also decreased the concentrations of pro-inflammatory cytokines, TNF-$\alpha$ and IL-6 in serum on oral administration of the amla fruit extract (50 mg/kg body weight) (Pradyumna *et al.*, 2013). Further, the Beta-glucogallin an aldose reductase inhibitor that catalyzes the reduction of toxic lipid aldehydes to their alcohol products and mediates inflammatory signals triggered by lipopolysaccharide (LPS) was isolated from *Emblica officinalis*. This molecule may be a potential therapy for inflammatory diseases (Chang *et al.*, 2013).

## Chemoprotective and Anticancer Effect

*Emblica officinalis* is a wonder berry known for the treatment and prevention of cancer (Rajarama and Siddiqui, 1964; Jeena *et al.*, 2001; Wiart, 2013). The crude extract of *Emblica officinalis* was reported to counteract hepatotoxic and renotoxic effects of metals due to anti-oxidant activity (Roy *et al.*, 1991). Triphala exhibits chemopreventive potential when included in the diet of the mice, results in the lowering of the benzo(a)pyrene [B(a)P] induced forestomach papillomagenesis this may be due to the increased antioxidant status in animals by Triphala (Deep *et al.*, 2005). Lipid-metabolizing enzymes, lipids and lipoproteins are reported to be associated with the risk of breast cancer. Kalpaamruthaa (KA) is a modified Siddha preparation containing *Emblica officinalis*, *Semecarpus anacardium* (SA) and honey. The elevated levels of free cholesterol, total cholesterol, triglycerides, phospholipids and free fatty acids and decreased levels of ester cholesterol in plasma, kidney and liver found in cancer suffering animals were reverted back to near normal levels on treatment with KA and SA (Veena *et al.*, 2006). Triphala significantly decreased lipid peroxidation and the activity of lactate dehydrogenase (LDH) in 1,2-dimethylhydrazinedihydrochloride (DMH) induced Endoplasmic reticulum stress (ER stress) in mouse liver (Sharma and Sharma, 2011). In one another study, methanolic extract of the fruit at a dose of 200mg/kg BW possesses optimum chemopreventive effect against DMBA-induced buccal pouch carcinogenesis (Krishnaveni and Mirunalini, 2012). Growth inhibitory activity of *Emblica officinalis* was primarily manifested through induction of apoptotic cell death and through inhibition of AP-1 further accompanied by suppression of viral transcription resulted in growth inhibition of cervical cancer cells (Mahata *et al.*, 2013). The effect of the *Phyllanthus emblica* against cancer was also tried to enhanced by creating synthesize silver nanoparticles by amla extract, and it was observed that the AgNPs (silver nanoparticle) capped with biomolecules of amla enhanced cytotoxicity in laryngeal cancer cells through oxidative stress and apoptotic function on Hep2 cancer cells (Rosarin *et al.*, 2013).

## Hepatoprotective Effect

*Emblica officinalis* fruits have been reported to be used for hepatoprotection in Ayurveda (Bhattacharya *et al.*, 2005). The protective effect and further inhibition of hepatic toxicity against ethanol induced rat hepatic injury was reported by some authors (Sultana *et al.*, 2005; Pramyothin *et al.*, 2006). The standardized herbal extracts showed reduction of lipid peroxidation and cellular damage against tert-butyl hydroperoxide (t-BH) induced toxicity (Hiraganahalli *et al.*, 2012). The fruit extract was reported as anti-hyperglycemic and hepato-renal in fluoride induced toxicity, it significantly reduced plasma glucose levels, SGOT, SGPT, ACP, ALP, hepatic G-6-Pase activity and increased hepatic glycogen content and hexokinase activity (Vasant and Narasimhacharya , 2012), fruit extract also found efficient in lessening intraperitoneally injected iron dextran-induced liver toxicity in Swiss albino mice (Sarkar *et al.*, 2013), protect liver against antituberculosis (anti-TB) drugs-induced hepatic injury due to its membrane stabilizing, antioxidative and CYP 2E1 inhibitory roles (Tasduq *et al.*, 2005). The extract of *E. officinalis* and Chyavanaprash showed hepatoprotective effect in carbon tetrachloride ($CCl_4$) induced liver injury in rats. Both extracts were found to inhibit these elevated levels of serum and liver lipid peroxides (LPO), glutamate-pyruvate transaminase (GPT) and alkaline phosphatase (ALP) significantly, showing that the extract could reduce the induction of fibrosis in rat's model (Jose and Kuttan, 2000).

## Cardioprotective, Cholesterol and Dyslipidemia

The *E. officinalis* showed cardioprotective effect against isoproterenol (ISP)-induced cardiotoxicity in rats, in this study the pretreatment with *E. officinalis* exhibited restoration of hemodynamic and left ventricular function along with significant preservation of antioxidants, myocytes-injury-specific marker enzymes and significant inhibition of lipid peroxidation the protection was attributed to its potent antioxidant and free radical scavenging activity which was evidenced by favorable improvement in hemodynamic, contractile function and tissue antioxidant status (Ojha *et al.*, 2012). The fresh fruit homogenate of showed adaptation on myocardial antioxidant system and oxidative stress induced by ischemic-reperfusion injury (IRI) against rats hearts by augmenting endogenous antioxidants and protects rat hearts from oxidative stress associated with IRI (Rajak *et al.*, 2004). Earlier, a human pilot study demonstrated a reduction of blood cholesterol levels in both normal and hypercholesterolemic men with treatment (Jacob *et al.*, 1988). It also decreased the low-density lipoprotein cholesterol and increased HDL cholesterol in ovariectomized rats fed chow or fructose. In ovariectomized and fructose-fed rats, it prevented insulin resistance aside from subduing the rise in triglycerides. Amla may be explored for its use in preventing dyslipidemia in postmenopausal women (Koshy *et al.*, 2012).

## Antimicrobial and Antimutagenicity

Amla has also been reported for its antimicrobial activities (Srikumar *et al.*, 2007; Rani and Khullar, 2004). The ether extract and 80 percent alcoholic extract of fruits acidified with hydrochloric acid, were found to have antibacterial activity. The other extract of acidified alcoholic extract showed the highest activity, inhibiting the growth of *M. pyogenes* var. *S. typhosa* and *S. paratyphi* at a concentration of 0.21mg /ml and that of *M. pyogenes* var. *albus*; *S. schottmmellari* and *S. dysenteriae* at a concentration of 0.42mg/ml (Khorana *et al.*, 1959).The potent antibacterial activity was reported against *Escherichia coli, K. ozaenae, Klebsiella pneumoniae, Proteus mirabilis, Pseudomonas aeruginosa, S. paratyphi A, S. paratyphi B* and *Serratia marcescens* (Saeed and Tariq, 2007). The chloroform and acetone extracts of Triphala showed inhibition of mutagenicity induced by both direct and S9-dependent mutagens (Kaur *et al.*, 2007). The fungal endophytes inhabiting *Emblica officinalis* showed antimicrobial and antioxidant activity (Nath *et al.*, 2012).

## Osteoarthritis

Osteoarthritis (OA) is a serious, degenerative disease. *Emblica officinalis* fruits have been reported as chondroprotective agent in osteoarthritis therapy (Sumantran *et al.*, 2008). The ayurvedic formulations (extracts of *Tinospora cordifolia, Zingiber officinale, Emblica officinalis*) were equivalent to glucosamine and celecoxib (Chopra *et al.*, 2013). There is preliminary evidence *in vitro* that its extracts induce apoptosis and modify gene expression in osteoclasts involved in rheumatoid arthritis and osteoporosis (Penolazzi *et al.*, 2008).

## Respiratory Diseases

Amla is especially valuable in tuberculosis of the lungs asthma and bronchitis. Pulmonary antioxidant defenses are widely distributed in lungs and include both enzymatic and non enzymatic systems. The primary non enzymatic antioxidants are membrane bound vitamin C and Vitamin E. Amla is the richest source of flavonoids and vitamin C. As an antioxidant, it is very effective in inhibiting lipid peroxidation by scavenging reactive species and free radicals, thus preventing tissue damage. Dietary supplement with amla protects against *K. pneumoniae* mediated respiratory tract infection by keeping a check on the induction of proinflammatory cytokine like TNF-α (Saini *et al.*, 2008). It has also shown hypotensive effect and also a synergistic cholinergic and synergistic histaminergic effect on MABP, HR and RR of anaesthetized male dogs (Geer *et al.*, 2005). In Turkey, the fresh fruit is used for inflammations of the lungs. The juice or extract of the fruit is mixed with honey and pipit added is given to stop hiccough and also in painful respiration. The expressed juice of the fruit along with other ingredients is used to cure cough, hiccough, asthma and other diseases (Jayaweera, 1980).

## Antidiabetic Effect

Due to its high vitamin C content, *E. officinalis* is effective in controlling diabetes. A tablespoon of its juice mixed with a cup of bitter gourd juice, taken daily for two months will stimulates the pancreas and enable is to secrete insulin, thus reducing the blood sugar in the diabetes. Diet restrictions should be strictly observed while taking this medicine. It will also prevent eye complication in diabetes (Sampat Kumar *et al.*, 2012). Diabetes mellitus increases the risk of cardiomyopathy and heart failure independent of underlying coronary artery disease. There may be a decrease in insulin sensitivity associated with autonomic dysfunction, hypertension and left ventricular (LV) hypertrophy. An increase in oxidative stress contributes to the characteristic morphological and functional abnormalities that are also associated with diabetic cardiomyopathy, the fruit juice was found beneficial for the treatment of myocardial damage associated with type 1 diabetes mellitus (Patel and Goyal, 2011). Another study was conducted where alloxan-induced diabetic rats were given an aqueous amla fruit extract. Decrease of the blood glucose, triglyceridemic levels and also an improvement of the liver function caused by a normalization of the liver-specific enzyme alanine transaminase activity was observed (Qureshi *et al.*, 2009). The hydro methanolic extract of *E.officinalis* leaves effectively normalized the impaired antioxidant status in streptozotocin induced diabetes in dose dependent manner than the glibenclamide-treated groups of rats. The extract exerted rapid protective effects against lipid peroxidation by scavenging of free radicals and reducing the risk of diabetic complications (Nain *et al.*, 2012). Aldose reductase (AR) has its involvement in development of secondary complications of diabetes including cataract. *E. officinalis* is proved as an important of AR. Exploring the therapeutic value of natural ingredients that people can incorporate into everyday life may be an effective approach in the management of diabetic complications (Suryanarayana *et al.*, 2004). The anti-diabetic effect of *E. officinalis* was also reported by Kalekar and his co worker through an insulin sensitizing effect (stimulation of glucose uptake into adipocytes) (Kalekar *et al.*, 2013).

## Indigestion, Anaemia, Jaundice and Dyspepsia

Clinical studies were conducted to investigate the effect of crude amla in gastritis syndrome. The crude amla was given in 20 cases in a dose of 3 gm, 3 times a day for 7 days. The drug was found effective in 85 per cent of the cases. It was observed that the drug did not have any significant beneficial effect in cases of hypochlorhydria. Only cases of hyperchloridia with burning sensation in abdominal and cardiac regions and epigastric pain were benefited (Singh and Sharma, 1971). The fresh juice of Amla is given as tonic, diuretic and anti-bilious remedy. It is also helpful in burning sensation, over thirst, dyspepsia and other complaints of digestive system. Use dried fruit with iron. Fermented liquor prepared from the root is used in jaundice, dyspepsia, cough, etc. Fruit possesses prokinetic and laxative activities in mice along with spasmodic effect in the isolated tissues of guinea pig and rabbit, mediated partially through activation of muscarinic receptors (Mehmood *et al.*, 2012). The therapeutic efficacy of amla in case of dyspepsia was evaluated with promising results in human subjects (Singh and Sharma, 1971; Chawla *et al.*, 1982).

## Anti-Ulcer and Wound Healing

A herbomineral formulation of the Ayurveda medicine named Pepticare, composed of *Emblica officinalis*, *Glycyrrhiza glabra* and *Tinospora cordifolia* was tested for its anti-ulcer and anti-oxidant activity in rats. Reports were made that Pepticare exhibit anti-ulcer activity, which can be attributed to its anti-oxidant property (Bafna *et al.*, 2005). Methanolic extract of Eo was studied against ulcer and similarly represent significant ulcer protective and healing effects was observed this might be due to its effects both on offensive and defensive mucosal factor (Sairam *et al.*, 2002). The healing activity of gallic acid enriched ethanolic extract (GAE) of fruits (amla) against the indomethacin-induced gastric ulceration in mice was investigated. The activity was correlated with the ability of GAE to alter the cyclooxygenase- (COX-) dependent healing pathways. Treatment with GAE (5mg/kg/day) and omeprazole (3mg/kg/day) for 3 days led to effective healing of the acute ulceration, while GAE could reverse the indomethacin-induced pro-inflammatory changes of the designated biochemical parameters. The ulcer healing activity of GAE was, however, compromised by coadministration of the nonspecific NOS inhibitor, N-nitro-L-arginine methyl ester (L-NAME), but not the i-NOS-specific inhibitor, L-N6-(1-iminoethyl) lysine hydrochloride (L-NIL). Thus the study suggested that the GAE treatment accelerates ulcer healing by inducing PGE (2) synthesis and augmenting e-NOS/i-NOS ratio (Chatterjee *et al.*, 2012*). E. officinalis* fruit extract promoted NO production, endothelial wound closure, endothelial sprouting, and VEGF mRNA expression. Therefore, it also proves useful in endothelial function and restoring wound healing competency (Chularojmontri *et al.*, 2013).

## Other Diseases

Ophthacare is a herbal eye drop preparation containing basic principles of different herbs *viz Carum copticum, Terminalia belerica, Emblica officinalis, Curcuma longa, Ocimum sanctum, Cinnamomum camphora, Rosa damascena* and *Meldespumapum*. It exhibits a beneficial role in a number of inflammatory, infective and degenerative ophthalmic (Biswas *et al.*, 2001). Body weight loss in extract administered irradiated animals was significantly less in comparison with animals who were given radiation only (Singh *et al.*, 2005). Amla churna produced a dose-dependent improvement in memory of young and aged rats. It reversed the amnesia induced by scopolamine and diazepam. Amla churna may prove to be a useful remedy for the management of Alzheimer's disease due to its multifarious beneficial effects such as memory improvement and reversal of memory deficits (Vasudevan *et al.*, 2007; Ali *et al.*, 2013).

## Conclusion

The use of medicinal plants in the management of various illnesses is due to their phytochemical constituents and dates back to historical age. While being exceptional for its ethnic, ethnobotanical and ethnopharmaceutical use, it is an important ingredient of many Ayurvedic medicines and tonics. Various extracts and herbal formulations of *Emblica officinalis* showed activities against various diseases and result is similar to standard drugs. It is one of the richest natural sources of Vitamin C and plays a vital role in preventing innumerable health disorders. It is considered to be a safe herbal medicine without any adverse effects. So it can be concluded that the Indian gooseberry is a traditionally and clinically proven fruit for both its application and efficacy.

## References

Alam, M.I, and Gomes, A. (2003). Snake venom neutralization by Indian medicinal plants (*Vitex negundo* and *Emblica officinalis*) root extracts. *J. Ethnopharmacol.*, **86(1):** 75-80.

Ali, S.K., Hamed, A.R., Soltan, M.M., Hegazy, U.M., Elgorashi, E.E., El-Garf, I.A., and Hussein, A.A. (2013). *In-vitro* evaluation of selected Egyptian traditional herbal medicines for treatment of alzheimer disease. *BMC Complement Altern Med.*, **13(1):** 121.

Asmawi, M.Z., Kankaanranta, H., Moilanen, E. *et al.* (1993). Antiinflammatory activities of *Emblica officinalis* Gaertn leaf extracts. *J Pharm Pharmacol.*, **45:** 581–584.

Bafna, P.A., and Balaraman, R. (2005). Anti-ulcer and anti-oxidant activity of pepticare, a herbomineral formulation. *Phytomedicine.*, **12(4):** 264-70.

Bhattacharya, A., Chatterjee, A., Ghosal, S., and Bhattacharya, S.K. (1999). Antioxidant activity of active tannoid principles of *Emblica officinalis* (amla). *Indian J Exp Biol.*, **37(7):** 676-80.

Bhattacharya, A., Kumar, M., Ghosal, S., and Bhattacharya, S.K.(2000). Effect of bioactive tannoid principles of *Emblica officinalis* on iron-induced hepatic toxicity in rats. *Phytomedicine.*, **7(2):** 173-5.

Biswas, N.R., Gupta, S.K., Das, G.K., Kumar, N., Mongre, P.K., Haldar, D., and Beri, S.(2001). Evaluation of Ophthacare® eye drops—a herbal formulation in the management of various ophthalmic disorders. *Phytotherapy Research.*, **15(7):** 618–620.

Blasdell, K.S., Sharma, H.M., Tomlinson, P.F., and Wallace, R.K.(2013). Subjective survey, blood chemistry and complete blood profile of subjects taking Maharishi Amrit Kalash (MAK). *Faseb Journal,* **5:** A1317.

Chang, K.C., Laffin, B., Ponder, J., Enzsöly, A., Németh, J., LaBarbera, and D.V., Petrash, J.M.(2013). Beta-glucogallin reduces the expression of lipopolysaccharide-induced inflammatory markers by inhibition of aldose reductase in murine macrophages and ocular tissues. *Chem Biol Interact.*, **202(1-3):** 283-7.

Chatterjee, A., Chatterjee, S., Biswas, A., Bhattacharya, S., Chattopadhyay, S., and Bandyopadhyay, S.K. (2012). Gallic acid enriched fraction of *Phyllanthus emblica* potentiates indomethacin-induced gastric ulcer healing via e-NOS-dependent pathway. *Evid. Based Complement Alternat Med.,* 487380. doi: 10.1155/2012/487380.

Chaudhuri, R.K., Guttierez, G., and Serrar, M. (2003). Low Molecular-Weight Tannins of *Phyllanthus emblica*: A new class of anti-aging ingredients. *Proceedings Active Ingredients Conference, Paris.*

Chaudhuri, R.K. (2004). Standardised extract of *Phyllanthus emblica*: A skin lightener with anti-aging benefits. *Proceedings PCIA Conference, Guangzhou, China.*

Chawla, Y.K., Dubey, P., Singh, P. *et al.* (1982). Treatment of dyspepsia with Amalaki (*Emblica officinalis*) an Ayurvedic drug. *Indian J Med Res.,* **76:** 95–98.

Chopra, A., Saluja, M., Tillu, G., Sarmukkaddam, S., Venugopalan, A., Narsimulu, G., Handa, R., Sumantran, V., Raut, A., Bichile, L., Joshi, K., and Patwardhan, B. (2013). Ayurvedic medicine offers a good alternative to glucosamine and celecoxib in the treatment of symptomatic knee osteoarthritis: a randomized, double-blind, controlled equivalence drug trial. *Rheumatology (Oxford).*

Chularojmontri, L., Suwatronnakorn, M., and Wattanapitayakul, S.K. (2013). *Phyllanthus emblica* L. enhances human umbilical vein endothelial wound healing and sprouting. *Evid Based Complement Alternat Med.* doi: 10.1155/2013/720728.

Deep, G., Dhiman, M., Rao, A.R., and Kale, R.K. (2005). Chemopreventive potential of Triphala (a composite Indian drug) on benzo(a)pyrene induced forestomach tumorigenesis in murine tumor model system. *J. Exp Clin Cancer Res.,* **24(4):** 555-63.

Dhanalakshmi, S., Devi, R.S., Srikumar, R., Manikandan, S., and Thangaraj, R. (2007). Protective effect of Triphala on cold stress-induced behavioural and biochemical abnormalities in rats. *Yakugaku Zasshi.,* **127(11):** 1863-7.

El-Mekkawy, S. (1995). Inhibitory effects of Egyptian folk medicines on human immuno deficiency virus (HIV) reverse transcriptase. *Chem Pharm Bull,* **43:** 641- 648.

Geer, M.I., Zia-ul-Arifeen, S., Ahmad, D.B., Moinuddin, G., Ahmad, A.K., and Devi, K. (2005). Hypotensive potential of aqueous extract of *Emblica officinalis* on anaesthetized dogs. *JK-Practitioner*, **12(4):** 213-215.

Golechha, M., Bhatia, J., and Arya, D.S. (2012). Studies on effects of *Emblica officinalis* (Amla) on oxidative stress and cholinergic function in scopolamine induced amnesia in mice. *J Environ Biol.*, **33(1):** 95-100.

Gupta, M., Banerjee, D., and Mukherjee, A. (2013). Evaluation of analgesic, antipyretic and anti-inflammatory effects of methanol extract of traditional herbal medicine using rodents. *J Pharmacognosy Phytother.*, **5(6):** 106-113.

Hiraganahalli, B.D., Chinampudur, V.C., Dethe, S., Mundkinajeddu, D., Pandre, M.K., Balachandran, J., and Agarwal, A. (2012). Hepatoprotective and antioxidant activity of standardized herbal extracts. *Pharmacogn Mag.*, **8(30):** 116-23.

Ihantola-Vormisto, A., Summanen, J., Kankaanranta, H. *et al.* (1997). Anti-inflammatory activity of extracts from leaves of *Phyllanthus emblica*. *Planta Med.*, **63:** 518–524.

Jacob, A., Pandey, M., Kapoor, S., and Saroja, R. (1988). Effect of the Indian gooseberry (amla) on serum cholesterol levels in men aged 35-55 years. *Eur J Clin Nutr.*, **42 (11):** 939–44.

Jagetia, G.C., Malagi, K.J., Baliga, M.S., Venkatesh, P., and Veruva, R.R. (2004). Triphala, an ayurvedic rasayana drug, protects mice against radiation-induced lethality by free-radical scavenging. *Journal of Alternative and Complementary Medicine.*, **10(6):** 971-8.

Jayaweera, D.M.A. (1980). Medicinal Plants used in Ceylon Part 2. *National Science Council of Sri Lanka. Colombo.*

Jeena, K.J., Kuttan, G., and Kuttan, R. (2001). Antitumour activity of *Emblica officinalis*. *J Ethnopharmacol.*, **75:** 65–69.

Jose, J.K., and Kuttan, R. (2000). Hepatoprotective activity of *Emblica officinalis* and Chyavanaprash. *J. Ethnopharmacol.*, **72(1-2):** 135-40.

Kalekar, S.A., Munshi, R.P., Bhalerao, S.S., and Thatte, U.M. (2013). Insulin sensitizing effect of 3 Indian medicinal plants: an *in vitro* study. *Indian J Pharmacol.*, **45(1):** 30-3.

Kaur, S., Arora, S., Kaur, K., and Kumar, S. (2001). The *in vitro* antimutagenic activity of Triphala—an Indian herbal drug. *Food Chem. Toxicol.*, **40(4):** 527-34.

Khan, K.H. (2009). Roles of *Emblica officinalis* in medicine - A Review. *Botany Research International*, **2(4):** 218-228.

Khorana, M.L., Rao, M.R.R., and Siddiqui, H.H. (1959). Antibacterial and antifungal activity of *Phyllanthus emblica* Linn. *Indian. J. Pharm.*, **21:** 331.

Kim, H.J., Yokozawa, T., Kim, H.Y., Tohda, C., Rao, T.P., and Juneja, L.R. (2005). Influence of amla (*Emblica officinalis* Gaertn.) on hypercholesterolemia and lipid peroxidation in cholesterol-fed rats. *J Nutr Sci.*, **75(2-3):** 65-69.

Koshy, S.M., Bobby, Z., Hariharan, A.P., and Gopalakrishna, S.M. (2012). Amla (*Emblica officinalis*) extract is effective in preventing high fructose diet-induced insulin resistance and atherogenic dyslipidemic profile in ovariectomized female albino rats. *Menopause.*, **19(10):** 1146-55.

Krishnaveni, M., and Mirunalini, S. (2012). Chemopreventive efficacy of *Phyllanthus emblica* L. (amla) fruit extract on 7,12-dimethylbenz(a)anthracene induced oral carcinogenesis—a dose-response study. *Environ. Toxicol. Pharmacol.*, **34(3):** 801-10.

Mahata, S., Pandey, A., Shukla, S., Tyagi, A., Husain, S.A., Das, B.C., and Bharti, A.C. (2013). Anticancer Activity of *Phyllanthus emblica* Linn. (Indian Gooseberry): Inhibition of transcription factor AP-1 and HPV gene expression in cervical cancer cells. *Nutr Cancer.*, **65(1):** 88-97.

McIntyre, A. (2005). Herbal treatment of children: Western and Ayurvedic perspectives. *Elsevier Health Sciences.* 225.

Mehmood, M.H., Rehman, A, Najeeb-Ur-Rehman, and Gilani, A.H. (2012). Studies on prokinetic, laxative and spasmodic activities of *Phyllanthus emblica* in experimental animals. *Phytother Res.*

Muthuraman, A., Shailja, Sood, S., and Singla, S.K. (2011). The antiinflammatory potential of phenolic compounds from *Emblica officinalis* L. in rat. *Inflammopharmacol.*, **19:** 327–334.

Mythilypriya, R., Shanthi, P., and Sachdanandam, P. (2007). Analgesic, antipyretic and ulcerogenic properties of an indigenous formulation—Kalpaamruthaa. *Phytother Res.* **21(6):** 574-8.

Nain, P., Saini, V., Sharma, S., and Nain, J. (2012). Antidiabetic and antioxidant potential of *Emblica officinalis* Gaertn. leaves extract in streptozotocin-induced type-2 diabetes mellitus (T2DM) rats. *J Ethnopharmacol.*, **142(1):** 65-71.

Nath, A., Raghunatha, P., and Joshi, S.R. (2012). diversity and biological activities of endophytic fungi of *Emblica officinalis*, an ethnomedicinal plant of India. *Mycobiology.* **40(1):** 8-13.

Nemmani, K.V., Jena, G.B., Dey, C.S., Kaul, C.L., and Ramarao, P. (2002). Cell proliferation and natural killer cell activity by polyherbal formulation, Immu-21 in mice. *Indian J Exp Biol.*, **40(3):** 282-7.

Ojha, S., Golechha, M., Kumari, S., and Arya, D.S. (2012). Protective effect of *Emblica officinalis* (amla) on isoproterenol-induced cardiotoxicity in rats. *Toxicol Ind Health.*, **28(5):** 399-411.

Patel, S.S., and Goyal, R.K. (2011). Prevention of diabetes-induced myocardial dysfunction in rats using the juice of the *Emblica officinalis* fruit. *Exp Clin Cardiol.*, **16(3):** 87-99.

Penolazzi, L., Lampronti, I., Borgatti, M., Khan, M., Zennaro, M., Piva, R., and Gambari, R. (2008). Induction of apoptosis of human primary osteoclasts treated with extracts from the medicinal plant *Emblica officinalis*. *BMC Complementary and Alternative Medicine.*, **8:** 59.

Perianayagam, J.B., Sharma, S.K., Joseph, A., and Christina, A.J. (2004). Evaluation of anti-pyretic and analgesic activity of *Emblica officinalis* Gaertn. *J.Ethnopharmacol.*, **95(1):** 83-5.

Pradyumna, R.T., Okamoto, T., Akita, N., Hayashi T, Kato-Yasuda N, and Suzuki K. (2013). Amla (*Emblica officinalis* Gaertn.) extract inhibits lipopolysaccharide-induced procoagulant and pro-inflammatory factors in cultured vascular endothelial cells. *Br J Nutr.*, **7:** 1-6.

Pramyothin, P., Samosorn, P., Poungshompoo, S., and Chaichantipyuth, C. (2006). The protective effects of *Phyllanthus emblica* Linn. extract on ethanol induced rat hepatic injury. *J Ethnopharmacol.*, **107(3):** 361-4.

Qureshi, S.A., Asad, W., and Sultana, V. (2009). The effect of *Phyllantus emblica* Linn on type — II diabetes, triglycerides and liver- specific enzyme". *Pakistan Journal of Nutrition.*, **8(2):** 125–128.

Rajak, S., Banerjee, S.K., Sood, S., Dinda, A.K., Gupta Y.K., Gupta S.K., and Maulik, S.K. (2004). *Emblica officinalis* causes myocardial adaptation and protects against oxidative stress in ischemic-reperfusion injury in rats. *Phytother Res.*, **18(1):** 54-60.

Rajarama, Rao, M.R., and Siddiqui, H.H. (1964). Pharmacological studies on *Emblica officinalis* Gaertn. *Ind. J.Exp. Biol.*, **2:** 29-31

Rosarin, F.S., Arulmozhi, V., Nagarajan, S., and Mirunalini, S. (2013). Antiproliferative effect of silver nanoparticles synthesized using amla on Hep2 cell line. *Asian Pac. J. Trop. Med.*, **6(1):** 1-10.

Roy, A.K., Haimanti, D., Sharma, A., and Talukder, G. (1991). *Phyllanthus emblica* fruit extract and ascorbic acid modify hepatotoxic and renotoxic effects of metals in mice" *Ind. J. Pharmacog.*, **29:** 117- 126.

Saeed, S., and Tariq, P. (2007). Antibacterial activities of *Emblica officinalis* and *Coriandrum sativum* against gram negative urinary pathogens. *Pak J Pharm Sci.*, **20(1):** 32-5.

Sai, R.M., Neetu, D., Yogesh, B., Anju, B., Dipti, P., Pauline, T., Sharma, S.K., Sarada, S.K., Ilavazhagan, G., Kumar, D., and Selvamurthy, W. (2002). Cyto-protective and immunomodulating properties of Amla (*Emblica officinalis*) on lymphocytes: an *in-vitro* study. *J. Ethnopharmacol.*, **81(1):** 5-10.

Saini, A., Sharma, S., and Chhibber, S. (2008). Protective efficacy of *Emblica officinalis* against *Klebsiella pneumoniae* induced pneumonia in mice. *Indian J Med Res.*, **128:** 188-193.

Sairam, K., Rao, C.V., Babu, M.D., Kumar, K.V., Agrawal, V.K., and Goel, R.K. (2002). Antiulcerogenic effect of methanolic extract of *Emblica officinalis*: an experimental study. *J Ethnopharmacol.*, **82(1):** 1-9.

Sampath, K.P., Bhowmik, D., Dutta, A., Yadav, A., Paswan, S., Srivastava, S., and Lokesh, D. (2012). Recent trends in potential traditional Indian herbs *Emblica officinalis* and its medicinal importance. *J Pharmacognosy Phytochem.*, **1(1):** 24-32.

Sarkar, R., Hazra, B., and Mandal, N. (2013). Amelioration of iron overload-induced liver toxicity by a potent antioxidant and iron chelator, *Emblica officinalis* Gaertn. *Toxicol Ind Health.*

Scartezzini, P., and Speroni, E. (2000). Review on some plants of Indian traditional medicine with antioxidant activity. *Journal of Ethnopharmacology.*, **71**: 23–43.

Sharma, A., and Sharma, K.K. (2011). Chemoprotective role of triphala against 1,2-dimethylhydrazine dihydrochloride induced carcinogenic damage to mouse liver. *Indian J Clin Biochem.*, **26(3)**: 290-5.

Singh, B.N., and Sharma, P.V. (1971). Effect of *Amalaki* on amalapitta. *J. Res. Ind. Med.*, **5**: 223-229.

Singh, I., Sharma, A., Nunia, V., and Goyal, P.K. (2005). Radioprotection of Swiss albino mice by *Emblica officinalis*. *Phytother Res.*, **19(5)**: 444-6.

Srikumar, R., Parthasarathy, N.J., Shankar, E.M., Manikandan, S., Vijayakumar, R., Thangaraj, R., Vijayananth, K., Sheela, D.R., and Rao, U.A. (2007). Evaluation of the growth inhibitory activities of Triphala against common bacterial isolates from HIV infected patients. *Phytother Res.*, **21(5)**: 476-80.

Srikumar, R., Parthasarathy, N.J., and Sheela, D.R. (2005). Immunomodulatory activity of triphala on neutrophil functions. *Biol Pharm Bull.*, **28(8)**: 1398-1403.

Sultana, S., Ahmad, S., Khanand, N., and Jahangir, T. (2005). Effect of *Emblica officinalis* (Gaertn) on CCl induced hepatic toxicity and DNA synthesis in Wistar rats. *Indian J Exp Biol.*, **43(5)**: 430-6.

Sumantran, V.N., Kulkarni, A., Chandwaskar, R., Harsulkar, A., Patwardhan, B., Chopra, A., and Wagh, U.V. (2008). Chondroprotective potential of fruit extracts of *Phyllanthus emblica* in osteoarthritis. *Evid Based Complement Alternat Med.*, **5(3)**: 329–335.

Suryanarayana, P., Kumar, P.A., Saraswat, M., Petrash, J.M., and Reddy, G.B. (2004). Inhibition of aldose reductase by tannoid principles of *Emblica officinalis*: implications for the prevention of sugar cataract. *Mol Vis.*, **12(10)**: 148-154.

Tasduq, S.A., Kaisar, P., Gupta, D.K., Kapahi, B.K., Maheshwari, H.S., Jyotsna, S., and Johri, R.K. (2005). Protective effect of a 50 per cent hydroalcoholic fruit extract of *Emblica officinalis* against anti-tuberculosis drugs induced liver toxicity. *Phytother Res.*,**19(3)**: 193-7.

Thakur, R.S., Puri, H.S., and Husain, A. (1989). Major Medicinal Plants of India. Central Institute of Medicinal and Aromatic Plants, Lucknow, India.

Treadway, L. (1994). Amla traditional food and medicine. *Herbal Gram.*, **31**: 26.

Vasant, R.A., and Narasimhacharya, A.V. (2012). Amla as an antihyperglycemic and hepato-renal protective agent in fluoride induced toxicity. *J Pharm Bioallied Sci.*, **4(3)**: 250-4.

Vasudevan, M., and Parle, M. (2007). Effect of Anwala churna (*Emblica officinalis* Gaertn.): an ayurvedic preparation on memory deficit rats. *Yakugaku Zasshi.*, **127(10)**: 1701-7.

Veena, K.P., Shanthi, S.P. (2006). The biochemical alterations following administration of Kalpa amrutha and *Semecarpus anacardium* in mammary carcinoma. *Chem Biol Interact.*, **161(1):** 69-78.

Wiart, C. (2013). Note on the relevance of *Emblica officinalis* Gaertn. for the treatment and prevention of cancer. *Eur J Cancer Prev.*, **22(2):** 198.

# Index

## Z